This book is the first to describe the development of scientific activity in the Antarctic (as distinct from exploration) in all its aspects. Coverage spans three centuries, starting with Halley who laid the foundations of geophysics which was to be the principal driving force behind Antarctic science for most of its history. Although early researchers built up a picture of the main features of the Antarctic environment, the idea of science specific to the continent emerged only later. As the main disciplines of oceanography, earth sciences, the sciences of atmosphere and geospace, terrestrial biology, medicine, and conservation developed, the clear interactions between them within an Antarctic context led to the emergence of the holistic view of Antarctic science which we hold today.

A History of Antarctic Science

Studies in Polar Research

This interdisciplinary series, aimed at all scientists with an interest in the world's cold regions, reflects the growth of research activity in the polar lands and oceans and provides a means of synthesizing the results. Each book in the series covers the present state of knowledge in a given subject area, resulting in a series which provides polar scientists with an invaluable, broad-ranging library.

Other titles in this series:

The Antarctic Circumpolar Ocean
Sir George Deacon

The Living Tundra
Yu. I. Chernov, transl. D. Love

Arctic Air Pollution
edited by B. Stonehouse

The Antarctic Treaty Regime
edited by Gillian D. Triggs

Antarctica: The Next Decade
edited by Sir Anthony Parsons

Antarctic Mineral Exploitation
Francisco Orrego Vicuna

Transit Management in the Northwest Passage
edited by C. Lamson and D. Vanderzwaag

Canada's Arctic Waters in International Law
Donat Pharand

Vegetation of the Soviet Polar Deserts
V. D. Aleksandrova, translated by D. Love

Reindeer on South Georgia
Nigel Leader-Williams

The Biology of Polar Bryophytes and Lichens
Royce Longton

Microbial Ecosystems of Antarctica
Warwick Vincent

Chronological List of Antarctic Expeditions
Robert K. Headland

The Age of the Arctic
G. Osherenko and O. R. Young

Subantarctic Macquarie Island
P. M. Selkirk, R. D. Seppelt and D. R. Selkirk

The Growth and Decay of Ice
G. H. S. Lock

Antarctic Fish and Fisheries
Karl-Hermann Kock

Frontispiece; Portrait by Thomas Murray of Edmond Halley, Fellow of the Royal Society and Captain in the Royal Navy. Halley commanded the first voyages ever undertaken specifically for a scientific purpose and in 1700 penetrated into Antarctic waters. He is generally regarded as the founder of geophysics, a science which has several times provided the impetus for Antarctic researches. (Courtesy of the Royal Society.)

A History of Antarctic Science

G. E. FOGG

CAMBRIDGE UNIVERSITY PRESS
Cambridge, New York, Melbourne, Madrid, Cape Town, Singapore, São Paulo

Cambridge University Press
The Edinburgh Building, Cambridge CB2 2RU, UK

Published in the United States of America by Cambridge University Press, New York

www.cambridge.org
Information on this title: www.cambridge.org/9780521361132

First published 1992
This digitally printed first paperback version 2005

A catalogue record for this publication is available from the British Library

Library of Congress Cataloguing in Publication data

Fogg, G. E. (Gordon Elliott), 1919–
A history of Antarctic science / G. E. Fogg.
p. cm. – (Studies in polar research)
Includes bibliographical references and index
ISBN 0 521 36113 3
1. Antarctic regions – Discovery and exploration. 2. Scientific
expeditions – Antarctic regions. I. Title. II. Series.
G860.F64 1992
919.8′904 – dc20 91-48063 CIP

ISBN-13 978-0-521-36113-2 hardback
ISBN-10 0-521-36113-3 hardback

ISBN-13 978-0-521-67337-2 paperback
ISBN-10 0-521-67337-2 paperback

Science must be tempered with humanity, and the best way of doing this is to explain its organic development, and also to show all that was really great, beautiful and noble in these civilizations of old, all that our conceited scientists and inventors have too often forgotten and disdained. We must teach reverence for the past, – not simply for its own sake or because it is still full of treasures, – but for the sake of the present which it will aid us to appreciate, and for the sake of the future in which it will help us to walk with dignity. (George Sarton, 1919)

Contents

Foreword

A theme throughout this book is the interaction of science and politics. In the beginning it was a secret instruction to look for a southern continent, added to his purely scientific remit to observe the transit of Venus in Tahiti, that led to the circumnavigation of Antarctica by James Cook. Claims to portions of the continent were made by explorers and scientists on behalf of nearly all the nations which sent the expeditions which followed his in the next 170 years. The resultant bickering over sovereignty was only allayed by an inspired collaboration of diplomatists and scientists in framing the Antarctic Treaty of 1961. This was a remarkable political innovation, enabling co-operation between nations of opposing ideologies in living and researching in a hostile environment. In the aftermath of the Falklands conflict we were able to strengthen Britain's presence in the South Atlantic by increasing our scientific effort. This paid off remarkably quickly in a totally unexpected way with the discovery by the British Antarctic Survey of the ozone 'hole' over Antarctica in the austral spring. This brought home to the whole world the potentially dangerous changes in the environment which mankind's activities are bringing about and led to the first measures to control pollution on a global scale. Antarctica's greatest value to us undoubtedly lies in the information which comes out of it about the functioning of our planet. I hope that this book, by one who has himself worked in the Antarctic, will help in better appreciation of the unique importance of Antarctic science for the management of our environment and in promoting co-operation between scientists and politicians in solving the problems which it presents.

Margaret Thatcher

MARGARET THATCHER
April 1991

Preface

One person cannot have the expertise to assess properly developments in all the diverse disciplines which are concerned with the Antarctic. The difficulty of tracing their history might be to some extent avoided by stopping at the point at which things become really complicated – some 30 years ago – but this would be to leave the tree without its fruits and, much as the professional historian of science may disapprove, the attempt has been made to bring this account as nearly up to date as possible. In spite of the problems, I am convinced that the history of science in Antarctica is worth writing about and that it should be attempted by a single author. From its beginnings, almost alone among scientific enterprises, research in southern latitudes has extended over a wide range and it seems particularly desirable for practical management as well as for scientific understanding that this holistic approach should be maintained. Multi-author treatments are ill-suited to do this; an inspired collaboration between two or three specialists would be best but failing this a synthesis is only to be achieved by the individual mind. The result may be only a rough sketch of what is desirable but one hopes it may provide a foundation on which others may build.

The interactions between politics and Antarctic science invite particular attention and it is most appropriate that the Foreword to this book has been provided by the Rt. Hon. Margaret Thatcher, OM, PC, FRS. I am deeply grateful to her for her encouragement.

Among the many who have helped with advice and criticism I am particularly grateful to Dr D. W. H. Walton, Chairman of the Editorial Board of Studies in Polar Research, whose comments, based on his wide knowledge of the Antarctic and of books, have been invaluable in the shaping of this one. My gratitude is also due to those who have generously given of their time in reading and commenting on individual chapters – Professor A. Clarke, Dr C. S. M. Doake, Dr J. R. Dudeney, Dr R. I. Lewis Smith and Dr M. R. A. Thomson of the British Antarctic Survey, and Dr

M. Deacon of the Department of Oceanography, University of Southampton. Many others have helped, more than they perhaps realize, not only by advice and discussion of particular points but by remarks in casual conversation. These include J. Caffin, Dr J. Davenport, Professor S. Z. El-Sayed, T. D. Fogg, Sir Vivian Fuchs, FRS, Dr D. J. Grove, R. K. Headland, Dr F. Jacka, A. G. E. Jones, Dr R. M. Laws, FRS, Dr G. A. Llano, Dr A. McConnell, Sir John Mason, FRS, Dr G. de Q. Robin, A. Savours (Mrs Shirley), Professor W. D. P. Stewart, FRS, Dr C. W. M. Swithinbank and R. Thomson. They have steered me through many crevassed areas and must not be blamed if I have come to grief in others which I might be expected to have navigated on my own.

Librarians are a friendly and helpful tribe and there have been no exceptions among those whom I have encountered. The Wolfson Marine Science Library of the University College of North Wales and the libraries of the Royal Society and the Scott Polar Research Institute have between them met the bulk of my needs. Others that I have used are the Bodleian Library, the British Library, the City of Westminster Reference Library, the National Meteorological Library, the Science Museum Library and those of the Antarctic Division of the New Zealand DSIR, the British Antarctic Survey, the Canterbury Museum in Christchurch, New Zealand, the Linnean Society, the Marine Biological Association, the Natural History Museum and the Royal Institution. I am most grateful for having been allowed to use these libraries and to their staffs for their great help. Not least, my thanks must be expressed to my daughter, Mrs Helen Dwyer, for her able assistance in locating sources and checking references.

This research was greatly assisted by the award of an Emeritus Fellowship by the Leverhulme Trust and by a grant from the Royal Society, to both of which bodies I am extremely grateful.

My thanks are also due to those, to whom acknowledgement by name is made in appropriate places in the book, who have generously allowed the use of illustrations and quotations. Much of the expert photography necessary was done by William Rowntree, ARPS. I am also grateful to Camilla Adamkiewicz, Phyllis Ellis, Margaret Sands and Anne Sands for the processing of the manuscript.

Finally, my thanks go to my wife for her understanding support and tolerance of my preoccupation over many years with things remote from the domestic scene.

G. E. Fogg
Menai Bridge, Isle of Anglesey
March 1991

A note for the reader

Maps showing the locations of some of the geographic regions of Antarctica, island groups and the principal bases mentioned in the text will be found in Figs. 6.5 (p. 170) and 6.9 (p. 184). Otherwise the map of Antarctica in *The Times Atlas of the World*, 7th comprehensive edition, 1985, is recommended for general purposes, and for more detail the map issued by the American Geographical Society in 1970 should be consulted. It has not been possible to give information about every Antarctic expedition but some particulars of all those involved in science between 1872 and 1925 will be found in chapter 4. An invaluable *Chronological List of Antarctic Expeditions and Related Historical Events* has been compiled by R. K. Headland (Cambridge University Press, 1989). Dealing with events after the International Geophysical Year of 1957–58 becomes difficult without the resort to the use of acronyms. A list of these abominations together with some abbreviations that may not be familiar is given below.

AWS	Automatic weather station
BANZARE	British, Australian and New Zealand Antarctic Research Expedition
BAS	British Antarctic Survey
BIOMASS	Biological Investigation of Marine Antarctic Systems and Stocks
BP	Before present
CCAMLR	Convention for Conservation of Antarctic Marine Living Resources
COSPAR	Committee on Space Research
CRAMRA	Convention for the Regulation of Antarctic Mineral Resource Activities

CSAGI	Comité Spéciale de l'Année Geophysique Internationale
CTD	Conductivity-Temperature-Depth (oceanographic instrument package)
EPOS	European 'Polarstern' Study
FAO	Food & Agriculture Organization
FGGE	First Global GARP Experiment
FIBEX	First International BIOMASS Experiment
FIDS	Falkland Islands Dependencies Survey
GARP	Global Atmospheric Research Programme
IAPO	International Association of Physical Oceanography
IBEA	International Biomedical Expedition to the Antarctic
IBP	International Biological Programme
ICSU	International Council of Scientific Unions
IGY	International Geophysical Year
IMS	International Magnetospheric Study
IOC	Intergovernmental Oceanographic Commission
IUBS	International Union of Biological Sciences
IUCN	International Union for the Conservation of Nature
IUPS	International Union of Physiological Sciences
LMT	Local magnetic time
LST	Local solar time
MNAP	Managers of National Antarctic Programmes
NASA	National Aeronautics & Space Administration (USA)
NOAA	National Oceanic & Atmospheric Administration (USA)
NSF	National Science Foundation (USA)
POLEX-South	South Polar Experiment
SCALOP	Standing Committee on Antarctic Logistics & Operations
SCAR	Scientific Committee on Antarctic Research
SCOR	Special Committee for Oceanic Research
SCUBA	Self-Contained Underwater Breathing Apparatus
SIBEX	Second International BIOMASS Experiment
SMS	Seiner Majestäts Schiff (German Navy)
SPA	Specially protected area
SSSI	Site of Special Scientific Interest
TAE	Commonwealth Trans-Antarctic Expedition
TOMS	Total Ozone Mapping Spectrometer
UGO	Unmanned Geophysical Observatory

USARP	United States Antarctic Research Programme
UT	Universal time
VLF	Very low frequency
WMO	World Meteorological Organization

1

Introduction

From a strictly scientific point of view there is nothing radically different about Antarctica to set it apart from the rest of the natural world. One might think that there, if anywhere on the surface of the planet Earth, would be the place to test the validity of the Copernican principle that physical laws remain the same everywhere in the solar system. However, while it is true that the geophysical and meteorological situations which it presents are unparalleled anywhere else and that its living organisms are remarkable in their successful adaptation to some of the most extreme conditions found on the planet, the same scientific laws appear to operate in Antarctica as elsewhere. Antarctic science is not *sui generis* or self-contained but has grown out of mainstream science, on which it still depends completely and to which it contributes knowledge essential for the understanding of our world as a whole. To a large extent, then, the history of science in Antarctica is the history of science in general but nevertheless the Antarctic does have several features of peculiar interest to the historian of science which deserve, but have not hitherto received, special consideration and which justify the writing of this book.

Firstly, it should be said that although such diverse topics as ionospherics, plate tectonics, penguin physiology and human psychology are included, it is customary and justifiable to speak of Antarctic science in the singular rather than in the plural. Obviously this situation at least partly originated with, and is still sustained by, the necessity for scientists working in this remote part of the world to share common logistic support. All organizations, whether national or international, concerned with Antarctic research include both physical and biological work in their remit. But beyond this, in an intellectual climate which has become predominantly specialist and reductionist, Antarctic science has maintained a characteristic holistic flavour which it acquired early on in its development. Just as at the end of the nineteenth century a discussion on Antarctic research ranged from terrestrial magnetism to geology, meteorology and biology (Murray, 1898) so there still appear texts, such as those of Quam (1971), Fuchs & Laws (1977) and Walton (1987), which bring together the whole range of scientific activities in the Antarctic. Such an outlook is rare and, one may think, of great value in present day science.

Another remarkable feature is the close involvement of Antarctic science with politics. Antarctic scientists have often been used as political instruments (Beck 1986) and it would be unrealistic for them to think that their work can be isolated from the spheres of interest of economics, law and politics (Vicuña, 1983). This contrasts with the situation in science in general in which, until recently, most major advances have been made by individuals little concerned with world affairs. Antarctic science is different because support on a national scale and therefore involvement with politics has usually been necessary for exploration and investigation to be possible; Antarctic science has always been 'big science' in the sense of Price (1986). This was so at the very beginning. The voyages of Halley, Cook and Bellingshausen had respectable scientific objectives but they were relatively as expensive as ventures into space are today and were funded by governments which saw that they might serve purposes other than the acquisition of knowledge for its own sake. Apart from the prestige which geographical discovery brought to a nation, information and experience useful in navigation and commerce was gained as well as the possibility of the annexation of potentially valuable territories. This marriage of Antarctic science and politics has lasted but with the passage of time the gentler, dependent, partner has assumed a certain moral ascendancy and may even be said to be moulding her lord's character for the good. The Antarctic Treaty of 1961, which arose out of the needs of scientific research, was unprecedented as a way of resolving differences between nations. It may well serve as a model for more co-operative attitudes between governments in the future and it may be the saving for posterity of the last great wilderness on earth. If a survey of the origins of this singular accord will help to ensure its development in the future then its history should be explored from every angle.

Another interest in the history of Antarctic science lies in the special qualities there may have been in the scientists themselves. The heroic images of Scott, Shackleton, Amundsen and Mawson are still vivid, and some more recent explorers of Antarctica have seen themselves as following in this tradition; witness book titles such as *Antarctica my Destiny, Assault on Eternity* and *A World of Men*. The general public encourages this by looking on anyone who visits Antarctica as a hero but with modern facilities and survival techniques there is little call for heroism; most scientists are not inclined that way and most national Antarctic organizations positively discourage any such tendencies. Heroics and science, whilst not necessarily antagonistic, are certainly not synergistic. As Medawar (1984, p. 263) has written, the work of scientists 'is in no way made deeper or more cogent by privation, distress or worldly buffetings'. The horrific winter journey undertaken by Wilson, Cherry-Garrard and Bowers in 1911 to obtain early stage embryos of emperor penguins, called for more determination and endurance of physical hardship than any other research venture has ever done but, sadly, yielded little of scientific value. The off-hand receipt of the eggs by British Museum (Natural History) scientists, which so outraged Cherry-Garrard, was insensitive but in retrospect is justified by the scientific outcome (Parsons, 1934). On the other hand, tenacity of purpose is a virtue in scientists as well as in explorers. The 35lbs (15.9 kg) of geological specimens with which Scott and his companions burdened themselves on their final march to their deaths, although possibly clung to in desperation 'to salvage something from defeat at the Pole' (Huntford 1979, p. 556), did prove to be of considerable value (Tingey, 1983).

Science and technology are always interdependent but the dependence of Antarctic research on technical advances has been absolute. James Cook's delineation of the Antarctic continent would not have been possible without ships capable of navigating unknown and dangerous waters, and had he not succeeded, in an empirical fashion, in keeping scurvy at bay during his long periods at sea. Later, radio, air transport, ice-breakers, satellite communications and numerous other products of technology have played their indispensable roles in enabling the ever more demanding requirements of science to be met in an extreme and hostile environment.

The propinquity between the different scientific disciplines brought about by shared logistic support has not necessarily led to the development of projects spanning different sciences – a geophysicist and an ornithologist together on an Antarctic base may learn something of each other's subjects and profit from discussion but no radically new science is likely to emerge – however, the indirect benefits have been enormous. Thus, for nearly two centuries the study of terrestrial magnetism has been a principal object of Antarctic expeditions but naturalists, who would have been quite unsuccessful in finding support for a venture to the Antarctic on their own, have been taken along and Antarctic biology has developed as a result. Equally, biologists provided the justification for the *Discovery* Investigations but a proper understanding of the ecology of whales required a background of physical oceanography, and major contributions to our knowledge of ocean circulation followed.

In assessing the effects of these and other influences on Antarctic science the historian has a wealth and variety of material, perhaps unique in science, on which to draw. In pursuit of objectivity scientists avoid discussion of personal motives and feelings in their formal publications and these consequently give little idea of how their researches actually originated and proceeded (Price, 1964; Medawar, 1984, p. 132). However, because of the money involved, officials and committees concerned with Antarctic expeditions have been more than usually prolific in production of memoranda, minutes and instructions. On the other hand, for the individual scientist a visit to Antarctica is an experience which he feels compelled to record in diaries, letters, travelogues or, sometimes, poetry and paintings. These reveal that chance, caprice, friendship, desire for fame or profit, or striving for some visionary ideal may play as great a part in shaping the course of research as they do with other human affairs.

The possible ramifications of a history of science in Antarctica are thus immense but two limits can be drawn. One is geographic: the Antarctic will be taken as that area of the earth's surface within the Antarctic Convergence, or, as it is often called, the Polar Front.[1] This is a discontinuity in the Southern Ocean, encircling the Antarctic continent in a line undulating around 50° S, where cold northward-moving surface water dips beneath warmer water. It varies little in position from year to year and climatically and biologically as well as oceanographically forms a sharp boundary delimiting a region with characteristic Antarctic features. Included within it are islands, such as South Georgia, Kerguelen and Heard, which together with Marion and Macquarie Islands lying just northward of it, may be classified as sub-Antarctic (Holdgate 1964b). It should be noted that this definition of the Antarctic is different from that adopted in the Antarctic Treaty, namely the land area and ice shelves south of latitude 60° S. Then, secondly, science may be defined as our attempts to understand the natural world – not the mere accumulation of facts about it but their assimilation into abstract concepts from which predictions can be made. Thus, although exploration and scientific research are closely interrelated in the Antarctic, most geographical discovery, the history of which has been the subject of many books (Hayes, 1928; Sullivan, 1957; Kirwan, 1962; Quartermain, 1967; Bertrand, 1971; Fuchs, 1982), can be left aside.

Law (1985) distinguished eras in Antarctic venture – trading, imperialistic, scientific and exploitive – but all four of these motivations have operated to varying degrees throughout man's involvement with the

Antarctic. In the trading era, which Law puts as ending around 1890, the expedition led by Ross in 1839–43 was conceived, planned and executed as a purely scientific venture, whereas in the post-imperialistic, scientific, era after 1930 science has been carried out by some nations to serve motives which can be fairly described as imperialistic. For the purposes of this book Law's categorization is too simplistic, as any rigid categorization will be with this particular subject. The historiographer of science distinguishes between a horizontal approach, in which the development through time of a given narrow topic is studied, and a vertical approach, in which the historian starts with a perspective that is more interdisciplinary in nature and the science which is in focus is seen as one element in the cultural and social life of a particular period (Kragh, 1987). What is being attempted here is a vertical approach to the study of a number of intertwining horizontal elements.

If, following Harrison (1987), we liken science to a stream flowing through history, then Antarctic science might be seen as a river beginning in limestone country. Its source – the genius of Halley – is definite and already a copious outflow but this immediately disappears underground to re-emerge and disappear again, each time reappearing – as in the episodes of Cook, Bellingshausen and Ross – as a deeper and more strongly flowing stream. After its final resurgence at the limestone escarpment – the heroic age of Scott, Shackleton and Amundsen – it is an obviously continuous river but it quickly becomes deltaic, splitting into half-a-dozen or so flows – different disciplines – which, although running in separate channels, nevertheless share the same source and are ineluctably interdependent. In the limestone phase – to persist with the simile for a little longer – the course of the stream is determined by the terrain, the science in this early phase being ancillary to and shaped by other activities, the object of which was not understanding of the Antarctic itself. It was around the turn of the nineteenth century that the idea of polar science began to emerge and, like a river crossing an alluvial plain, came to determine its own course. In the heroic age scientific research was the avowed object of most expeditions even though patriotism and adventure were the forces that raised funds and brought in the volunteers in their thousands. In giving an account of the history of science in the Antarctic it is possible in the early stages to adhere reasonably well to chronological order and deal with the subject in successive episodes but, especially after the International Geophysical Year of 1957–58, the complexity and scale of Antarctic operations escalated so much that even an approximately diachronic treatment becomes unmanageable. In considering these later stages, therefore, it seems best to deal separately with the different areas of Antarctic investigation, always bearing in mind that they are subject to common constraints and influences and are interdependent in a special way that is not general in the world of science.

Endnote

1 The terms Antarctic Convergence and Polar Front are not exactly equivalent. For discussion of this see Gordon (1971b) and Deacon (1984, pp. 114–19).

2

The science of the early explorations

2.1 The scientific and technological background

In the fifteenth century European shipwrights had developed the wooden full-rigged sailing ship to a form which was to persist, relatively unchanged, for another 400 years. These ships had sufficient capacity to undertake prolonged voyages of exploration and the divided sail plan enabled them to be managed safely on the high seas and off unknown shores (Naish, 1957). Courage and fortitude was needed to sail them but it was not lack of these which delayed the beginning of organized exploration in the Antarctic. For a long time there had been the idea of a *Terra Australis Incognita*, such as pictured in the maps of Orontius Finaeus (1531) and Ortelius (1573), which arose from a feeling that the two hemispheres must counterbalance (Rainaud, 1893; Bushnell, 1975). There are some striking similarities between the general outline of this continent in Orontius Finaeus's map and the modern map but the suggestion of Weihaupt (1984) that there was actual knowledge of the broad configuration of Antarctica before the mid-sixteenth century strains credibility. The figment of a large and populous southern continent persisted until nearly the end of the eighteenth century and provided incentive for exploration into the middle southern latitudes but there was little to take men further south. Such probing into Antarctic waters as had been made were not encouraging; there seemed little of commercial value beyond latitude 55° S and no large populations of infidels to arouse the powerful urge of Spain and Portugal to evangelize. Nor had science developed to a point where it demanded observations in those parts.

However, by the latter half of the seventeenth century science was evolving rapidly – in Pledge's (1939) view this was the single most important episode in the history of science. It took place mainly in the Low Countries, France and southern England, and advance was largely in mathematics, mechanics, optics and astronomy, with a major preoccupation in the problems of navigation. At a time of increasingly active exploration it was important for both political and commercial reasons to be able to locate places precisely on the world map. Latitude could be determined accurately by measuring the meridian altitude of the sun but means of finding longitude at sea were inadequate. Observatories were established in Paris in 1669 and at Greenwich in 1675 for the purpose of making celestial observations which would enable longitude to be ascertained at sea. The most promising idea at this time was to use the moon's movement across the heavens as a clock that would permit comparison of local time with that at some reference point and so make possible calculations of longitude. Another idea was that magnetic variation might be used. These lines of investigation led to the first major sea voyage to be undertaken specifically for a scientific purpose – the voyage by Edmond Halley in HMS *Paramore* in 1698 to 1699. In a second voyage, in 1699 to 1700, Halley went just south of the Polar Front but for weightier reasons than this it can be taken as the definitive starting point for a history of science in the Antarctic.

2.2 Edmond Halley

Edmond Halley (Fig. 2.1) perhaps does not receive the sustained regard due to one who has been described as 'the second most illustrious of Anglo-Saxon philosophers' (Anon. 1880). He was most notably a pioneer in both geophysics and stellar astronomy but had remarkably wide interests. Without his persistence and very practical help with the publication, Newton's *Principia* would never have appeared (Armitage, 1966). He was entirely free from jealousy and was much more enthusiastic about the success of the *Principia* than of anything that he himself had accomplished (Biog. Brit. 1757, quoted in MacPike, 1932).

Not much personal information about Halley has survived.[1] He was born probably in 1656, son of a prosperous London soap manufacturer. In 1673 he entered Queen's College, Oxford, but left in 1676, without taking a degree, to sail for St. Helena in the South Atlantic to catalogue and chart the stars of the southern celestial hemisphere which are invisible from Europe. The idea of doing this necessary work seems to have been his own and just as remarkable is the practicality with which he planned and obtained support for the venture. His catalogue, the earliest containing telescopically determined star positions to appear in print, was published in 1679. From our point of view two important outcomes of the expedition were that it turned Halley's attention southwards and that it gave him practical seagoing experience.

Fig. 2.1 Edmond Halley (1656–1742) at the age of 80 – he holds a diagram illustrating his theory of terrestrial magnetism. Attributed to Michael Dahl. (Courtesy of the Royal Society.)

The determination of longitude had great fascination for Halley and much of his astronomical work was devoted to this end (Cotter, 1981). However, his first recorded scientific observation, in 1672, was on magnetic variation; terrestrial magnetism continued to interest him for most of his life. In 1683 he devised a model of terrestrial magnetism which supposed the earth to have four magnetic poles and in 1692 he extended it to describe secular changes in the magnetic field (Barraclough, 1985). The idea of a voyage to map variation and test various methods for determining longitude came from Benjamin Middleton, who in 1693 requested assistance from the Royal Society in procuring a suitable vessel. Halley was involved from the start and the proposal was that he should accompany Middleton on the voyage. The justification and general arrangements for the voyage were set out thus:

> Whereas this Honble Society has at all times, according to Its Institution, appeared ready and willing to promote any Designe tending to the Advancement of Usefull Arts or the Discovery of Nature. And whereas the variations of the Magneticall Needle are as yett unknowne to Us in all that vast Tract of Sea betweene America and China, being neare halfe the Globe, none of our Journalls giving the Least Account thereof. And whereas the Vibrations of the Pendulum in Clocks are found to be Swifter and Slower in differing Latitudes, hindring the Discovery of the Longitude at Sea by that meanes, unless this Difference be adjusted by Accurate Observation in Severall Places. And whereas there has of late been Severall other Methods thought of for discovering the Longitude at Sea by the Motion of the Moone and other Celestiall Bodys, which have not as yett been effectually put in Practice, Soe that it cannot be as yett concluded how farr the Same may be relyed on for the use of Navigators. It is therefore most humbly prayed that this Honble Company would please to Lend their Assistance and Good offices to Obtaine of their Ma^{tys} a vessell which may be Secure in all weathers, but not exceeding. 60. Tunns burthen for a voyage to be undertaken by Benjamin Middleton Esqr and Edmond Halley in Order to discover what may be Learnt in the aboves^{d} Particulars: the designe being to compass the Globe from East to West through the great South Sea. And the said Benj: Middleton for promoting the said Undertaking does oblige himself to goe the Voyage and to Victuall and Man the said Vessell at his owne proper Costs and Charges, and Likewise to render an Account of his Proceedings to the Rt. Hon^{bles} the Lords of the Admiralty and to this Society. And the Care of Making the Necessary Observations is undertaken by the s^{d} Edmund Halley, whose Capacity for Such Purposes is Supposed to be Sufficiently knowne to this Honble Company.
>
> (Royal Society, *Collectanea Newtoniana*, vol. IV. no. 425, March 1693.)

A present-day scientist can but wistfully admire the brevity and elegance of this proposal and the dispatch with which it was considered. The Royal Society gave its support, the Lords of the Treasury reported favourably, Queen Mary II gave encouragement, and within four months the Admiralty ordered the building of a vessel of 'about Eighty Tuns Burthen' specially for the voyage (Thrower, 1981). The vessel, to be named the *Paramore*, was of the type known as a pink, particularly suitable for carrying stores and for navigation in shallow water. An unexplained delay followed the launching and Halley became involved with other things. A vexatious task as Comptroller of the Mint at Chester occupied two years and then he became involved with the Tsar Peter (the Great), who was then 20 years of age. The Tsar was visiting England to improve his knowledge of seamanship and shipbuilding, for which this country had a reputation second to none. He was also interested in science and Halley was recommended to him as a source of information. The two got on extremely well; the Tsar, finding Halley's conversation both instructive and entertaining, 'treated him with great distinction, admitting him to the Familiarity of his Table' (memoir attributed to M. Folkes, quoted in MacPike, 1932).

Halley had received a Commission as Master and Commander of the *Paramore* in 1696 and this was renewed in 1698. In addition to his experience at sea he had studied tidal phenomena and, of course, was an expert on navigation but nevertheless it was remarkable, even for those days, for a civilian to be commissioned to command a man-of-war on a long voyage. By this time Middleton had faded from the scene and Halley became responsible for finding security for the crew's wages. In his instructions from the Admiralty the variations of the compass and accurate determination of the latitude and longitude of ports of call are laid down as the objects of the voyage, which was now to be confined to the Atlantic and to include, if possible, determination of the position of the coast of the *Terra Incognita* which was supposed to lie between the tip of South America and the Cape of Good Hope (Thrower, 1981, p. 268).

The *Paramore* sailed on 20th October 1698 and reached Fernando Loronho, off the coast of Brazil, by 8th February 1699. Halley having found his crew insubordinate, then sailed north to Barbados, where he came to the conclusion that it was best to return to England. His difficulty originated with Edward Harrison, lieutenant and mate who, although Halley had not realized it at the beginning of the voyage, held him responsible for the cool reception of a paper on the problem of longitude which he, Harrison, had submitted to the Royal Society in 1694 (Armitage, 1966, p. 140). Harrison thought that as a professional naval officer he knew better than Halley and criticized him in front of the crew. There is no evidence that Halley was in any way remiss in his handling of the vessel and from what we know of his straightforward and cheerful character it seems unlikely that he was unable to command the respect of the men under him. When the *Paramore* got back to England on 22nd June 1699 Harrison went before a court-martial but, to Halley's dissatisfaction, received no more than a reprimand.

In spite of all this, the scientific programme was successful and Halley's

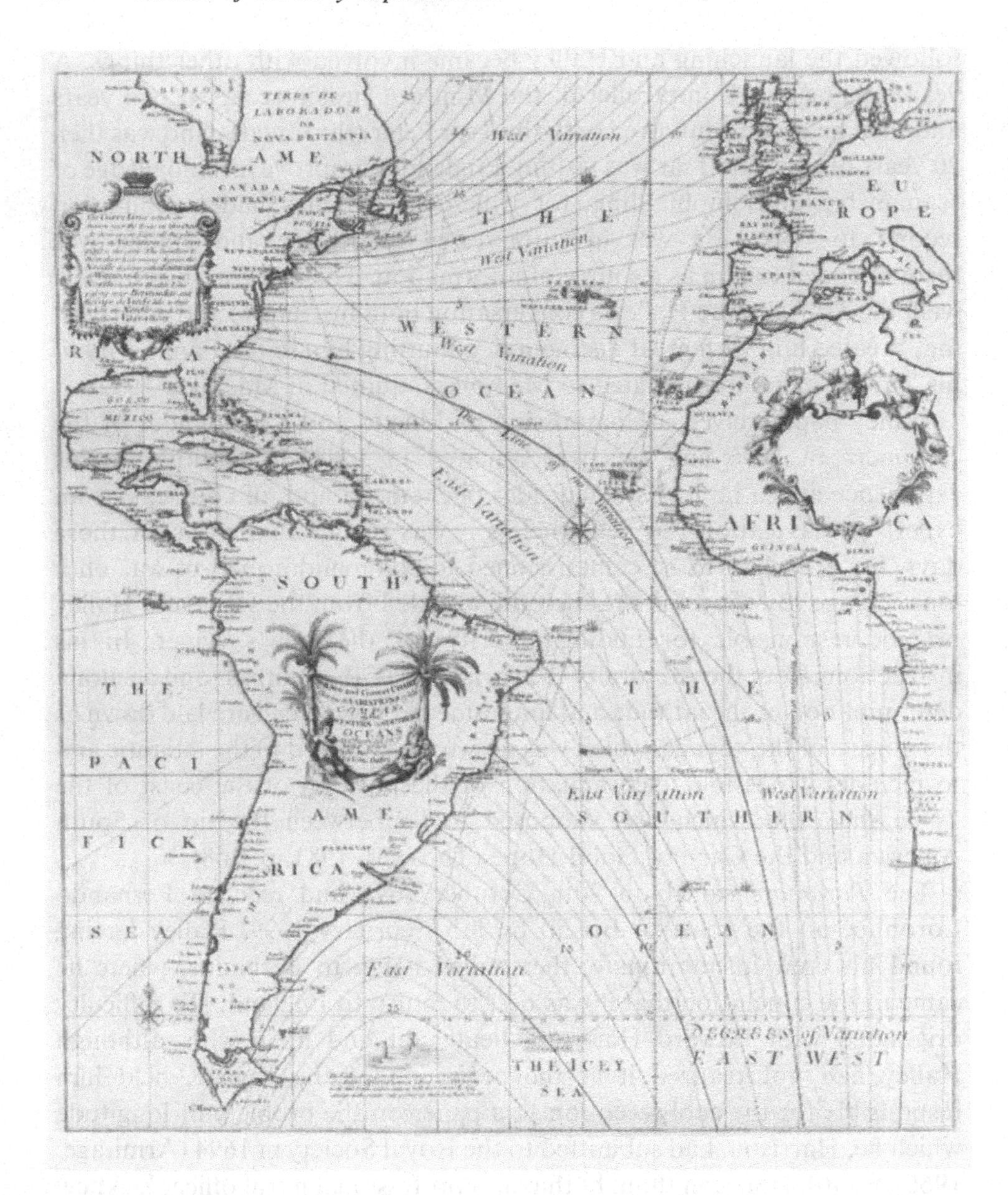

Fig. 2.2 Halley's Atlantic Chart, 1701, showing magnetic variation. This is an early version lacking the dedication to William III. The inscription to the east of Tierra del Fuego reads 'The SEA in these parts abounds with two sorts of Animalls of a Middle SPECIES between a Bird and a Fifh, having necks like Swans and Swimming with their whole Bodyes always under water only putting up their long Necks for Air.' (Reproduced by permission of the Royal Geographic Society, London.)

wish for a second voyage was granted. The *Paramore* sailed again on 16th September 1699, this time without a professional officer higher in rank than mate, and the voyage was a harmonious one. Its cruise track is given in Fig. 2.2. On December 29th 1699 the *Paramore* left Rio de Janeiro to sail south on a mission for which Halley had specifically requested permission:

> To Mr Secretary Burchett
> These
> humbly present
> Honoured S[r]
>
> I entreat that in the orders their Lopps please to give me, it may be specified that I endeavour to make discovery of the South unknown lands, between Magellan Streights and the Cape of Good Hope, between the Latitudes of 50° and 55° South, if I meet not with the Land sooner.
>
> I am
> Your Honours most obed[t] Serv[t]
> Edm. Halley
> Sept. 12° 1699 (Thrower, 1981, p. 299).

Besides making magnetic and astronomical observations Halley recorded weather conditions, currents, colour of the sea, and flora and fauna. On 21st January he noted in his journal (Thrower, 1981) at about 44° 22′ S 49° 29′ W that 'the Sea appear'd very white and abundance of small Sea Foule about the Shipp and severall beds of weeds drove by the Ship of which we took up some for a Sample being what none of our people had Seene eles where'. The whiteness of the sea was very likely caused by a bloom of coccolithophorids, such as often occurs in this sea area, and the drifting weed was no doubt the giant kelp, *Macrocystis pyrifera*, which grows in profusion around South Georgia and the Falklands. On 27th January at latitude 50°45′S he mentioned two kinds of penguins but it is not possible to determine the species from his descriptions. For the next day there are puzzling references to birds and large animals:

> We have had a Continuall thick ffog for this 24 hours, which obliged us to goe away with our fore top saile only, on the Capp; and to Sound every two hours, We have had Severall of the Diveing birds with Necks like Swans pass by us, and this Morning a Couple of Annimalls which some supposed to be Seals but are not soe; they bent their Tayles into a sort of Bow thus and being disturb'd shew'd very large Finns as big as those of a Large Shirk The head not much unlike a Turtles. This morning it was very cold and the Themometer at but 4 above frezing in my Cabbin.
> (Thrower, 1981, p. 159).

The bird with a neck like a swan is pictured on his Atlantic chart (Fig. 2.2) and is somewhat reminiscent of a cormorant. It may have been a king penguin but is not really identifiable. The seal-like creature may have been a killer or bottle-nosed whale (Harrison Matthews, quoted by Ronan, 1968 and Thrower, 1981). The fog and the drop in temperature indicate that he had crossed the Polar Front which at the present day is at approximately 51° S at 41° W, the longitude of the *Paramore* on that day.

On 1st February, Halley attained his furthest south, 52°24′ S, and saw his first icebergs, which were of characteristic tabular form. He described them in a letter to Burchett:

> in Lattd: 52½° and 35° west Longitude from London, we fell in with great Islands of Ice, of soe Incredible a hight and Magnitude, that I scarce dare write my thoughts of it, at first we took it for land with chaulky clifts, and the topp all covered with snow, but we soon found our mistake by standing in with it, and that it was nothing but Ice, though it could not be less then 200 foot high, and one Island at least 5 mile in front, we could not get ground in 140 fadtham. Yet I conceive it was aground, Ice being very little lighter then water and not above an Eight part above the Surface when it swims; It was then the hight of Summer, but we had noe other signe of it but long Days; it froze both nigth and day, whence it may be understood how these bodies of Ice are generated being allways increased and never thawing. The next day February the 2d. we were in Imminent Danger to looss our ship and lives, being Invironed with Ice on all Sides in a fogg soe thick, that we could not see it till was ready to strike against it, and had it blowne hard it had scarce been possible to escape it: Soe I stood to the Northward to get clear of it, which in the Lattd. of 50° I did, and their saw the last Ice. (Thrower, 1981, p. 306.)

Halley's estimates of longitude, which were by dead reckoning, have been revised by Barraclough (1985) and he was probably some 4° further east at this point than he thought. The *Paramore* was then at least 200 nautical miles away from South Georgia and in any case throughout this period there appears to have been fog which would have obscured the island. South Georgia had probably been sighted by Antoine de la Roche, a London merchant, in 1675 and a translated precis of his account of this was published in Spain in 1690 (Headland, 1984). Halley appears not to have seen this report and, if he had, its lack of information about latitudes and longitudes would have given him no particular reason for searching in that region. As it was, his concern for the health and safety of his crew made him decide to head north again. The *Paramore* arrived at Plymouth on 27 August, 1700 having, unusually for those times, lost only one life.

The most important immediate result of this voyage was the publication in 1701 of a chart of the Atlantic showing magnetic variation (Fig. 2.2). Later Halley was to incorporate this into a world chart which appeared in various editions and revisions, and remained in use until the end of the eighteenth century. The idea of finding longitude by magnetic variation had by then been recognized as impracticable but, of course, there was still the need for accurate variation charts from which to correct compass bearings (Cotter, 1981). The Atlantic chart is of particular cartographical significance as it is believed to be the earliest printed map with isolines of variations; indeed isolines have also been called Halleyan lines.

His observations of declination and inclination, after correction of their reported positions to allow for error in estimation of longitude, were considered reliable enough by Bloxham *et al.* (1989) to use in a study of geomagnetic secular variation.

Thus Halley, in the course of his second voyage, penetrated deliberately if marginally into the Antarctic region. Some of his observations on animals are ambiguous – judging from his indifference to Leeuwenhoek's discoveries Halley was little interested in biology (Palm, 1989) – but other things match precisely with what we should expect in that sea area today. In view of his brief experience and the absence of any soundings, his conclusion that the large icebergs which he saw were ice islands may be cxcused.

After this Halley made no more excursions south. He became Astronomer Royal in 1720 and remained active as a scientist until he died in 1742 at thc age of 86. In the end it was his indirect contributions to Antarctic science that were most important. He carried out original work on geomagnetism and was the first to connect this with the aurora (Ronan, 1968). He also studied the behaviour and cause of trade winds and made attempts on a scientific basis to determine the age of the earth. He stepped out of his study to make the whole world his observatory and can justly be regarded as the founder of the science of geophysics (Armitage, 1966; Ronan, 1968). As we shall see, this branch of science has subsequently on more than one occasion provided the major impetus for large scale Antarctic investigations. Apart from this he had shown that it was worth spending public money on an expensive expedition mounted specially for a scientific purpose. There was also his advocacy of using the transit of Venus as a means of estimating 'astronomy's most important constant' – the sun's distance from the earth – and his publication in 1716 of a programme for taking advantage of the transits which he predicted would occur in 1761 and 1769. This programme aroused international interest and provided the initial impetus that launched Captain James Cook on his voyages and circumnavigation of Antarctica. The Antarctic voyage of Bellinghausen was undertaken in emulation of Cook, but perhaps there was a more basic link between this first Russian venture to the far south and Halley. Halley presumably described his plans for the voyage of the *Paramore* to Peter the Great and the idea that naval vessels should be used for scientific expeditions was perhaps implanted in the minds of Russia's rulers at this point.

It was entirely appropriate that the Royal Society should have named the base on the Brunt Ice Shelf Halley Bay which was set up in 1956 for the International Geophysical Year and that this name should be perpetuated by the British Antarctic Survey station in the same area.

2.3 *Terra Australis Incognita* and the theoretical geographers

Nevertheless, for the most part of the eighteenth century it was the lure of a southern continent extending into temperate latitudes rather than anything else which provided the most powerful motive for exploration in the southern hemisphere. Fortuitously, isolated sightings of capes and coastlines approximately fitted the imagined outlines of the continent as shown on contemporary world maps (Rainaud, 1893). The Dutch navigator Jakob Roggeveen voyaged into the Pacific in search of a part of the continent, Davis Land, and his furthest south of 62°30′, achieved in 1722, matched with a large southerly embayment in the Cape Horn region of the map of the supposed continent. The abundance of birds and ice which he encountered were taken to indicate proximity of land and the idea of the austral continent was not put in doubt. Bouvetøya (54°25′ S) was discovered in 1739 by Bouvet de Lozier, who had been commissioned by the Compagnie Français des Orientales to establish a port of refreshment on the continent for their vessels *en route* to India and China. Not being successful in circumnavigating it, Bouvet persisted that what he had seen was part of an extensive coast, so again belief in a continent was not questioned although it had to be admitted that it was nearer the pole than hitherto thought. Yves-Joseph de Kerguelen-Trémarec was likewise looking for the continent when in 1772 he came upon the group of islands now named after him. A second visit convinced even him that they fell far short of his expected paradise and the proponents of the idea of a temperate southern continent had to accept defeat.

During this period theoretical geographers were also preoccupied with *Terra Australis Incognita* (Rainaud, 1893). Pierre Louis Moreau de Maupertuis, mathematician and astronomer, put a reasonable case for its investigation. He did not attach much importance to the argument that a continent must be there to counterpoise those in the northern hemisphere but in 1756 pointed out that an enormous area of the southern hemisphere was unknown and that if there was a continent, because of its isolation things on it might be very different from what they are elsewhere. He was not without practical experience in exploration, having led an expedition to Lappland to measure the length of a degree along the meridian as a test of the Newtonian hypothesis that the Earth is an oblate spheroid, and thought that ice need not be too much of an obstacle to access if the techniques of Arctic whalers were employed. Another Frenchman, Phillipe Buache, one of the founders of physical geography, in a paper concerned with the framework of the physical globe presented to the Académie des Sciences in 1754, linked up the mountain chains of the southern hemisphere with islands discovered in the higher latitudes. He supposed the same general structure for the Antarctic as for the Arctic, that is, a nearly land-locked polar sea with large rivers, like those of Siberia, acting as sources of ice. Charles de Brosses, in his *Histoire des navigations aux terres australes* published in 1756, besides giving a catalogue of 47 voyages, discussed the greater cold of the Antarctic as compared with the Arctic. This he attributed to a displacement of the Earth's axis in the recent past from a position in which Labrador and Iles Kerguelen were the poles, ignoring the fact that everywhere in the southern hemisphere the ice is in closer proximity to the equator than it is in the north. Like all his contemporaries he associated ice formation with land. He suggested that temperatures might actually rise towards the poles on account of the continuous sunshine in summer. He proposed that just as Magellan had encircled the Earth between parallels, a circumnavigation along a great circle through the poles was desirable. Long before this was accomplished by the Transglobe Expedition of 1979–82, the error of his reasoning about temperature was apparent.

2.4 The voyages of James Cook

We must now return to the transit of Venus. For various reasons, among which were inaccuracies in the determination of longitude, observations in 1761 were conflicting and disappointing (Woolley, 1969). Since the next pair of transits was not due until 1874 and 1882, the second of the pair to occur in the eighteenth century, due to take place in 1769, was the subject of anxious planning by astronomers. Arrangements were made to observe it from as many places as possible and, in Britain, Dr Thomas Hornsby, Professor of Astronomy at the University of Oxford, reminded the Royal Society of its duty in this respect. The Society responded enthusiastically and petitioned King George III for support for an expedition to the South Pacific so that the base line for observations might be extended as much as possible. The petition was granted immediately. Following the Peace of Paris in 1763, England was in a phase of enterprise and expansion – this was the time of the agricultural revolution and the apogee of the East India Company – and, as Hornsby had pointed out (Beaglehole, 1974), there might well be commercial advantage in a settlement in the South Pacific.

The Royal Society's choice of a leader for the expedition was Alexander Dalrymple, a man of ability and with knowledge of eastern seas, but vain, sometimes foolish, and obsessed with searching for the mythical continent of the map-makers. The Admiralty found this suggestion 'totally repugnant to the rules of the Navy' – memory of the trouble but not the achievement that arose from Halley's commission still lingered. Dalrymple thereupon declined to have anything further to do with the voyage. Instead, James Cook, a marine surveyor of known ability who had already published a paper in the *Philosophical Transactions of the Royal Society of London*, was chosen. He was then low down in the naval hierarchy but this does not imply that the Admiralty was regarding the expedition as of little importance. Much thought was given to what might be the most suitable vessel and the one of their choice, the barque *Endeavour*, was fitted out with the utmost care (Beaglehole, 1974). There is no need here to give biographical details of Cook[2] but a little should be said about the scientists he took with him. The Royal Society chose Charles Green to have the responsibility of making the astronomical observations. He had been assistant to the Astronomer Royal and was of the utmost use to Cook in fixing positions, and contributed greatly to the scientific success of the voyage (Beaglehole, 1968). Joseph Banks, then 25 years old but already a Fellow of the Royal Society of two year's standing, was of an entirely different sort. A young man of independent means, he had become attracted to botany and decided that the voyage of the *Endeavour* would give him the opportunity of something grander and botanically more rewarding than the grand tour of Europe conventional among his peers. The Royal Society gave him strong support and he and Cook took to one another immediately. Along with Banks, Cook accepted a retinue of eight, with great tolerance, because the *Endeavour* was already crowded beyond her capacity. This included two draughtsmen, an assistant naturalist and Daniel Solander, a Swede, who as a pupil of Linnaeus was the ablest botanist resident in Britain at the time. These scientists were equipped with the very best instruments available at the time. Of the books they took that by de Brosses seems to have been particularly influential (Carr, 1983).

The *Endeavour* left Plymouth in August 1768 and returned to London in July 1771. Conditions in Tahiti were perfect for observing the transit but the actual measurements did not prove of as much value as was hoped because the planet's atmosphere made it difficult to decide the exact moment of contact. In addition Cook had been given secret instructions to search for a continent south of Tahiti. This yielded negative results but otherwise, of course, the voyage was immensely successful. New Zealand was circumnavigated and the eastern coast of Australia discovered and accurately charted. This voyage marked the culmination of the use of astronomical data for fixing longitude, as envisaged by Halley. Beside this there was a vast mass of new plants and ethnological information. One is tempted to write about the *Endeavour* at length but for most of this glorious voyage she was north of 40° S and no direct contribution to our knowledge of the Antarctic came from it. Nevertheless, the indirect contribution to Antarctic science must be judged to be considerable. The voyage was a resounding triumph, in the estimation of their Lordships of the Admiralty, and no less so in the view of the general public; it reinforced the idea that an expensive expedition undertaken for scientific purposes could be of great value to a nation. It was no innovation to have had scientists aboard; de Brosses had regarded these as necessary members of a voyage of discovery and Bougainville, who was much influenced by him, had taken a botanist and an astronomer on his circumnavigation of 1766 to 1769, but those on the *Endeavour* acquitted themselves particularly well, indeed their activities were integral to the voyage, and a pattern was set that was to prove marvellously fruitful in the future. It was on this voyage that Cook eliminated the possibility of an extensive land mass in the temperate part of the south Pacific, thus clearing the way for exploration at higher latitudes. With Beaglehole (1968) we can regard the *Endeavour* voyage as giving Cook the experience, transforming his competence into genius, which enabled him to carry out the great work of his next circumnavigation. It may be doubted whether like things would have happened had Dalrymple been in charge.

Cook had already conceived the idea of a second voyage to the South Seas and sketched a plan for its execution by the end of the *Endeavour* cruise. The plan was a masterly one, making use of the prevailing winds and having Queen Charlotte's Sound in New Zealand as a base for refreshment to enable long sweeps at high latitudes to be made in the summer months. This plan was accepted by the Admiralty without question and two barques similar to the *Endeavour*, eventually to be named the *Resolution* and the *Adventure*, were purchased for him. Their fitting out was done under

Cook's supervision and so high did he stand in the Admiralty's estimation that his requests were invariably granted immediately. He evolved his own method of getting things done quickly – going to the appropriate government office, he would explain his needs and write the necessary formal applications on the spot (Beaglehole, 1969, p. xxvi). The expedition had the support of the Board of Longitude, which had been established by Act of Parliament in 1714. The Board was to bear the cost of an observer to accompany each ship and, together with the Royal Society, provide the instruments needed for carrying out and testing methods for determination of longitude. William Wales and William Baly, chosen by Maskelyne, the Astronomer Royal, went as the observers. The instruments with which they were supplied were of the best and are listed by Rubin (1982a). A prize of up to £20,000, an enormous sum in those days, had been offered for the discovery of a means of determining longitude to within 30 nautical miles after a voyage of six weeks, and timepieces which offered the possibility of doing this were among the equipment. A copy by Larcum Kendall of a watch made as long ago as 1762 by John Harrison worked satisfactorily, and Harrison was eventually awarded the prize after a dispute in which the Royal Society played a not very creditable part.

Things did not go smoothly when it came to arrangements for naturalists to go on the voyage. Banks was invited to take part by Lord Sandwich, First Lord of the Admiralty, and accepted with alacrity for himself and Solander. But, encouraged by public adulation, Banks had become too big for his boots. He imagined himself standing at the South Pole and turning on his heel through 360° of longitude but seems not really to have appreciated the essentially geographic objects of the voyage. He assumed that it was being arranged for him, that Cook was no more than the ship's master, and that he should have command of the expedition. A retinue of draughtsmen, an artist, secretaries, servants, two horn-players and, as it later emerged, a girl friend, was planned and Cook, his good nature overcoming his better judgement, agreed to the *Resolution* being altered to accommodate this party and their impedimenta. The potentially disastrous effect of the extra superstructure was plain as soon as the *Resolution* was under sail and Cook immediately asked that it be removed. When it had been and Banks found out, he ordered his retinue and equipment out of the ship in high dudgeon. The Admiralty accepted this as resignation from the expedition and Lord Sandwich, by now embarrassed by his friend's egotism, quietly but firmly saw to it that the withdrawal was final. It is a pity that we have to leave one who was to be a uniquely influential figure in the British scientific establishment – he has been described as 'a one-man benevolent research council'[3] – in this unfavourable light, but although the rift with Cook was not permanent, Banks played no further important part in Antarctic matters.

At one point in the negotiations there was a suggestion that Joseph Priestley should join the expedition as astronomer and we may reflect for a moment on the consequences for science had this happened. Priestley had already made the crucial experiments which led to the discovery of oxygen and photosynthesis. Would his genius have been able to exercise itself under shipboard conditions or might it even have developed further with enforced abstinence from the distraction of politics? He was a man of the laboratory rather than of the field and maybe would have done no more than Wales or Baly.

The substitute for Banks was Johann Reinhold Forster, born near Danzig but domiciled in England since 1766. Having started life in the Church he had travelled widely in Russia, and was accomplished in many fields. In 1767 he succeeded Priestley at the Warrington Academy as tutor in modern languages and natural history. By 1772 he had published a translation into English of Bougainville's *Voyage autour du monde*, in the preface to which he boldly outlined what a well-qualified philosopher might accomplish on a voyage of discovery (Hoare, 1982, p. 47). In the same year he had been elected to the Fellowship of the Royal Society, his supporters including Banks, Solander and Daines Barrington, a lawyer, antiquary and naturalist with influential connexions. Forster had angled for a berth on the *Resolution* but Banks and Solander gave him no encouragement. When Lord Sandwich decided that Banks should not go, he, on Barrington's advice, picked on Forster as the best man to fill the gap and by some adroit manœuvres secured both a parliamentary grant for him and royal approval of his appointment (Hoare, 1982). Forster thus became the first government-paid naturalist on a British exploring expedition but he seems to have been given no written instructions about his duties (Lenz, 1980). His own great ability was supplemented by that of his son, George, a promising naturalist and draughtsman, whom he insisted should accompany him. Father and son worked together in close harmony but there was to be little accord between Forster senior and the rest of the *Resolution's* company. Beaglehole (1969) has no good words for him as a man and considers that Forster exploited his son, but Lenz (1980) attributes the hostility which he aroused to his outspoken comments on inefficiency and despotism. He seems to have lacked tolerance and humour, and to have been liable to take anything less than complete compliance with the demands that he made for his scientific work as evidence of enmity.

Nevertheless, he was the most learned of the scientists associated with Cook. There seems to be no list of books taken aboard the *Resolution* but that Forster took many with him is evident from his journal (Hoare, 1982, p. 63). He was an avid collector and his library was eventually deemed to be the most outstanding one in private hands in Germany.

The *Resolution* and *Adventure* sailed from Plymouth on 13 July 1772. The cruise included three ice-edge sweeps – in the Atlantic-Indian Ocean sector from December 1772 to March 1773, in the Pacific sector from November 1773 to February 1774, and in the Atlantic sector in January and February 1775 – the great achievement of which was the delimitation of the Antarctic continent. Furthest south was reached on January 30th 1774. An extract from Cook's journal for that day may be given, not because a 'furthest south' is of particular scientific significance but because the attainment of it led Cook to express rather more of his thoughts than was his wont. After commenting on the continuous mass of ice that faced him and the 'ice-mountains' (i.e. ice-islands or icebergs) which lay behind it, he went on:

> it must be allowed that these prodigeous Ice Mountains must add such additional weight to the Ice fields which inclose them as must make a great difference between the Navigating this Icy sea and that of Greenland. I will not say it was impossible any where to get farther to the South, but the attempting it would have been a dangerous and rash enterprise and what I believe no man in my situation would have thought of. It was indeed my opinion as well as the opinion of most on board, that this Ice extended quite to the Pole or perhaps joins to some land, to which it had been fixed from the creation and that it here, that is to the South of this Parallel, where all the Ice we find scatered up and down to the North are first form'd and afterwards broke off by gales of Wind or other cause and brought to the North by the Currents which we have always found to set in that direction in the high Latitudes. As we drew near this Ice some Penguins were heard but none seen and but few other birds or any other thing that could induce us to think any land was near; indeed if there was any land behind this Ice it could afford no better retreat for birds or any other animals, than the Ice it self, with which it must have been wholy covered. I who had Ambition not only to go farther than any one had done before, but as far as it was possible for man to go, was not sorry at meeting with this interruption as it in some measure relieved us, at least shortned the dangers and hardships inseparable with the Navigation of the Southern Polar Rigions; Sence therefore, we could not proceed one Inch farther to the South, no other reason need be assigned for my Tacking and Standing back to the north, being at this time in the Latitude of 71°10′ S, Longitude 106°54′ W. (Beaglehole, 1969, p. 323.)

Forster's entry for this day is brief and colourless. He was exasperated by the search for a continent and a few days earlier had given vent in his journal to a tirade, obviously against Cook, about those who 'against humanity and reason' insisted on continuing it. However, Forster and Cook were agreed that the ice must have reached as a solid mass to a distance of 20° from the Pole, being broken by storms and eroded by melting at its edge in the summer, to be extended again in winter, and that the presence of land was not essential for its formation. The positions of the ice-edge as determined by the voyage of the *Resolution* accord well with what would be expected today in a bad year (Rubin, 1982a). The *Resolution* was probably only 100 nautical miles or so from land at its furthest south but the conditions that day were against any sighting and neither Cook nor any of his officers made unequivocal claim to have done so (Jones, 1982, p. 55). This was the nearest the *Resolution* came to the continent. Later, after his fruitless search for Bouvetøya, which showed that this could not be part of a continent, Cook wrote:

> That there may be a Continent or large tract of land near the Pole, I will not deny, on the contrary I am of opinion there is, and it is probable that we have seen a part of it. The excessive cold, the many islands and vast floats of ice all tend to prove that there must be land to the South and that this Southern land must lie or extend farthest to the North opposite the Southern Atlantick and Indian Oceans, I have already assigned some reasons, to which I may add the greater degree of cold which we have found in these Seas, than in the Southern Pacific Ocean under the same parallels of Latitude. (Beaglehole, 1969, p. 643.)

He had got the rough outline of the continent correctly. Forster (1778) in his comments on the coldness of the Antarctic as compared with the Arctic hit on an explanation, which is the opposite of Cook's and which is incorrect. He assumed that air temperatures depend on reflection of the sun's rays and the vast areas of ocean, absorbing rather than reflecting light, would thus result in lower temperatures. As elsewhere, Forster showed that he was not a physicist.

Apart from geographical discovery, this voyage contributed much to other scientific knowledge of the Antarctic region (Rubin, 1982a). What they found may seem commonplace and obvious now but it must be remembered that Cook and his companions were venturing into the entirely unknown.

Determinations of magnetic variation and dip, which previously with Halley and afterwards with Ross, provided the principal justification for sailing to the Southern Ocean, were included by the Board of Longitude among the duties which Wales was to carry out. At this time the effects of iron aboard ship were not really appreciated – Cook had the rather endearing habit of keeping the keys of the leg irons in the binnacle (May quoted by Jones, 1982) – and although Wales noted that the variation altered by as much as 4° according to whether the ship was on the port or starboard tack he did not relate it to shifting of the distribution of iron. Frequent determinations of variation and a few of dip were made but call for no comment in the Antarctic context. The aurora australis was reported probably for the first time, from 57°8′ S 80°59′ E on February 17 and its similarity to the aurora borealis was recognized (Forster, 1778) but the *Resolution*'s scientists were evidently not aware of Halley's suggestion that such displays are related to geomagnetism.

Wales, in addition to his other duties, had been specifically instructed by the Board of Longitude to make measurements of the temperature and saltiness of sea-water. For temperature measurements they had an apparatus, based on that invented by Stephen Hales some 20 years before, in which the thermometer was enclosed in a container fitted with valves so that it would retain a sample of water from the deepest point to which it was lowered. Four sets of measurements made in the Antarctic showed deep water (100 fathoms, 183 m) to be slightly warmer than that at the surface (Rubin 1982a). Wales did not try to explain this and Richard Watson (1782), who was aware that cooling of surface water would increase its density and so cause it to sink, in commenting on his results did not link the temperature inversion with salinity differences. Wales was not provided with equipment for determining saltiness and, although some samples for analysis were taken on the way home from the Cape of Good Hope, none were taken from the Antarctic. Currents were measured in calm weather by log line from a small boat held by a kedge or other heavy weight lowered to 100 fathoms (183 m) or so, a method obviously unreliable if there were subsurface currents. Cook concluded that in the Southern Ocean the set of the current was always northerly and it was this that brought ice from the south. His log records changes in air temperature and abundance of sea-birds at latitudes corresponding well with the present position of the Polar Front, and the sea temperature measurements just mentioned are also in accord with this (Rubin, 1982a). He did not, however, connect these phenomena with the pattern of currents to get an inkling of the grand design of water circulation in the southern hemisphere. Forster (1778), however, did recognize the implications of variations in sea temperatures and water movements between the northern and southern oceans. No deep soundings were made during the cruise.

Ice was obviously a subject for observation and speculation, apart from being a constant danger (Herdman, 1959). Cook had met ice before off the coast of Nova Scotia, Canada, and two of his crew had experience of Greenland ice. The dogma of the time was that ice is only formed near land but both Cook and Forster seem to have revised their initial acceptance of this view in the course of the voyage. Cook speculated on the effect of cold on sea-water and did not doubt that it could freeze but thought that this was unlikely in the open sea and that ice might form at the sea surface from falling snow in areas sufficiently protected by land, thus laying a foundation for its accumulation without the sea-water itself freezing. Forster went further and envisaged thick ice building up in the open sea by rafting of new ice as well as by accumulation of snow. Forster records (Hoare, 1982, p. 214) an experiment carried out by Cook which showed that a contraction in volume takes place when ice melts and he related this to the lesser specific gravity of ice as compared with water. Was cold a real substance which enters the water and expands it when it freezes, Forster wondered. The finding that ice collected at sea yields sweet drinking water was put to good use (Fig. 2.3) and Cook was puzzled as to what had happened to the brine which must have been formed when sea-water froze. Icebergs, referred to as 'ice-islands', were under continuous scrutiny. Estimates of size were entered up in logs; different people made widely different guesses but Wales seems to have taken the trouble to use his sextant to get their measure. Cook made an interesting calculation on the first Antarctic leg of the voyage which showed that icebergs covered one thirty-ninth of the sea area traversed. The continuous breakup and erosion of bergs and their occasional sudden disintegration were recorded by Wales and others. Forster realized that icebergs move only under the influence of currents or very strong winds and that they may become grounded. He attributed their blue colour to reflection of the colour of the sea but noted without further comment that the colour was often evident 20 or 30 feet (6 or 9 m) above the water and that ice is still blue when the sea looked like 'tincture of verdigrease'. Forster, early on in his journal, and Cook, near the end of his, put forward similar ideas on the origin of the tabular bergs characteristic of the Antarctic. Both rejected Buache's hypothesis that the ice is formed by freezing of river water but considered, rather, that it arises from the accumulation of snow in sheltered bays, imagining the ice cliffs which they had seen in such bays extending until the whole was filled. Neither,

Fig. 2.3 The *Resolution* taking aboard ice, 9 January 1773; copper engraving after W. Hodges from Cook, *A Voyage towards the South Pole* (1777) vol. 1, p. 37.

apparently, had heard of glaciers. Cook, incidentally, was badly out in doubting 'if ever the Wind is violent in the very high Latitudes' so that the sea there would freeze over and snow accumulate more readily. Forster adduced the stratification which he had seen in icebergs as evidence that deposition of snow played a part in their formation. Cook supposed that eventually large pieces of these expanses broke off under their own weight whereas Forster thought that they were freed in warm summers by water from the land melting them along their margins and buoying them up. These ideas are not too far off modern ones on the formation of tabular bergs from shelf ice. Cook did not arrive at the idea that the weathering of tabular bergs which, of course, he had seen in action, produces bergs of other shapes but suggested that irregular stretches of coast act as templates for these.

The only landing made from the *Resolution* in the Antarctic region was a brief one on South Georgia, the first ever on typically Antarctic terrain, and Forster had minimal opportunity to use his considerable knowledge of geology. In Possession Bay he noted that the rock was a bluish grey slate laid down in horizontal strata and that there appeared to be no other minerals.

Opportunities for botany were as brief as those for geology. Forster (1778) recorded only two species from South Georgia, one new which he named *Ancistrum*. A specimen of this collected in Possession Bay by the Forsters survives in the Natural History Museum, London but it is now known as *Acaena ascendens* (Fig. 2.4). The other, a 'well known grass' called by him *Dactylis glomerata*, was probably *Poa flabellata*, the tussock grass. Forster remarked on its growth habit, providing shelter for seals, penguins and shags, and its role in soil formation. The identity of a third plant, a 'Plant like Moss which grows on the rocks', mentioned by Cook but not by Forster is obscure (Greene, 1964). Forster surmised that the South Sandwich Islands must be incapable of supporting any plant life.

However, Forster's forte was zoology and particularly with birds he made a major contribution. Out of the total of around 44 birds now recorded as breeding in the Antarctic and sub-Antarctic he saw about 31, of which 14 were then species new to science and first described by him. As an example of his ornithological activities here is an excerpt from his journal.

> Jan. ye 13th [1773] I went out to shoot a bird we saw swimming on the Sea. I winged it & it got some shot into the neck, so that we got the bird alive. It proved to be the *Quackerbird*, which I found to be a kind of Albatross, never described before, & I called it therefore *Diomedea palpebrata*. It is of a dusky grey, its head, wings and tail black, the bill of a

Fig. 2.4 Specimens of the greater burnet, *Acaena adscendens*, collected by J. R. and G. Forster, Possession Bay, South Georgia, 17 January 1775. This is the only plant material known to have survived from the first collecting expedition on South Georgia. (Courtesy of the Natural History Museum, London.)

> jetty black, the feet of a dusky blue: the (rachis) shafts of the Quill & tail feathers white; the Eye which is of a hazel colour, is surrounded with a white except before: the Tail wedgeshaped. (Hoare 1982, p. 213.)

This bird, the light-mantled sooty albatross, is now known as *Pheobetria palpebrata* (Forster). The value of the elder Forster's bird descriptions was enhanced by his son's illustrations (Fig. 2.5). The elder Forster saw the southern sea elephant, *Mirounga leonina*, and the southern fur seal, *Arctocephalus tropicalis gazella*, on South Georgia but was not able to add much to previous knowledge of these beasts. Other seals were seen swimming but for obvious reasons he was unable to describe them. There was a similar difficulty with whales and although he shot and killed what was possibly a Pacific pilot whale in New Zealand waters, the ship was

Fig. 2.5 Drawing of the chinstrap penguin, *Pygoscelis antarctica* (Forster), by G. Forster. (Courtesy of the Natural History Museum, London).

going at $3\frac{1}{2}$ knots and it was lost (Hoare, 1982, p. 295). Neither krill swarms nor Antarctic fish seem to have attracted his attention.

The *Resolution* returned to England on 29 July 1775. This time it was Cook who got the great acclaim, while Forster, who had imagined himself getting as much public recognition as Banks did after the *Endeavour* voyage, was left out (Lenz, 1980). Moreover, there was frustration in store for him over the publication of the history of the voyage. Forster was under the impression, rightly or wrongly, that the Admiralty had agreed to him writing the history and benefiting financially from its sale. The Admiralty did not see it this way, Forster refused to accept any compromise, and after much acrimony it was Cook's journal that was polished up to become the official account. *A Voyage towards the South Pole and Round the World*, appeared in 1777 and is a classic of Antarctic and Pacific exploration. Prohibited from publishing himself before this appeared, Forster resorted to George, who managed to write his *A Voyage Round the World*, which perforce had to appear without illustrations, and have it published six weeks before Cook's volumes appeared. This initiated a new and popular style of travel writing but contributed little new on the scientific side. The elder Forster was then able to devote his time to his own *Observations made during a Voyage Round the World*. This was a general scientific and philosophical account containing much relating to the Antarctic – although he made evident his abhorrence of this part of the world – and even more on the anthropology of the Pacific islands.

Cook is generally regarded as a supreme navigator rather than a scientist. Certainly he was self-taught and without formal training as a scientist but in the eighteenth century this was of no consequence. In electing him to their Fellowship in 1776, the Royal Society recognized him as a scientist of distinction but in awarding him their highest honour, the Copley medal, they did so for not the best reason. This medal was given for a paper on the health of seamen, a subject near to the heart of the President, Sir John Pringle, a medical man (Pringle, 1776) but here Cook had done no more than apply thoroughly and intelligently empirical rules mostly recommended by others (Cook, 1776). James Lind in 1747 had carried out what was probably the first controlled trial in clinical nutrition and shown clearly that oranges and lemons are effective cures for scurvy. Cook was eminently successful in holding scurvy at bay on the ships under his command but he seems not to have heard of Lind's work and achieved his results by a variety of different measures, not one of which he could be sure was effective (Carpenter, 1986). Forster (1778), who had his own elaborate ideas about scurvy being caused by an imbalance in food of the acids, phlogiston and alkalies, gave a nauseating account of food and hygiene on eighteenth century voyages which increases one's admiration for Cook's practical achievement, however lacking it was in scientific basis. Cook himself thought this was his finest accomplishment:

> But, whatever may be the public judgement about other matters, it is with real satisfaction, and without claiming any merit but that of attention to my duty, that I can conclude this Account with an observation, which facts enable me to make, that our having discovered the possibility of

> preserving health amongst a numerous ship's company, for such a length of time, in such varieties of climate, and amidst such continued hardships and fatigues, will make this Voyage remarkable in the opinion of every benevolent person, when the disputes about a Southern Continent shall have ceased to engage the attention, and to divide the judgement of philosophers. (Cook, 1777, vol. II, p. 292.)

We must look on Cook's control of scurvy as an essential prerequisite without which his other, far more outstanding, work would have come to naught. In his passion for exact observation and measurement and in his systematic testing of geographical hypotheses he was clearly a scientist of a high order. It must also be remembered that he and Banks laid the foundation for anthropological studies in the Pacific.

Cook had the scientist's curiosity and a broadening of his attitude to learning is evident from his journals. At first he was scornful of the botanizing of Banks and Solander (Beaglehole, 1974, p. 161) but later he supported their work with enthusiasm (Ibid, p. 329). His troubles with Banks over the second voyage seemed not to have led to a souring of his relations with scientists but Forster was an awkward character, both during the voyage and afterwards when it came to publication. Cook's journals make no mention of difficulties or arguments with Forster but his alleged outburst 'Curse the scientists and all science into the bargain!' when his third voyage was in preparation may have expressed something of his feelings (Beaglehole, 1969, p. xlvi). It is significant that he took no naturalist with him on his third voyage. This was to the end of the earth opposite to the Antarctic, so it does not concern us greatly. However, it was when the *Resolution* touched at Kerguelen on this voyage that the surgeon, Anderson, added to knowledge of sub-Antarctic botany by describing, among other things, the Kerguelen cabbage (see Fig. 3.12) giving it the name *Pringlea* in honour of the President of the Royal Society. Nearly four years later, towards the sombre ending to this voyage, the friendly relations established in Kamchatka may perhaps have helped to implant the image of the recently assassinated Cook in the Russian consciousness and so paved the way for Bellingshausen.

William Wales, with his high intellectual standards, contributed much to the success of the second voyage by reliable observation and measurement. The report which he wrote with Baly, the *Original Astronomical Observations, made in the course of a Voyage towards the South Pole, and Round the World* (1777) is a valuable repository of data but adds little in interpretation or speculation. He did, however, lay *Terra Australis Incognita* to rest with the words:

> the notion which some persons have got concerning the necessity of a counterpoise, is so very unphilosophical, that I am much surprised how so many ingenious Gentlemen have happened to adopt it.
>
> (Wales & Baly, 1777 p. 8.)

The elder Forster's book perhaps goes too far in the other direction but should not be dismissed in such severe terms as those used by Beaglehole (1969). Forster was an expert taxonomist and made full use of his time on the *Resolution*. His views on ice formation and distribution were similar to those of Cook; they evidently discussed such matters together in depth (Hoare, 1982, p. 451) so that it is difficult to apportion credit for ideas. Lenz (1980) argues that it was Forster's criticisms of the establishment which led to him being underrated in Britain. Certainly his reputation in Germany, to which he returned in 1780, was, and still is, greater. His son George was to have an influence on the young Alexander von Humboldt. Hoare (1982) suggests that it was partially through the elder Forster that the theoretical implications of Cook's voyage were transmitted into the corpus of scientific knowledge. His *Observations* was one of the most important of the early systematic treatments of geography and he can be looked on as the forerunner for some of the early nineteenth century interests in the geo- and life sciences.

> The risk one runs in exploreing a coast in these unknown and Icy Seas, is so very great, that I can be bold to say, that no man will ever venture farther than I have done and that the lands which may lie to the South will never be explored. (Beaglehole 1969, p. 637.)

These words coming from Cook must have been discouraging. There were, indeed, no officially sponsored voyages of scientific investigation in the Southern Ocean for over 40 years after his second voyage. During this period Europe was preoccupied with the Napoleonic Wars and after the Battle of Waterloo (1815), although there was an upsurge of interest in exploration, commercial and political eyes were directed more towards the Arctic and the North-west Passage than to the Antarctic (Gough 1986). There was much activity of an unofficial sort in the far south, however, and this must be discussed later, but the next venture in the style of Halley and Cook was from Russia and did not sail until 1819.

2.5 The voyage of Thaddeus Bellingshausen

During the eighteenth century Anglo-Russian contacts in science had been promoted, in particular by Daniel Dumaresq, a Channel Islander, who resided in St. Petersburg in 1747–57 and again in 1764–66. In 1763 he was elected as an honorary Foreign Member of the Imperial Academy of Sciences (which had been founded in 1726) and as a Fellow of the Royal Society. Although his interests extended widely, including the exploration of Kamchatka, experiments on freezing, magnetism and the transit of Venus (Appleby, 1990), he evidently did not touch on anything specifically related to the Antarctic. The connexion between the Russian and British navies, which, as we have seen, began with Peter the Great, had also continued to be close. Adam Ivan Krusenstern, trained as a cadet in the British Navy and in 1803 was the first Russian to circumnavigate the world. He was the initiator of a phase of great activity in exploration by Russia. An excellent seaman, he was far sighted and firm of purpose, and inspired a number of Russian voyages, notably those of Golovin and Kotzebue. Russia needed to develop a supply route to its far eastern territories and so it was important to gain experience in sailing southern seas and to find possible bases and ports of refreshment. There was also exploration in the Bering Straits area, in which, incidentally, rivalry with Britain did not preclude exchange of charts and information between the hydrographers of the two countries (Gough, 1986). In 1819 expeditions were dispatched to both the north and south. The preliminary memoir to the account of the Antarctic voyage begins thus:

> Emperor Alexander Pavlovich (Alexander I) of glorious memory, desiring to help in extending the fields of knowledge, ordered the despatch of two expeditions, each consisting of two vessels, for the exploration of the higher latitudes of the Arctic and Antarctic Oceans. Following His Majesty's commands, announced on the 25th of March 1819, two vessels were selected for exploration in the Antarctic Sea, viz. the sloops *Vostok* (East) and *Mirnyi* (Peaceful), under the command of Captain (2nd rank – now Vice-Admiral) Bellingshausen. (Debenham, 1945, p. 1.)

Alexander I, a contradictory character of general, rather than thorough, education, who managed to combine advanced liberal ideas with thoroughgoing autocracy, did indeed take a personal interest. He followed his grandmother, Catherine II, in wanting to extend Russia's influence in the world but by peaceful methods. He had enlightened ministers in charge of marine affairs, in particular the Marquis de Traversey, and it was they who saw the advantage of such expeditions for training in seamanship and navigation and provided the detailed instructions. These instructions were modelled on those for British expeditions and show the influence of Cook. A Russian translation of his book *Voyage towards the South Pole and Round the World* had been published in St. Petersburg in 1796 and there was great admiration for him in Russian naval circles.

Thaddeus Fabian von Bellingshausen[4] was born in 1779. He had served as a junior officer with Krusenstern in his circumnavigation, and in June 1819, when he was offered the command of the Antarctic expedition, was Captain in charge of the frigate *Flora*, carrying out a hydrographic survey of the Black Sea. After the Antarctic voyage he was to attain the rank of Admiral and when he died in 1852 was Governor of the port of Kronshtadt in Russia. He was undoubtedly a superb seaman and navigator and from his account of the voyage seems to have been an educated, thoughtful and observant person. His fellow officer Lieutenant Mikhail Lazarev, in command of the *Mirnyi*, was born in 1788 and had spent five years in the British Navy. He too was a fine seaman and an educated man.

The great object of the expedition was exploration in the closest possible vicinity to the South Pole but it was also to gather on the way as much information as possible about the activities of other nations. Imperial impatience is evident in the hurry with which it was dispatched. Bellingshausen had little more than a month for preparation and the Imperial Academy of Science had to excuse itself on grounds of lack of time from drawing up instructions for the scientists. The Minister of Naval Affairs, however, dashed off a set of instructions for a comprehensive and exhaustive programme which might have kept a number of savants busy for their lifetimes; it included beside the usual geodetic, astronomical and meteorological work, a proposal that balloons should be used to determine wind direction at levels above the sea surface, and plans for research on human anatomy should it be possible to obtain bodies for dissection. I. M. Simanov, a professor of astronomy, was taken as astronomer. Scientific instruments were purchased in London at the beginning of the voyage but a ready-made pendulum clock was not to be had so gravity measurements were foregone. The two naturalists appointed, Mertens of Halle and Kuntze of Leipzig, declined the invitation. Bellingshausen did not learn of this until he reached the rendezvous in Copenhagen and since, even with the help of the now aged Sir Joseph Banks, he was unsuccessful in recruiting anyone in London, the expedition was without expertise in natural history. Bellingshausen felt this lack keenly and considered that the plea by Martens and Kuntze of a lack of notice was unreasonable since all they required for their work was books, which could have been got in Copenhagen or London. In his account of the voyage he several times draws attention to missed opportunities for observations on plants and animals. In one of his rare criticisms of authority he recorded that two Russian students of natural history were willing to join him but were not permitted to do so because of a prejudice in favour of foreigners. Both ships

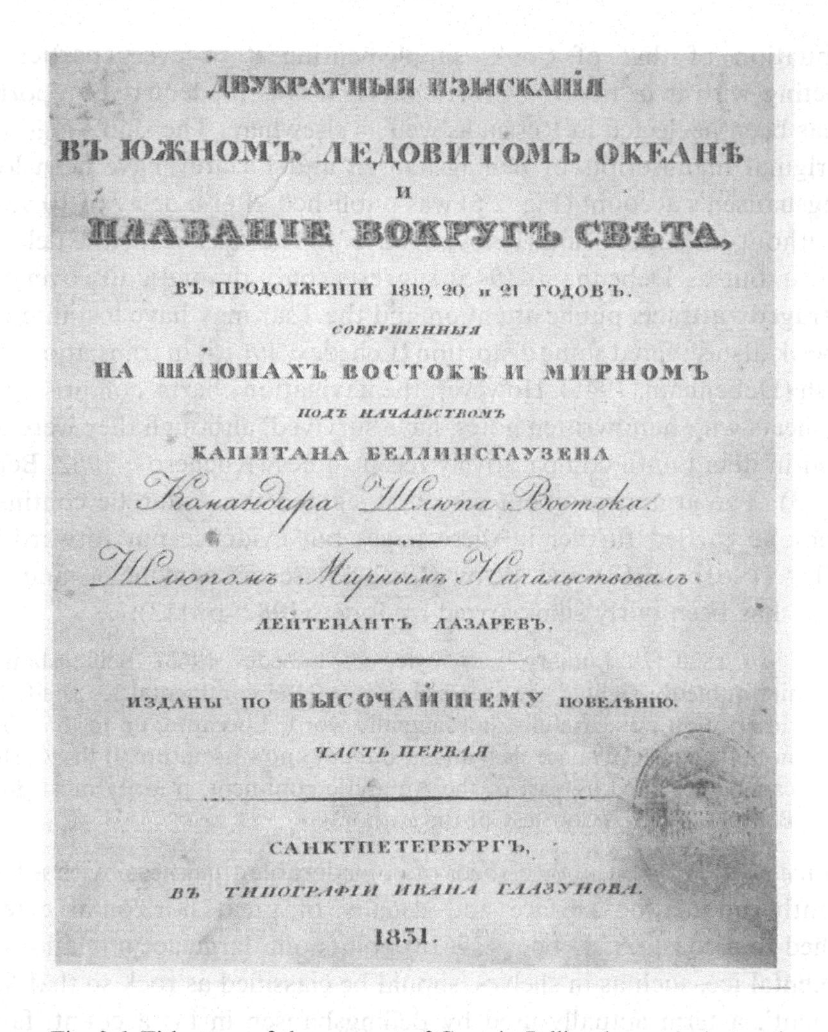

ДВУКРАТНЫЯ ИЗЫСКАНІЯ

ВЪ ЮЖНОМЪ ЛЕДОВИТОМЪ ОКЕАНѢ

И

ПЛАВАНІЕ ВОКРУГЪ СВѢТА,

ВЪ ПРОДОЛЖЕНІИ 1819, 20 и 21 ГОДОВЪ.

СОВЕРШЕННЫЯ

НА ШЛЮПАХЪ ВОСТОКЪ И МИРНОМЪ

ПОДЪ НАЧАЛЬСТВОМЪ

КАПИТАНА БЕЛЛИНСГАУЗЕНА

Командира Шлюпа Востока.

Шлюпомъ Мирнымъ Начальствовалъ

ЛЕЙТЕНАНТЪ ЛАЗАРЕВЪ.

ИЗДАНЫ ПО ВЫСОЧАЙШЕМУ ПОВЕЛѢНІЮ.

ЧАСТЬ ПЕРВАЯ

САНКТПЕТЕРБУРГЪ,

ВЪ ТИПОГРАФІИ ИВАНА ГЛАЗУНОВА.

1831.

Fig. 2.6 Title page of the account of Captain Bellingshausen's Antarctic voyage, 1819–21. (Courtesy of the Scott Polar Research Institute.)

had surgeons but their interest in natural history does not seem to have extended beyond stuffing birds.

The choice of ships might also have been better had more time been allowed for consideration. The *Vostok* and the *Mirnyi* were ill-matched in speed and throughout the account of the voyage the strain of keeping contact is evident. Some strengthening was added but neither vessel had been built for work amongst ice. The two ships sailed from Kronshtadt on 14 July 1819 (new style; Bellingshausen's account used the old calendar).

The circumnavigation by the *Vostok* and *Mirnyi* ranks among the great voyages of all time. In the words of Mill (1905) it 'was a masterly continuation of that of Cook, supplementing it in every particular, competing with it in none'. Unfortunately, it was inadequately reported and has been neglected in Russia as well as elsewhere. The ship's logs and the original manuscripts of Bellingshausen and Lazarev have been lost. Bellingshausen's account (Fig. 2.6) was published after a delay of 10 years and without his supervision (Belov, 1962–63). The reason for the delay is not clear but as Debenham (1945) suggests, only dramatic discovery or high tragedy attracts public attention and the Tsar may have lost interest. The book also suffered some distortion (Lebedev, 1961a) in translation into English (Debenham, 1945). However, the navigation charts, comprising 15 large sheets with handwritten notes, have survived, although they were not studied in detail until comparatively recent times (Kucherov, 1962; Belov 1962–63). The argument about who first sighted the Antarctic continent will not be carried further in these pages but evidence put forward by Lebedev (1961b, 1963) and Belov (1962–63) seems convincing and the position has been fairly summarized by Jones (1982, p. 113):

> In 1820 [28 January, new style, see Lebedev 1963], Bellingshausen undoubtedly sighted the seaward edge of the continental ice shelf. His description was carefully and factually worded because, up to that time, nobody had seen an ice shelf and he did not know its nature. If the ice shelf can be accepted as part of the Antarctic continent, priority must go to Bellingshausen. (Courtesy of the author)

An *ice shelf* is a floating ice-sheet of considerable thickness, with a level or gently undulating surface and usually of great horizontal extent, attached to a coast. As Lebdev (1961a) points out, large accumulations of continental ice, such as in shelves, should be classified as rock so that 'ice-continent', a term actually used by Bellingshausen in his account, fairly signifies a continent (but see Armstrong, 1971).

The *Vostok* and *Mirnyi* returned to Kronshtadt on 5 August 1821 (new style), the former badly strained but both with crews in good health. The geographical achievements in Antarctica had been the tighter circumscription of the still hypothetical continent and the discovery of Peter I Island (the first land to be sighted within the Antarctic Circle) and Alexander Island. Possibly with Buache's ideas (p. 17) in mind and certainly influenced by Humboldt's writings, Bellingshausen speculated on submarine mountain chains and in supposing that the South Sandwich Islands formed part of the same range that surfaced further west as South Georgia identified part of that major physical feature, the Scotia Ridge. Following the same train of thought he looked, fruitlessly, for a continuation southwards from Macquarie Island of what is now known as the

Macquarie Ridge. Had he arrived a little later in the season, when ice conditions might have been more favourable (Rubin, 1982b), it would have led him to the Balleny Islands and, possibly the Ross Sea.

Apart from this, Bellingshausen added much to the still scanty knowledge of the Antarctic environment but we need not be concerned with the details of his observations on ice, sea temperatures, atmospheric pressure, winds, weather, surface currents and the aurora (see Rubin, 1982b for summaries of these). It should, however, be noted that he gave more precision to Cook's observation of the important fact that south of 60°, winds are generally easterly. He accumulated more magnetic data than appeared in his book and passed much of it on to Gauss, who published it in 1840 (Armstrong, 1971). As Belov (1966) points out, his estimate of the position of the South Magnetic Pole at 76° S 142.5° E is close to that found by Ross some 20 years later and the honour of locating it must thus be shared. One of the few occasions when there was a near coincidence of the tracks of the *Vostok*, from December 1819 to 23 January 1820, and *Resolution*, from 14 January to 11 February 1775, afforded an opportunity to compare year to year variations in conditions in the vicinity of the South Sandwich Islands (Ruben, 1982b). Having found considerably more ice than was reported by Cook, Bellingshausen rightly concluded that 'these fields of solid ice are not permanent and consist of pieces of floating ice' (Debenham, 1945, p. 394). He thought, incorrectly, that icebergs were formed by the addition to sea ice by freezing from below and snow from above but correctly surmised that irregular bergs were produced by uneven erosion and capsizing of tabular bergs. He carried out experiments on the freezing of sea-water, finding that it froze less solidly and compactly than did fresh-water and that the ice formed from it yielded fresh-water on melting. He found that the specific gravity of water at the surface among ice floes in the vicinity of the South Sandwich Islands was 1.0997 whereas that of water brought up from 256 fathoms was 1.1009 but left this finding without comment (Debenham, 1945, p. 100). He was abreast of the literature on ice. Nairne (1776) had reported on the freezing of sea-water and shown by determinations of density that water from the ice it produces approximates to fresh-water whereas the residue of water which it leaves after freezing is a brine of density greater than that of sea-water. Nairne also determined the freezing point of sea-water as 28.5 °F. There was also a comprehensive account of sea ice available from Scoresby (1815). Although Scoresby was clear that some icebergs were formed on coasts he too thought that large bergs could be formed well away from land by accretion of precipitation on sea ice.

Bellingshausen was the first to record volcanic activity in the Antarctic region. A 'crater, from which a thick stinking vapour was continually rising' was seen on Zavodovski Island. On landing, there were doubts as to whether the stink came from the crater or the penguins but half way up the mountain the ground was found to be warm. Earthquake shocks were noted from both the *Vostok* and the *Mirnyi* in the vicinity of Macquarie Island.

We have already noted that Bellingshausen was handicapped by lack of a naturalist. He kept records of birds, seals and whales encountered and some of them were sketched by the artist Mikhailov. Inevitably, identifications were sometimes imprecise. An animal that from the description must have been leopard seal is left simply with the designation 'a species of seal' (Debenham, 1945, p. 113). Likewise the crabeater seal was pictured recognizably in Bellingshausen's *Atlas* but not named (Bertram, 1940). What must have been the first emperor penguin to be seen by man was confused with the king penguin (Debenham, 1945, p. 388). Diatoms were seen in sea ice but their nature not understood (Debenham, 1945 p. 377). A landing was made on King George Island in the South Shetlands but 'they observed no flora except the moss' (Debenham, 1945, p. 428). Nevertheless, Bellingshausen was an acute observer and curious about the creatures he came across. He commented on the adaptation of penguins to life in the sea and noted that their food was 'shrimps', that is krill (Debenham, 1945, p. 118). He estimated the swimming speed of dolphins (Debenham, 1945, p. 145) and realized that albatrosses must roam more widely than almost any other kind of bird (Debenham, 1945, p. 175). Under his leadership a good naturalist might have accomplished much.

In the course of this splendid voyage in the Southern Ocean the loneliness of those who sailed in the *Vostok* and the *Mirnyi* was briefly relieved by encounters with sealers on South Georgia, Macquarie Island and in the South Shetlands. We must now go back a little in time to consider the activities of these rather differently motivated groups of explorers of the Antarctic.

2.6 Explorations by sealers

Seafaring men seeking quick profits had not been put off by Cook's forbidding picture of Antarctica. Already towards the end of the eighteenth century there was a lucrative business in fur and oil from seals in the southern South American region. The slaughter was such that populations were quickly depleted and shipowners had to be prompt in following up any hint of new sealing beaches. Cook's brief mention of fur seals on South

Georgia, was sufficient to send sealers in that direction. The ships that they used were of types already tested in the Arctic (Falla, 1964). The first to go to South Georgia seems to have been the *Lord Hawkesbury*, which sailed from London in 1786 under the command of Thomas Delano (Headland, 1989). A leopard seal skull collected on South Georgia, probably during the 1789–90 season, was presented to the museum of the Royal College of Surgeons, London, by a Mr Kearne, a whaler, and was later described by Home (1822). American sealers soon followed the British (Bertrand, 1971). Soon the Americans established a flourishing trade with Canton, the source of furs being the Antarctic.

The slaughter of seals was on a large scale and thorough, a vessel often finishing with a cargo of several tens of thousands of skins. Beaches were completely cleared but the average sealer seems to have had faith in an inexhaustible supply of seals. A few realized that this was not so and that there was need for regulation. As early as 1788 M. Leard, a master in the Royal Navy, suggested to Lord Hawkesbury, President of the Council for Trade and Foreign Plantations, that licences should be issued to limit the number of crews sealing on a beach and that killing of females with young should be prohibited (King, 1964–65). He also pointed out that it was wasteful not to make use of the oil in seal blubber. Nothing came of this proposal, perhaps because it would produce diplomatic complications with Spain, in whose South American territories many of the sealing beaches were.

It was not in a sealer's interests, of course, to let anyone else know of its whereabouts if he found a new sealing ground and ship's logs were usually sketchy at the best. Added to this, sealers usually carried no navigational instruments apart from a magnetic compass so that positions were fixed very roughly (Mitterling, 1959, p. 65). In these circumstances it is not surprising that there has been controversy as to who discovered what. Bertrand's (1971) reconstruction of the cruise track of the *Hersilia*, the first American vessel, as far as we know, to arrive at the South Shetland Islands, is illustrative of the ambiguities and uncertainties encountered. However, we do not have to discuss priorities here and can simply note some of the generally accepted credits. Macquarie Island was discovered in July 1810 by Frederick Hasselburgh, master of the brig *Perseverance* out of Sydney, Australia. William Smith, master of the merchant brig *Williams* of Blythe, England, made the first certain sighting of the South Shetlands in February 1819 when forced by strong head winds off the usual course around Cape Horn. The value of his discovery was immediately realized by American sealers but Smith resolutely refused to divulge to them the position of the new land. The *Williams* was later chartered in Valparaiso, Chile, by the British commanding officer in the Pacific, Captain William Shirreff, who after previous doubts was persuaded of the potential importance of the new discovery, and with Edward Bransfield as master and Smith as pilot, returned to survey the islands (Miers, 1820). It was on this voyage that Bransfield in January 1820 sighted the Trinity Peninsula on mainland Antarctica. A year later Nathaniel Palmer in the sloop *Hero*, tender to an American sealing fleet, after making what was probably the first visit to the flooded crater of Deception Island, saw the same part of the continent (Mitterling, 1959). On his return to the South Shetlands Palmer fell in with the *Vostok* and the *Mirnyi*. No doubt the Russians and the Americans found each other difficult to understand but discrepancies between Bellinghausen's laconic report of this encounter (Debenham, 1945) and Fanning's (1924) more romantic version do not help to clear up the confusion about this phase of Antarctic history. There seems little doubt, however, that the first landing on the Peninsula was made in February 1821 by Captain John Davies from the *Cecilia*, tender to the American sealer *Huron*, probably at Hughes Bay 64°20′ S 61°15′ W. He used the word 'continent' to describe that upon which he had set foot (Mitterling, 1959; Bertrand, 1971). The South Orkney Islands were discovered by George Powell in the British sloop *Dove*, sailing in consort with Palmer in the American sloop *James Monroe*, in December 1821 (Powell, 1822; Mitterling, 1959).

Most sealers had no thought beyond collecting as many seal skins as possible and such accounts as they gave of the Antarctic are of little use. Despite the huge numbers slaughtered, few skins or skulls found their way to museums, and sealers' records refer simply to 'seals' without differentiating species. Miers (1820), who did not himself visit the South Shetlands, concluded from a description given to him in Valparaiso by the mate of the *Williams* that the rock of these islands might be 'chlorite-slate or schistose horneblende'. He quoted statements that trees could be most distinctly seen on one island through the telescope and that there was an abundance of land-birds and freshwater ducks. In addition to seals and sea-otters there was an animal differing from the sea-otter which he imagined might prove to be a variety of *Ornithorhyncus*, the duckbilled platypus. Bransfield's instructions for the survey of the South Shetlands included orders to observe, collect and preserve every object of natural science. This order was disregarded but the surgeon on the *Williams*, Adam Young, who landed on King George Island, produced a credible list of seals and birds, and noted that the only vegetation was a stunted grass and a species of moss (Young,

1821). The occurrence on the South Shetlands of a grass, assumed now to be *Deschampsia antarctica*, is mentioned in other records from around this time (Smith, 1981). Natural history specimens from the South Shetlands were taken to the US and some were presented to Dr Samuel L. Mitchill, a physician who was an influential figure in American science in the early nineteenth century. It is reported that they included quartz, amethyst, porphyry, onyx, flint, zealite, pumice and pyrites but no scientific description was published (Bertrand, 1971, p. 108). A new species of lichen found in the South Shetlands, *Usnea fasciata*, was, however, described by the well-known American botanist, Dr John Torrey (Torrey, 1823). The rock and mineral specimens seem to have found their way into the collection of the New York Lyceum of Natural History and presumably were lost when the premises of the Lyceum were destroyed by fire in 1866 (Bertrand, 1971, p. 109). Captain Winship of the ship *O'Cain* of Boston, USA, brought back specimens of shells, minerals and coal from the South Shetlands which were presented to a Mr Topliff and examination by experts invited but no publication in the scientific literature resulted (Bertrand, 1971, p. 115). Powell made observations and recorded them. In his notes accompanying his chart of the South Orkneys (Powell, 1822) he mentioned that on three occasions he found sea temperatures at around 170 fathoms (311m) depth to be higher than those at the surface and gave a systematic table of barometric pressures and air and sea temperatures, extending over nearly four months.

Few sealers published general accounts of their voyages. Among those who did was Captain Benjamin Morrell, whose *Narrative of Four Voyages* (1832) is generally dismissed as inaccurate and exaggerated but anyway contains little of scientific import (Mitterling, 1959; Bertrand, 1971). Before that, however, Weddell had published in 1825 his *A Voyage towards the South Pole*, one of the classics of Antarctic exploration.

James Weddell was born in 1787.[5] In 1819 he became commander of the brig *Jane* and sailed to the South Shetland Islands, which he explored and surveyed in addition to taking a disappointing cargo of seal skins. His fame rests on his next voyage, in 1822–24, during which the *Jane*, accompanied by the cutter *Beaufoy*, attained a furthest south of 74°15′, which was to remain unsurpassed for nearly 90 years, in the sea which now bears his name. This voyage is the one described in his book.

Weddell was a man of independent spirit and of enquiring and accurate mind. He seems to have been self-taught but wrote excellent English, was well read, had intellectual friends, and became a Fellow of the Royal Society of Edinburgh. His exploring proclivities were encouraged by the owners of *Jane* and *Beaufoy* and he took with him a selection of the best chronometers available as well as other instruments. He also had the essential attribute of being a good leader who could keep his men going cheerfully under conditions of hardship.

Weddell's discovery of open water extending far south in longitude 34° W was lucky because of an unusually ice-free season and, realizing that he would be challenged, he got his Chief Officer and two seamen to swear to the accuracy of his log before the Commissioners of His Majesty's Customs. Nevertheless, his claim was doubted by Dumont d'Urville who failed to penetrate the ice in this region in 1838. Weddell kept an accurate record of the distribution of ice and noted an iceberg containing black earth. He measured sea surface temperatures almost daily until both his thermometers were broken. Based on such observations and those of recent Arctic explorations he included as an appendix to his book a dissertation on the state of the poles. In this he pointed out that a formula produced by Mayer of Göttingen for calculating temperatures at different latitudes took no account of the effects of local features such as rocky or mountainous land and that the mean annual temperature at 61° S calculated from this formula, namely around 44 °F, was considerably higher than the 35 °F which he himself measured in January, the warmest time of the year. This discrepancy he attributed to the effect of the barren land of the South Shetland archipelago. His own observations showed little decrease in temperature in the open sea going south from the South Shetlands and at 74° S the formula gave a more realistic result. Leslie (1820, p. 459) had shown by calculation that

> the intensity of the cold would not be sensibly augmented in penetrating from the Arctic Circle to the Pole. The existence of an open sea towards the extreme north is hence not improbable.

Starting from his own observations of open water in the far south, Weddell reached a similar conclusion for Antarctica, arguing that because, as he believed, field ice is only formed in the proximity of land and since no land had so far been found within 20° of it, that there must be open sea around the South Pole. Any ice which formed in winter, he thought, would be melted as a result of the sun's continuous presence for two months in the summer at an elevation of more than 20° above the horizon. The chart which he presented shows latitudes south of about 77° devoid of both land and permanent ice. He gently dismissed the idea put forward by one St. Pierre, 'whose writings in general do honour to his head and heart' and who 'treated the subject of Polar temperature in a course of reasoning truly

Fig. 2.7 Weddell seal, *Leptonychotes weddelli*, engraving after a drawing by J. Weddell (1825), *A Voyage towards the South Pole*, opp. p. 22. (Courtesy of the Scott Polar Research Institute.)

inventive'. St. Pierre held that at the poles there are enormous pyramids of ice, the annual melting of which produces a flow of melt-water towards the equator – one occasion on a sufficient scale to produce Noah's flood. Such large accumulations of ice, Weddell thought, would be of such an altitude that they would remain permanently frozen. Moreover, the northward flowing current in the Southern Ocean was not of the magnitude postulated and icebergs were never of the height attributed to them by St. Pierre.

Weddell made measurements of magnetic declination and for South Georgia his value fits well on the curve showing the change in declination with time from the first observations in this locality by Halley to the present day (Simmons, 1987). He looked in vain for the aurora australis seen by Forster but thought that it might not have been visible because darkness was never complete during that part of his voyage. While attempting an observation of the altitude of the sun from a mountain near Adventure Bay, South Georgia, he noticed that the mercury in his artificial horizon was in tremulous motion, suggesting a seismic disturbance. He noted volcanic activity in the South Shetlands and that the rock there consisted of quartz with iron pyrites and copper pyrites. He mentioned taking rock samples when he was in Tierra del Fuego for later identification in Edinburgh but it is not clear whether he took samples from the Antarctic too.

Weddell has achieved a place in the biological literature by having a seal named after him. This was first described by Robert Jamieson from six skins collected at Saddle Island, South Orkneys on January 13, 1823. One of these skins, now the type specimen of *Leptonychotes weddelli*, the Weddell seal, described by Lesson in 1826 is still in the Edinburgh Museum, Scotland. The picture of this animal given in his book (Fig. 2.7) is perplexing; the Weddell seal does have a relatively small head but is not nearly so microcephalic as he depicted. The legend on the plate states 'Drawn from nature by I. Weddell' but in the text he says that it is a drawing of the one deposited in the Edinburgh Museum. Possibly he did the drawing in Edinburgh and the specimen failed to refresh his memory of the living animal. The name he gives the beast – the sea leopard – is an error but there is some suggestion in his drawing of the true sea leopard. Nevertheless, Weddell's observations on seals were on the whole accurate and penetrating. He gave good accounts of the habits of the sea elephant and of the fur seal. He realized the effect on the fur seal populations of wholesale extermination by the sealers and suggested that by not killing females until the young were able to take to the water the Antarctic sealing grounds might be managed to provide a sustainable yield of 100,000 furs per year. Writing some 85 years later about Weddell's observations on the king penguins, Murphy (1948) said 'The details of his study have long been overlooked, or perhaps disbelieved, by ornithologists, but they actually comprise the best account of the bird's life history that has yet been published. Nothing in my own observations would lead me to change a line of Weddell's almost forgotten history'. He also gave a good description of the courtship behaviour of the albatross.

Weddell's career after the voyage described in his book is obscure and he died in London at the age of 47.

Captain Edmund Fanning, a native of Stonington, Connecticut, USA, and one of the first sealers in the Antarctic, also left an account of his voyages (Fanning, 1924). Like Weddell he was an educated man, taking books to sea with him and using a 'microscopic glass' to look at plankton. He was a careful observer and in his book gave a good description of the macaroni penguin. He recorded catching 'cod' – presumably *Notothenia rossii* – in the kelp beds of South Georgia and finding that they were larger the deeper the water, an observation in accord with what we now know of the life history of this fish. His sense of duty to the owners of his ships limited his explorations and he made no particularly outstanding discoveries. A plan put forward by him for a voyage of exploration in far southern waters had the moral, if not financial, support of the US government but had to be abandoned in June 1812 with the declaration of war against

Great Britain (Bertrand, 1971, p. 30). Later, however, he was one of the prime movers behind the first overt American exploring expedition to the Antarctic. Together with Jeremiah N. Reynolds, of whom more in chapter 3, he succeeded in raising sufficient funds privately to send two sealing brigs, the *Seraph* and the *Annawan*, down south after an attempt to get the government to support an expedition had failed (see p. 59). These vessels, commanded by the veteran sealing captains, Benjamin Pendleton and Nathaniel Palmer, sailed in October 1829 with the object of extending Palmer's discoveries south of the South Shetlands (Mitterling, 1949, p. 97; Bertrand, 1971, p. 144). The expedition, which returned in 1831, was unsuccessful in achieving its geographical objectives, partly because the crew were counting on their share of sealing profits and threatened to mutiny unless sealing was done, but it did result in the first papers of the 'professional' sort with which we are familiar today to come out of the Antarctic.

The scientific programme of the expedition was sponsored by the Lyceum for Natural History of the City of New York. Private citizens were invited to lend charts, books and instruments, and the *Annawan* sailed well equipped with instruments and several hundred books. The 'scientific corps' consisted of James Eights of Albany, New York (Fig. 2.8), John Frampton Watson, Jeremiah Reynolds and two unidentified assistants (Bertrand, 1971, p. 147). It was James Eights who made the outstanding scientific contribution (Clarke, 1916; Calman, 1937; Martin, 1940; Mitterling, 1959; Bertrand, 1971; Hedgpeth, 1971; Miller & Goldsmith, 1980). Born in 1798, he had trained as a physician and it is not known how he came to be selected as a naturalist on the expedition. It took him to Tierra del Fuego, Staten Island and the South Shetlands, including Deception Island. He was an acute and accurate observer and a good draughtsman (Fig. 2.9). He collected 13 cases of specimens, including rocks, lichens and marine animals, and published five papers between 1833 and 1852 on his Antarctic observations and these collections. His papers[6] are the only substantial records of the expedition.

In one of his papers Eights (1846) discussed the transport of rock and animals by drifting ice, thereby anticipating Darwin by six years (see p. 50). He used the word 'tabular' to describe the characteristic shape of Antarctic icebergs, apparently for the first time. He held that these always originate from land and in suggesting that glaciers gradually encroach on the sea and project over it, expressed clearly the essence of the modern view on the nature of ice shelves and the way in which tabular bergs are formed from them. Other forms of iceberg he thought must result from uneven erosion

Fig. 2.8 Dr James Eights at 25; watercolour (1820) by E. Emmons. (Collection of the Albany Institute of History & Art, gift of John M. Clarke.)

and capsize but as we have seen, Bellingshausen had already realized this. Seeing a drifting berg running aground then swinging around and freeing itself he speculated on the effects on the sea bottom of such a massive impact and what its ultimate geological expression might be. In another paper (Eights, 1833b) he described the geology of the South Shetlands, the main formation which he saw being basalt resting on strata of argillaceous conglomerate. Both rocks contained inclusions of quartz, in its various forms, plus calcite, barytes, pyrites and malachite. He also found a fragment of carbonized wood embedded in the conglomerate – the first

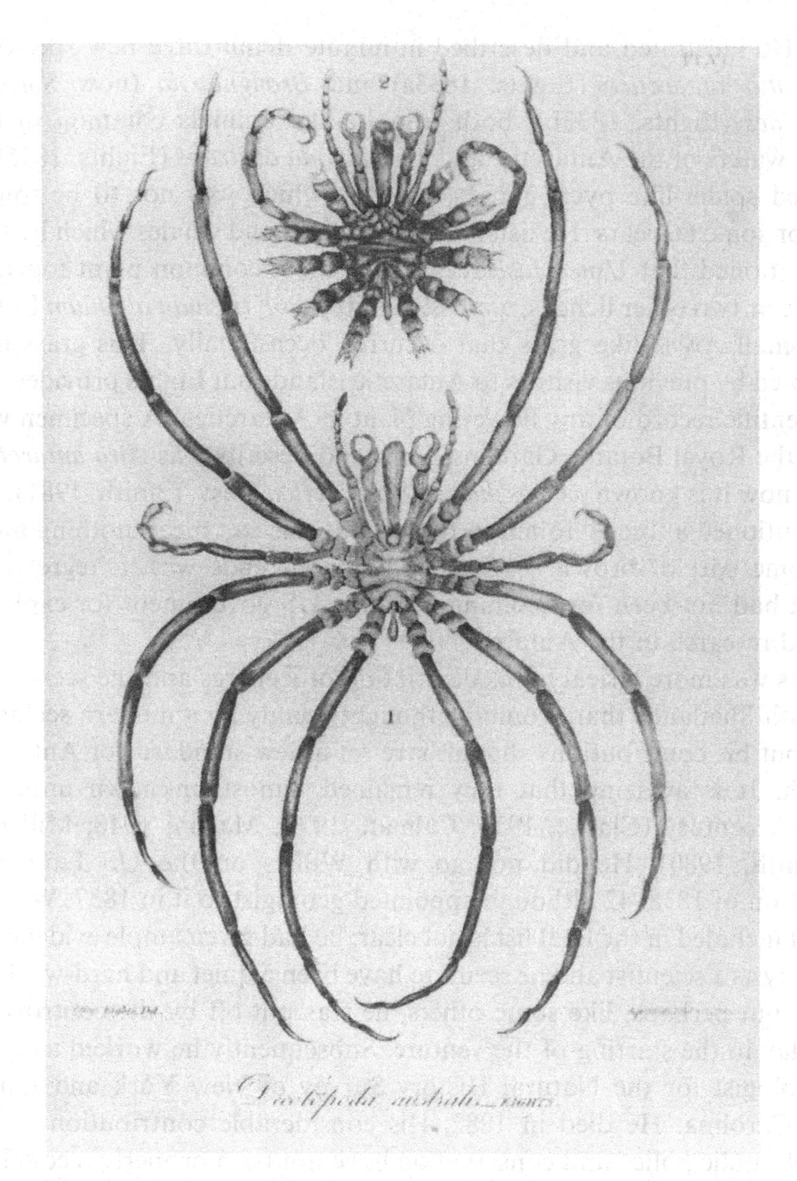

Fig. 2.9 The pycnogonid, *Decolopoda australis*, Eights (1835) ***Boston Journal of Natural History***, **1** (2), plate VII. Fig. 1, dorsal view; Fig. 2, ventral view with the greater parts of the legs removed. (Courtesy of the Natural History Museum, London.)

fossil to be found in the Antarctic. Rounded pieces of granite, he presumed, since he recovered one a foot in diameter from an iceberg, must have been transported from land further south.

Eights was undoubtedly the pioneer in the study of Antarctic invertebrates. He illustrated and described in minute detail three new species – *Glyptonotus antarctica* (Eights, 1833a) and *Brongniartia* (now *Sorolis*) *trilobitoides* (Eights, 1833b), both trilobite-like animals common in the shallow waters of the Antarctic, and *Decolopoda australis* (Eights, 1835), a 10-legged spider-like pycnogonid (Fig. 2.9) which was not to be found again for some 66 years. He listed the birds, seals and whales which he saw and mentioned that *Usnea fasciata* was the most common plant together with one or two other lichens, a moss, possibly *Polytrichum alpinium* Linn., and a small *Avena*-like grass that occurred occasionally. This grass had been noted by previous visitors to Antarctic islands but Eights provided the first scientific record of any flowering plant in Antarctica. A specimen was sent to the Royal Botanic Gardens, Kew, and described as *Aira antarctica* Hook.; now it is known as *Deschampsia antarctica* Desv. (Smith, 1981). He also mentioned a 'fucus' found in the sea but this can mean nothing more than some sort of brown seaweed. This paper ends with a regret that support had not been forthcoming from the US government for exploration and research in the Antarctic.

Eights was more lyrical in his description of icebergs and the scenery of the South Shetlands than would be thought seemly by a modern scientific editor but his contributions should have set a new standard for Antarctic research. It is amazing that they remained almost unknown until the twentieth century (Clarke, 1916; Calman, 1937; Martin, 1940; Miller & Goldsmith, 1980). He did not go with Wilkes on the US Exploring Expedition of 1838–42 although appointed geologist to it in 1837. Why he was not included in the final list is not clear; he had given ample evidence of his ability as a scientist and he seems to have been a quiet and hard-working person, but perhaps, like some others, he was put off by the controversy and delay in the starting of the venture. Subsequently he worked as artist and geologist for the Natural History Survey of New York and visited North Carolina. He died in 1882. His considerable contribution in the fields of public policy and conservation have not been properly recognized (Miller & Goldsmith, 1980).

By 1829 no fur seals were to be found in the South Shetland Islands and the same sorry outcome was soon repeated on islands in other sectors of the Southern Ocean (Headland, 1989, gives a statistical summary of Antarctic sealing). As the fur trade ceased to be profitable, attention was switched to the less rewarding business of exploiting elephant seals for their oil but populations of these animals likewise declined, by around 1870, to levels too low to bother with. Nevertheless, hope was slow to die and the search for new sealing grounds continued up to the middle of the nineteenth century. Some of this exploration, however, was done for its own sake without great expectation of immediate financial gain. Foremost in this was the British firm of Enderby. The founder of this firm, Samuel Enderby, was inquisitive about the unknown regions which his ships visited and one of his three sons, Charles, continued this interest and was a member of the Royal Geographical Society from its foundation in 1830. The Enderbys chose men of some education and naval experience to captain their vessels and encouraged them to explore. This went beyond financial prudence and brought eventual ruin to the firm (Jones, 1969) but the results in terms of geographical discovery were great.

The voyages of John Biscoe in the brig *Tula*, accompanied by the cutter *Lively*, took place in the years 1830 and 1832. Biscoe's first approach to the continent was at the beginning of 1831 when he sailed mostly far to the south of Bellingshausen's track (although he does not seem to have known this) between the longitudes of 1° E and 50° E. Horsborugh (1830), hydrographer to the East India Company, had predicted from the occurrence of icebergs at unusually low latitudes that there must be a large land mass near the pole in this sector. Biscoe proved him right and named his discovery Enderby Land. After a pause for recuperation in Tasmania, where he encountered James Weddell, Biscoe resumed his circumnavigation at high latitudes and on the west coast of the Antarctic Peninsula discovered the island which he named in honour of Queen Adelaide. He also found what is now called Anvers Island but believed it to be part of the continent. As far as sealing was concerned the voyage was a failure. He lost many of his crew through scurvy and then desertion of most of those who remained alive and the *Lively* was finally wrecked in the Falklands but to have achieved what he did with sailing vessels of only 150 and 50 tonnes is unsurpassed. He was awarded a Royal Premium, equivalent to a gold medal, by the Royal Geographical Society and dispatched on a second expedition, but this came to naught. He died destitute in 1843 (Savours, 1983).

Biscoe's great contribution was to confirm that major land masses exist in the Antarctic. He was a careful observer and his log, a full detailed narrative, contains records of birds, seals, whales, ice conditions and auroral displays but it did not add much to scientific knowledge of these things. Biscoe was not a scientist and clung obstinately to his belief that icebergs were formed at sea and that there was no land continent only a huge solid mass of sea-ice in spite of Weddell's report of an earth-contaminated berg and the evidence of his own eyes.

It is probable but not certain that the *Magnet*, master Peter Kemp, was an Enderby ship. Sailing south from Kerguelen in 1833 he sighted Heard Island, being the first to do so, and then that part of the Antarctic continent now known as the Kemp Coast. Although the chart of this voyage survives, Kemp's journal has been lost. Another Enderby voyage in search of land was that of the cutter *Sabrina* and the schooner *Eliza Scott* under the commands of H. Freeman and John Balleny, respectively, in 1838–39 (Jones, 1969). Sailing south from New Zealand they achieved a higher latitude, 69°02′ S, than anyone else had done in these longitudes and discovered the Balleny Islands. A hazardous and momentary landing by Freeman on Young Island, which like the others was evidently volcanic, enabled the collection of a few stones. These later proved to be scoriae and basalt, with crystals of olivine. An appearance of land was seen from 65°10′ S 117°04′ E, off what is now known as the Sabrina Coast. An iceberg with a block of rock attached to it was encountered at 61° S, 450 nautical miles from the nearest suspected land. Charles Darwin (Darwin, 1839) appended a note to Balleny's report (Balleny, 1839) commenting on the transporting power of ice. Darwin had taken the trouble to interview a mate from the *Eliza Scott* about the rock on the iceberg and concluded that the bottom of the Antarctic sea and the shores of its islands must be scattered with masses of foreign rock – the counterpart of the erratic boulders of the northern hemisphere. He made reference to an American naturalist, who can only have been James Eights, who had already arrived at this conclusion although the idea had been suggested by a Swedish mineralogist, Daniel Tilas, in 1742. Darwin made much of this role of icebergs and thereby perhaps delayed for some time the recognition of glaciers as an important agency for transporting erratics (Mills 1983). The *Sabrina* was lost without trace in the Southern Ocean but the *Eliza Scott* got back to London in time to inform Ross, who was about to sail for the Antarctic (see p. 76), of the discovery of the Balleny Islands. Again, the main value of the voyage had been to confirm the existence of the Antarctic continent and to add to general Antarctic experience rather than to contribute scientifically.

2.7 William Scoresby: pioneer polar scientist

Nothing has been said yet about whalers in Antarctic waters. A few whalers turned their efforts to seals in the boom years of sealing but during the period that is being considered there was no other reason for them to go to the far south. Whales were still abundant in more accessible waters and the big whales of the Antarctic could not be caught and handled by existing methods. But note must be taken of a redoubtable whaler/scientist, who although he sailed only in Arctic seas and had little immediate influence on exploration or science in the far south, was a pioneer among polar scientists and made studies of a kind which eventually became one of the major interests of Antarctic research.

This was William Scoresby the younger. Born in 1789, he accompanied his father, a whaling master of almost legendary reputation, on a voyage to Greenland seas at the age of 11. Only five years later he was mate on the whaling ship *Resolution* when with the elder William Scoresby in command she sailed to 81°30′ N, the most northerly latitude then reached by any ship. The next year he went to the University of Edinburgh, where he was influenced by Thomas Hope (Professor of Chemistry), John Playfair (Professor of Natural Philosophy and pupil of the famous geologist James Hutton), John Leslie (Playfair's successor, a mathematician interested in meteorology) and Professor Robert Jameson, (Professor of Natural History, a mineralogist who became a particular friend and mentor). These were outstanding men and, indeed, at that time Edinburgh was a scientific centre second to none. He was also befriended by Banks, now Sir Joseph, in his 30th year as President of the Royal Society and one of the great Englishmen of the period. Scoresby came to correspond regularly with Banks who provided him with apparatus such as a device for deep water sampling. In 1811 Scoresby was in command of the *Resolution* and he continued as a whaling master until 1823. During his voyages he observed everything around him – he experimented, pondered and recorded in meticulous detail – as well as keeping a paternal eye on his crew and catching more whales than any other Whitby whaler. His scientific work was embodied in several papers and the two volumes of *An Account of the Arctic Regions*, published in 1820. He was elected a fellow of the Royal Society in 1824, shortly after he had left the sea to study theology at Queens' College, Cambridge, and enter the Church. With all this to do he still found time to develop improved magnetic needles for ships' compasses, to collaborate with a then obscure young physicist, James Prescott Joule, as well as to maintain a wide interest in matters scientific, as for example, through the British Association, of which he was a founder member. He visited the United States twice and in 1856 undertook a voyage to Australia and back to test his ideas on the magnetism of iron ships. He died the following year.

An Account of the Arctic Regions has been described as 'one of the most remarkable books in the English language' (Harmer, 1928) and without doubt it set up milestones not only in Arctic geography and polar science generally but in oceanography as well. Even in small matters it is admirable; unlike most scientific books of the time it gives exact references to the literature. The first volume gives an account of the geography of the Arctic regions, the results of investigations on the polar seas, observations on ice and meteorology, and descriptions of Arctic animals. The second volume is concerned with the history and practice of whaling. In a chapter entitled *Hydrographic Survey of the Greenland Sea* Scoresby gave the results of his temperature measurements, which showed that deep water was often warmer than that at the surface and he suggested that a branch of the Gulf Stream reached the Arctic and there sank below the colder but less dense water of the Greenland Sea. He described the distribution of waters of different colour, distinguishing blue, more transparent, water from the green, more turbid, water in which whales were abundant and mentioned that there is sometimes a sharp front between them. He found that innumerable minute organisms were present in the green water, gave remarkably accurate drawings of some of the bigger ones and descriptions that show that he had recognized chains of diatoms. He knew that the larger of these organisms were the food of whales and surmised that these animalcules in turn subsisted on the smaller ones, 'thus producing a dependant chain of animal life, one particular link of which being destroyed, the whole must necessarily perish' – here we have the foundations of biological oceanography. He was emphatic that ice can be formed in open water and did not preclude the possibility that large icebergs might originate away from land so that, in his view, the occurrence of icebergs in the Southern Ocean need not be taken as evidence of land around the pole. He discussed seasonal and secular changes in Arctic sea-ice. Banks was particularly interested in his report that in 1817 the coast of West Greenland was free of ice and pointed out the value of knowledge of ice distribution for the understanding of climate. Scoresby's drawings of ice crystals provide an example of his care and patience in observation. A section on *Effects of the Ice on the Atmosphere, and of the Ice and Sea on each other* deals with the very modern topic of air–sea interactions. His discussion of air temperatures follows the pattern of those of Leslie and Weddell but in noticing that the sun's rays are reflected by snow without producing any material elevation of temperature whereas under the same conditions pitch may be melted, he had found the key to the problem of polar heat budgets without realizing it. He regretted that he did not get any opportunity of investigating the 'agitation of the magnetic needle' by the aurora. He also considered the physiological effects of cold and the phenomenon of wind-chill.

To us, Scoresby's book seems to be a remarkably accurate and comprehensive manual of polar science which could have provided a firm basis for the rapid advance in understanding of the Antarctic. It was a best seller and Scoresby did not lack contacts with Antarctic explorers. He is quoted by Weddell in his book (1825), and perhaps they encountered each other at meetings of the Royal Society of Edinburgh; he met Simanov, the astronomer who accompanied Bellingshausen, when he visited Paris in 1824; and he was in correspondence with James Ross who was to lead the 1839–43 expedition of the *Erebus* and *Terror*. Nevertheless, his immediate impact, even in Arctic exploration, was small. His book contains a discussion of the practicalities of overland travel in the Arctic and the sensible suggestion of overwintering in the ice to provide a forward base for extended exploration. He proposed that the government should fit out an expedition to search for a North-west Passage, the expenses of which might at least partially be defrayed by taking whales. This was first ignored then later taken up enthusiastically by both the Royal Society and the Admiralty but with the offer of a subordinate position for Scoresby which he felt bound to refuse (Stamp & Stamp, 1976, p. 66; Hall, 1984, p. 201). He met with a similar rebuff in the field of magnetism, being excluded from the Admiralty Committee for the Improvement of Ships' Compasses. On Antarctic exploration he had no perceptible influence in his lifetime.

The reason for this may not be far to seek. Scoresby was a whaler and his exploration and research were private ventures conducted within limits set by his obligations to the ship owners and his crew. In passing, because he set out so clearly the position in which sealers in the Antarctic must also have found themselves, it is worth quoting from his journal:

> My situation, whenever researches were in view, was therefore a very delicate one; and even when quite compatible with the chief design of the voyage was not without anxiety. Most of our insurance, indeed, was effected under the permission 'to touch at or land upon, any coast whatever and for any purpose whatever'. But some of the underwriters, I understand, had intimated their design of withstanding any claim on the part of the owners of the Baffin, if losses or damage was sustained other than in the direct pursuit of the whale fishery. To the honour and liberality of several of the underwriters of Liverpool, however, I ought to mention that they gave me every encouragement to pursue my researches, even with larger sums than otherwise they would have done, and this at the lowest premium. (Stamp & Stamp, 1983, p. 91).

No doubt his reputation as a superb seaman accounts for the liberality of the underwriters but the point is that in the Navy he might have found more scope for exploration. However, his experiences when he had served as a volunteer naval rating in the seizure of the Danish fleet in 1807 had prejudiced him. The Navy felt that it had a monopoly of Arctic exploration and Sir John Barrow, Second Secretary to the Admiralty, appears to have been determined to maintain this and, perhaps, in particular to exclude Scoresby (Stamp & Stamp, 1976). After Bank's death Scoresby had no powerful friend in the Royal Society to take his side and Barrow saw to it that he did not receive support for exploration. Happily, there was a change in attitude to Scoresby and towards the end of his life he was consulted by the Admiralty about the search for Franklin and was lent instruments for his work on magnetism but the monopolistic attitude of the Navy towards polar exploration was to persist in Britain for better or for worse until the end of the nineteenth century.

Scoresby did not come fully into his own until well into the twentieth century. When in 1917 an Interdepartmental Committee was set up to look into research and development in the Falkland Islands Dependencies the worth of his work was recognized (Interdepartmental Committee, 1920, p. 13). Later, when the *Discovery* Investigations which resulted from this Committee's recommendations were under way, a specially built vessel, the Royal Research Ship *William Scoresby*, was to contribute notably to Antarctic science.

Endnotes

1 This account of Halley's life is based on Armitage (1966), Cotter (1981), MacPike (1932) and Thrower (1981).

2 See Beaglehole (1974), Rubin (1982a), Stamp & Stamp (1978), Woolley (1969).

3 Sir David Smith in the Sir Joseph Banks Memorial Lecture to the Royal Society of Tasmania (in co-operation with the Royal Society). 7 November 1988.

4 Information about Bellingshausen and his associates is taken largely from the Introduction in Debenham (1945).

5 For information about Weddell see the Introduction by Sir Vivian Fuchs to the 1970 reprint of Weddell (1825).

6 The Antarctic papers of James Eights have been reprinted in Quam (1971).

7 For information about Scoresby see the Introduction by Sir Alister Hardy to the 1969 reprint of Scoresby (1820) and Stamp & Stamp (1976, 1983). Also to be noted is Scoresby, W. (1980) *The Polar Ice and the North Pole*, Caedmon Reprints.

3

The national expeditions of 1828 to 1843

3.1 The scientific and social background

We now return from the *ad hoc* investigations associated with scaling to the mainstream of science in the early nineteenth century. Science was expanding apace and also changing in nature and outlook. The descriptive studies characteristic of the eighteenth century were being replaced by work which was more analytical and experimental, concerned more with underlying structures and processes than external features. Not unrelated to this was a change in those who pursued science, from being mostly gentlemen amateurs to professionals. Following the death of its President, Sir Joseph Banks, in 1820 and after a short interregnum the Royal Society of London chose a professional scientist, Sir Humphrey Davy, to be his successor, and physical science came to predominate over biology. Nevertheless, its transformation, from a club which admitted aristocratic amateurs and patrons of science and learning, to the society restricted to scientists of distinction which it became by the end of the nineteenth century, was slow (Hall, 1984). Long before that happened it was no longer possible to remain in the forefront and at the same time be abreast of anything more than a small segment of the whole field of scientific endeavour. Accordingly, specialist societies were formed; for example, in London there were the Linnean Society (1799), the Geological Society (1807), the Royal Astronomical Society (1820) and the Royal Geographical Society (1830). At the same time the general public was becoming more interested in science, perhaps as its practical applications became evident in such things as the steam engine. To meet popular demand for information and to maintain contact between scientists of different disciplines the Gesellschaft Deutscher Naturforscher und Ärtze was founded in 1822, to be followed by the British Association for the Advancement of Science in 1831 and the American Association for the Advancement of Science in 1848. The industrial revolution had created a

Fig. 3.1 Alexander von Humboldt in 1848; drawing by Rudolf Lehmann. (Courtesy of the Trustees of the British Museum.)

favourable climate for the British Association and it quickly began to have an important influence on the direction of science. Industrialization was slower to establish itself elsewhere but the French Revolution released a burst of creative activity, French governments were more supportive of science, and Paris, rather than London, became its chief centre up to about 1850 (Pledge, 1939).

Antarctic investigations were peripheral to these developments but one outstanding figure of the period is of special significance to us. This is Baron Alexander von Humboldt (1769–1859; Fig. 3.1) who had some connection with the Antarctic through George Forster (chapter 2). Later he heard of Bellingshausen's exploits from the astronomer Simanov and communicated an account of the voyage to the Literary Gazette (Humboldt, 1824). This connexion might have become more direct, for Humboldt was invited to join Bougainville on a circumnavigation, one objective of which was the South Pole, but Napoleon, needing to devote his financial resources to his Egyptian campaign, postponed the voyage indefinitely (Cawood, 1977). Instead, between 1799 and 1804 Humboldt travelled widely in South America and it was on the mass of data which he accumulated on his journeyings that he established the foundations of modern physical geography. His distinctive approach – empirical, quantitative and synthetic – differed from that of his great contemporaries and does not fit with the conventional present day subdivisions of science (see Nicolson, 1987). He has been described as the last complete scientist; he was an expert on geomagnetism, geology and botany, a pioneer in climatology, oceanography and biochemistry, and introduced techniques such as isothermal plotting, cross-section of land form, new methods of geological mapping and numerical analysis of data (Bowen, 1970). He saw nature holistically – 'Nothing can be considered in isolation. The general equilibrium, which reigns amongst disturbances and apparent turmoil, is the result of an infinity of mechanical forces and chemical attractions balancing each other out. Even if each series of facts must be considered separately to identify a particular law, the study of nature, which is the greatest problem of *la physique generale*, requires the bringing together of all the forms of knowledge which deal with modifications of matter' (quoted by Nicolson, 1987). This seems a fair statement of the philosophy of Antarctic science at the end of the twentieth century but Humboldt more immediately provided the stimulus which revivified Antarctic studies in the 1830s. Magnetism at that time was a fashionable topic; instruments for studying it were being developed to a high degree of precision (O'Hara, 1983; McConnell, 1985; Multhauf & Good, 1987). Oersted in Copenhagen, Denmark, and Ampére and others in France had discovered the laws relating electrical current and magnetism. Humboldt had been introduced to terrestrial magnetism by the nautical astronomer J. B. de Borda and it was his pleas for international collaboration in the establishment of a world-wide network of magnetic observatories that provided the initial stimulus for the expeditions which we are about to consider (Cawood, 1977) as well as the pattern for subsequent polar years and the International Geophysical Year. His observation that following the auroral display on 21 December 1806, the horizontal magnetic intensity the following morning was distinctly lower was a landmark in auroral research (Schröder 1984).

Interest in exploration was high in the early nineteenth century and the discoveries made by the Enderby captains were received with acclaim but British public attention and government money were for some time directed to the North-west Passage where rivalry with Russia provided a powerful political motive (Kirwan, 1962). As the fur seals were exterminated, so the number of voyages going south of 60° S diminished drastically; Headland (1989) lists 94 such voyages between 1800 and 1830 but only 13 between 1830 and 1843. In the 22 years after Bellingshausen's return there was only one scientific voyage, that of Foster, to the far south, apart from the sealing expedition led by Pendleton which, as we have seen, had little scientific impact at the time. Then came six years in which three major expeditions, sponsored respectively by the governments of France, the US and UK, sailed for the Antarctic. The forces which launched them were much the same as those which had operated in the eighteenth century – national pride and international rivalry, the needs of navigation and in particular the charting of magnetic declination, and the urge to collect specimens – with only a tincture of new scientific ideas.

3.2 The United States exploring expedition

These three ventures overlapped but we may start with the US Expedition, since it follows on most closely to the expeditions described in the previous chapter. There are many accounts of the events preceding its sailing[1] so that here a summary should suffice as background for the more scientific aspects. In 1818 one John Cleves Symmes, residing in Kentucky, put forward a theory that the earth is hollow with holes at the poles allowing communication with concentric spheres within (Fig 3.2). There is no suggestion that he got this idea from Halley, who to explain the phenomena of terrestrial magnetism had supposed that the earth might consist of concentric, but certainly not separated, spheres (Mitterling, 1959). He prudently accompanied this pronouncement with a certificate of his sanity. Some took the theory seriously and his plea for an expedition to look into the matter was supported by New England merchants who, keen to promote any exploration that might find new sealing and whaling grounds, sent petitions to Congress. Wilkes (Morgan *et al.*, 1978, p. 358) saw this as the origin, and thus Symmes as the initiator, of the expedition which he was to lead. One of Symme's most active supporters for a while was the ambitious Jeremiah N. Reynolds who we encountered in the previous chapter (see also Bartlett, 1940; Mitterling, 1959). They parted after a year and Reynolds, happily dropping the 'holes in the poles' idea, conducted his own campaign for an Antarctic expedition, in which he was

Fig. 3.2 Wooden model as used by John C. Symmes to illustrate his theory of the hollow earth. (Courtesy of the Library, The Academy of Natural Sciences of Philadelphia.)

supported by the newspapers. His proposals were considered by the government and he was employed by the Secretary of the Navy, Southard, to collect information on regions most in need of exploration; his report was published in 1828. Reynolds was no scientist and Wilkes (Morgan *et al.*, 1978 p. 322) thought him entirely unfitted for the task. However, as we have seen, financial aid for an expedition was refused, the expedition led by Pendleton in 1829 to 1831 was privately funded and unsuccessful in the eyes of Pendleton himself and his agent Edmund Fanning. After the expedition's return Fanning and Reynolds separately renewed their importuning of Congress, Fanning arguing the impractibility of a privately sponsored expedition. Petitions went to the House Committee on Commerce and this reported favourably, Reynolds being invited to give an address in the chamber of the House. His eloquent presentation of wellworn arguments for a voyage of exploration and scientific investigation helped to convince

Congress, which on May 18, 1836, authorized a sum of $300,000 for 'a surveying and exploring expedition to the Pacific Ocean and the South Seas'. President Jackson supported the idea but the new Secretary of the Navy, Dickerson, whom Reynolds had antagonized, was less than whole-hearted in organizing it. The fitness of the vessels assigned to the expedition was questioned and by the time alterations recommended by a board of enquiry had been carried out much of the appropriation had been spent and still they were not ready to sail. Eventually four naval vessels, the sloops of war *Vincennes* and *Peacock*, the brig *Porpoise* and the store ship *Relief*, together with two pilot boats, *Sea Gull* and *Flying Fish*, made up the expeditionary fleet.

Selection of personnel for the expedition was equally plagued with indecision. After four other naval officers had declined the appointment, Lieutenant Charles Wilkes, then aged 40, was put in command on March 20, 1838. There was widespread disapproval in the Navy of such a junior officer receiving the appointment but it seems in retrospect that only he was capable of carrying the expedition through to the successful conclusion it eventually reached. As he himself remarked (1845 I, p. xiv) 'The very state of things that brought the Expedition into general disrepute, was of great advantage to me, for I was left to perform my duties unmolested'. Wilkes had been connected with the expedition since the beginnings, having been made responsible for the selection of instruments and visiting England, France and Germany to procure them. He had entered the navy as a midshipman at the age of 20 but had a natural inclination towards science and had become head of the Naval Department of Charts and Instruments in 1833. In character he was energetic and decisive, quick-tempered, stubborn, something of a martinet and determined to make a success of the expedition come what may. He set the date of departure as 10 August 1838, and managed to sail only eight days after that, fully aware that the vessels in his charge were ill assorted and shoddily refitted. Wilkes summed up his attitude to the expedition thus:

> I was a long time in suspense what course it became me to follow and I communed with myself under very distressing thoughts. I weighed well the different points – the imbecility of the Administration; its failure to secure the reliance that should be at all times paramount in securing the great objects entrusted to it by the People; and its disregard of all common sense as well as knowledge of control over Officers and men. So far as they were concerned I was turned loose to manage everything and to assume the necessary responsibility required. If this had been solely the reason I had to consider I should have thrown up the command and declined to act under so flagrant an act of their public duties, but I had the cause of my

> Country at heart, and the disgrace which had attended the getting up of the Expedition and its failure and folly, as well as the honest expectations of the whole country, made me consider the whole in another light. I had made up my mind the Expedition should not fail in my hands and believed I could carry it out to a successful issue. (Morgan *et al.*, 1978, p. 376).

These were no idle words; faced with a situation of risk and danger he invariably put the honour of the US before safety. He had not even received proper instructions:

> I had organized the whole out of the wreck, if I may so call it, of the first attempts and had drawn up my instructions in full which were copied and issued to me under the Signature of the then Secretary of the Navy, Mr. Paulding, with some few additions as to the conduct and demeanour of the Crews, and closing with a paragraph that the successful accomplishment of the duties would entitle us to the thanks and rewards of our country. (Morgan *et al.*, 1978, p. 376).

The administration had started off properly by consulting learned individuals and societies about the science that should be pursued and the experts who should do it. Admiral Krusenstern contributed a memorandum on navigation (Wilkes, 1845, pp. 354–9). The American Philosophical Society provided detailed advice covering astronomy, physics, oceanography, magnetism, auroras, meteorology, zoology, botany, geology, mineralogy, and, as the expedition was visiting other regions as well as the Antarctic, philology, ethnography and medicine. The value of worldwide magnetic observations was taken for granted; already at this date the US had several magnetic observatories (Multhauf & Good, 1987). The advice was all sound and useful but contained little that was innovative. Among the books that were suggested for taking were those by Weddell, Lyell and Humboldt (Conklin, 1940) but not that of Scoresby, although this had been received and favourably reviewed in the US 10 years previously (Griscom, 1823). The instructions prepared by the Academy of Natural Sciences of Philadelphia have been lost but a sidelight on equipment is that although the expedition was well provided with surveying instruments, there was a shortage of microscopes for the naturalists to the extent that a member of the Academy thought it necessary to lend his own (Rehn, 1940).

Neither the Society nor the Academy were willing to suggest particular scientists for the expedition but the former grimly called attention 'to the importance of so multiplying the members as to provide for the losses which may be looked for from various causes' (Conklin, 1940). The selection of the scientific corps was attended by confusion and ill-feeling. As we have seen, James Eights was chosen as geologist but not included in the final list. Jeremiah Reynolds hoped for the position of secretary to the expedition but, much to the relief of the scientists, failed to get it (Tyler, 1968, p. 8). Matthew F. Maury, later to become one of the founders of the science of oceanography, was selected but opted out because he felt his field of interest overlapped too much with that of Wilkes, whom he regarded as unfitted to lead the expedition (Tyler, 1968, p. 29). Asa Gray was given a position but resigned it when offered a professorship of botany (Eyde, 1985). Several disgruntled scientists and artists found themselves with their careers interrupted and out of pocket because they had purchased equipment against salaries which began only on the day of sailing (Tyler, 1968). We need not spend time on the scientists who did sail because apart from the artist/zoologist Titian Ramsay Peale, none of them was involved in the Antarctic legs of the cruise, Wilkes making himself and his officers responsible for the physical studies which were carried out.

Relations between naval and scientific personnel were not altogether happy. Wilkes considered science as secondary to survey and meteorology and did not always appreciate what was needed for his naturalists to do their work properly. For example, he forbade the taking of zoological specimens below decks because of the smell, thus severely limiting the examination and drawing of the material. It was more peaceable aboard the *Peacock*, commanded by Captain W. L. Hudson, than on the *Vincennes*, Wilkes's flagship. Nevertheless, Peale, who was on the *Peacock*, complained of lack of assistance – the British and French expeditions would be much better off in this respect he thought – and of not being given proper opportunities to collect (Poesch, 1961) – but this has been a perennial complaint from ship-borne naturalists. The officers thought of the scientists as passengers, ranking below them irrespective of age or knowledge, and the scientists were upset by the secrecy which surrounded so much of the operations. On the scientists' side the geologist, Dana became increasingly sceptical of the competence of the officers in making meteorological observations (Tyler, 1968, p. 37). One of Wilkes's techniques for collecting data must have been somewhat irksome; as suggested originally by the American Philosophical Society his officers were required to keep journals in which objects of interest, however small, were to be recorded and these had to be shown to the commander of their vessel every week.

In a cruise of nearly four years the Wilkes Expedition spent a total of not quite three months south of 60° S. In late February 1839, a bad time of the year, Wilkes in the *Porpoise* accompanied by the *Sea Gull* sailed from Tierra del Fuego to explore to the south-east of Palmer Land, whilst the

Peacock, commanded by Hudson, with the *Flying Fish* were dispatched west to the longitude where Cook had achieved his furthest south. Wilkes took no naturalists with him but made meteorological observations and measurements of wave characteristics. So late in the season the weather was bad and no landings could be made except from the *Sea Gull* in the harbour of Deception Island. There a search was made for a self-registering thermometer left by Foster in 1829 (see p. 74). It was not located but a note left in a bottle led to it being found later by a sailing vessel captained by James Smyley in February 1842, when it read 5° below zero Fahrenheit (Bertrand 1971). This same Smyley, incidentally, recorded what seems to have been the first volcanic eruption on Deception Island in historic times (Bertrand, 1971). Things were quieter for the *Sea Gull*. Hot springs were seen but no temperatures taken. Peale volunteered to go on the *Peacock*, although with little expectation of encountering much interesting natural history. His forte was preservation of conservation and specimens – his father was the founder and proprietor of the Philadelphia Museum and he had been given to understand by no less a person than President Jackson that one of his chief objectives should be the collection of material for the projected Smithsonian Institution – but the few bird specimens he was able to collect from Antarctic waters presumably perished when the *Peacock* was wrecked towards the end of the cruise. His records of bird, seal and whale sightings presented little that was new. He did, however, make accurate drawings of swimming penguins. Discolouration of water at the ice edge and of the ice itself, undoubtedly due to diatoms, was seen and put down as a 'deep earthy stain' (Wilkes, 1845, I p. 414). The *Peacock* and *Flying Fish* were forced to return by ice closing in after the latter had reached 70° 14′ S. Measurements of water temperatures were made from the *Porpoise* and the *Peacock* but when deep water soundings and temperature measurements were attempted the copper wire, seemingly being used for the first time for this purpose, parted and the apparatus was lost from both ships (Wilkes 1945 I, pp. 139 & 152).

The major Antarctic venture during the Wilkes Expedition was in January and February 1840 when the *Vincennes*, *Peacock*, *Porpoise* and *Flying Fish* cruised along the coast of the continent in the sector 100° E to 160° E. The distance of the land was underestimated, giving rise to controversy into which it is not necessary to go here, but there is no doubt that it was sighted (Hobbs, 1940; Bertrand 1971). The naturalists had been left behind but there would have been little for them to do on this leg since there were no landings. Observations of magnetic variables and meteorology were made at Cambridge, Massachusetts, USA, to parallel those made throughout the period of the cruise. Examples of the Antarctic observations are given in Fig. 3.3. Internationally agreed term days for magnetic observations were kept whenever possible during the expedition and the data obtained put the south magnetic pole at 70° S 140° E (cf. the positions estimated by Bellingshausen, p. 37, and Ross, p. 86). It was in this sector that there was contact between the expeditions of Wilkes, d'Urville and Ross. However, there was no co-operation except that Wilkes generously gave Ross a copy of the chart of his recent discoveries, but that only led to unfortunate misunderstandings.

The aftermath of the Wilkes Expedition was as acrimonious as its inception. Its return was greeted with indifference; the country was in a deepening economic depression and interest in natural history was giving way to specialization in science. Peale, whom Poesch (1961) described as having had 'a kind of joyful excitement about the world and God's creatures, great and small', was bitter about the intrusion of what he called 'closet philosophy' into his field. Wilkes was brought before a court-martial on petty charges, preferred by his own officers, of oppression, injustice to his men, illegal and severe punishment of savages, falsehood (which included falsely stating that the Antarctic coast had been sighted) and scandalous conduct. No doubt he brought this on his own head by his high handedness but it is satisfactory to record that he was acquitted and subsequently rose to high rank in the US Navy. He was as indefatigable in getting reports of the expedition written and published as he had been in commanding it on the seas. An immense number of specimens had been collected. These were not all looked after as they should have been on arrival in the US and much was lost (Poesch, 1961; Viola, 1985) but since little of the material came from the Antarctic this need not concern us. There was delay and argument over the writing of reports and when they eventually appeared it was in extremely limited editions (Poesch, 1961). The final tally was the five-volume *The Narrative of the United States Exploring Expedition* (1845) and 20 volumes of scientific reports (Bartlett, 1940). The title page of the copy of the *Narrative* consulted by Poesch bore a note saying 'Nothing has been used in its preparation that is not STRICTLY AMERICAN' and for a time it was forbidden to consult European scholars in the preparation of reports (Poesch, 1961), but such manifestations of misplaced patriotism cannot obscure the magnificence of the final achievement (Viola & Margolis, 1985). The Royal Geographical Society recognized this by awarding Wilkes its gold medal.

This was the one and only major exploring expedition conducted under canvas by the US. It signalled the emergence of this young nation into the

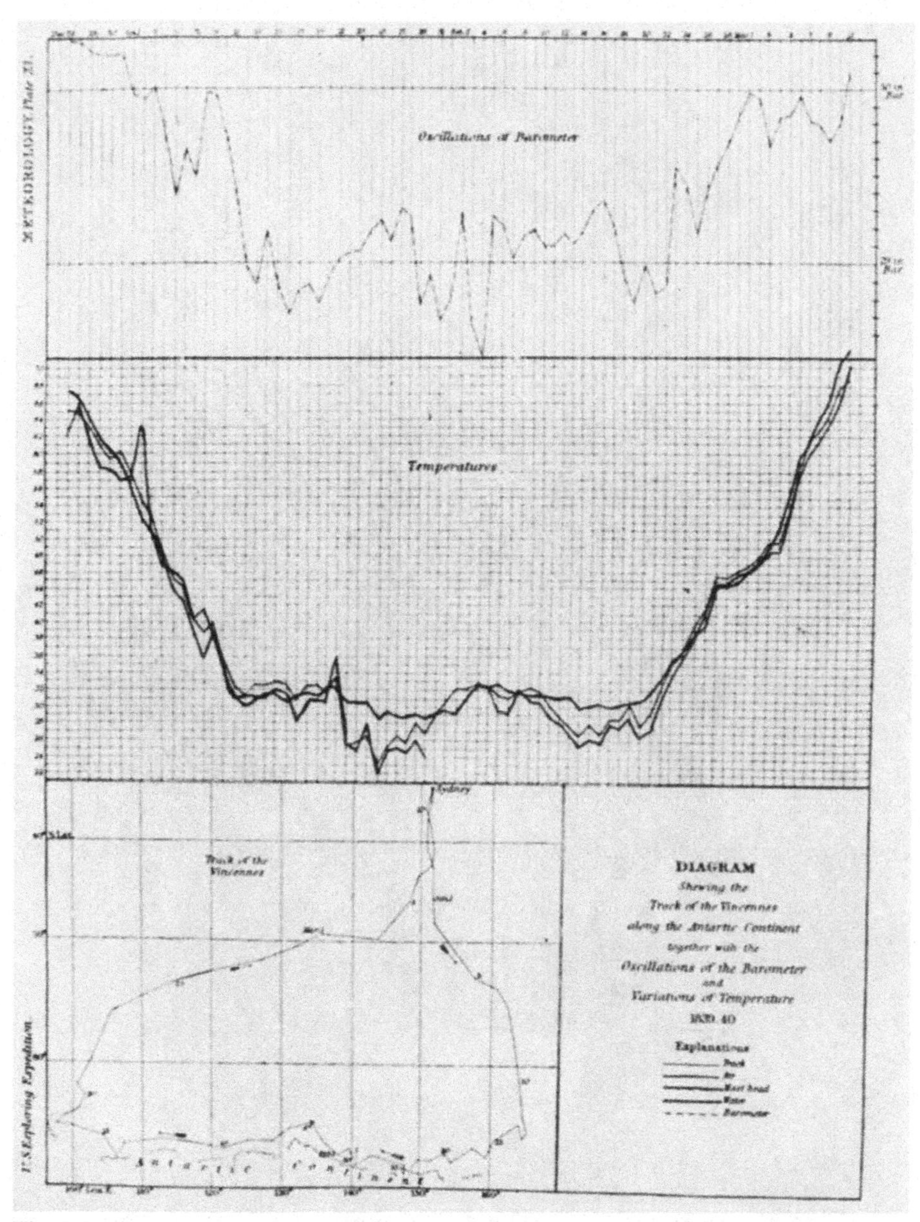

Fig. 3.3 Temperature and pressure measurements and track of the *Vincennes* off Wilkes Land, December 1839 to March 1840; from Wilkes, 1851, *United States Exploring Expedition, 1838 to 1842*, vol. XI, *Meteorology*, plate XI.

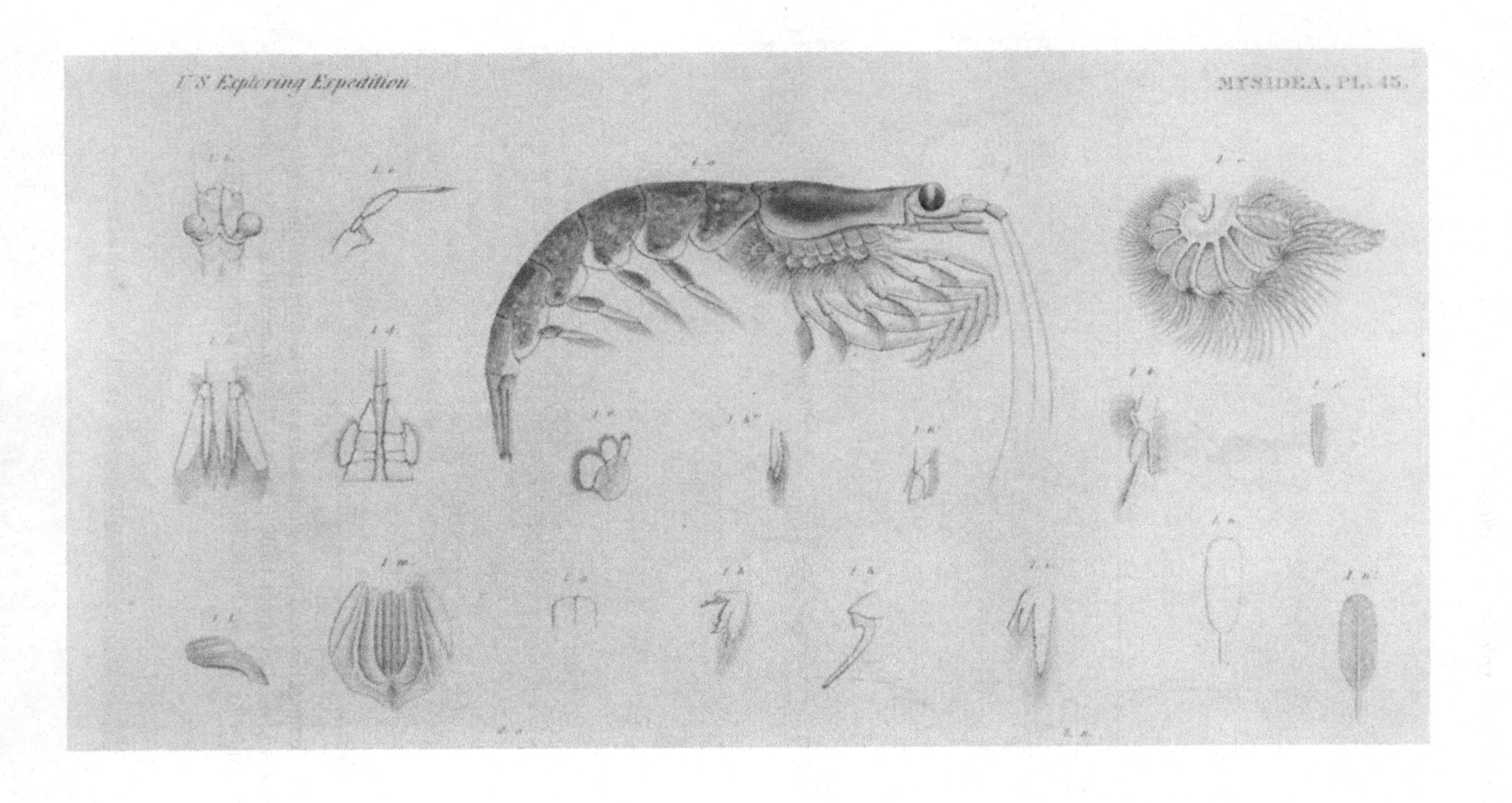

Fig. 3.4 The first illustration of krill, *Euphausia superba*; from J. D. Dana, 1855, plate 43 in Folio Atlas accompanying *Crustacea* in *US Exploring Expedition*, vol. XIII, part 1. (Courtesy of the Natural History Museum, London.)

world as a major naval power. Equally, however, it brought to the notice of the nations the scientific capacity of the US and in the long run it was to establish US interests in Antarctica. Its immediate contribution to Antarctic science was to provide evidence that there is land of continental extent within the Antarctic Circle. Apart from that it added to the knowledge of hydrography and meteorology in the Southern Ocean. The magnetic results were mostly left unpublished but have been assembled by Ennis (1934) from Wilkes's fragmentary notes. They are important for establishing the secular variations in terrestrial magnetism during the nineteenth century. Photometric measurements over several days showed the intensity of light to be less in the far south than in the far north – an important point. Geologically it contributed little immediately but the member of its scientific corps who was to attain greatest eminence, James Dwight Dana, made observations in the Pacific which are fundamental to the modern theory of plate tectonics (Appleman, 1985) which plays so great a part in present day Antarctic geology. It was Lieutenant Totten of the *Porpoise* who in February 1840 collected the type specimen of *Euphausia superba* Dana, the Antarctic 'krill' (Fig. 3.4) which, as Wilkes (1845, II, p. 327) already realized, is the main food of penguins and whales. Otherwise the contribution to Antarctic biology was slight. None of the supposedly new species of birds which Peale described from Antarctic waters (Bartlett, 1940) have survived and no fish were collected south of the Polar Front (Fowler, 1940). An engraved copper plate of southern hemisphere fish (Fig. 3.5), from which prints were evidently not published, shows the high standard that was achieved.

3.3 The French expedition

Meanwhile, the re-established monarchy in France had launched an expedition south with far more dispatch and no argument[2]. Captain Jules-Sébastien-César Dumont d'Urville proposed a new voyage of exploration to the Ministre de la Marine in January 1837 and sailed in September of the same year with the corvettes *Astrolabe* and *Zélée*. d'Urville, born in 1790 and accepted into the Navy at the age of 17, was a gifted linguist and scholar, interested in astronomy, geology, entomology, botany and, above all, ethnology. It was he who, while on duty with the fleet in the eastern Mediterranean, saw the recently unearthed Venus de Milo and secured it for France. He took part in the voyage round the world of the *Coquille* in 1822–25, publishing a book on the flora and fauna of the Falkland Islands as a result and in 1826–29 he commanded the *Coquille*, renamed the *Astrolabe*, in a second circumnavigation. Weddell's exploits were known in

Fig. 3.5 Engraved copper plate by J. H. Richard illustrating berycoid fish collected by the US Exploring Expedition from southern hemisphere waters, 1838–42. This and other plates were intended for a work on the Expedition fishes by L. Agassiz which was never published. (Courtesy of the National Anthropological Archives, USA.)

France by 1825 and both Humboldt and F. Arago expressed regret that d'Urville had not attempted to follow them up on this voyage (d'Urville, 1847 II, p. 19). d'Urville has been described as cold, aloof and unlikable but his writings give evidence of a humorous and thoughtful, if crusty, individual. Whether his personality was the cause of it or not, he fell from favour with the establishment after his second voyage and passed a period of seven years unregarded, poor and ill with gout. His proposal for a third voyage of exploration was made in desperation but it was favourably received by the new Ministre de la Marine, Vice-Admiral Rosamel, and then by King Louis Philippe. d'Urville's object in making the proposal was to further his studies in Pacific ethnology and he had no thought of visiting the Antarctic, having no liking for navigation in the ice, but in granting him two ships for the expedition the King expressed a wish that it should start with a venture towards the South Pole. It is not clear whether this idea was the King's. He had evidently seen the books of Weddell and Morrell (d'Urville, 1847, I, Introduction) but also he was on friendly terms with Humboldt and the suggestion may have come from this source. In informing d'Urville of the support of the government, Rosamel stated that the object was not only prosecution of hydrography and science but to further the interest of French commerce, especially of the sperm whale fishery which in the south seas at that time was dominated by the French. The Antarctic component seems to have been added for the glory of France. d'Urville was greatly taken aback by this Royal request; he regarded Morrell as mendacious and Weddell's honesty as unproven; accident or loss in the Antarctic might prevent him from carrying out the studies which were near to his heart. However, he knew that expeditions south were going forward from both the US and the UK and, realizing its popular appeal, he accepted the mission.

Thus resolved d'Urville went forward energetically with his preparations. To ensure the co-operation of his officers and crew he persuaded the King to give a premium of a hundred gold francs to each man on reaching 75° S with an additional 20 for every degree beyond that. In a preparatory visit to London, where he was received politely but with hints that the Antarctic was a British preserve, d'Urville was assured that Weddell was 'a true gentleman' and that his reports could be trusted. The secretary of the Royal Geographical Society, Captain Washington, whom he met on this visit, had written anonymously urging on the President and Council of that society the desirability of Antarctic exploration and d'Urville presented a full translation of this document to the Paris Geographical Society at the same meeting at which he announced his own plans. The Academy of Sciences was approached on his behalf for advice but he was not on good terms with some of its members and he took exception to the instructions which they provided for physical observations to be made during the voyage and especially in the far south. Arago, who had been charged with drawing up directions for the work on magnetism declined to do so and Savary, who replaced him on the Commission, perfunctorily referred d'Urville to the instructions prepared by Arago for the voyage of the *Bonité* in 1836. Obviously d'Urville's preference was for the biological sciences but even here the Academy, in his view, responded in an unencouraging manner. However, both Humboldt and Krusenstern sent their good wishes for the success of the voyage.

The *Astrolabe* was well proven in exploration and the *Zélée* was a similar frigate but neither was strengthened against ice and their gun-ports were points of weakness in facing Antarctic storms. Their complement included Vincendon Dumoulin (hydrographer), Hombron, Le Breton, Le Guillou, Jacquinot (surgeons and assistant surgeons also acting as naturalists), and Coupvent Desbois (a sub-lieutenant who assisted Dumoulin in physical observations). Some of the sailors had seen service in the Arctic.

d'Urville's first main task after leaving Toulon on 7 September 1837 was to follow Weddell's track. Sailing due south of the South Orkneys he encountered dense pack-ice and after being trapped for a horrifying five days was glad to make his escape, then, after more fruitless searching for a way south, to abandon the attempt. From his subsequent experience d'Urville realized that pack-ice is extremely changeable and that it was possible that Weddell had struck a good year and actually got as far south as claimed, but he remained suspicious. His fellow captain, C.-H. Jacquinot of the *Zélée*, was more prepared to admit that Weddell had found open sea where they had found impenetrable ice. The two ships then sailed west and discovered new land around the tip of the Antarctic Peninsula. For nearly two years after that d'Urville engaged in work in Oceania. However, he determined on another attempt on the Antarctic. This was beyond his instructions but he was sensitive to innuendo following his failure to follow Weddell. The French, he decided, must have the glory of being first at the South Magnetic Pole. Sailing south from Tasmania they sighted land in January 1840 and then followed the coasts which he named as Adélie and Clarie westwards as far as ice conditions would permit over 11° of longitude. A landing was made on a small island, completely lacking vegetation, from which geological specimens (granite/gneiss) were collected. The magnetic pole, estimated by Dumoulin as being around 72° S 138° E, was clearly inland and out of their reach. The French have ranked d'Urville with Cook and, especially when one bears in mind that throughout this voyage he was ill and in constant pain, his achievement was outstanding.

The *Astrolabe* and *Zélée* arrived back in France in early November 1840. The account of their voyage, *Voyage au pole sud et dans l'Océanie*, was published in 10 parts accompanied by illustrations in the *Atlas* and followed by other volumes dealing with scientific results. Before this d'Urville together with his wife and son perished in a railway accident in 1842.

d'Urville's discovery of the Adélie and Clarie coasts helped to substantiate the existence of a large land mass near the pole and determined the future location of a major part of France's scientific effort in the Antarctic. The magnetic determinations suffered as the two ships had not been supplied with matched instruments and those for measuring dip and intensity could not be used satisfactorily at sea. In the published results only compass declinations are reported and those from the *Astrolabe* alone, the observations from *Zélée* being deemed unreliable (Dumoulin & Coupvent-Desbois, 1842). Among the other observations made, that of continual easterly winds in the high latitudes attained was one of the first records of this important feature of atmospheric circulation. Barometer, temperatures, sea states and currents were also noted. Equipment invented by Biot for sampling water at depth was tried but did not work, evidently because of a defect in the particular piece of apparatus. Transparency of sea water was measured by noting the depth at which an earthenware plate became invisible – a device now known as the Secchi disc after the Italian astronomer who used it in the Mediterranean in the 1860s – but the use of a polarizing viewer to increase the accuracy of the end point was unsatisfactory (d'Urville, 1847, vol. I, pp. 3 and 174) and few determinations of transparency seem to have been made. Ice formations were described; d'Urville was at first inclined to explain the irregularity of ice-fields as arising from disruption of smooth ice by seismic action (d'Urville, 1847, vol. II, p. 171) but this hypothesis later seems to have been abandoned. He commented on the uniformity in form and height of tabular icebergs and, once they had seen an ice shelf (which he called a *barrière de glace*), both he and his officers had no doubt that these bergs were produced by the breaking off of pieces of the shelf. They were, however, puzzled as to how ice shelves themselves originated although d'Urville realized that they must have solid land as a basis even though they might extend some distance away from it. Grange (1848) in writing the official account of this part of the scientific work argued that tabular bergs are too enormous and uniform

Fig. 3.6 *Lithodes* sp., a crab of sub-Antarctic waters collected on the voyage of *Astrolabe* and *Zélée*, 1837–40; from J. B. Hombron and H. Jacquinot, *Zoologie, Atlas*. (Courtesy of the Scott Polar Research Institute.)

to have been formed on land, in which, of course, he was in part right. He did, however, take a useful step forward in discussing Antarctic ice in relation to Scoresby's observations in the Arctic and the work of de Saussure, Forbes and others in glaciers in the Alps. Forbes (1846) by experiment and observation had established a theory of viscous flow of glaciers based on the supposition that ice became plastic under pressure. The geological report (Grange, 1854) says little relating to the Antarctic. The volumes on zoology and botany were accompanied by atlases of beautiful illustrations (Fig. 3.6); perhaps the most notable discovery was that of the crabeater seal, *Lobodon carcinophagus*. A recommendation by the Academy of Sciences (1837) that marine zooplankton should be sampled by means of nets of bolting silk does not seem to have followed but some dredging of benthic organisms was carried out.

3.4 Geodesy and the visit of HMS *Chanticleer* to Deception Island

In the first third of the nineteenth century British effort in exploration was largely absorbed by the Arctic but while this retarded further ventures south it eventually provided the experience and the men, for one of the most successful of Antarctic expeditions. Before this, however, there was a voyage to the Antarctic which, because it made no new geographical discoveries, attracted little attention.

At the beginning of the century surveying and cartography were demanding a firm geodetic basis. A trigonometric survey of England had begun in 1783, primarily to establish geodetic connection between Greenwich and Paris. A principal method in physical geodesy has been gravimetric determinations by means of the seconds pendulum and in 1816 a prayer from the House of Commons to the Prince Regent resulted in the Royal Society setting up a committee to oversee pendulum measurements at the trigonometrical stations established in Britain.[3] In 1818, John Pond, the Astronomer Royal, drew the attention of this committee to the value of pendulum measurements in different parts of the globe, which seemed 'the more necessary from the circumstance of the French Government having directed the Members of their Academy to commence a similar operation'.[4] There was a quick response when this was referred to the Admiralty and accommodation was found on HM sloop *Isabella*, about to sail for Baffin Bay, in response to Scoresby's initiative. One object of this cruise was to enable Captain Edward Sabine, a young artillery officer, to carry out pendulum measurements in the Arctic. This was momentous for Antarctic research in that the *Isabella* was commanded by John Ross who was accompanied by his 18-year-old nephew, Midshipman James Clark Ross, of whom there will be more to be said shortly. Sabine also took part in the voyage to East Greenland and Spitzbergen of HMS *Griper* in 1825 when with Henry Foster, one of the leading scientific officers of the Navy, he carried out magnetic, astronomical and pendulum observations. In 1828 Foster was given command of HMS *Chanticleer* and charged with continuing in the South Atlantic his observations to determine the specific ellipticity of the earth. In welcoming 'the liberal and enlightened views which have actuated His R.H. the Lord High Admiral in meeting the outfitting of an enterprise solely and simply for the promoting of scientific research' the Council for the Royal Society went on to say 'perhaps the most important of all the southern stations is that of New South Shetlands; and yet more so any other lands in still higher southern latitude.'

Instructions for the *Chanticleer* were drawn up by the Board of Longitude (Day, 1967) and it was suggested by the Royal Society that in addition to the pendulum measurements and establishing precise latitudes and longitudes, Foster should make magnetic and meteorological observations and collect natural history specimens. The *Chanticleer* was too small to accommodate a naturalist and, as was becoming usual, the surgeon, Dr William Webster, was given the job (see Jones, 1974, for an account of his life). The *Chanticleer* sailed in the spring of 1828 and reached Deception Island in early January 1829, staying there nearly two months and leaving the self-registering thermometer already mentioned and the name Pendulum Cove as momentos of her visit. On the return voyage Foster was drowned in a river while carrying out a geodetic traverse of the Isthmus of Panama and it was Webster who wrote the published account of the voyage (Webster, 1834). A detailed appendix to this account gives the scientific results. Besides the pendulum data it contains careful accounts of the geology and thermal phenomena of Deception Island, some general observations on its plants and animals, and descriptions, but no illustrations, of dissections of leopard seals and penguins. The first algae to be collected in the Antarctic, *Cystophaera jaquinotii* and *Desmarestia* were brought back by the *Chanticleer* (Hooker, 1847).

3.5 'The magnetic crusade'

The *Chanticleer* returned to England in 1831. In this same year James Clark Ross reached the North Magnetic Pole by sledge and the British Association for the Advancement of Science was founded. Four years later at the Association's meeting in Dublin, Ireland, Edward Sabine, now a powerful figure in both the Association and the Royal Society, in a report on terrestrial magnetism pointed out that south latitudes were more accessible along certain meridians than previously supposed and that magnetic observations should be made in those regions.[5] Responding to his enthusiasm the Association resolved,

> That a representation be made to Government of the importance of sending an expedition into the Antarctic regions, for the purpose of making of observations and discoveries in various branches of science, as Geography, Hydrography, Natural History, and especially Magnetism, with a view to determine precisely the place of the Southern Magnetic Pole or Poles, and the direction and inclination of the magnetic force in those regions.[6]

It was perhaps Sabine, also, who suggested to Humboldt that he should write to the President of the Royal Society on the subject (Cawood, 1979). A weighty letter, dated 23 April 1836, resulted and appealed for the setting up throughout the British Empire of a series of magnetic observatories similar to that which the Tsar had established in Siberia, and for the dispatch of an expedition primarily for magnetic studies in the Antarctic. The Royal Society referred this letter to G. B. Airy and S. H. Christie, both experts on terrestrial magnetism, who strongly endorsed the idea of international co-operation both in this field and science generally (Cawood,

1977; O'Hara, 1983). A committee was set up, a government grant was obtained for the purchase of instruments and an offer accepted from the Master General of the Ordance for magnetic observations to be made at stations occupied by Officers of the Engineers,[7] but things got bogged down. Sabine renewed his exhortations at the British Association meeting in Liverpool in 1837; the most important blank to be filled in magnetic observations was in the southern hemisphere and this could only be done by a naval expedition led by an officer of scientific ability and experience. He went on: 'I need hardly say that the country possesses a naval officer in whom these qualifications unite in a remarkable degree with all others that are requisite'[8] and his audience was in no doubt that he was referring to Captain James Ross.

The anonymous pamphlet in the form of a letter signed A. Z. of 1837 addressed to the Royal Geographical Society urging the importance of an Antarctic expedition has already been mentioned and seems to have been of little avail, for that Society took no part in the campaign. However, A. Z., coming out into the open as Captain Washington, secretary to the Royal Geographical Society, made a plea for Antarctic exploration at the British Association's meeting in Newcastle in 1838, demanding that 'that individual who has already planted the "red cross of England" on one of the northern magnetic poles' should be selected to lead an expedition and forstall the Americans. Already before this outburst of patriotic fervour the Council of the Association had decided on action and a committee appointed to put a memorial embodying the resolutions about an expedition before government did so in September 1838. We may note that scientists lacked secretarial help even more in those days than they do now, for Sir John Herschel, a key figure in negotiations with the government had himself to make copies of the memorial by hand.[9] Wiederkehr (1985) considers that it was a meeting between the German physicist Wilhelm Weber and Sir John on the occasion of the coronation of Queen Victoria which sparked off this activity. Measurements from the southern hemisphere were needed by Weber's colleague, Gauss, to verify his general theory of terrestrial magnetism and Britain as the leading sea-power of the day was in the best position to provide them. The Prime Minister, Lord Melbourne, referred the British Association memorial to the Royal Society. This august body might have taken umbrage at this presumption of the young Association but fortunately did not and thereafter matters moved apace. The Royal Society committee appointed two years before to consider Humboldt's letter excused the delay, quite reasonably, on the grounds that methods had needed evaluating and that the delay had 'given time for maturer consideration and it now seems that more continuous and systematic observations involving more costly instruments and means are called for'.[10] The committee's report detailed the desirable objectives of an expedition and Council, approving its recommendations, recollected that the government had supported a similar expedition by the 'illustrious Halley' and dispatched a deputation led by its President, to communicate this to the Prime Minister. He decided that the expedition should take place, the Chancellor of the Exchequer allocated ample funds and the Admiralty provided two ships HMS *Erebus* and HMS *Terror*. The administration was to change in 1841 but Sir Robert Peel, Lord Melbourne's successor, continued support for the expedition. A letter dated 13 June 1839, from the Secretary to the Admiralty, Sir John Barrow – still enthusiastic about polar exploration by the Navy – to the Secretary of the Royal Society gave the news that the right man had been appointed to lead it and requested the Society's help in planning the scientific work.[11]

It had been touch and go for had there been any postponement, Ross would have accepted another command.[12] Council resolved that Barrow's letter be referred to the Committees of Physics, Meteorology, Geology and Mineralogy, Botany and Vegetable Physiology, and Zoology and Animal Physiology. Further a circular was sent to *inter alia*, Humboldt, Gauss, Weber, Arago – these four were the leading authorities on terrestrial magnetism – the Minister of Public Instruction at St. Petersburg, Krusenstern and de Saussure, advising them of what was afoot and requesting their co-operation in arranging for observations to synchronize with those of the expedition.[13] The specialist committees reported on 25 July 1839, and their recommendations were incorporated into an 85-page booklet for the use of the expedition. This was a useful document – so much so that an extended version of the sections dealing with physics and meteorology was issued for general use in 1840 with another, revised, edition in 1842 (Royal Society, 1840). It emphasized the importance of measurements made synchronously over as much of the globe as possible. Observation days should include the *terms* (four times a year) of the German Magnetic Association, which, under Gauss and Weber at Göttingen had become the leading organization in the field. During these terms, observations were to be made hourly for not less than a week. Among the oceanographic work that was recommended were temperature measurements at depth, the daily release of drift bottles to track currents, measurements of the transparency of the sea using Mr. Talbot's sensitive paper, and deep-sea sounding (a method of echo sounding was suggested but evidently not used). Geological collections should be representative and not merely of rarities. 'The vegetation of the Antarctic regions and of the most southern countries which the expedition may visit, should be an object of especial attention, for however sterile and uninviting a place may appear to be, it is most desirable to know exactly what plants these regions produce' (Royal Society, 1840). The zoologists requested specimens – young and eggs at different stages as well as adults – of the great penguin and directed attention to the smaller oceanic crustacea as a prolific and hitherto unexplored field of investigation. In the Antarctic, said the instructions '*every thing* is of interest'. However, they did not include any recommendations for work on the physical and physiological effects of cold such as had been made to Mr George Fisher when in 1821 he was about to go to the Arctic seas under the command of Captain Parry.[14]

We might pause at this point to pay a tribute to Edward Sabine as the originator and main driving force behind the successful launching of the expedition. This was part of the 'Magnetic Crusade' which was described at the time by William Whewell as 'by far the greatest scientific undertaking the world has ever seen' (Cawood, 1979). Although an army man, rising eventually to the rank of Lieutenant General, Sabine saw little active service (Stone, 1984). He had only a minimum of scientific training but achieved an early reputation as an explorer and geophysicist and his lifelong preoccupation was collecting magnetic data. He might well have been the inspiration for Gilbert's modern Major-General (Hall, 1984). His wife too had scientific interests and was responsible for the translation into English of works by Humboldt and, probably, Gauss (O'Hara, 1983). As a scientific administrator he was effective, with a gift for promoting harmony and co-operation but, although he was President of the Royal Society from 1861 until 1871, he was perhaps not a very profound scientist and was accused by Babbage, a man with a chip on his shoulder, of inaccurate calculation (Stone, 1984; Hall, 1984). He was responsible for reducing and writing up the magnetic data obtained by the Ross expedition and, through Herschel, provided Gauss with much of the information he needed.

3.6 The Antarctic voyage of HMS *Erebus* and HMS *Terror*

James Clark Ross (Fig. 3.7) had spent 15 summers and eight winters in the Arctic and as a seaman and navigator was superb. He had an enquiring mind and early on had shown an aptitude for science. Besides being proficient in geomagnetic observations he had a bent towards natural history, in which his ability had been recognized by election to the Linnean Society in 1823 (Ross, 1982). Apart from Ross and the officers who assisted him in magnetic and other physical observations, the surgeons were

Fig. 3.7 Sir James Clark Ross; portrait by Stephen Pearce. A Fox dip circle is shown to Sir James's right. (Courtesy of the National Maritime Museum, London.)

responsible for the science, there having been no question that full-time naturalists should accompany the expedition. Among the medicals it is Joseph Dalton Hooker (Fig. 3.8) who stands out as a scientist. Through his father, William Jackson Hooker, at that time Regius Professor of Botany at the University of Glasgow, Scotland, he had met Ross and it had been agreed that he should be taken on the Antarctic expedition if he completed a medical course so that he could qualify as an assistant surgeon. He was 22 when he joined the *Erebus*, already seriously devoted to botany and

Fig. 3.8 Joseph Hooker as a young man. (Reproduced with the permission of the Director and Trustees of the Royal Botanic Gardens, Kew.)

grudging time spent on other things. He eventually succeeded his father as Director of the Royal Botanic Gardens, Kew, and became one of the great scientific figures in Victorian England (Bower, 1913; Turrill, 1963). Towards the end of his long life he was a powerful influence in the planning of the *Discovery* expedition of Scott. The assistant surgeon on the *Terror*, David Lyall, also contributed to the botanical work and John Robertson, the surgeon of the *Terror*, was officially the zoologist of the expedition. Other aspects of natural history were in the hands of Robert McCormick, the surgeon of the *Erebus*. He had been a shipmate of Charles Darwin for a short time on the *Beagle* but Darwin wrote of him as 'a philosopher of rather an antient date; at St. Jago by his own account he made general remarks during the first fortnight & collected particular facts during the last' (Keynes, 1979, p. 59). On his side McCormick came to feel, undoubtedly with justification, that his position as naturalist on the *Beagle* had been usurped. He was intensely interested in geology and ornithology. In the latter, the wildfowling element seems often to have predominated over the science but it did provide the type specimen of the Antarctic skua, *Catharacta maccormicki.* He was hard-working but excitable and sensitive, and was probably justified in his criticism of Ross's inflexibility, which kept a medical officer aboard each ship under circumstances when they might without undue risk have been ashore collecting natural history specimens (McCormick, 1884, p. 162). Hooker too, although he got on well with his captain and was appreciative of the facilities given to him for his work, thought Ross somewhat autocratic. There might have been conflict between Hooker and McCormick, who was officially in charge of all natural history, but fortunately McCormick was little interested in plants and was therefore content to leave them entirely to Hooker.

In equipping the expedition there was much experience gained in the Arctic on which to draw. The *Erebus* and *Terror* were vessels of a type already thoroughly tested in the ice. Both were bomb vessels (Pearsall, 1973); exceptionally strongly built to carry heavy mortars, they had additional strengthening for the Antarctic, providing a mass of wood seven feet in thickness at the bows, and although they took a tremendous battering in the pack they returned home essentially unscathed. Steam power was considered but deemed to be of too recent introduction to be trusted in the Antarctic (Mill, 1905, p. 208). Among the provisions were tinned meats, soups and vegetables but there seems to have been no lead poisoning from solder on the tins as occurred on the ill-fated Franklin expedition which sailed to the Arctic in 1845. With the variety of foods available scurvy was successfully held at bay; 2455 lbs (1.4 tonnes) of pickled walnuts – extremely rich in vitamin C as we now know – taken as 'medical' comforts may have contributed appreciably to this. A warm-air stove invented by Mr. Sylvester was to prove speedy and effective in circulating air through the ships and drying them out. Warm clothing of the best quality was taken and issued gratis to the crews. The scientific equipment was also of the best. The magnetic and meteorological apparatus has been listed by Savours & McConnell (1982) but the most important item from our point of view was the statical 'dipping needle deflector' which had been described by R. W. Fox in 1843 (Fig. 3.9). This was capable, when mounted on a gimballed table, of measuring declination, dip

Fig. 3.9 The Fox dip circle; this particular instrument was made by W. George, Falmouth, *c*. 1850. (Courtesy of the Trustees of the Science Museum, London.)

and intensity with acceptable accuracy on a vessel at sea. It was infinitely handier than the more precise instruments used by Gauss and Weber, giving Ross a greater advantage over his rivals and then being widely used for the next 50 years (McConnell, 1985; Multhauf & Good, 1987). Duplicate sets of magnetic instruments were provided for each vessel and to minimize interference the after part of the ship was fastened entirely with copper and composition metals. Hooker, lecturing in 1846 (Fig. 3.10), said:

> I believe no instruments, however newly invented, was omitted, even down to an apparatus for daguerreotyping and talbotyping, and we left England provided with a register for every known phenomenon of nature, though certainly not qualified to cope with them all.[15]

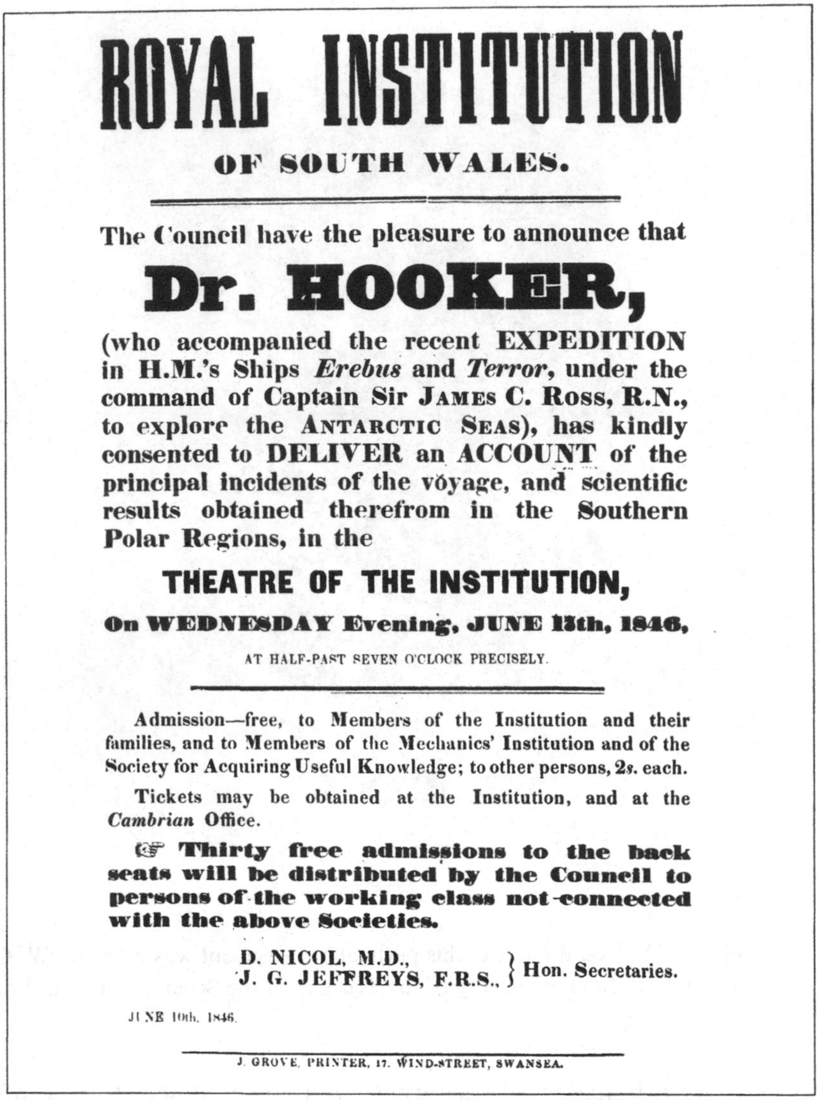

ROYAL INSTITUTION

OF SOUTH WALES.

The Council have the pleasure to announce that

Dr. HOOKER,

(who accompanied the recent EXPEDITION in H.M.'s Ships *Erebus* and *Terror*, under the command of Captain Sir JAMES C. ROSS, R.N., to explore the ANTARCTIC SEAS), has kindly consented to DELIVER an ACCOUNT of the principal incidents of the voyage, and scientific results obtained therefrom in the Southern Polar Regions, in the

THEATRE OF THE INSTITUTION,

On WEDNESDAY Evening, JUNE 15th, 1846,

AT HALF-PAST SEVEN O'CLOCK PRECISELY.

Admission—free, to Members of the Institution and their families, and to Members of the Mechanics' Institution and of the Society for Acquiring Useful Knowledge; to other persons, 2*s*. each.

Tickets may be obtained at the Institution, and at the *Cambrian* Office.

☞ Thirty free admissions to the back seats will be distributed by the Council to persons of the working class not connected with the above Societies.

D. NICOL, M.D.,
J. G. JEFFREYS, F.R.S., } **Hon. Secretaries.**

JUNE 10th, 1846.

J. GROVE, PRINTER, 17, WIND-STREET, SWANSEA.

Fig. 3.10 Poster advertising a lecture on the Antarctic by Joseph Hooker, 17 June 1846. (Reproduced with the permission of the Director and Trustees of the Royal Botanic Gardens, Kew.)

They evidently found it difficult to cope with the photographic apparatus as there is no record of it having been used. Hooker also mentioned that Captain Ross took an excellent library of scientific works.

Unlike the expeditions of d'Urville and Wilkes this one had the Antarctic as its main objective and altogether it spent nine months south of 60° S. Before that, however, a magnetic observatory was established in Hobart,

Tasmania. This was near a point of maximum intensity and continued observations for 14 years. The instruments set up there were used to check those taken on the ships (Savours & McConnell, 1982). The first incursion further south was between November 1840 and April 1841 and penetrated through the pack-ice into the open waters of what was afterwards called the Ross Sea, discovering Victoria Land, Ross Island with its active volcano, Mount Erebus and the Barrier (later the Ross Ice Shelf). These major geographical discoveries would not have been made had Ross gone directly for the South Magnetic Pole as d'Urville and Wilkes did. Ross scorned to follow in their tracks and decided – he had been given permission to exercise his own judgement by the Admiralty – to go south along a more easterly meridian, 170° E, where in the summer of 1839 Balleny had reached 69° S and found open sea. Thus, it was shown that Antarctic pack-ice could be mastered and the route, which led into the heart of Antarctica and eventually to the South Pole, was discovered. Had d'Urville or Wilkes made an attempt in the same sector they would not have succeeded because their ships could not have withstood the ice. Ross's following two ventures south were anticlimaxes. The expedition reached home in September 1843. Its return seems to have aroused little public interest but Ross was shortly afterwards awarded the Founder's Medal of the Royal Geographical Society and received the Gold Medal of the Royal Geographical Society of Paris from the hands of d'Urville.

The major scientific achievement was the provision of definitive charts of magnetic declination, dip and intensity in the southern hemisphere, replacing conjecture with observation (Fig. 3.11). This was the greatest work of its kind so far attempted; the results were more reliable and detailed than those of either d'Urville or Wilkes, having been meticulously corrected for deviation and checked daily between the two ships. They were presented by Sabine in a series of lengthy papers in the *Philosophical Transactions of the Royal Society of London* between 1842 and 1868, the last appearing after Ross's death in 1862. These reports also included results obtained on a voyage in 1845 by HMS *Pagoda* south of the Cape of Good Hope to the ice off Queen Mary Land at 68° 10′ S 35° E, which was made to fill in the major gap left by the Ross expedition (Sabine, 1846; meteorological observations given by Clerk, 1846). Gauss's theory, as provisionally applied, had failed to produce for the southern hemisphere the characteristic features of the magnetic field such as were well established for the north. Now there were sound and substantial data to provide the numerical coefficients needed to revise the Gaussian theory. The necessity for such correction is illustrated by comparing the position predicted by Gauss for the South Magnetic

(a)

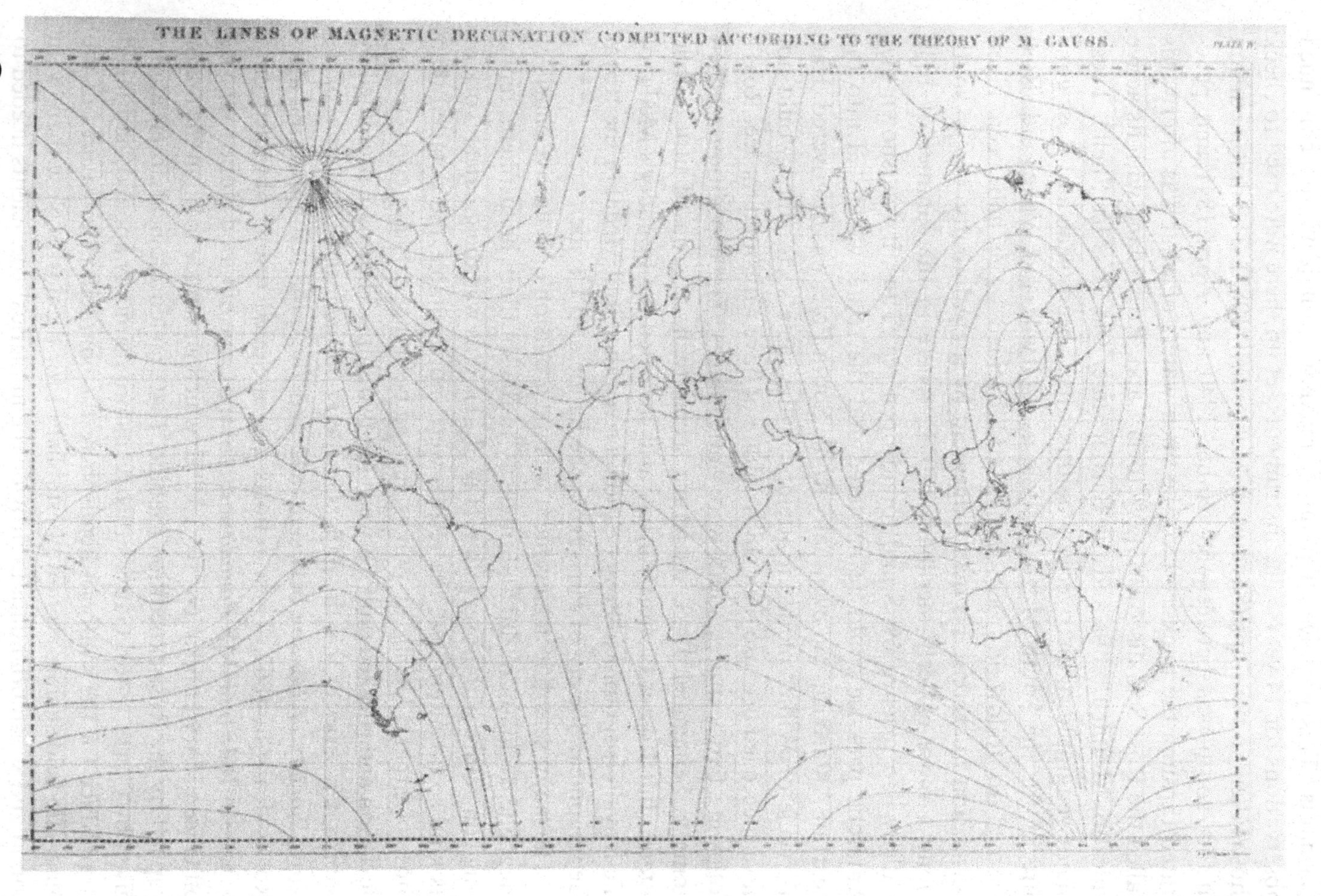
THE LINES OF MAGNETIC DECLINATION COMPUTED ACCORDING TO THE THEORY OF M. GAUSS.

(b)

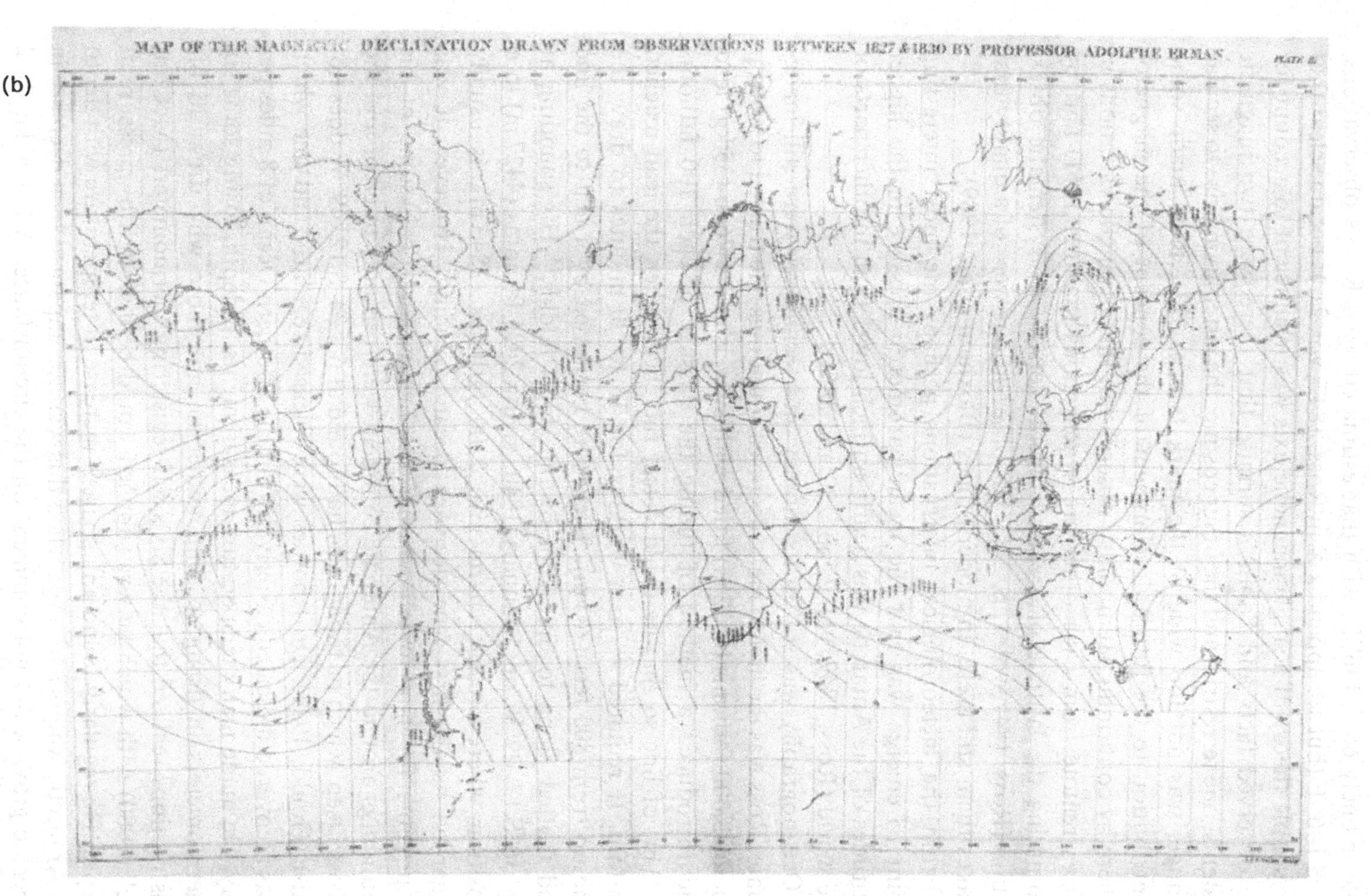

MAP OF THE MAGNETIC DECLINATION DRAWN FROM OBSERVATIONS BETWEEN 1827 & 1830 BY PROFESSOR ADOLPHE ERMAN

Fig. 3.11 World charts of magnetic declination: (a) as predicted by Gauss, Royal Society 1842, and (b) as observed. (Courtesy of the Royal Society.)

Pole, namely 66° S 146° E, with that estimated from Ross's observations, 75° S 146° E (but see appendix X in Ross, 1847, vol. 2). It became clear that the simple theory of a single magnetic axis was insufficient to account for the observed facts and Sabine's surmise that there were two magnetic systems, one terrestrial and the other cosmic, the latter giving rise to secular change, was not too wide of the mark (see p. 322). Ross made some contribution to the future development of the concept of the magnetosphere by confirming Humboldt's observation that magnetic variations in places remote from each other are synchronous (Sabine, 1851) and by estimating the latitude of greatest auroral activity by systematic observation (Ross, 1847, vol. 1, p. 266). His observations were of a quality that makes them still useful in modern times (Bloxham *et al.*, 1989).

From the meteorological observations perhaps the most interesting point to emerge was that the mean barometric pressure in the higher latitudes of the Antarctic was nearly an inch lower than in other parts of the world (Ross, 1847, vol. 2, pp. 229 and 385).

Oceanography was still in its infancy and techniques were still inadequate. Ross was the first to make deep-sea soundings. He trusted to a hemp line of 3,600 fathoms (6572 m) fitted with swivels and a 76 lb (35 kg) lead and introduced a method of timing the passage of each 100 fathoms (183 m) of line so that he could detect the check when the lead reached bottom. It required 40 men working for 16 to 20 minutes to draw the line up from 800 feet (243 m). They were harnessed by a belt to the line and trotted along to time kept by the ship's fiddler.[15] His sounding at 27° 36′ S 17° 29′ W on 3 January 1840, giving 2425 fathoms (4427 m) in a position where present charts give 2100 fathoms (3843 m), is usually regarded as the first true deep-sea ocean sounding. However, it was impossible for him to tell whether the line was vertical, some of his soundings around the Antarctic continent were certainly in error and the 'Ross Deep' which he thought he had found at 68° 34′ S 12° 49′ W does not exist (Gould, 1924). Drift bottles were released regularly and one which was recovered indicated easterly drift in the Southern Ocean of 8 miles per day. He made many temperature measurements at depth, losing some six thermometers in doing so, but neither he nor those who drew up his instructions realized that the Six's self-registering thermometers (McConnell, 1980) with which he was supplied would be distorted by pressure at depth. He regularly found an apparent increase in temperature with depth in Antarctic waters and, even when this was accompanied by a decrease in specific gravity, did not comment on the circumstance. A letter dated 12 October 1839 from Humboldt to the First Lord of the Admiralty giving various suggestions for the scientific work of the expedition pointed out the fallacy of the general assumption that sea-water has the same point of maximum density as fresh-water (Deacon, 1971). This letter was included in the extended 1840 edition of the instructions but Ross did not see it until he had returned from the expedition (Ross, 1847, vol. 2, p. 24). The effect of pressure on thermometer readings had actually been demonstrated as long ago as 1833 but the paper had been passed over (Deacon, 1971). Following an idea which d'Urville had put forward after his 1826–29 voyage, Ross thought that what he called 'the circle of mean temperature' where the mean temperature of the sea obtained throughout its depth, could be distinguished around 56° S. Although this concept was based on faulty observations it is interesting that Ross's circle corresponds well with the Polar Front (Deacon, 1971). Analysis by Forchhammer (1865) of three samples of sea-water collected in the Antarctic gave widely differing values for total salt content and ratio of sulphate to chloride. Assuming that there was no contamination or evaporation of the samples, this might be accounted for by the samples being taken from pack-ice.

In glaciology the major discovery was the enormous ice shelf in the Ross Sea. Ross showed that it was floating at its seaward edge (Ross 1847, vol. 1, p. 222) and it was obvious to him as to d'Urville that it must be the source of tabular bergs. This was clear, too, to the armourer of the *Erebus*, James Savage, who dictated the following to his shipmate, C. J. Sullivan:

> The Fragments as I call the floating Islands though Large Enough to build London on their Summit must through a Long Succession of years have parted from the Barrier they never could accumulate to Such an Enormous hight otherwise.[16]

McCormick was assiduous in collecting rocks – shooting penguins in order to recover the pebbles from their crops being one method. Had he but realized it, his specimens, which included gneiss, granite, quartzite and slate, provided evidence of the continental character of Antarctica. However, they lay unexamined in the British Museum (Natural History) and by the time a petrological description of them was published (Prior, 1898) the material collected by HMS *Challenger* had made the point. McCormick gave an account of the geology of Kerguelen to the Royal Society (McCormick, 1841), one of the most interesting findings being that of fossil trees, but as Prior (1898) remarked, his writings 'in most cases resolve themselves into exasperating (from a petrological point of view) descriptions of birds, for the doctor appears to have been a more enthusiastic ornithologist than geologist'. In appendices to the account of the voyage he

Fig. 3.12 *Pringlea antiscorbutica*, the Kerguelen cabbage, from plates XC–XCI in J. D. Hooker (1847) *The Botany of the Antarctic Voyage*. (Courtesy of the Director and Trustees of the Royal Botanic Gardens, Kew.)

described the geology of the Antarctic continent and southern islands as well as of other places visited. The active vulcanism evident in Mount Erebus and the basalt and lava he found on Possession Island led him to the conclusion, pointed out as erroneous by Ferrar (in Scott, 1905), that the whole range of Antarctic mountains seen were of volcanic origin.

The botanical results, embodied in Hooker's massive *Flora Antarctica* (1847), are still basic for the plant taxonomist but are, however, more concerned with the southern temperate zone than the true Antarctic. The floras of the sub-Antarctic islands visited were described with reasonable completeness (Fig. 3.12) and he gave a good account of the biology of tussock grass. The two indigenous flowering plants of the Antarctic, *Colobanthus quitensis* and *Aira* (*Deschampsia*) *antarctica*, were described, the latter, he noted, occurring further south than any other flowering plant and having been found in the South Shetlands by Eights. He recorded the enormous growths of *Macrocystis* and *Durvillea* surrounding the islands and the floating masses of these brown seaweeds occurring nearly as far south as open water was free of icebergs down to 64° S. No vestige of plant life was found in the Ross Sea area but a landing on Cockburn Island in the Erebus and Terror Gulf yielded a collection of several different lichens and algae. Hooker noted that rocks in the sun became quite warm, providing sufficiently good conditions, he thought, to enable growth of resistant cryptogams such as were found.[17] He concluded that the ability to resist low temperatures was not correlated with toughness since delicate plants such as *Ulva* and *Calothrix* were able to survive the cold. Hooker made pioneer studies in plant geography (Bower, 1913), using the ratio of natural orders to number of species as an index of the variety in the flora of a particular place. He found this index to decrease towards the poles but to be higher in the Arctic than at equivalent places in the Antarctic. He noticed similarities in the floras of circum-Antarctic lands, for example, the occurrence of *Nothofagus* in Australia and South America, and favoured the idea of land bridges to account for such facts of plant distribution.

The instructions for the expedition had called attention to zooplankton and Humboldt in his 1839 letter recommended that observations should be made on the colour of the sea and its microscopic animals. These could be concentrated by filtration on fine linen and samples collected at different latitudes and longitudes. Sampling was carried out routinely by Hooker, who made drawings and preserved samples. Some of his drawings, (now in the Natural History Museum, London, see Fogg, 1990) show clearly recognizable diatom species. Samples, collected by tow-net, by filtration, from ice and the guts of salps, were sent to the eminent German proto-zoologist C. G. Ehrenberg. From these samples seven new genera and 71 new species (Ehrenberg, 1844) were identified. Many of the diatom species described (Fig. 3.13) are still accepted although Ehrenberg regarded them as animals. Ross, who had an otherwise clear idea of the food chain in Antarctic seas, was content to regard this as a region without plants and with minute infusorial animals as the eventual source of nourishment. During the voyage discoloured ice was frequently seen and at first accounted for as

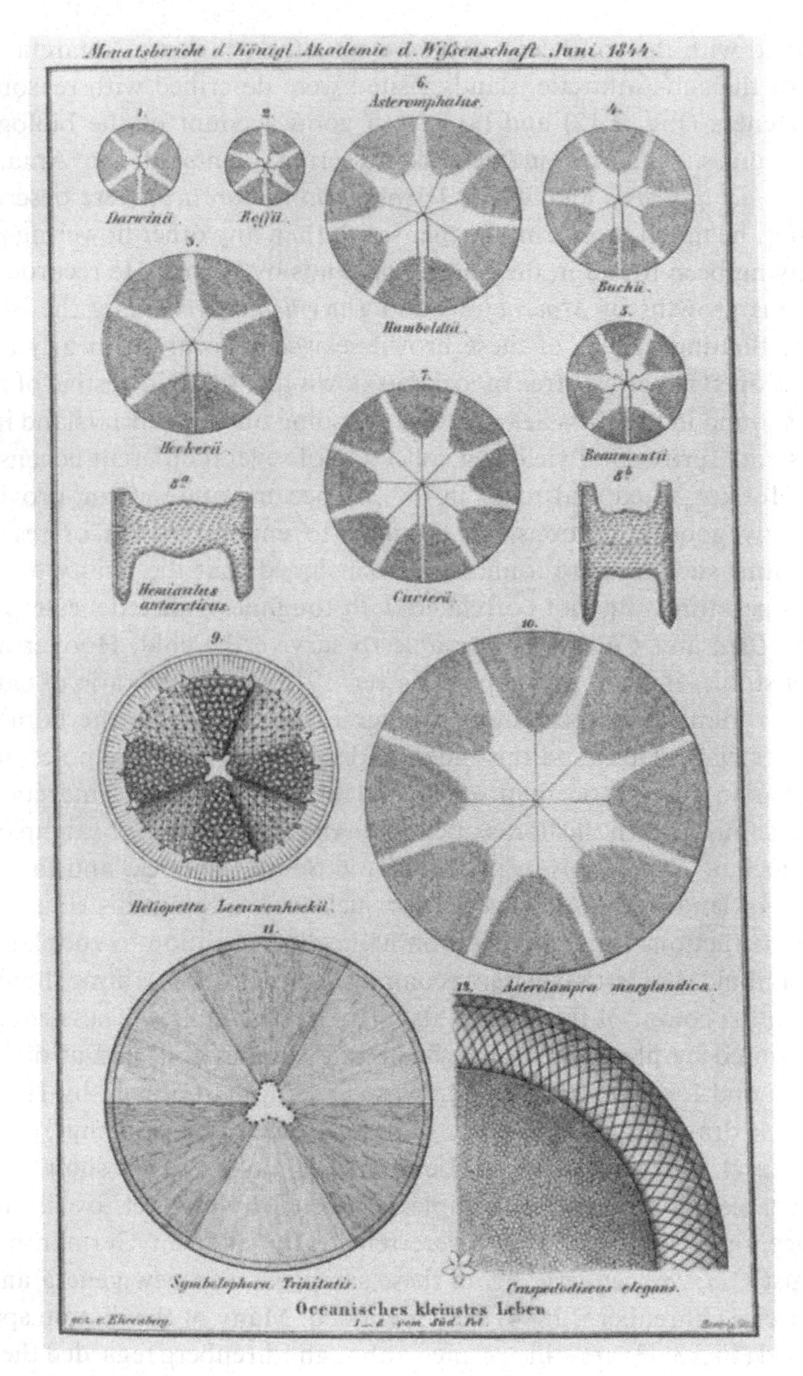

Fig. 3.13 Plate illustrating Antarctic diatoms collected by Joseph Hooker and described by C. G. Ehrenberg, 1844. (Courtesy of the Royal Society.)

being contaminated with volcanic ash but Hooker's examination of melted ice samples showed what we now know, that the colouration is produced by diatoms. An important point was noted when returning north after the second excursion south:

> The sea was remarked to have assumed its oceanic light blue colour, from which we inferred that the ferruginous animalculae, which give a dirty brownish tint to the waters of the southern ocean, prefer the temperature which obtains in the vicinity of the pack; for here, as in the arctic regions, our approach to any great body of ice was invariably indicated by the change in colour of the sea. (Ross, 1847, vol. 2, p. 214).

This distribution of phytoplankton is characteristic of the Southern Ocean. It was Hooker, when he came to write his section on Diatomaceae for his *Flora Antarctica* (1847), who first expressed the true significance of phytoplankton and emphasized the point that diatoms are widely distributed throughout the oceans. The scientific study of phytoplankton as a basic factor in biological oceanography and limnology may be said to have started with Hooker's observations (Taylor, 1980; Fogg, 1990).

Ross intended to work on the larger planktonic organisms himself and Hooker collected material for him. Sadly, nothing came of his devotion to duty. Ross retained the collection but did not work on it and after his death it was found to be in such a condition as to be useless (Ross, 1982). So was lost an opportunity to initiate an important branch of biological oceanography (Davenport & Fogg, 1989).

The zoological material collected by dredging suffered much the same fate. Hooker's drawings depict ophurids, nudibranchs, bryozoa and a pycnogonid (Davenport & Fogg, 1989) whilst an appendix to the account of the voyage gives a description of the corals collected in Antarctic seas. It is mentioned that some came from great depths and it is a pity more was not made of this. Around this time Edward Forbes had put forward the idea that no animal life is found in the sea below 300 fathoms (548 m). Ross was aware that he had disproved this (Deacon, 1971, p. 282) but did not publish and Forbes's idea, becoming dogma, delayed the advance of marine biology by several decades.

The larger animals fared better. A grant of £1,000 was made towards the publication of *The Zoology of the Antarctic Voyage of H.M. Ships Erebus & Terror*, which was edited by Sir John Richardson, naval surgeon, Arctic explorer and authority on fish, and J. E. Gray, Keeper of Zoology in the British Museum (Richardson & Gray, 1844–75). It appeared in parts between 1844 and 1875 and is a rather uneven work, not confining itself to *Erebus* and *Terror* material, but, at its best, excellent (Fig. 3.14). In the

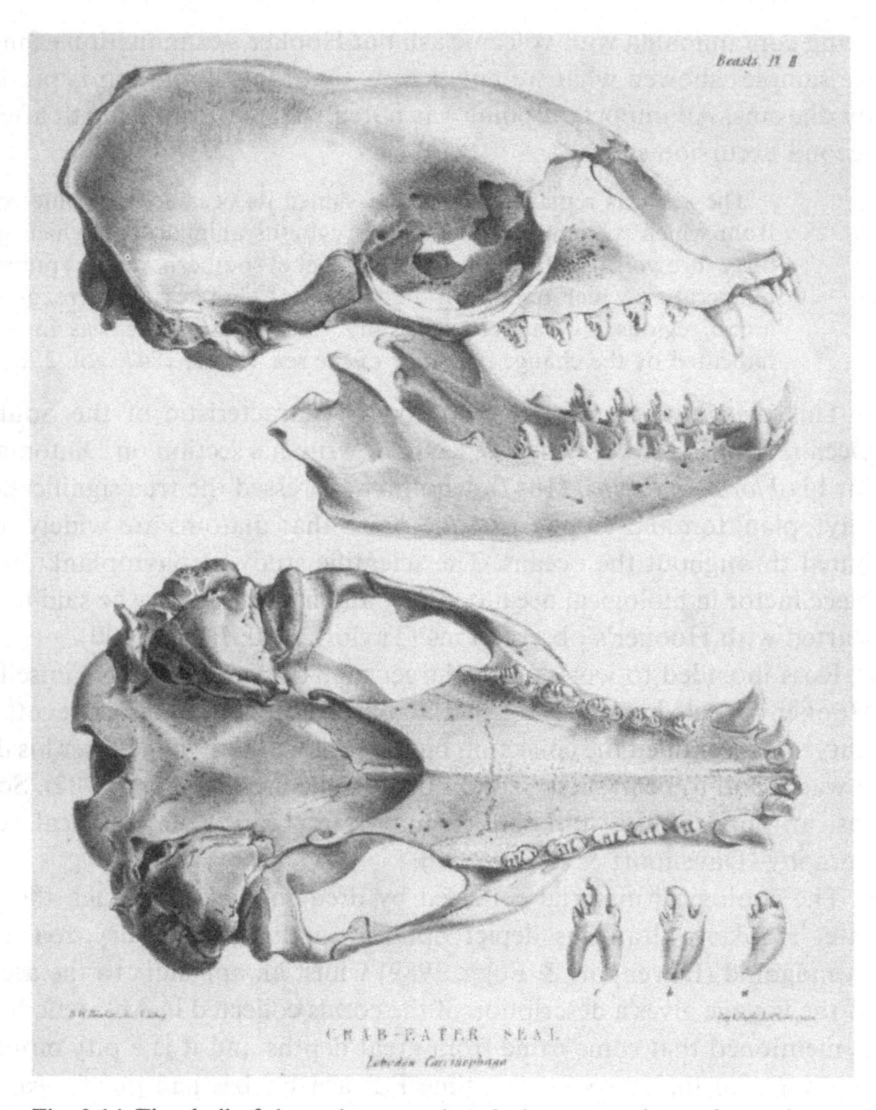

Fig. 3.14 The skull of the crabeater seal, *Lobodon carcinophagus*, from *The Seals of the Southern Hemisphere*, plate II, in Richardson & Gray, 1844–75. (Courtesy of the Scott Polar Research Institute.)

section on *The seals of the southern hemisphere* there is the first description of the rare Ross's seal, *Ommatophoca rossi* Gray. The section *On the cetaceous animals* gives scarcely any information on material from the expedition. That on *Fishes*, however, gives almost the first account of fishes from south of 50° S and was based largely on the rather decayed material brought back by Ross. The *Birds* section included descriptions of six species of penguin, including, notably, the emperor. The instructions for the expedition had been insistent that specimens of this bird, which was confused with the king penguin, should be collected. Ross's naturalists found it a difficult bird to kill and eventually resorted to a tablespoonful of hydrocyanic acid, but specimens were brought back and enabled Gray to determine that it was a distinct species, to which he gave the name *Aptenodytes forsteri* Gray.

3.7 Comment on the mid-nineteenth century expeditions

The three major expeditions which have been discussed show many similarities. All were government sponsored, conducted in naval vessels, and, although scientific societies were consulted about programmes of work, the execution of the science was directed by naval officers. In each, the study of terrestrial magnetism was a major object and in this field their programmes of work were similar. Any idea of scientific investigation of the Antarctic was secondary to this. To a greater or lesser extent inspiration for each expedition can be traced back to Alexander von Humboldt and the scheme of international co-operation proposed by the German Magnetic Association was followed by all three. There, however, international co-operation virtually stopped. There was mutual respect and some exchange of information between the leaders but in each expedition patriotic rivalry ran high and misunderstandings arose too easily. This is exemplified by the episode when the French *Astrolabe* and the US *Porpoise* passed within hailing distance of each other at 64° 48′ S but did not speak. Had there been more consultation and co-operation between the three nations more might have been accomplished. None of them undertook any appreciable exploration on land, although Ross had the necessary experience and would have attempted to get to the South Magnetic Pole had a landing near it been possible. All three expeditions escaped without major disaster in Antarctic waters but returned with tales of hairbreadth deliverance from destruction. Of the three, the expedition by Ross was undoubtedly the best organized and greatest in achievement. In conception and execution it showed Victorian England at her best and it had the same degree of technical superiority over its rivals that in 70 years time the Norwegians were to have over the British on the Antarctic ice-fields.

Endnotes

1 Apart from Wilkes's *Narrative* (1845) and *Autobiography* (ed. Morgan *et al.*, 1978) see Mill, 1905; Conklin, 1940; Rehn, 1940; Haskell, 1942; Mitterling, 1959; Kirwan, 1962; Tyler, 1968; Bertrand, 1971; and Viola & Margolis (eds.) 1985.

2 d'Urville's *Voyage au Pole Sud* (1847) has recently been translated into English by Rosenman (1987). A short historical appraisal of the discovery of Terre Adélie is given by De Pradel de Lamase (1950).
3 Royal Society Council Minutes 28 March 1816.
4 Ibid. 28 January 1818.
5 British Association, Report **5**, 61–90.
6 British Association, Report of 5th Meeting, XX–XXI.
7 Royal Society Council Minutes 13 March 1837.
8 British Association, Report **6**, 1–85.
9 Royal Society archives, Herschel Letters, Sabine to Herschel 23 November 1838.
10 Royal Society Council Minutes, 22 December 1838.
11 Ibid. 13 June 1839.
12 Royal Society archives, Herschel Letters, No. 25.
13 Royal Society Council Minutes, 27 June 1839.
14 Ibid. 12 April 1821.
15 Royal Botanic Gardens Kew, J. D. Hooker papers 351–81.
16 Ibid. Narrative written by C. J. Sullivan for James Savage Sailor on board HMS *Erebus*, Rio de Janeiro, 19 June 1843.
17 Ibid. Antarctic Journal of J. D. Hooker, p. 223.

4

Averted interest and consolidation

4.1 The mid-nineteenth century view of Antarctica

By the middle of the nineteenth century a picture of Antarctica was beginning to emerge. It was established that there was a substantial land mass around the South Pole. Magnetic phenomena had been charted and some idea of geology obtained. Salient facts about ice, ocean currents, winds and atmospheric pressure had been noticed – indeed, even at this stage a master-mind might have constructed a rough model of the dynamics of ocean and atmosphere around Antarctica. Many of the larger plants and animals had been described and the basic role of plankton in the Southern Ocean was appreciated. Yet there matters rested for almost 50 years. In the period between Ross's return and the sailing of the *Challenger* – 29 years – perhaps no more than 20 voyages went south of 60° S (Headland, 1989).

The main cause of this 'age of averted interest' was, of course, that there were no tangible prospects of colonization nor commercial incentives to attract governments to further ventures south. Sufficient had been learned about terrestrial magnetism for navigational purposes and whales were for the time being numerous enough in more accessible waters to satisfy demand, particularly after the introduction in the 1860s by Sven Foyn of techniques which allowed the large rorquals to be caught. Interest in exploration was still intense but in England the urge towards polar regions was largely absorbed by attempts to find the North-west Passage and in search for those lost in trying to find it. For Britain and Europe generally, exploration in the Arctic and in Africa provided ample excitement for those intrigued by unknown lands and the US was preoccupied with exploring her own western territories. There were no major developments in communications or transport to make journeys to places as remote as the Antarctic any easier or safer.

Nor was there much demand from scientists for further work in the

Antarctic. Both Ross and Hooker writing immediately after their voyage of 1839–43 make it evident that they found no scientific problems in the Antarctic itself calling for urgent attention and that they did not expect anyone to visit again the regions into which they had penetrated. It was not that science generally was quiescent; this was the time of the birth of thermodynamics, of electromagnetic theory, of the atomic theory and organic chemistry, of evolution and the cell theory, but those fields of science which now depend on information from Antarctica were insufficiently developed. There was a lull in geomagnetic studies from 1860 to 1900 (Multhauf & Good, 1987; Bloxham *et al.*, 1989). Synoptic meteorology developed apace in the period 1850 to 1875 with the setting up of national networks of weather stations and international co-ordination established by discussion at international congresses but the idea that events in the Antarctic should be taken into account was slow to take hold. In geology the concept of uniformitarianism, put forward by Lyell in his *Principles of Geology* published between 1830 and 1833, was yet to lead to interest in the Antarctic as a place to study the geomorphological effects of ice. Interest in glaciology itself was satisfied by investigations in the Swiss Alps (e.g. Whymper, 1870). The efforts of Humboldt to advance physical geography in its widest sense culminated in the publication of his *Kosmos* between 1845 and 1862 and might have led to continued interest in the Antarctic but after his death in 1859 his holistic approach became unfashionable (Bowen, 1970).

4.2 Maury's campaign for an expedition south

A lone voice raised at this time on behalf of Antarctica was that of Matthew Maury, who, it will be recalled, had declined to go on the Wilkes Expedition.[1] Since 1844 he had been Director of the US Naval Observatory and from the data on winds and currents he was able to collect, he compiled charts and sailing directions which were of immense value to shipping. He was active in organizing international co-operation in making observations and played a leading part in the first International Maritime Meteorological Congress, held in Brussels in 1853. He realized that the high southern latitudes held the key to understanding the weather of the southern hemisphere and successive editions of his textbook *The Physical Geography of the Sea* show his increasing preoccupation with the Antarctic (Maury, 1855, 1883; Bertrand, 1971). However, he was handicapped by the paucity of information from this region and his ideas on the circulation of the atmosphere and ocean were somewhat confused (Deacon, 1971). His advocacy of an expedition to the Antarctic fell on deaf ears in the US, where the clouds of civil war were gathering, but elsewhere was listened to with the respect due to a man of his great achievements, but no action. Addressing the Royal Geographical Society in 1860 he outlined meteorological differences between Arctic and Antarctic and urged the importance, for commercial as well as scientific reasons, of further research in southern regions. This, he thought, was a task for the British. In the discussion which followed his address his challenge was evaded except by the Hydrographer, Captain J. Washington – who had been so active in arguing for Antarctic exploration 25 years before – and who said little more than that the next expedition ought to be the responsibility of the US. The following year Maury circulated a detailed plan for an international Antarctic exploring expedition to the Ministers in Washington representing Austria, France, UK, Italy, the Netherlands, Portugal, Russia and Spain (Bertrand, 1971). None of these countries took any action but it may be that this initiative of Maury's paved the way for the flurry of activity at the end of the century.

4.3 The rise of oceanography and *Challenger*'s incursion into Antarctic waters

Although Maury's advocacy was ineffectual at the time, it was, in fact, marine science that led to the next positive move. Oceanography in the mid-nineteenth century was in the same incoherent state as Antarctic science and for much the same reasons. Government support was needed for research at sea, this had not been forthcoming and those who once had an interest in marine science – Humboldt, the Russian physicist Emil Lenz, Scoresby and Sabine – had turned to other fields. Nevertheless, with the backing of the Royal Society, two biologists, Charles Wyville Thomson and William B. Carpenter, were given the opportunity to carry out scientific work on HMS *Lightning* off the north-west coast of the British Isles in 1868. This not only produced evidence casting doubt on the dogma that no life existed in the depths of the sea but found some surprising temperature distributions. These results encouraged the Royal Society to promote more oceanographic cruises, for which the Admiralty provided HMS *Porcupine* in 1869 and 1870. The existence of life in the ocean depths was proved beyond doubt and Carpenter carried out important work on temperature structure which led him to revive Humboldt's idea that polar cold provided the main motive power for a flow of water towards the equator (Carpenter, 1868; Deacon, 1971).

Carpenter, at this time was vice-president of the Royal Society, together with the Hydrographer, George H. Richards, played the crucial role in the launching of the cruise of HMS *Challenger*, which is generally taken as the beginning of the modern science of oceanography (Rice, 1986). The former was anxious to promote more work on deep-sea fauna and on temperature distributions, the latter needed more deep-sea soundings in connexion with the rapidly expanding business of laying submarine cables. In a letter dated 15 June 1871, to the Secretary of the Royal Society, Carpenter pointed out that other nations were contemplating physical and biological work on the deep sea and that the British government ought to take the lead and support a comprehensive investigation of the oceans.[2] The Council, under the Presidency of Sabine, responded by setting up a committee charged with making recommendations within five months. Before it reported, Carpenter had prepared the ground and ascertained that the government would look favourably on the scheme. The principal recommendation of the report was that a scientific circumnavigation should be directed to:

> Investigate the Physical Conditions of the Deep Sea, in the great Ocean basins – the North and South Atlantic, the North and South Pacific, and the Southern Ocean (as far as the neighbourhood of the great ice-barrier); in regard to Depth, Temperature, Circulation, Specific Gravity, and Penetration of Light; the observations and experiments upon all these points being made at various ranges of depth from the surface to the bottom.[3]

Studies were also to be made on seawater chemistry (Fig. 4.1), bottom deposits and the distribution of living organisms. The expedition should leave in the latter half of 1872 and full publication of the results should be regarded as an integral part of the project. These recommendations were agreed without undue delay by the Admiralty.[4] In further instructions it was emphasized that the southern ice-barrier should be approached as closely as safety would permit, that deep-water fauna at the margin of the Antarctic sea should be examined, that the botany and zoology of the sub-Antarctic islands were of interest, and that the study of temperatures at depth in the Southern Ocean was expected to yield particularly important results.[5]

All had gone smoothly and although Burstyn (1968) has identified a number of factors, including a favourable political climate and a new government financial structure, contributing to the success of the application there seems no doubt that it was Carpenter's personal acquaintance with senior ministers that was crucial. However, Deacon (1971) regards this apparent open-handedness on the part of government towards science as deceptive. The *Challenger* expedition was a single high-profile project likely to bring credit to the government without involving it in continued expense. At about the same time an application to the Treasury for support

Fig. 4.1 The chemical laboratory aboard HMS *Challenger*, from Thomson & Murray, 1885, p. 12. (Courtesy of the Royal Society.)

of a much less expensive, more directly useful, but continuing project on tides, was turned down flat. Later, when the publication of the *Challenger* reports proved to be more costly than first thought the Treasury was reluctant and niggardly.

Although the Royal Society saw the advantages of extending the *Challenger* investigations into Antarctic waters there was evidently no enthusiasm for more ambitious ventures into this region. The President of the Royal Geographical Society, seeking the Royal Society's support for renewed exploration in the Arctic, argued that such exploration would provide the men versed in the skills of ice navigation necessary if the Royal Society intended to organize an Antarctic expedition in connexion with the transit of Venus in 1882. Council resolved to give scientific advice to Arctic expeditions if necessary but was not to be drawn further.[6]

Challenger sailed in December 1872 under the command of George Nares, an Arctic explorer of distinction. Carpenter at the age of 59 had decided not to go to sea again and the six civilian scientists, appointed by the Royal Society, were headed by Wyville Thomson, then Professor of Natural History at the University of Edinburgh. With him were John Murray (an energetic natural historian who later was to prove a powerful advocate for Antarctic research) and the chemist J. Y. Buchanan, who also had an interest in meteorology, but no physicist accompanied the expedition.

The *Challenger* was in Antarctic waters in early 1874. After calling at Kerguelen and Heard Islands she reached the edge of the pack in mid-February and worked her most southerly station at about 65° S. She was not strengthened against ice and after following the edge of the ice eastwards for about two weeks she turned north for Melbourne at the beginning of March – the first steamship to have crossed the Antarctic Circle. Her return to England in May 1876 excited little public attention. After some argument the collections were taken to Edinburgh, rather than the British Museum in London, for distribution to experts for study, and the official scientific reports of the expedition appeared in 50 volumes, a total of about 30,000 pages, between 1880 and 1895. Despite the foresight of the Royal Society in drawing attention to the importance of full publication of results there was much trouble over this as described by Burstyn (1968) and Deacon (1971). Suffice it to say here that the Treasury considered that the expedition had cost enough, that the first editor, Wyville Thomson, was driven to his death by the worry of the rigid accountability forced upon him, and that the project was only brought to completion by the determination and private means of John Murray. Burstyn (1968) summed it up by writing, 'The *Challenger Report* is a magnificent monument to the vision of William Carpenter, the practicability of George Richards, the martyrdom of Wyville Thomson, and the industry of John Murray'.

The achievements of the *Challenger* expedition have been assessed elsewhere (e.g. Deacon, 1971; Schlee 1973). The hydrographic data largely supported Carpenter's ideas but there was no one involved with the expedition with sufficient knowledge of physics to unravel fully the complex situation that had been found. The meteorological observations confirmed the existence of a permanently high atmospheric pressure over the Antarctic continent. Major contributions came from deep-sea dredgings; calcareous *Globigerina* ooze disappeared south of Kerguelen (i.e. at the Polar Front) to be replaced by siliceous diatom ooze in correspondence with the abundance of these organisms in the plankton of Antarctic waters. The sea bottom was found to be strewn with rocks, dropped from icebergs, of different sorts from those found on the sub-Antarctic islands and which

Fig. 4.2 Georg von Neumayer. (Courtesy of the Alfred-Wegener-Institut, Bremerhaven.)

must have been derived from land further south. A varied benthic fauna was dredged up from 1675 fathoms (3058 m). South of 43° S dredgings brought up 830 animals (protozoa not included) belonging to 398 species of which 326, or nearly all those described, were new to science (Murray, 1894a). Existing information on the natural history of the sub-Antarctic islands was added to. These findings, some of which will be discussed further in later chapters, all added to the general picture of Antarctica.

4.4 Neumayer and the growth of German interest in the Antarctic

During the period of the *Challenger* cruise there were other ventures south that should be noted briefly. One was the voyage of the

Grönland, captained by Eduard Dallman, to the South Shetlands and the west coast of the Peninsula as far down as the Biscoe Islands. Germany, from whence the *Grönland* sailed, was at that time the leading power on the continent of Europe, both politically and scientifically, and had a persistent advocate for Antarctic research in Georg von Neumayer (Fig. 4.2). Inspired by the exploits of Wilkes, d'Urville and Ross, Neumayer had resolved to devote himself to the study of terrestrial magnetism and the science of the sea. He served before the mast, spent some time in Australia as a gold-digger, knew Humboldt and had the great chemist Liebig as a sponsor. While carrying out magnetic and meteorological observations at the Flagstaff Observatory in Australia (1858–64), funds for which he had obtained from King Maximilian II of Bavaria, he followed Maury's plan in collecting data on the Southern Ocean. This observatory he regarded as a possible base for a venture to the Antarctic but nothing came of this (Neumayer, 1895, 1901). Returning to Europe he became Director of the Deutsche Seewarte at Hamburg which he built up into a first-class oceanographic institution. He was chosen to lead an expedition to the Antarctic as a preliminary to a projected transit of Venus expedition in 1874 but this was postponed at the outbreak of the Franco-Prussian War (Neumayer, 1895, 1901) and then the death of his patron, Admiral Tegethoff, put an end to the idea. Drygalski (1904) noted that while Neumayer was an enthusiastic campaigner in support of Antarctic exploration he enviced surprise and almost revulsion when the call came for action to realize the plans. Nevertheless, it was evidently through his influence that the Society for Polar Navigation in Hamburg, founded to promote whaling and sealing in the Arctic, had sent Dallman on a voyage of exploration south, the Society's director, Albert Rosenthal, meeting much of the cost (Neumayer, 1895; Mill 1905). This expedition made some geographic discoveries but achieved little of scientific note (Anon. 1875). A German expedition to observe the transit of Venus went ahead without Neumayer, sailing in the frigate *Gazelle* to Kerguelen. Astronomically, if not meteorologically, the Indian Ocean sector of the Antarctic was the best place to see this transit and there were parties from Britain and the US on Kerguelen for the same purpose. The weather was favourable during the first part of the transit and the Americans successfully used a photographic method of recording it. The Germans did some surveying and the Assistant Surgeon of the American party, Dr Jerome H. Kidder, a competent amateur naturalist, made collections of geological and zoological specimens, which went to the Smithsonian Institution, and of plants, which went to Professor Asa Gray at Harvard (Bertrand, 1971). Otherwise, the three

Fig. 4.3 Lieutenant Karl Weyprecht.

expeditions, which were all naval, concentrated on astronomy and contributed little to Antarctic science.

4.5 Weyprecht and the First International Polar Year

The second transit of this pair, which took place in 1882, was marked by more general scientific activity which showed a growing awareness of the need for intensive collaborative research in polar regions as well as in astronomy. Before this, Purnell (1878) tried to arouse interest in a joint venture to be supported by Australia and New Zealand, pointing out that these two countries were best situated to explore the Ross Sea area, and to contribute to science and open up commercial possibilities at relatively little cost. However, the new attitude seems to have begun with Lieutenant Karl Weyprecht (Fig. 4.3), co-leader of the Austro-Hungarian North Pole Expedition of 1872–74, who, when the expedition's ship the *Tegethoff* became beset off Franz Josef Land, passed his time by thinking about the general problems of polar research. On his return he put his ideas before the Austrian Royal Geographical Society and then the Royal Geographical Society in London (Weyprecht, 1875), when he said:

> The key to many secrets of Nature, the search for which has now been carried out for centuries (I need only refer to magnetism and electricity,

> the greatest problems of meteorology) is certainly to be sought for near the Poles. But as long as Polar Expeditions are looked on merely as a sort of international steeple-chase, which is primarily to confer honour upon this flag or the other, and their main object is to exceed by a few miles the latitude reached by a predecessor, these mysteries will remain unsolved . . .
>
> Decisive scientific results can only be attained through a series of synchronous expeditions, whose task it would be to distribute themselves over the Arctic regions and to obtain one year's series of observations made according to the same method.

He was restating something that had already been implicit in Halley's work on geomagnetism and argued more recently by Maury and Neumayer (Baker, 1982) but he persisted until he had achieved practical expression of the ideal of co-ordinated polar studies. In a paper read at a meeting of the Association of German Naturalists and Physicians held in Graz in 1875 he stated six principles of Arctic exploration:

> 1. Arctic exploration is of the greatest importance for a knowledge of the laws of nature.
> 2. Geographical discovery carried out in these regions has only a serious value in as much as it prepares the way for scientific exploration as such.
> 3. Detailed Arctic topography is of secondary importance.
> 4. For science the Geographical Pole does not have a greater value than any other point situated in high latitudes.
> 5. If one ignores the latitude the greater the intensity of the phenomena to be studied the more favourable the place for an observational station.
> 6. Isolated series of observations have only a relative value.
>
> (quoted from Baker, 1982)

This was the first explicit distinction drawn between polar exploration for its own sake and scientific investigation, and, of course, it applies equally well for the Antarctic. He went on to suggest a chain of five or six stations around the North Pole to make simultaneous observations using similar instruments in various branches of physics and meteorology and also botany, zoology and geography. There might be another station with the same programme in the Antarctic. The proposal was submitted, without arousing much interest, to various academies of science but was eventually taken up at the International Meteorological Congress in Rome in 1879. There, a resolution was adopted recognizing the great importance of synchronous meteorological and magnetic observations and calling on governments to support such work. The International Meteorological Committee sent out invitations to an International Polar Conference. This was held in Hamburg from 1 to 5 October, 1879, under the chairmanship of Neumayer and was attended by representatives from Austria, Denmark, France, Germany, The Netherlands, Norway, Russia and Sweden, with Belgium, UK, Portugal and Spain sending apologies for absence. The conference drew up a programme, based on that submitted by Weyprecht and his sponsor and patron Count Wilczec to the International Meteorological Congress, which provided for a minimum of eight circumpolar stations in the Arctic to operate for one year from autumn 1881 to autumn 1882. An International Polar Commission was established with Neumayer as its first president. A second meeting of the Conference, held in August 1880, decided to postpone the programme for a year but it was launched in 1882 with the participation of 11 nations and the establishment of 12 major stations in the Arctic, mostly sponsored by governments. Before that, in 1881, Weyprecht had died of tuberculosis at the age of 41 (Baker, 1982; Barr, 1985; Bretterbauer, 1985).

The First International Polar Year finally included two expeditions south, a plan by Lieutenant Bove of the Italian Navy for a scientific voyage to the Antarctic foundering through lack of funds (Mill, 1905). Of those that went, that by the French to the Cape Horn area produced an impressive body of work but lies outside our province although it may be noted in view of our present preoccupation with the carbon dioxide concentration in the atmosphere that they found this to be appreciably less in this region than in the northern hemisphere. The other was a German expedition under the leadership of Dr K. Schrader to South Georgia. This was sponsored by the German Polar Commission, clearly under the influence of Neumayer. South Georgia was selected as a site because it is on a longitude almost diametrically opposed to that of the magnetic observatory established by Neumayer in Australia (Headland, 1984; Barr 1985). The scientists accompanying Schrader were Dr P. Vogel (physicist and mathematician), Dr K. von den Steinen (medical officer and zoologist), Dr H. Will (botanist), E. Mosthaff (engineer and draughtsman) and Dr O. Clauss (physicist). The expedition was disembarked from SMS *Moltke* in Royal Bay at the southern end of South Georgia and dwelling and observatory huts erected. The scientific programme was carried through for the full year, giving the first complete set of meteorological data for over a year to be obtained for the Antarctic. Because of fluctuations of temperature above and below zero and gales, difficulties were experienced with the mountings for the magnetic instruments but these were overcome and a full set of measurements synchronizing with those made at the Arctic stations was made. Almost miraculously, considering South Georgia's weather, a period of clear skies coincided exactly with the transit of Venus. Measurements of glacier movement were made and tidal observations showed anomalies which later proved to have followed the eruption of Krakatoa on 27 August, 1883. Von den Steinen carried out investigations of the sublittoral marine fauna and made a superb pioneer study of the birds (in Neumayer, 1890–91; Barr, 1984, 1985; see Fig. 10.3) and Will provided the first extensive description of the flora, with autecological notes including a comprehensive account of tussock grass (in Neumayer, 1890–91; Greene, 1964). There was deliberate introduction of domesticated plants and animals but they did not survive.

The results of the First International Polar Year were impressive, although in meteorology its contribution was not as useful as it might have been because of the wide spacing of the stations, nor were all of the magnetic results published or fully evaluated. However, fortuitously, the year had been one of considerable sunspot activity with two exceptional magnetic storms and it provided incontrovertible evidence of a connexion between the aurora and magnetic variation (Barr, 1985). It linked Antarctic investigations with those in the Arctic and, above all, it set an example that was to be followed and developed in the future with immense gain to science.

4.6 Reconnaissances by whalers

The final meeting of the International Polar Commission was held in 1891 but interest in the Antarctic had been revived and the decade was to be one of activity culminating in the period which has come to be known as the heroic age of Antarctic exploration. However, while the scientists were still talking, the whalers, realizing that northern waters were being fished out, went on expeditions south on their own account. These expeditions were that of the *Balaena, Active, Diana* and *Polar Star* out of Dundee in Scotland from 1892–93, two by the Norwegian schooner *Jason* under the command of Captain C. A. Larsen and sponsored by a Hamburg firm between 1892 and 1894, and that of the *Antarctic* sponsored by the veteran Norwegian whaler, Svend Foyn, and led by Captain L. Kristensen and H. J. Bull in 1894 and 1895. The first two of these went to the Peninsula area and the other to the Ross Sea. All three went in search of right whales, by then almost exterminated in northern waters, which Ross had reported as being abundant in Antarctic waters. None were found. It seems unlikely that Ross, who was familiar with the right whale in the Arctic, could have been mistaken and it is possible that hunting of this species in southern temperate waters had already reduced the stock in the southern hemisphere as a whole to a very low level by the 1890s (Bull 1896), but Racovitza (1903) considered that Ross was in error.

The attitude to science on these expeditions varied from just tolerant to actively hostile. The Arctic explorer, Leigh Smith, was instrumental in getting two surgeons, William S. Bruce and C. W. Donald, both interested in natural history, appointed to go with the Dundee fleet. The Royal Geographical Society and the Meteorological Office provided instruments for precise navigation and meteorological observation (Mill, 1905) and H. R. Mill gave scientific instructions (Bruce, 1896). Nevertheless, Bruce's account of the expedition is full of frustration; commerce, he wrote, was the dominating note and no concessions were made to science. Thus, when the bucket which he used for sampling surface water was lost it was not replaced and sampling for salinity determinations had to be completely stopped thereafter. His rock samples consisted of little more than pebbles recovered from the crops of penguins or from icebergs. W. G. Burn Murdoch, an artist who was on the *Balaena* with Bruce, summed up the situation vividly:

> Perhaps it seems incredible that gentle, loving law-abiding Bruce and the writer if possible more so, could have lain swinging in our hammocks under an inverted whale-boat watching the blue lace-lined seas slipping past and planning how to tip the old man overboard or slit his throat. But we were too timid. This little confession may let scientists know of Bruce's bitter disappointment at having his scientific work hindered at every turn, mocked and jeered at, and will show how greater scientific results were not obtained. In fact, it may make some wonder at the amount of work he did bring home. (Brown, 1923, p. 56.)

Larsen, who was probably stimulated to go south by Nansen, was himself scientifically inclined but he had to remind himself 'we were not sent out for scientific exploration, but for whale and seal hunting' and the temptation to investigate had to be resisted (Larsen, 1894). Bull, a resident in Australia, had to go back to his native Norway for financial backing for his expedition. There, Svend Foyn gave immediate and unstinted support but Foyn, Bull (1896) recorded, 'looked upon scientific men and theorists with a feeling of suspicion, mingled with some disdain'. Scientific societies in Australia were enthusiastic and lent books and charts but a suggestion that one or more young scientists should join the expedition was not followed up and Bruce, who had obtained leave to participate, could not get to Melbourne in time. So the venture was conducted on strictly commercial lines. There were, nevertheless, some results of scientific value from these expeditions. Not the least of these was that Bruce became inspired to do more work in the Antarctic, and later on led his own, very fruitful, scientific expedition in the *Scotia*. Otherwise there was no particularly significant scientific outcome of his 1892–93 cruise (Bruce, 1896).

Larsen found fossil molluscs and coniferous wood of Lower Tertiary age on Seymour Island in the same region (Larsen 1894). Eight's discovery 60 years before of fossil wood in the South Shetlands having been overlooked, this aroused great interest as evidence that the Antarctic had not always been as cold as it is now (Murray, 1894a,b). Bull's expedition to the Ross Sea found lichens on Possession Island and Cape Adare. That this was the first record of plant life at this latitude can probably be put down to the fact that Hooker was prohibited by his Captain from landing on Possession Island after he had fallen in the sea so it was McCormick, who had no interest in plants, who went ashore as the only naturalist. Bull also collected rock samples (reported on by J. J. H. Teall in an appendix to Bull, 1896) and was the first to find mummified seals. Perhaps his major contribution was to show that it was not too difficult to penetrate into the Ross Sea and to make a landing, and to have identified a site at which an expedition might overwinter.

4.7 Growing interest among scientists

Meanwhile there was much talk about Antarctica among the scientists of several nations. In Germany, where interest had been stimulated by its own recent expeditions to the Peninsula and South Georgia, Neumayer was still active. The *Deutsche Geographentag*, which began in 1881, provided an annual platform for discussion on Antarctic matters. At its fourth meeting, in 1884, a resolution calling for an expedition to investigate the physical geography of the Antarctic included a recommendation that the *Geographentag* should work towards this in collaboration with the Polar Commission but this came to nothing. A rumoured German-American expedition to be directed by Neumayer and financed by a German-American millionaire did not materialize (Mill, 1905). Between 1865 and 1893 a total of 26 papers on the theme of Antarctic exploration was published by German authors (Neumayer, in Murray, 1894a). At the eleventh *Geographentag* in 1895 a session, at which Neumayer, Drygalski and Vanhöffen spoke, was devoted to the Antarctic and led to a renewal of the call for a scientific expedition and the setting up of the *Deutsche Südpolar Kommission*. This decided that the sector between 60° and 70° E, for which the physical circumstances seemed good but which remained unexplored, should be the objective. An expedition with two wooden steamships, one of which should overwinter, lasting at least three years and staffed by competent scientists, was proposed (Neumayer, 1895, 1901).

At the British Association meeting in Aberdeen in 1885, Admiral Sir Erasmus Ommannay, who had participated in the search for Franklin, drew attention to the neglect of Antarctica as a field for exploration. This stimulated interest as far away as the colony of Victoria in Australia. Under pressure from the Royal Geographical Society of Australasia and the Royal Society of Victoria, the colonial government agreed to pay premiums to whalers or sealers bringing back cargo from south of 60° S but was not able to finance an expedition itself (Mill, 1905). Murray (1886) made a proposal to involve two naval vessels and a guarantee of £150,000 to ensure proper equipment and support. This was endorsed by the Royal Society of Edinburgh and the Scottish Geographical Society but a high-powered committee set up by the British Association to consider Antarctic exploration took no effective action. The Government of Victoria decided to provide £5,000 if the Imperial Government did likewise. The Royal Society of London and the Royal Geographical Society supported the idea and the Colonial Office recommended that the money should be made available. However, the Treasury, perhaps sensing that the support was not altogether wholehearted and fearing that this was but the forerunner of further, more costly, ventures, rejected the application (Mill 1905).[7] Britain having failed them, the Australians looked elsewhere. At a meeting of the Australasian Association for the Advancement of Science in Christchurch, New Zealand, in 1891, T. Kirk had delivered a paper on the botany of the Antarctic islands and, at the same meeting, a committee was appointed to consider the question of Antarctic exploration. There was a definite prospect of support, the multi-millionaire Oscar Dickson of Stockholm, a generous supporter of polar research (Frängsmyr, 1989), having offered to equip an expedition subject to there being an Australian contribution of £5,000. It was suggested that the expedition should be commanded by Baron A. E. Nordenskiöld, the Arctic explorer, or perhaps Nansen, and sail under the Swedish flag. This proposal, too, came to nothing.

Undefeated, proponents of Antarctic exploration continued their campaign. In 1891 the geographical section of the British Association heard the case put forward again (Morgan, 1891). The following two years saw the departure of the whaling expeditions already described. Then in 1894 Murray, addressing the Royal Geographical Society, gave a comprehensive review of the state of Antarctic science and concluded it with a plea for the renewal of exploration in this region. Again he suggested that two vessels of the Royal Navy were needed and that their complement should include civilian scientists. He emphasized that what was now required was not a dash for the South Pole but 'steady, continuous, laborious, and systematic exploration of the whole southern region with all the appliances of the modern investigator'. He had a distinguished audience and several of them gave their views at length in the discussion which followed the address. Hooker pointed out the part which had been played by chance in past exploration and suggested a preliminary circumnavigation to probe for weak places in the pack ice. The Duke of Argyll suggested that the study of the operation of a true ice-sheet would be of great value to geological science. Nares reminded the audience that if the Navy undertook the exploration there would be a saving in that the ships and men would have to be paid for anyway. Nevertheless, the First Sea Lord was quick to suggest that the expense should be properly borne by the education vote. The President, Clements Markham, concluded by expressing the hope that an expedition need not involve the wholesale slaughter of seals and referred to the efforts of the British Association and its sister organization in Australasia to promote further investigation in the field. Neumayer supported Murray in a letter in which he illustrated in precise terms the importance of further studies in the Antarctic for meteorology – pointing out that there was not yet a single measurement of winter temperatures – terrestrial magnetism and geology.

The head of steam that had been generated began to have practical effects at the sixth International Geographical Congress, held in London in 1895. At this meeting there was a lengthy address in convoluted German from Neumayer in which he reviewed the main aspects of Antarctic science, outlined the plans of the proposed German expedition, and concluded with the hope that there might be international collaboration (Neumayer, 1895). This was followed by a discussion; 'The exploration of the Antarctic regions', it was agreed, 'is the greatest piece of geographical exploration still to be undertaken' and 'in view of the additions to knowledge in almost every branch of science which would result from such a scientific exploration the Congress recommends that the scientific societies throughout the world should urge in whatever way seems to them most effective, that this work should be undertaken before the close of the century.'

A member of the newly elected Parliament went so far as to promise that the British Government would 'freely grant any expenditure that may be necessary for such a grand national undertaking'. It is scarcely necessary to say that this did not happen or that, in spite of Sir Clement's wistful

Fig. 4.4 Henryk Arçtowski in the laboratory on *Belgica*; from Cook, 1900, facing p. 208.

cxpression of preference for a committee of one, a committee of national representatives was appointed to consider what steps ought to be taken.

4.8 The voyages of the *Belgica, Valdivia* and *Southern Cross*

While the major proponents became involved in protracted scheming and political manoeuvring, three lesser expeditions went south and returned. The first of these was already in preparation before the Geographical Congress took place. This was the expedition in the *Belgica* under the leadership of a young Belgian naval lieutenant, Adrien de Gerlache, and was the first truly scientific expedition to be specifically directed to the Antarctic since the days of Ross. It was Gerlache's own idea and had to be financed mainly by public subscription organized by the Brussels Geographical Society. The Belgian Government – preoccupied with colonization of the Congo – gave only meagre support and the final sum collected was barely adequate. The ship's complement was polyglot and included notably Roald Amundsen (see Fig. 11.1), then unknown, and Dr Frederick Cook (see Fig. 11.1), an American with Arctic experience, as surgeon. No written instructions were provided either by the government or the learned body and no programme for the voyage was drawn up (Arçtowski, 1901b). The science depended on Henryk Arçtowski (Fig. 4.4), a Polish geologist, and Emil Racovitza, a Roumanian zoologist (see Fig. 11.1). The *Belgica* entered Antarctic waters at the beginning of 1898 and carried out a survey of the west coast of the Peninsula until she was entrapped by ice on 2 March. Whether Gerlache intended to get beset or not is obscure but 'the first Antarctic night' which followed was terrible: one man died, two went mad, and that matters were not worse was largely due to the good sense of Cook. Eventually, after drifting through 17 degrees of longitude the ship was liberated from the ice on 14 March 1898. The scientific results were considerable in spite of the inadequate resources and haphazard organization. For the first time a consecutive meteorological record extending over the winter south of the Antarctic Circle was obtained and showed a minimum temperature of −43° C. Arçtowski produced the first coherent account of the physical geography and petrology of the Peninsula in spite of difficulties:

> A few strokes of the oars brought us to the beach amid cries of 'Hurry up, Arçtowski!' I gave a hammer to Tellefsen, with orders to chip here and there down by the shore, while I hurriedly climbed the moraine, picking up specimens as I ran, took the direction with my compass, glanced to the left and the right, and hurried down again at full speed to get a look at the rock *in situ*; meanwhile Cook had taken a photograph of the place from the ship – and that is the way geological surveys had to be carried out in the Antarctic. (Arçtowski, 1901a.)

He discussed the possibility that the Peninsula was a continuation of the Andes through the Scotia Arc, citing his own bathymetric observations as evidence against there being a more direct connexion across the Drake Passage. He also found evidence pointing to more extensive glaciation in the Peninsula area in the past and made observations on ice formations. Racovitza made general studies of flora and fauna (see p. 342 and Fig. 10.2), discovering for the first time the smaller terrestrial animals, mites and collembola, which are so abundant among mosses and lichens. The scientific results of the *Belgica* expedition were published in a series of extensive reports and briefly summarized in appendices in the book by Cook (1900).

The *Valdivia* expedition, like that of the *Challenger*, was concerned with the deep sea. It was sponsored by the German Government and led by Professor Karl Chun of Leipzig (a zoologist). It included a large staff of scientists and was certainly the most comfortable and well-found expedition of its sort to date. After sailing from Hamburg in August 1898 it visited first the Firth of Forth for special advice from Sir John Murray. Its main contribution to Antarctic geography was the rediscovery and accurate determination of the position of Bouvetøya. *Valdivia* penetrated as far south as the ice edge at 64° 15′ S 54° 20′ E, soundings in this area showing unexpectedly great depths averaging about 3,000 fathoms. Dredging brought up rocks of obviously continental origin. The *Valdivia* called at Kerguelen and returned to Hamburg at the end of April, 1899. Its results were published in 24 volumes between 1902 and 1940. Most of these dealt with the zoology of temperate and tropical waters but one volume, by Karsten, described and illustrated the Antarctic phytoplankton.

The third of this group of expeditions was that under Carsten Borchgrevink in the *Southern Cross* to Cape Adare in 1898 to 1900 – the first overwintering on land within the Antarctic Circle. Borchgrevink, a Norwegian living in Australia, had accompanied Bull on his visit to the same area in 1895 and afterwards gave himself much of the credit for that expedition. He spoke at the International Geographical Congress, bringing a breath of realism into the predominantly academic proceedings but evidently making few friends. He was unsuccessful in raising funds in Australia but was able to persuade the publisher of the magazine *Tit-Bits*, Sir George Newnes, to provide for a small expedition. This infuriated Sir Clements Markham who considered it a diversion of support from the project which the Royal Geographical Society had in hand. At Newnes' insistence the *Southern Cross* sailed under the British flag but all except three of the complement were Norwegian. The scientists were Louis Bernacchi from Tasmania (physicist), William Colbeck, a sublieutenant in the Royal Naval Reserve (magnetic observer) and Nikolai Hanson from Norway (zoologist), who died during the winter. The meteorological observations, mostly made at two-hourly intervals and extending over an entire year, were important in giving the first detailed picture of the climate of the maritime Antarctic continent (see Fig. 9.1 and p. 291). The prevailing ESE and SE winds indicated the existence of an anticyclone extending over much of the continent with a corresponding flow of air towards the South Pole at upper levels (Bernacchi, in Murray, 1901). The magnetic, auroral, geological and zoological observations were unremarkable (Borchgrevink, 1901) and some of the natural history notes and specimens were lost or destroyed (Evans & Jones, 1975). A landing was made on the Ross Ice Shelf and a furthest south of 78° 50′ S was attained. It was only slowly that the geographical establishment in Britain came to acknowledge the achievements of this pioneering expedition (Evans & Jones, 1975).

4.9 Naval tradition versus science: the *Discovery* expedition

In 1900 the geographical and scientific establishment was absorbed in the final preparations for its own expedition after a bitter squabble as to what sort of person should lead it. Even five years later Mill was coy in writing about the events that led to Captain Robert Falcon Scott of the Royal Navy being chosen but the causes of the trouble now seem clear. Following his election as President of the Royal Geographical Society in 1893 Sir Clements Markham, who had accompanied Nares on his Arctic expedition in 1875, made it his main business to promote another Antarctic expedition. The Society appointed a special Antarctic Committee which drew up a report, recommending a naval expedition very much along the lines proposed by Murray in 1886, and this was adopted by the Council. Markham, anxious to muster all possible support, then took a step he was later to regard as disastrous and invited the Royal Society's participation. The Royal Society, consistent with its policy of not getting involved in exploration for exploration's sake, was evasive in its first response and made its own enquiries of the First Lord of the Admiralty about support for another magnetic survey in southern regions.[8] It received no encouragement but Markham was undeterred and, as we have seen, the International Geographical Congress in 1895 was strongly supportive. He continued his lobbying and, following a lunch on 7 February, 1897, to which he invited the First Lord to meet Fridtjof Nansen – recently back from his epic drift in the *Fram* across the Arctic Ocean – and John Murray, he received an assurance that if there were an expedition the Admiralty would take great interest and lend instruments (Markham, 1986). However, the Prime Minister, Lord Salisbury, after delaying a reply to a request for support until June 1898, was definite in his refusal to dispatch a government expedition. Before this Markham had decided that his only hope was to raise funds privately. This was a slow business but a private donation of £25,000 in March 1899 gave the necessary fillip and as a result of a deputation representing both Societies putting the case to the First Lord of the Treasury, Mr Balfour, government support of £45,000 was promised, provided that this was matched by an equal sum from other sources (Fig. 4.5). This was perhaps not a sudden conversion to altruistic support of science – Germany, expanding in every field, was voting funds for Antarctic exploration and could not be left unchallenged (Drygalski, 1904). Detailed planning could now go forward. This was at first in the hands of a joint committee of the Royal Geographical Society and the Royal Society which, having 33 members, was unwieldy and fractious. It was soon contracted to an executive committee of two members each from the two societies.

THE DAILY GRAPHIC, FRIDAY, JUNE 23, 1899.

Fig. 4.5 Mr Balfour, First Lord of the Treasury, receiving a deputation on behalf of the proposed National Antarctic Expedition, 22 June 1899. (Courtesy of the Director and Trustees of the Royal Botanic Gardens, Kew.)

Meanwhile, the zoologist P. L. Sclater had proposed to the Royal Society a discussion meeting on Antarctic science and this was held on 24 February 1898. Such was the interest that it went on until nearly midnight – something unprecedented in the history of the Society. The main speaker was John Murray who, after pointing out that without knowledge of the Antarctic we cannot understand the phenomena with which we are surrounded even in the habitable parts of the globe, reviewed the main problems needing to be solved (Murray, 1898). Nansen was present but is not on record as contributing to the discussion which followed but among those who did were the Duke of Argyll, Sir Joseph Hooker, Professor Neumayer, Sir Clements Markham, J. Y. Buchanan, Sir Archibald Geikie and Professor D'Arcy Thompson – all distinguished scientists with something pertinent to say. Neumayer began by remarking that politics had interfered in an unusual manner to retard progress. He emphasized the need for involving all branches of science in Antarctic research but thought that the positive advantages would be few at first. He went on to make a remarkably perspicacious statement on the importance of studies of geomagnetism and atmospheric electricity in the Antarctic (p. 290). That undeservedly forgotten character Buchanan, the chemist on the *Challenger* and the first appointed lecturer in geography at Cambridge, gave a graphic picture of atmospheric circulation over Antarctica. Markham was dismissive (Markham, 1986, p. 8) but if any one event can be taken as signalling the beginning of Antarctic science as a coherent field this discussion surely was it. It looked forward as well as bringing together what was already known and it adumbrated a firm scientific basis for the great expeditions that were to come in the next 15 years.

The difficulty was that Markham, the driving force behind the British expedition that was to come, was living in the past not only as regards the science but also in techniques of polar exploration. He stated his position clearly in his personal narrative of the origins of the expedition:

> Its main object would be the encouragement of maritime enterprise, and to afford opportunities for young naval officers to acquire valuable experiences and to perform deeds of derring doe. The same object would lead to geographical exploration and discovery. Other collateral objects would be the advancement of the sciences of magnetism, oceanography, meteorology, biology, geology; but these are springes to catch woodcocks. The real objects are geographical discovery, and the opportunities for young naval officers to win distinction in time of peace.
>
> (Markham, 1986).

The expedition must, therefore, be a naval one or, if government was unenlightened enough to be obdurate on this point, there should be as strong a naval component as possible. His insistence of youth extended to the command:

> The inexperience and haste in decision of young leaders are disadvantages which sometimes accompany their youthful energy, but they alone have the qualities which ensure success.

Its leader, then, had to be a young naval officer. He was to have command of the entire operations, both at sea and on land, and while one of the civilian complement, which Markham particularized as a geologist, a biologist and a physicist, might be placed over the others as director, he would have only limited powers such as Wyville Thomson had on *Challenger*.

Already in 1887 Markham had identified Scott, then a midshipman aged 18, as a likely leader. Scott volunteered for the position of commander on 5 June 1899, after a chance encounter with Markham. The Admiralty's agreement to seconding two naval officers to the expedition enabled him to

Fig. 4.6 Dr E. A. Wilson at work on a drawing of paraselene on the British Antarctic Expedition 1910–13; photograph by H. Ponting. (Courtesy of the Scott Polar Research Institute.)

be appointed four days later. Scott was at one with Markham in insisting that he should have complete command and be consulted about all further appointments. As far as the scientists went these eventually were H. T. Ferrar (geologist), T. V. Hodgson (marine biologist) and L. C. Bernacchi (physicist and who had been on Borchgrevink's expedition). To these must be added the two surgeons, E. A. Wilson (Fig. 4.6) as zoologist and artist, and R. Koettlitz, who had Arctic experience, as bacteriologist and botanist. The Royal Society, however, had other ideas. Markham himself had proposed J. W. Gregory, Head of the Geological Department of the British Museum and subsequently Professor of Geology in the University of Melbourne, as director of the civilian staff. This was approved by the Royal Society members of the joint committee but they went further to maintain that while Captain Scott was necessarily supreme in all matters related to the safety of his ship and its crew, the scientific director should be totally in charge of the scientific work, including geographical survey, the captain carrying out his wishes so far as they were consistent with safety. Markham and Scott would not accept this nor would Gregory agree to limitation of his role as scientific leader. The Royal Society had a majority on the joint committee and the Royal Geographical Society side was split by an Admiralty faction which wanted a naval surveyor rather than an executive officer such as Scott as commander. The upshot of much dissention and vituperation which almost wrecked the expedition was that Markham had his way and Gregory resigned (Poulton, 1901). We need not go into the details of all this, which were chronicled in public in the columns of *Nature* (Anon. 1901a) and in private by Markham (1986). *Nature*, of course, took the scientists' side, eulogizing Gregory, scorning Scott as having neither scientific nor polar experience, and criticizing the Royal Society for being weak-kneed. Poulton's statement (Poulton, 1901) of the Royal Society's position seems at this date to be reasonable and temperate. Markham developed a persecution complex, seeing spite behind the most innocent of actions, but his obsession carried the day.

Thus the Royal Society's attempt to put the National Antarctic Expedition on a thorough-going scientific basis failed but we may wonder what difference it might have made had they succeeded. Much has been written about Scott but here we can take it that the real person was something between the ultimate English hero and the incompetent amateur that he has variously been made out to be (Pound, 1966; Huxley, 1977; Huntford, 1979; Young, 1980). He had no more scientific schooling than was usually given to a naval officer in the late nineteenth century, that is, elementary physics, astronomy and mathematics. Nevertheless his interest in science developed. He had undertaken with enthusiasm a two-year training course, beginning in September, 1891, on the Navy's newest weapon, the torpedo. The Torpedo School, HMS *Vernon*, taught a certain amount of physics and electricity and carried out research in a desultory way (Hackmann, 1984); Scott emerged with first-class certificates in all subjects. Before the expedition departed he had a course of instruction in magnetic determinations from Captain E. W. Creak FRS, visited Germany to learn about the preparations for the Drygalski expedition, and had talks with Nansen. H. R. Mill, the contemporary historian of Antarctic exploration, noted his quickness in picking up the essentials of a scientific problem (Huxley, 1977). Scott spent much of the expedition's resources on science. He was sympathetic to the aims of his scientists and, indeed, selected men for his staff who afterwards proved their worth as scientists in the wide world. Scientists under him, such as Wright and Priestley (1922) payed tribute to his scientific insight and contributions but this was in a eulogy written after his death and in an official publication. Mawson speaking in 1935, described the scientific programme of Scott's last expedition, on which he had declined to go because he wanted to follow his own ideas, as very complete and highly successful (Jacka & Jacka, 1988, p. xxxii). George Simpson of his second expedition, who subsequently became an eminent meteorologist, considered him as a fellow scientist and wrote:

> One thing which never fails to excite my wonder is Captain Scott's versatile mind. There is no specialist here who is not pleased to discuss his problems with him; and although he is constantly asserting that he is only a layman, yet there is no one here who sees so clearly the essentials of a problem . . . He is constantly stating new problems and he seldom comes in from a walk without having made some useful observation. I must say he often sees things which have a bearing on my work which I have passed over without noting their import. We have had many a long discussion on the cause of our blizzards. A few days ago I had an idea as to the mechanics of these remarkable phenomena, and it gave me a sleepless night working out the details. The next morning I talked the matter over with him exactly as I should with a trained meteorologist or a physicist. We have come to the conclusion that my theory is probably correct in its main lines, today he asked me whether I had realised that according to the theory, I was abstracting heat from the Polar regions, and carrying it to places with a higher temperature, which was hardly probable from thermodynamic principles. Such a question shows a very high power of looking at a problem in its widest aspect and would not have been made by one out of a hundred who had not specialised in physics.[9]

The University of Cambridge awarded Scott an honorary doctorate in science after his return from his first expedition and certainly the scientific achievements of his two expeditions fully justify this. Huxley (1977) aptly summed up Scott's activities as a scientist when she wrote 'Markham, in picking a good sailor, had chosen better than he knew'.

Gregory, on the other hand, at 37 as compared with Scott's 33 years, was already a scientist of distinction when he was considered for the position of scientific director, indeed, he was elected FRS the year that the expedition sailed. He was far from being an armchair scientist, having explored in the Rift Valley in East Africa and made the first crossing of Spitsbergen. His ability to handle men in the field had been demonstrated in 1893 when under difficult conditions he led a party in the first survey of Mount Kenya.[10] He thus had excellent qualifications for the job. However, we cannot be sure that the science of the *Discovery* Expedition would have been better if he had directed it. Able young scientists often give of their best with less rather than more supervision and Scott's principle was 'to give each a maximum amount of freedom in his particular job' (Scott to Markham from Lyttelton, quoted by Huxley, 1977, p. 50). Ferrar, the geologist on Scott's first expedition, did good work and one wonders whether the physicist and the biologists would have done any better than they did with a geologist to oversee them. Gregory's apparent diffidence of manner concealed great tenacity of purpose; he does not seem to have been a person who could tolerate being told what to do and there might well have been difficulties, with serious consequences for the expedition, between him and the ship's captain, whoever he might have been. One cannot tell, but it may have been for the best to have had an undivided command even if that command lacked formal scientific experience. It was perhaps the exploration, rather than the science, that suffered from the rigidity of naval tradition – Gregory had realized that dogs were essential for polar travel whereas Scott, following Markham's prejudice, put his faith in man-hauling (Huntford, 1979).

The National Antarctic Expedition sailed in August 1901 in the *Discovery*, a wooden sailing ship with auxiliary power, purpose built in Dundee. H. R. Mill sailed with her as far as Madeira and George Murray as far as South Africa, both to give instruction in various branches of science. Murray had already put together his invaluable *Antarctic Manual* (1901), a compendium of articles on various aspects of polar science, excerpts from the records of previous Antarctic explorers and an Antarctic bibliography, for the use of the expedition. Of the *Discovery*'s library of some 1,200 books nearly half were scientific works.[11] The story of the expedition's sojourn at Hut Point on Ross Island has been often told (Scott, 1905; Armitage, 1905; Wilson, 1966; Pound, 1966; Huxley, 1977) and the scientific reports comprise six volumes published by the British Museum between 1907 and 1912 and three, with two accompanying volumes of photographs, sketches and panoramas, published by the Royal Society between 1908 and 1909. A main achievement was a description of the physical geography of the McMurdo Sound area. An idea was gained, first by observation from a captive balloon and then from a sledge journey across it to a farthest south of 82° 17′ S, of the extent of the Ross Ice Shelf, and it was recognized as being afloat. The balloon, the first to be used in the Antarctic, seems to have been taken at Sir Joseph Hooker's suggestion (Hooker, in Murray, 1898) – he evidently remembered the frustration of not being able to see over the Barrier. Armitage, Scott's second in command, led a party up the Ferrar Glacier and was the first to reach the polar plateau. A diversion on another journey on the Ferrar Glacier the following season led to Scott's discovery of Taylor Valley, the first of the snow-free 'oases' unique to the Antarctic, which have lately featured so much in Antarctic research. The detailed magnetic records once again showed a synchrony of disturbances in the Arctic and Antarctic (Chree, 1909). There was some criticism of the Meteorological observations from the Director of the Meteorological Office (National Antarctic Expedition, 1908; see p. 292), which Scott took amiss (Huxley, 1977), but these were of detail rather than substance. Vast collections of biological material were made by Wilson and Hodgson, the descriptions of them eventually filling five of the volumes published by the British Museum. The *Discovery* returned to England in September, 1904, having made the longest stay in the Antarctic of any vessel to date.

4.10 The *Gauss* expedition

After the Seventh International Geographical Congress in Berlin in 1899 it had been confirmed between Markham and Professor Erich von Drygalski (Fig. 4.7), leader of the proposed German expedition, that the two expeditions should avoid overlap, the Germans going to the quadrant between 0 and 90° E containing Enderby Land (Markham, 1986), as the Deutsche Sudpolar Kommission taking the advice of Neumayer had decided in 1895. An account of the consultations between the Germans and British was given by Drygalski (1904). Unlike the British arrangements those in Germany were going forward smoothly. When Scott paid his visit he was dismayed by the contrast; the Germans were unhampered by unwieldy committees and warring factions, their government had provided funds without argument and the whole enterprise was under the control of one man, albeit a scientist not a naval officer (Huxley, 1977, p. 38). Drygalski had led a four-year expedition to Greenland and in 1899, at the age of 34, had become Professor of Geography at Berlin. The promotion of the expedition was greatly helped, both financially and politically, by Admiral Graf von Baudissin of the German Admiralty and Graf von Posadowsky, the Imperial Home Secretary. The Kaiser gave his approval

Fig. 4.7 Erich von Drygalski. (Courtesy of the Afred-Wegener-Institut, Bremerhaven.)

in April, 1899. The ship, the *Gauss*, was built along the lines of Nansen's *Fram* and sailed from Kiel in Germany in August 1901. Drygalski (1904) was of the opinion that more could be achieved through the independent involvement of free individuals than by rigid adherence to a set plan and, being given a free hand, chose his staff accordingly. It included as naturalist Ernst Vanhöffen, who had been in Greenland and also on the *Valdivia* cruise; Hans Gazert, as surgeon; Emil Philippi, who had worked with John

Murray on deep-sea deposits, as geologist; and Friedrich Bidlingmaier, as magnetician and meteorologist. The captain of the *Gauss*, Hans Ruser, was subordinate to Drygalski. These arrangements worked well, Drygalski was a cheerful character who handled his men with a light touch and the *Gauss* was a happy, although not altogether well designed, ship.

After calling at Kerguelen to establish a geomagnetic and meteorological base she sailed south-east and sighted land at around 66° S 90° E, becoming beset soon after this. The winter was taken up with scientific observations which were made with assiduity; on one term day readings were taken every 20 seconds for 24 hours. Sledge journeys were made to the adjacent land. Drygalski used a balloon to survey the terrain from 1,600 feet (488 m), maintaining contact with the ground by telephone (see Fig. 5.7). Another technical innovation was to record penguin noises on an Edison phonograph. In the spring, after attempts to free the *Gauss* from the ice by blasting and sawing had failed, Drygalski had the idea of hastening melting by spreading dark coloured material on it. A trail of rubbish was laid over the 2,000 feet (610 m) to the nearest open water. This was effective in creating a channel and the ship was eventually freed to reach home by November, 1903. Permission to spend a second season in the Antarctic, in spite of a full report on the achievements of the expedition, written up on the voyage from the south, having been sent home from Cape Town, was refused. The Kaiser was disappointed that so little new territory had been discovered and in the eyes of the general public Drygalski's achievements were outshone by those of Scott but an enormous amount of data had been amassed. This was published in a series of 20 volumes between 1905 and 1931.

4.11 The *Antarctic* expedition

Exploration of the Peninsula sector had been allocated to Sweden and was organized on the initiative of Otto Nordenskjöld (Fig. 4.8), the nephew of the discoverer of the North-east Passage around Siberia. He had led geological expeditions in Tierra del Fuego and the Yukon, and in 1901 he was a 32-year-old lecturer in the University of Uppsala. The expedition was funded from private sources and although Nordenskjöld felt that it did not compete in scale with the British and German expeditions he was nevertheless able to take a substantial team of scientists: S. A. Duse (cartographer), K. A. Anderson (zoologist), G. Bodman (hydrologist and meteorologist), E. Ekelöf (medical officer and bacteriologist), C. Skottsberg (botanist), and J. M. Sobral, an Argentine naval lieutenant who

Fig. 4.8 Otto Nordenskjöld, from Nordenskjöld & Andersson, 1905, frontispiece.

assisted with meteorological, magnetic, astronomical and hydrographical work. Nordenskjöld's ship, the *Antarctic*, was captained by C. A. Larsen who had already sailed in Antarctic waters.

The *Antarctic* sailed from Gothenburg in Sweden in October, 1901, and arrived in the South Shetlands in January the following year. The story of the tribulations of the next two years would not be appropriate here; suffice it to say that the *Antarctic* was crushed in the ice off the east coast of the Peninsula and the successful rescue of the scattered survivors by an Argentine naval vessel involved a series of highly unlikely coincidences (Nordenskjöld & Andersson, 1905). In spite of all that, a full scientific programme was carried through. The longest series of meteorological observations so far, with hourly readings over most of the period, was obtained by the party under Nordenskjöld on Snow Hill Island. The usual magnetic determinations were made, observing the internationally agreed terms. Numerous important fossils were found (see Fig. 8.2) although many of them were lost with the *Antarctic*. Although tons of palaeobotanical specimens have been collected since from Mount Flora, in Hope Bay, no more species than the Swedish expedition found were discovered in the next 60 years (Adie, 1964). Geological studies were carried out on South Georgia and there were also hydrographical, glaciological and botanical studies (Nordenskjöld, 1908–20).

4.12 Scientific expeditions in the first quarter of the twentieth century

We have now reached the point where it becomes confusing to deal with separate expeditions. However, before turning to the modern period it will be useful briefly to annotate the expeditions of scientific importance which followed in the next 25 years (for more details see Bushnell, 1975; Headland 1989).

1902–4: the *Scottish National Antarctic Expedition* in the *Scotia* to the Weddell Sea and South Orkneys; sponsored privately and led by Dr W. S. Bruce (Brown, Pirie & Mossman, 1906; Bruce, 1907–15).

1903–5: The *French Antarctic Expedition* in the *Français* to the west coast of the Peninsula; supported by a government grant and private contributors, and led by Dr J.-B. Charcot. A primary aim had been to rescue the Nordenskjöld expedition but this was forstalled by Argentina. Charcot deliberately avoided the Ross Sea area so as not to get involved in international rivalry (Charcot, 1906).

1904: The occupation by Argentina of the meteorological station on Laurie Island, South Orkney Islands; this had been established by Dr Bruce, the Argentine Government accepting responsibility for it at his request.

1907–9: The *British Antarctic Expedition* in the *Nimrod* to the Ross Sea sector; sponsored privately and by the Australian and New Zealand governments and led by Lieutenant Ernest Shackleton. This was the expedition on which Shackleton found the way onto the polar plateau via the Beardmore Glacier, achieving a farthest south of 88° 23′ S and proving beyond all possible doubt the continental nature of Antarctica. Another party under Professor T. W. Edgeworth David reached the South Magnetic Pole (Shackleton, 1909; Murray, 1910).

The *Nimrod* expedition was the debut of perhaps the greatest of Antarctic scientists, Douglas Mawson (1882–1958; see Fig. 8.10). Born in England, he grew up in Australia at a time when interest in the Antarctic was intensifying. He trained as a geologist and following his mentor, Professor David, had the idea of taking a round trip on the *Nimrod* just to see the Antarctic coastline and ice-cap, and no more. To his surprise he found himself appointed by Shackleton as physicist for the entire duration of the expedition. After a few days of consideration he decided to acquiesce in this diktat. At first at Cape Royds he seems to have seen himself just as a scientist with a specific project but:

> it was only after his experiences during the ascent of Mount Erebus that he realised that there was a much wider scope of functions which he would need to fulfil in order to do justice to his scientific work. In other words, the work demanded of him – by his own character – now needed wider skills not only to make the best use of an entirely new kind of 'laboratory', but also to learn how to survive in it. A new responsibility would thus arise: to record experiences and observations of a more general value to humanity. He became a leader not for the sake of glory or power or even simply to set a record, but because of an overpowering conviction, justified as the facts were to prove, that he could do a better job than most others. And in this he differed significantly from the other great Antarctic leaders of his time – Shackleton, Scott and Amundsen.
>
> (Jacka & Jacka, 1988, p. xxviii.)

To resume our catalogue: 1908–1910: The *Second French Antarctic Expedition* in the *Pourquois-Pas*? to the west coast of the Peninsula; supported by government and private sources and again led by Dr J.-B. Charcot (Charcot, 1911).

1910–12: The *Norwegian Antarctic Expedition* in the *Fram* to the Ross Sea sector; sponsored privately and led by Roald Amundsen; a meticulously planned, professionally executed and successful dog-sledge journey to the South Pole with a minimum of science (Amundsen, 1912). Priestley & Tilley (1928) comment on the missed opportunity of extending knowledge of the geology of the Trans-Antarctic Mountains.

1910–13: The *British Antarctic Expedition* in the *Terra Nova* to the Ross Sea sector; sponsored privately with the support of the Royal Society and by contributions from Commonwealth governments and led by Captain R. F. Scott. This carried out valuable scientific work in many fields but was overshadowed by the tragedy of the death of Scott and his four companions on their return from man-hauling sledges to the South Pole, which they reached just after Amundsen (Huxley, 1913).

1911–12: The *Japanese Antarctic Expedition* in the *Kainan Maru* to the Ross Sea area; sponsored privately and led by Lieutenant Choku Shirase; chiefly notable in that there was no tradition of exploration in Japan at that time and that it was carried out, not unsuccessfully, without experience and with an unsuitable ship and equipment (Hamre, 1933; Asahina, 1973).

1911–12: The *Second German South Polar Expedition* in the *Deutschland* to the Weddell Sea; supported by a government grant and by private contributions and led by Dr Wilhelm Filchner; an attempt to establish a base on an ice shelf had to be abandoned and the *Deutschland* was trapped in the ice for nine months (Filchner, 1922).

1911–14: The *Australian Antarctic Expedition* in the *Aurora* to the Adelie Coast; supported by the Australian and British governments, scientific societies and private individuals, and led by Sir Douglas Mawson. This was the most thoroughgoing scientific expedition yet and was the first to use radio in the Antarctic (Mawson, 1915).

1912–13: Dr Robert Cushman Murphy of the American Museum of Natural History sailed south on the whaling and sealing brig *Daisy*. He made considerable contributions to knowledge of the birds and seals of South Georgia (Murphy, 1922, 1936).

1914–16: The *British Imperial Trans-Antarctic Expedition* in the *Endurance* and the *Aurora*; supported by government, the Royal Geographical Society and private donors, and led by Sir Ernest Shackleton; it failed in its objective of crossing the continent from the Weddell Sea via the pole to the Ross Sea, the *Endurance* being beset and crushed in the Weddell Sea, but it resulted in one of the most epic rescues of all time. In spite of everything some science was carried out (Shackleton, 1919; Wordie, 1918, 1921a,b).

1915–16: Sub-Antarctic circumnavigation by the US vessel *Carnegie*; the fifth cruise in a series conducting magnetic research (Bertrand, 1971).

1920–22: The *Graham Land Expedition*; supported from private funds with transport provided by various whalers and led by J. L. Cope; two of the four scientists landed at Andvord Bay on the west coast of the Peninsula elected to stay a second year to continue work on meteorology, tides, ice movements, geology and zoology (Bagshawe, 1939; Lester, 1923).

1921–22: Shackleton's voyage on the *Quest* was sponsored by a private benefactor with the object of exploring the coast line between Coats Land and Enderby Land. It continued after Shackleton's death at Grytviken, South Georgia, on 5 January 1922, doing some survey work but little science (Wild, 1923).

1923–24: The first whaling in the Ross Sea was undertaken by the Norwegian vessel *Sir James Clark Ross* accompanied by five catchers. The ship's doctor and a Swedish naturalist made a short journey on the Ross Ice Shelf and made geological, meteorological and zoological studies.

The USA is poorly represented in this list. The American Philosophical Society in 1909 proposed an expedition to explore Wilkes Land but, although President T. R. Roosevelt and his Secretary of the Navy were favourably disposed, no suitable vessel was available and the project came to naught. Peary, after his exploits in the north conceived an ambitious plan for a scientific expedition to winter at the South Pole but sufficient support, both scientific and financial was not forthcoming and this too failed (Gillmor, 1978).

4.13 The coming-of-age of Antarctic science

The expeditions centering on the 'heroic age' of Scott, Amundsen, Shackleton and Mawson were, with one or two exceptions, specifically directed to increasing knowledge of the Antarctic itself – a difference from the expeditions which took place before 1890. It had become recognized that geographic discovery and scientific exploration are different and require different kinds of expeditions. While these scientific expeditions did not lead to any striking advances in theory or practice, between them they built up a corpus of information obtained from many points around the continent and from the ocean surrounding it. There was now a framework defining the gaps in knowledge. The expeditions of this period were attended by much hardship and some tragedy but there emerged techniques for meeting the difficulties and hazards presented by the Antarctic. Perhaps the most important technical innovation in this period had been the use of liquid fuel and efficient portable stoves ('Primus' stoves) as introduced by Nansen for his journey across the Arctic Ocean (Nansen, 1897), which enabled extended overland journeys (Hatherton, 1986). A contrast with the expeditions prior to 1890 was that there was more dependence on private sponsorship than government support. The advent of the independent private person or group, engaged on projects of his own rather than those of governments, was a general feature of exploration in the second half of the nineteenth century (Kirwan, 1962). Spectacular escapes and tragedies caught the public imagination and Antarctic exploration became an exciting and patriotic thing for the wealthy to spend their money on and at the same time the influence of naval men in Antarctic expeditions waned. Although adventurers such as Shackleton played a major role, they mostly paid more than lip-service to science and many expeditions were run by men who were primarily scientists. In the countries of the British Empire, but not obviously in others, there was built up a cadre of able scientists imbued with enthusiasm for Antarctic science. This gave the Commonwealth leadership in Antarctic affairs which is still evident in the vastly different circumstances of the late twentieth century. At the beginning of the century some 10 nations were involved to a greater or lesser extent but neither the USA nor Russia was among them. Although rivalry between nations was a potent factor in getting expeditions sent out, there was some real international collaboration, as in the continuation of the system of simultaneous magnetic observations on term days and planning to the extent that efforts were made to avoid overlap between expeditions. Plans for a joint Swedish–British expedition, which Otto Nordenskjöld played a large part in drawing up, were abandoned on the outbreak of the First World War (Frängsmyr, 1989). Had this materialized it might have brought forward the establishment of a permanent scientific base in Antarctica by 25 years. Among others who recognized at this time that Antarctic science should be international and who actively promoted it to this end, was Professor Edgeworth David, a powerful voice in Australia. He gave encouragement to the Japanese expedition under Shirase at a time when its morale was at a low ebb, he gave unstinted congratulations to Amundsen on his attainment of the Pole – which the majority of his compatriots were inclined to regard as an act of trespass – and at the meeting of the British Association in Australia at the outbreak of the First World War he proclaimed with reference to a German delegate 'All men of science are brothers' (David, 1937). Antarctic science was consolidating into something resembling its present form.

Endnotes

1 For a brief biography and portrait of Maury see Wexler *et al.*, 1962.
2 Royal Society Council Minutes, 26 October 1871.
3 Ibid. 30 November 1871.
4 Ibid. 21 March 1872.
5 Ibid. 14 June 1872.
6 Ibid. 20 June 1872.
7 Ibid. 24 May 1894.
8 Ibid. 5 July, 25 October and 1 November 1894.
9 G. C. Simpson, 1912, Ms. diary in the Scott Polar Research Institute – Ms. 1097/49, p. 84.
10 J. W. Gregory. Obituary Notices of Fellows of the Royal Society, Vol. 1 (1932), pp. 53–9.
11 *Books in Discovery* 1901: National Antarctic Expedition Library: Scott Polar Research Institute archives.

5

The modern period – logistics and matériel

5.1 The inter-related growth of science and technology

From before the time taken as the start of this history up to the present, science, whether measured in terms of numbers of scientists or their publications, has grown exponentially with a remarkably uniform doubling time of somewhat less than 15 years (Price, 1986). Even the two world wars produced only minor deviations from the fitted curve. But, although there were no lasting changes in relative growth rates, there were nevertheless qualitative changes associated with these upheavals. Hall (1976), following Bernal (1965, p. 518), who considered the 'first large scale entry of industrial techniques and organization into physical science' as a significant turning point, distinguished the phase into which science entered after 1918 as that of the New Industrialism. To change to this new phase, however, occupied at least two decades rather than a point in time. Hackmann (1984) remarked apropos of their role in defence research that whereas scientists were employed as technical problem solvers in the First World War and continued in this subordinate position in the inter-war period, the Second World War irrevocably changed the relationship when scientists became involved in operational research. Antarctic science, while sharing in the general expansion, has been affected more than most other branches of science by the technological and social changes that took place. More than any other, Antarctic science is dependent on logistics, on the ability to place and maintain a scientist and his equipment in the right place at the right time. Expeditions to the Antarctic up to 1925 depended on techniques of transport, communication and survival, which had remained largely unchanged for 100 years. These were not necessarily inefficient; the two-man sojourn of T. W. Bagshawe and M. C. Lester off the Danco Coast in 1920–22 was primitive in the extreme (Fig. 5.1, Bagshawe 1939) but they collected more data per man than any other expedition (Wade 1945a) until the advent of computers and satellites. After 1925 the development of mechanized transport, the aeroplane, radio and technology based on better understanding of human physiology, were to make access to the Antarctic, travel within it and survival in its hostile environment, much less difficult.

5.2 Development of organization: the polar institutes

Equally with employment of new technology it was necessary to assemble, organize and assess available information about the Antarctic. Following an initiative by the Congress of World Economic Expansion held at Mons, Belgium, in 1905 a new International Polar Commission was set up in the following year. This held a congress in Brussels in 1908 which was attended by official representatives from some 12 countries. Among the reports it received was one from the International Polar Institute, a short-lived Belgian government-supported bibliographic agency established in 1907. Another meeting of the Commission was held in Rome in 1913 but little of note was achieved and the outbreak of war put a stop to any further activity (Panzarini, 1968).

A more effective step towards organization was taken with the foundation of the Scott Polar Research Institute in Cambridge in 1920. The idea of a centre for collecting and cataloguing information about polar regions was discussed between the two young geologists, Raymond Priestley and Frank Debenham, on Scott's last expedition. After the war they found themselves together in Cambridge and put the idea into practice with an attic in the Sedgwick Museum of Geology as their headquarters. With the support of James Wordie, another Cambridge geologist, who had been with Shackleton on the *Endurance* expedition, they were able to persuade the trustees of the Scott Memorial Fund, which had been set up to assist the dependants of those who had died on the return journey from the Pole, that a polar institute was a worthy object on which surplus funds could be spent. A condition of the transfer, made in 1926, was that a memorial building should be erected within 10 years. An additional gift from the Pilgrim Trust enabled this to be done and the Scott Polar Research Institute (Fig. 5.2) was officially opened by the Prime Minister, Stanley Baldwin, in 1934. The Institute, with Debenham as its part-time director, built up a collection of books and manuscripts, launched a journal, the *Polar Record* in 1931, and established links with polar scientists throughout the world. However, it was not until the 1960s, by which time it had been absorbed into the University of Cambridge as a sub-department within the Department of Geography, that it acquired laboratories of its own (King, 1980). An *Archiv für Polarforschung* was established privately in Kiel, Germany in 1926 by M. Grotewahl and produced a journal *Polarforschung* in 1931. The Norsk

Fig. 5.1 The hut at Water-Boat Point, Graham Land, 1921–22. (Bagshawe, 1939.)

Fig. 5.2 The Scott Polar Research Institute, Cambridge.

Polarinstitutt, was established by the Ministry of Commerce in Oslo in 1948. J.-B. Charcot did not live to see his dream of a French polar institute but Expeditions Polaires Français, a private logistics organization backed by considerable state funding, was founded by his protegé, Paul-Emile Victor in 1947 (Malaurie, 1989).

5.3 The Byrd expeditions and the general introduction of technology

The first major Antarctic programme after the First World War, the *Discovery* Investigations which began in 1925, was ship-borne except for work carried out at established whaling stations and laboratories in the United Kingdom and led to no important advances in logistic techniques for terrestrial expeditions. It was the first Byrd expedition (1928–30) that really began the introduction of modern technology. Although not the first Antarctic expedition to use the various contrivances that had become available, it was the first in using some of them effectively and extensively in exploration. The First World War had, of course, enormously stimulated the development of equipment and techniques of sorts useful in polar exploration. It took American enterprise to deploy them in the still largely unpredictable terrain of the Antarctic.

5.3.1 Ships

Of these, the first essential, little need be said at this point. Falla (1964) and Morley (1964) have given general accounts of the vessels used and their navigation. Rice (1986) gives descriptions of British ships. Something of the roles played by ships in Antarctic research will be found in chapter 7. After the era in which ships such as RRS *Discovery* and *Fram* were designed and built specially, there followed a period extending until after the Second World War when second-hand vessels were adapted or naval vessels employed. Byrd used the wooden sealer, *City of New York*, capable of punching through moderate ice, to accompany the much more vulnerable steel-hulled freighter, the *Eleanor Bolling*. Such ships were used primarily for transport and not as research platforms.

5.3.2 Electrical communication

The German expedition to South Georgia in 1881–82 used a telegraph line to signal time from the living quarters to the magnetic hut (Barr, 1985). An expedition at the beginning of the new century similarly used electrical communication to transmit time signals to instrument points (Fig. 5.3). The first use of the telephone to transmit speech in the Antarctic was on Scott's second expedition when Cape Evans and Hut Point, 15 miles

Fig. 5.3 C. S. Wright making transit observations on the British Antarctic Expedition 1910–13. Note the telephone line. (Courtesy of the Scott Polar Research Institute.)

apart, were linked with a bare aluminium wire and earth return (Huxley, 1913).

Marconi had introduced maritime radio in 1897 (the universal code call for help at sea was adopted in 1904) and Gerlache (1943, p. 242) first heard of it when he returned from being beset in the Bellingshausen Sea. Mawson in 1911 was the first to use radio in Antarctica, sending and receiving messages, via a station on Macquarie Island, when conditions allowed. Radio communications between field parties and base – which would have mitigated Mawson's ordeal when his companions perished – was not feasible. The application of the triode valve, invented by Lee de Forest in 1906, in the period 1912–15, together with the urgency of the war years gave radio communication great impetus (Tucker, 1978). Shackleton had two transmitting and receiving sets aboard the *Quest* in 1921–22 and *Discovery* was, as a matter of course, equipped with radio when she sailed south in 1925. Byrd, however, was the first to use radio in the Antarctic to maintain regular contact between base, field parties and aircraft (Fig. 5.4).

Fig. 5.4 The radio antennae at Little America in the winter night, 1928–30, from Byrd, 1930, facing p. 208. (Courtesy of Richard E. Byrd, III.)

Weather reports transmitted by radio from a geological party made an important contribution to the success of the flight to the Pole. Time signals transmitted to field parties eliminated a possible source of error in fixing positions and there is no need to enlarge on the general value of radio in increasing the efficiency and safety of work in the field. Communication with home also helped to maintain the morale of those working in Antarctica. There are more dubious advantages; Byrd's message to *Little America* saying that he was over the South Pole was picked up by an alert radio operator in New York and the news was immediately broadcast by loudspeaker in Times Square (Carter, 1979, p. 127). Scientists usually like to ruminate on their pronouncements before they are made public. News and entertainment from the outside world can be distracting and often seems vapid in Antarctica, and Byrd had some misgivings:

> When too much talk seems to be the cause of much of the grief in the world, no man could break the isolation of the Last Continent of Silence without a twinge of remorse. (Byrd 1935, p. 98.)

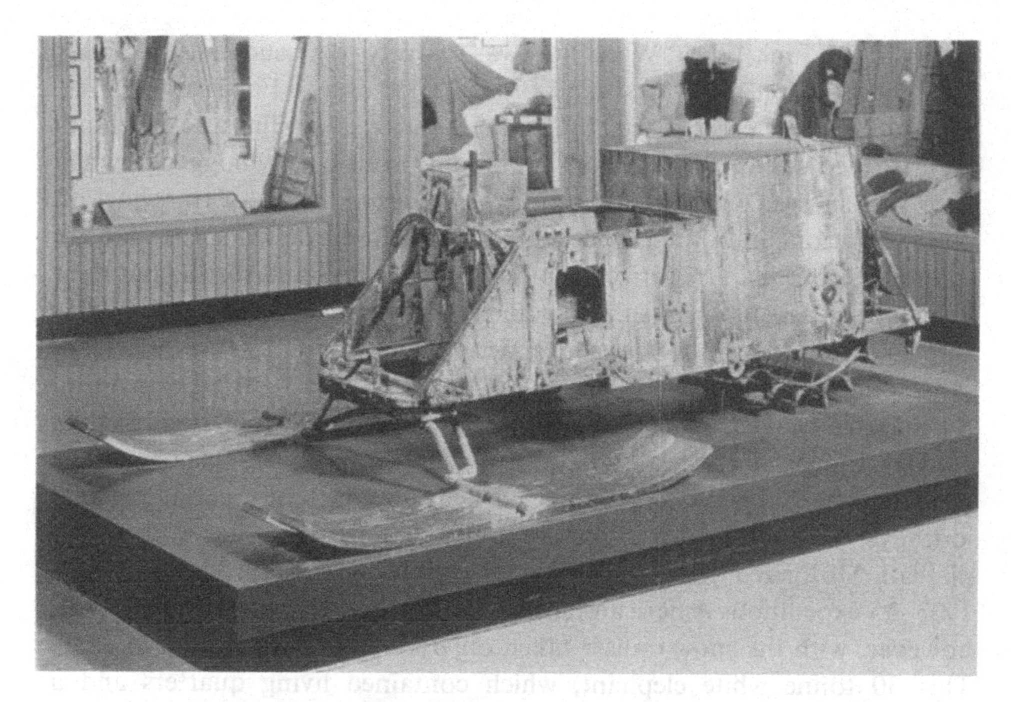

Fig. 5.5 Motor sledge taken to Antarctica by the Ross Sea party of the Imperial Transantarctic Expedition, 1914–16. (Courtesy of the Canterbury Museum, Christchurch, New Zealand.)

5.3.3 *Mechanized surface transport*

The internal combustion engine was another invention that was dramatically to alter the scale and scope of Antarctic science. The first gasoline-engined automobiles were made by Benz in 1885 and Daimler in 1886, and the Ford Model T motor car followed in 1908. Shackleton in 1907 took a motorcar and Scott in 1910 took motor sledges, to which he had devoted considerable thought and time in testing, to the Antarctic (Dibbern, 1976). Shackleton's Ross Sea party in 1914 also took a motor sledge (Fig. 5.5). These various vehicles performed satisfactorily only when the surface was suitable and although they showed that motorized transport had potentialities they were not of much help on the long journeys for which they were needed. Charcot had taken three lightweight crawler-type motor sledges with him on his 1908 expedition (these had been tested in Scott's company in winter in the French Alps) but they were never used, the terrain on the Peninsula being deemed unsuitable. Mawson in 1911 attempted to use a wingless aeroplane as a tractor but it could not be used in wind and soon came to grief. Although tracked vehicles had proved their worth in the First World War, Byrd was scarcely more successful than his predecessors when on his first expedition he tried using a Ford snowmobile. This had front wheels replaced by skis and double caterpillar tracks behind. It performed well on firm surfaces but became irretrievably bogged down in deep soft snow. Dog sledges remained a preferred means of transport on this expedition, and on many others to come. In crevassed areas they are still much safer than motorized vehicles; feeding dogs and sorting out their fights are pleasanter chores than engine maintenance work in subzero temperatures, and dogs provide men under isolated conditions an emotional satisfaction which machines cannot. However, on his second expedition Byrd used a heavy Cletrac crawler-type tractor, three Citroen tractors with crawler tracks at the rear, and two Ford snowmobiles. This time the tractors proved their worth, both in work around the base and on long field journeys, on which heavy seismic equipment was carried as far as the polar plateau. Motorized transport was firmly established in the Antarctic by this 1933–35 expedition. American faith in the big machine overreached itself, however, with the snow cruiser taken on Byrd's third expedition in 1939. This 30 tonne white elephant, which contained living quarters and a machine shop besides carrying food supplies and fuel for a year and a small aircraft on its roof, was designed by Dr Thomas C. Poulter, a physicist with an engineering bent who had been on Byrd's second expedition. It had worked well when tested on loose sand but was underpowered, and on snow surfaces it was three to five times too heavy for the tyres to support it, so that it sank under its own weight (Fig. 5.6, Carter, 1979; Freitag & Dibbern, 1986).

5.3.4 Aircraft

The introduction of air transport has been of profound importance for Antarctic research. Aircraft make travel in the Antarctic somewhat safer and enormously quicker, and by 1928 the various problems of operation and navigation under polar conditions were made manageable (see the various papers in Joerg, 1928). One obvious advantage is the extension of the range of view to a scale more matched to the immensity of the continent. At six feet (1.8 m) above ground the horizon is at a distance of 2.8 nautical miles (5.2 km), but at 10,000 feet (3,048 m) it is at 114.6 nautical miles (212 km) (*American Practical Navigator* quoted by Bertrand, 1971 p. 277). As already noted, both Scott (4 February, 1902) and Drygalski (29 March, 1902) used captive gas-filled balloons to increase their range of vision (Fig. 5.7). Wilson (1966, p. 111) regarded Scott's ballooning as 'an exceedingly dangerous amusement in the hands of such

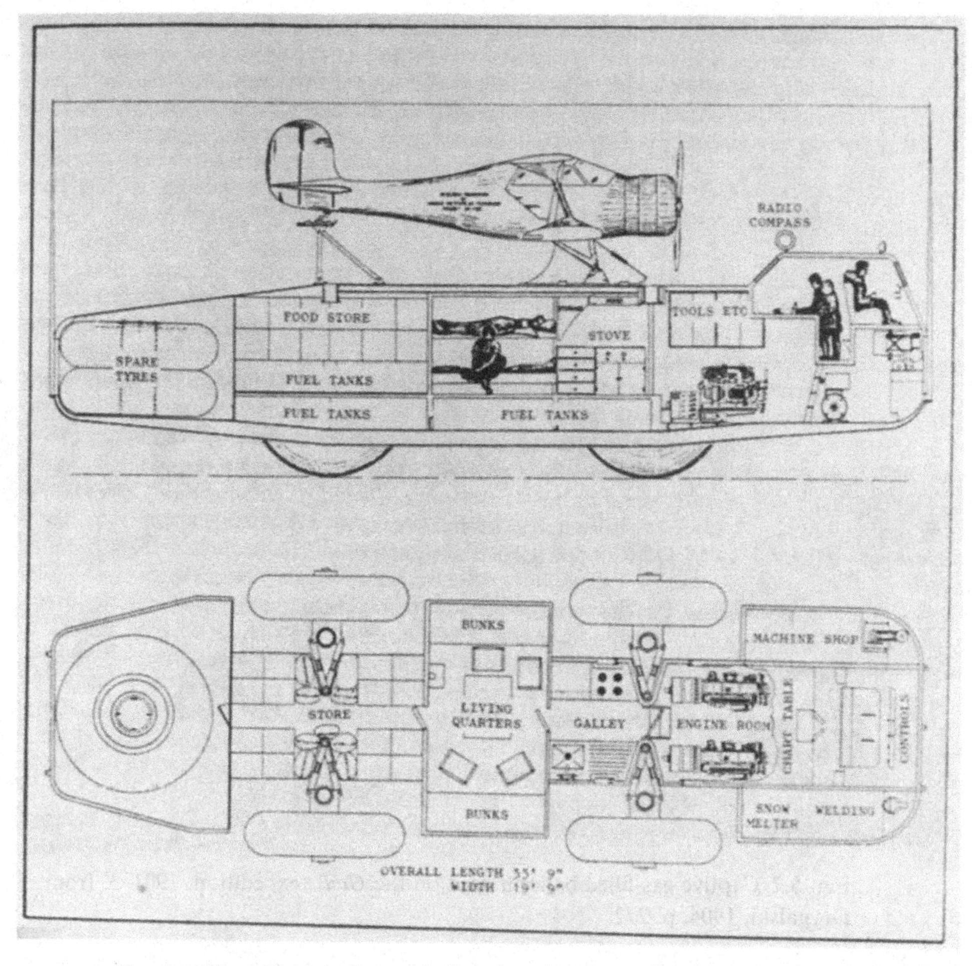

Fig. 5.6 The original plan and elevation drawings of the snow cruiser designed by T. C. Poulter and taken on the Second Byrd Expedition, 1933–35. (Courtesy of Dr Dean R. Freitag.)

inexperienced novices as we have on board' and the limited supply of gas-cylinders precluded both expeditions from making repeated ascents. The Inter-Departmental Committee (1920) in making the recommendations which led to the *Discovery* Investigations considered the use of an aeroplane but, while recognizing the potentialities, concluded that especially for a sea-borne programme it was of doubtful practicality. The first aeroplane flight in the Antarctic was made by Sir Hubert Wilkins from a base on Deception Island in a Lockheed Vega monoplane in November 1928. His 10-hour flight down the east coast of the Peninsula to an estimated 71° 20′ S demonstrated that aircraft could be used successfully in Antarctica.

Fig. 5.7 Captive gas-filled balloon used on the *Gauss* expedition, 1901–3, from Drygalski, 1904, p. 272.

Although numerous geographical discoveries were made, mapping of them could be approximate only.

In 1926 the dirigible *Norge* had made a successful transpolar flight in the Arctic. Although such lighter-than-air craft have certain advantages for long voyages the height of the continent and the violence of its weather precluded their use in the Antarctic (Nobile, 1928).

Within 15 months of the Wilkins flight, three more expeditions were using planes in the Antarctic; besides Byrd with his three at *Little America*, Captain Hjalmar Riiser-Larsen used a seaplane from the *Norvegia* in the vicinity of Bouvetøya and Mawson also used a ship-based plane for the exploration of Mac.Robertson Land, although high winds prevented it from making long flights. Bertrand (1971) notes that most of the aviators involved in these four expeditions had experience of flying in the Arctic. There were no catastrophes, although meteorological forecasts were sketchy at best, and Byrd's flight over the Pole was a spectacular triumph. Of more value scientifically was his use of a plane to place a party in the field to make a geological reconnaissance of the Rockefeller Mountains (Gould, 1931b). On his second expedition Byrd again took several aircraft, including a Kellett autogyro, the first vertical take-off machine to operate in the Antarctic, which was used for atmospheric sounding flights – an advance on the kite which Drygalski had used. After this came the first flight in 1935 across Antarctica, made by Lincoln Ellsworth. Until their exploring expeditions ceased in 1937, the Norwegian whalers continued to use small seaplanes for surveying and the British Graham Land Expedition, with their tiny DeHavilland Fox Moth, surveyed the west coast of the Peninsula, disproving Wilkin's idea that it was separated into islands by straights. We may note that all this pioneering in the use of aircraft, like most logistic innovations in the Antarctic, was done by private expeditions; 'governments generally prefer to follow well-trodden paths' (Swithinbank, 1988b). The German state-supported expedition of 1938–39 used aircraft primarily to establish territorial claims in what is now called New Schwabenland. This expedition was the first to use photogrammetry for mapping in the Antarctic. None of these expeditions used aircraft to any great extent to transport men and materials for scientific work in the field.

5.3.5 Aerial photography

These exploits left no doubt about the practicality and value of aircraft in geographical discovery. To extend this to accurate survey it was necessary to have appropriate photographic arrangements and reliable ground control. The aerial camera, developed for military purposes in the First World War, provided a permanent record of detail vastly superior to notes and sketches. Taken obliquely in overlapping series, a continuous strip of stereoscopic pairs is obtained, giving three-dimensional views from which relative elevations may be determined. On the first Byrd expedition the photographer, Colonel Ashley C. McKinley (Fig. 5.8), manually recorded time and latitude of the plane for each exposure. From such material approximate maps could be prepared but for accurate mapping it is essential that there should be features on the photographs, the positions of which have been determined by conventional techniques on the ground. Much of the aerial photography carried out on these early flights over Antarctica was unsuitable for cartography. Thus, although the 1939–41 US Antarctic Service Expedition produced large numbers of photographs, there was the limitation of having only one aerial camera so that overlapping groups could not be obtained (Shirley, 1945) apart from a general absence of ground control.

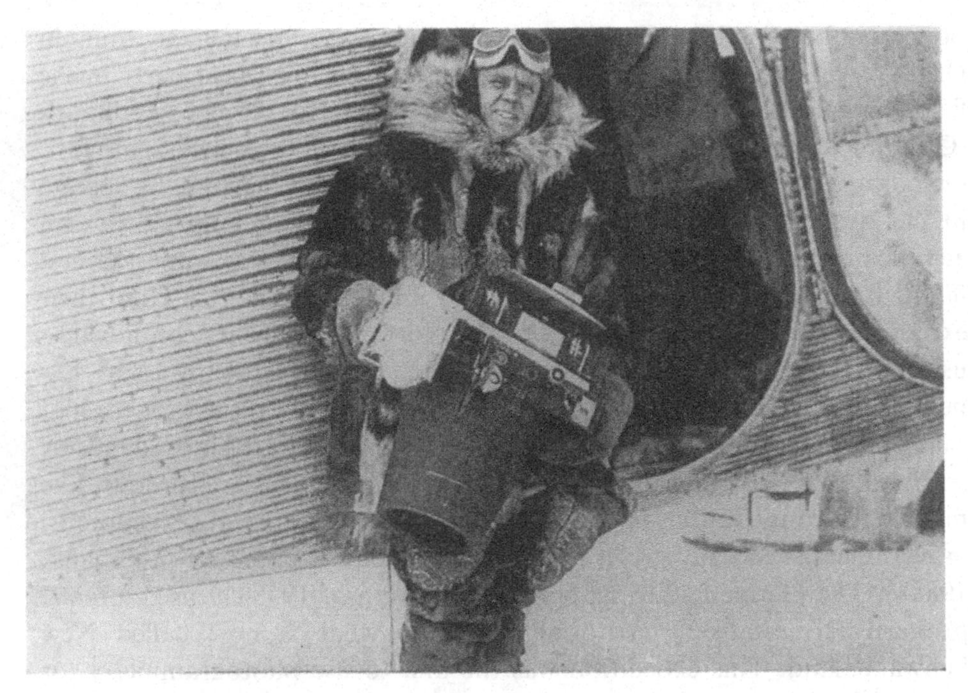

Fig. 5.8 Captain Ashley C. McKinley, aerial surveyor on the First Byrd Expedition, with his aerial camera; from Byrd, 1930, facing p. 36. (Courtesy of Richard E. Byrd, III.)

Proper co-ordination of aerial and ground survey was, in fact, achieved only after many years. The thousands of aerial photographs taken during *Operation Highjump* (1946–47) were again of little cartographical value because of the absence of ground control although this was provided for Wilkes Land by *Operation Windmill* the following year. In 1955–57 there was an expedition under the auspices of the Falkland Islands Dependencies Survey to undertake air photography and land-based control to provide accurate maps of the Peninsula, a specialized task entrusted to Hunting Aerosurveys Limited (Mott, 1986).

5.3.6 Laboratories

A requirement for most kinds of research, one that often causes quarrels and resentments between scientists but which is rarely mentioned in reports, is dedicated space. For magnetic research, for example, it is quite essential to have a place with an even temperature, absence of vibration and without unwanted magnetic influences. On the expeditions of the heroic age, special accommodation of one sort or another was constructed to provide this (Fig. 5.9). Otherwise, scientific work was done

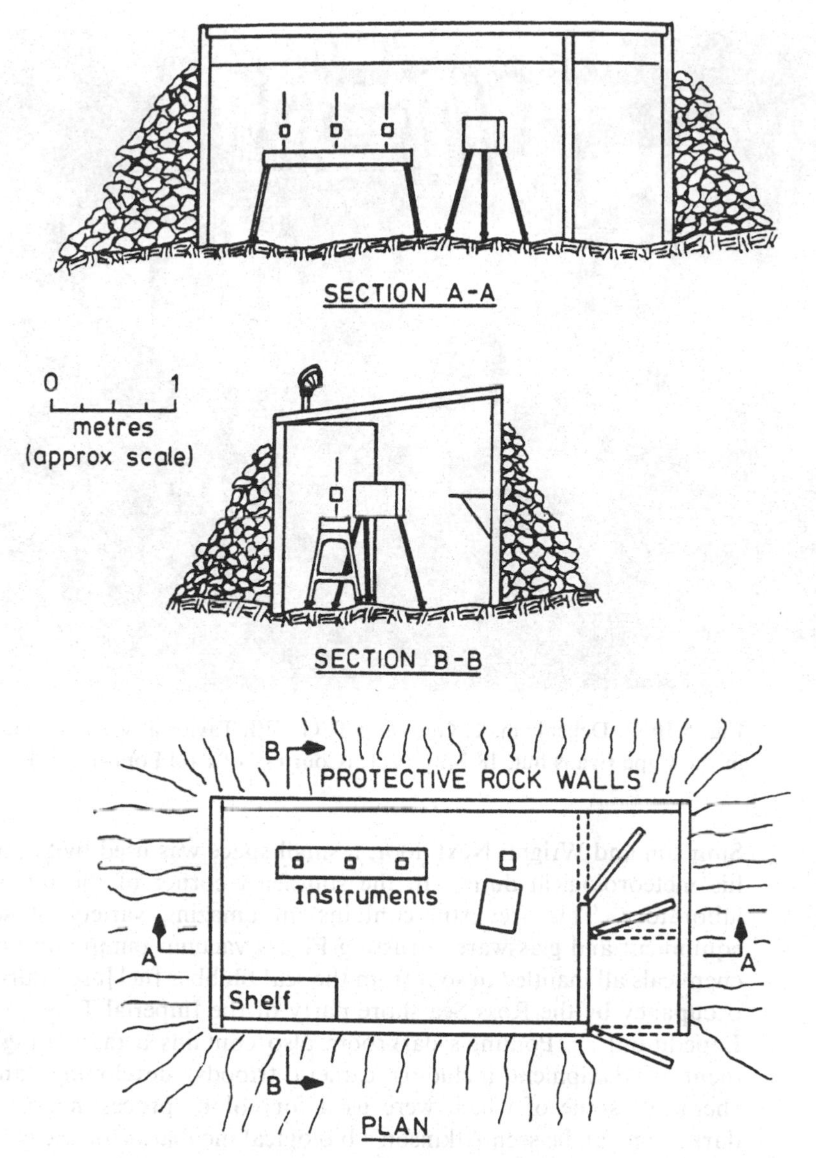

Fig. 5.9 The magnetograph hut at Cape Denison, Adélie Land, 1911–14; from E. N. Webb, 1975. (Courtesy of the Australian Antarctic Division.)

in the general living area with, at the most, a small bench set aside or a corner partitioned off (Fig. 5.10). In his account of the historic huts of the Ross Sea area, Harrowfield (1981, p. 57–61) gives a picture of the situation in the Cape Evans hut that housed Scott's second expedition:

> Nelson's biological work bench is still supported by two cases stencilled with his name, while on the other side are bunks once occupied by

Fig. 5.10 F. Debenham, T. Gran and T. Griffith Taylor at work in their cubicle in the Cape Evans hut, 18 May 1911. (Courtesy of Paul Popper Ltd.)

Simpson and Wright. Next door, a small space was used by Simpson for his meteorological duties. In the southeast corner of the hut was the laboratory. This area still contains an amazing variety of scientific equipment and glassware, a rusting Fleuss vacuum pump and bottles of chemicals all mantled in soot from the seal blubber fuel [used during later occupancy by the Ross Sea shore party of the Imperial Trans-Antarctic Expedition] . . . Ponting's darkroom also contains a fascinating assortment of equipment including camera tripods, developing tanks and chemicals some of which were used for colour processing. Above the darkroom can be seen Atkinson's biological incubator for the cultivation of micro-organisms and a small table by the darkroom was used by Meares during the winter of 1911 for making dog harnesses.

Better working conditions were provided on some of the expeditions following the First World War. The *Discovery* Investigations had a purpose-built marine biological laboratory on King Edward Point, South Georgia, from 1925 to 1931 (Headland, 1984) as well as shipboard laboratories (Fig. 5.11). Byrd's first expedition had several buildings, one of which contained designated quarters for geologist, meteorologist and physicist. However, the radio laboratory was no more than a 'corner of the

Fig. 5.11 Wide-angle view of the upper laboratory, RRS *Discovery*, 1925; from Bernacchi, 1938, p. 128. (Courtesy of Blackie & Son Ltd.)

administrative building' (Byrd, 1930) and Hanson's pioneer work on the reflecting layers of the ionosphere (Fig. 9.13) was done in improvised accommodation. Byrd's second expedition had a 'science hall' equipped 'to delve into twenty-two branches of science' (Byrd, 1935). On the other hand, the impecunious British Graham Land Expedition reverted to the ways of the heroic age, Bertram having to build himself a lean-to against the side of the living hut in which to carry out his dissection of seals (Rymill, 1939). Parenthetically, we may note that he also had to construct his own balance from a wooden beam pivoting on a pebble, ground to a knife-edge (Bertram, 1987).

5.3.7 *Techniques for living*

There was gradual improvement, rather than striking advance, on the first half of the century in techniques for living in the polar environment. Food on base had already been adequate and varied in the heroic age but remained spartan on sledging journeys until after the Second World War. Concentrated orange juice was taken as a source of vitamin C on Byrd's first expedition and that and halibut liver oil for vitamin A were included in the sledging rations for the British Graham Land Expedition. Light, loosely fitting, wind-proof clothing, the value of which had been recognized by Amundsen, was in general use. The string vest, which provides both insulation and ventilation, and so allows tolerable alternation of inactivity and intense activity without major adjustment of clothing, was devised on a scientific basis out of experience on the British Graham Land Expedition for the armed forces of the Second World War. It has become a standard item in Antarctic dress as well as a popular symbol of 'manliness' (Bertram, 1987). Siple (1945) attempted a scientific approach to the selection of clothing for cold climates based on the physiological characteristics of thermal output and regulation of body temperature, the rate at which the atmosphere absorbs heat, and the insulating properties of clothing. The walls of huts incorporated conventional insulation but reflecting aluminium foil was also used to reduce heat loss (Rymill 1939). These improvements were mainly empirical and *ad hoc* rather than the result of systematic application of science.

5.4 Post-Second World War developments

Once again, the urgent demands of war produced equipment and techniques adaptable to polar conditions and the Americans were able to make massive improvements in Antarctic logistics. Ice-breakers had been in use in the north since the late nineteenth century and the contribution which they could make in Antarctica had been envisaged in 1901 (Anon, 1901b) but the US ice-breakers *Northwind* and *Burton Island*, sailing with *Operation Highjump*, were the first to operate in the Southern Ocean (MacDonald, 1973). Without them the thin-skinned steel vessels that made up the rest of the fleet would have been unusable. The submarine *Sennet*, on the other hand, was worse than useless for surface operation in pack ice. The wide deployment of meteorological observers made possible, for the first time, twice-daily synoptic weather charts for the Antarctic. The combination of wheel and ski landing gear and jet-assisted take-off were important advances in flying techniques. A helicopter was flown for the first time in Antarctica but it was not until *Operation Windmill* that these aircraft were used as primary means of transportation. Hayter's (1968) account of his year as leader at Scott Base 1964–65 gives a vivid picture of the indispensable and dominating role which aircraft have come to play in modern Antarctic science, the complex logistics of their use, and of the amazing skill and dedication of the men who fly them. For aerial photography, trimetrogon cameras, which at regular predetermined intervals took three pictures simultaneously – one vertically beneath the plane and the other two obliquely to the port and starboard horizons – were used. Another innovation was the successful use of an airborne magnetometer, adapted from an antisubmarine airborne magnetic detector, for geological exploration. The snow tractors and powered sledges, 'Weasels', were more powerful and efficient than those used on earlier expeditions, and, thanks largely to Paul Siple, who since going with Byrd on his first expedition, had emerged as one of the foremost of Antarctic scientists, clothing, tents and insulation were all much better (Bertrand, 1971). Wet suits for diving were tested for the first time in the Antarctic but not put to any scientific use (Rose, 1980, p. 190).

5.5 Developments following the International Geophysical Year

Eleven nations – Argentina, Australia, Belgium, Chile, France, Japan, New Zealand, Norway, the USSR, the UK and the US – participated in the Antarctic programmes of the International Geophysical Year (IGY). The number of stations in Antarctica rose from 20 to 48, and the numbers of the wintering population from 179 to 912. The IGY Antarctic stations, their co-ordinates, altitudes, wintering numbers and scientific disciplines have been listed by Law (1959). The summer population was perhaps as many as 5000 (Sullivan, 1961). Again there were great advances in logistics and polar technology, mainly emanating from the US which in 1955–56 dispatched a naval task force consisting of three ice-breakers, three cargo vessels, an oil tanker and two oil barges under the code name *Operation Deep Freeze I* to establish the two coastal stations which would provide the bases for further operations. This was distinct from the US IGY organization and was not flexible enough for the liking of IGY scientists. It represented the national interest in exploration and mapping and had Paul Siple as its chief scientist (Siple, 1959; Crary, 1982). Its first, pioneering, task was the preparation of a hard-ice runway on McMurdo Sound to enable large transport planes with conventional landing gear, C-124 Globemasters, to fly in from New Zealand. At McMurdo they were loaded up with the heavy cargo needed for the construction of the Pole station. Transport of this by sledge was considered inpracticable and the Globemasters had to airdrop it at the Pole – an excessively wasteful procedure but undoubtedly the only one that could make possible the achievement of this stupendous task (Fig. 5.12). Accompanying such operations there was inevitably some loss of life. Anderson (1974) catalogued US aircraft losses from 1946 to 1973 in Antarctica, amounting to 20 helicopters and 30 fixed wing aircraft destroyed with 29 US scientists and support staff killed. This paper, incidentally, gives a good general account of US aircraft in Antarctica. The Soviets did not use aircraft for long-distance heavy transport during IGY; probably the elevation of their station near the Pole of Inaccessibility (*c.* 4000 m) would have precluded

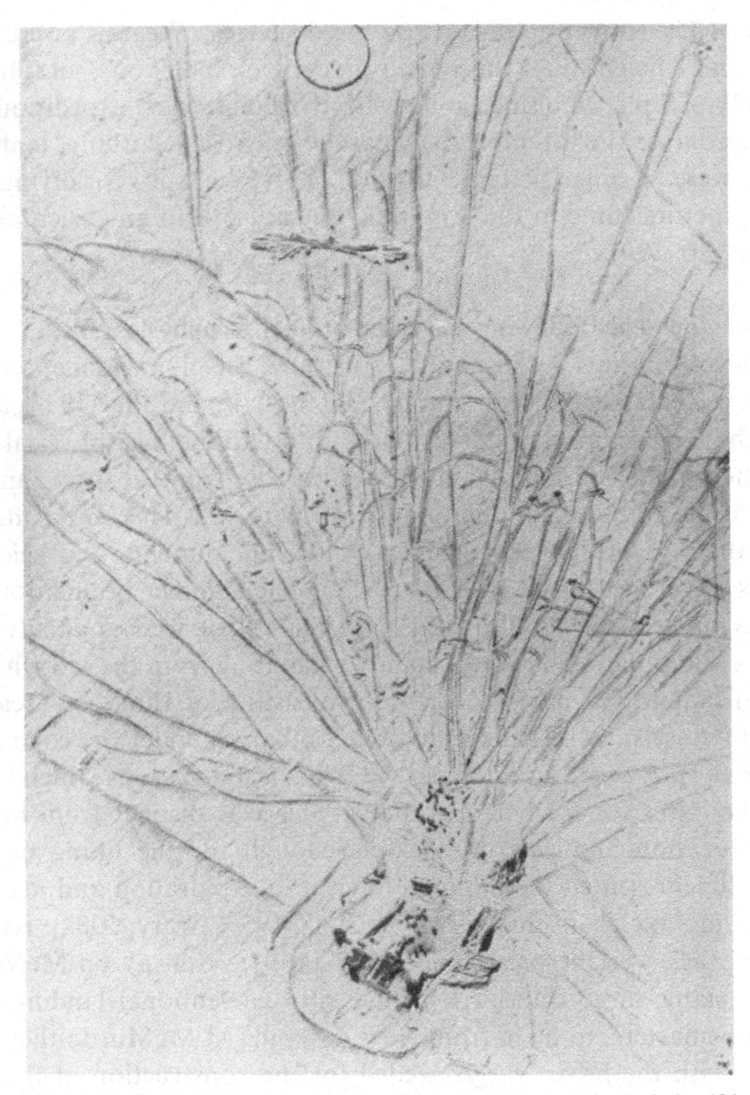

Fig. 5.12 Vertical air photograph of the South Pole. The Pole itself is marked off by a circle of oil drums. Below, 750 m away, is Amundsen-Scott Station. The horizontal marks just below the polar circle were made in an effort to retrieve a tractor which plunged 10 m into the snow when its parachute was severed during an air drop. From Siple, 1959, between pp. 320 and 321, US Navy photograph.

both air-dropping and aircraft take-off (Sullivan, 1961). Motorized surface transport was used extensively by all participants – most spectacularly perhaps by Sir Edmund Hillary who reached the Pole from the Ross Sea on standard Ferguson tractors fitted with tracks for use in snow. There was wry comment by Americans that in spite of the power of their D-8 tractors, weighing 50 tonnes or more each with their loads, the pace was still no more than Scott achieved by man-hauling (Sullivan, 1961) but, of course the amount of material shifted was enormously greater. Crevasse detectors made the passage of heavy tractors somewhat safer but nevertheless crevasse fields severely hampered progress. One of the heaviest items of equipment carried was a deep-drilling rig taken from McMurdo to Byrd Station. The Soviet expedition, led by Hero of the Soviet Union, M. M. Somov, who had directed the activities of the North Pole Drifting Station No. 2, was supplied by the diesel-electric sister ships, *Ob* and *Lena*, and relied heavily on tractors on land (Lebedev, 1959). The deep soft snow encountered on the way to the centre of the continent presented an almost insurmountable obstacle to these tractors, which sometimes sank to a depth of 5 ft (1.5 m) and which, despite superchargers, lost power as the plateau rose towards 13,123 ft (4000 m). In these circumstances Sovetskaya was set up 400 miles (741 km) short of the mark but from the scientific point of view this made little difference. Living conditions at this station were the most extreme encountered during the IGY, with temperatures down to −89.5° C it was only possible for men to work outside for 10 minutes at a time even with electrically heated suits. The 'thermal boot' developed by the US, and zip-fasteners, were innovations in clothing that were to pass into general use. Centralized weather information and forecasting has already been mentioned. Reliable telecommunication was an advance of enormous benefit made possible by co-ordination of radio transmissions through a complex system of nodal links. There were inevitably accidents, and some were fatal. The Japanese supply ship, the *Soya*, became trapped almost every time it ventured into the ice. The Japanese also lost a hut and the scientific records it contained by fire so that their station, *Syowa*, had to be closed midway through IGY (Sullivan, 1961). However, such happenings gave an opportunity to demonstrate that co-operation and mutual assistance between nations were realities in the Antarctic.

5.6 Ships in the modern period

Ships remain essential although mostly serving for supply rather than research. Ice-strengthened vessels, many of them chartered from the Danish shipping line Lauritzen and including such famous names as *Kista Dan, Magga Dan, Nella Dan* and *Perla Dan*, have played notable roles. However, after some 70 years there has been a return to specially designed ships. The 38 m US ship *Hero*, commissioned in 1968 and operational until 1984, was a shallow draught wooden vessel designed for inshore as well as

Fig. 5.13 *Polarstern*. (Courtesy of the Alfred-Wegener-Institut, Bremerhaven.)

offshore biological research and ketch-rigged to reduce roll and to allow silent operation when needed, for example, in bio-acoustic studies. The British Antarctic Survey's (BAS) 99 m RRS *Bransfield*, launched in 1970, is primarily a logistic support vessel with limited research capability. The West German 118 m *Polarstern* (Fig. 5.13) is a double-hulled ice-breaker, commissioned in 1982, built for research as a main function and capable of working and overwintering in the pack-ice zone. Besides having built-in laboratories and equipment for oceanographic work she can take up to 12 lab-containers fitted out for specialist purposes. She also functions as a supply vessel and helicopter carrier.[1] The new BAS logistics support and marine research vessel, RRS *James Clark Ross*, at present (1990) under construction is a 99 m ship capable of breaking 0.8 m thick level first-year ice at a constant speed of 2 knots. She is to be equipped with facilities for oceanography, marine biology and marine geoscience. Her quietness is an essential feature. A comprehensive Ethernet data system will link all laboratory, working and accommodation areas, and via satellite communication systems, shore-based mainframe computers (Drewry, 1990).

5.7 Building technology

Little need be said about this since if accomodation and facilities are adequate it makes no difference to the science how this is achieved. The laboratories of some present-day Antarctic bases such as those at McMurdo and the biological laboratories on Signy Island, compare favourably in available space and facilities with any elsewhere. It is now possible to place cabins fitted out as specialized laboratories wherever they are required. For temporary buildings unconventional materials have sometimes been used. The Transglobe Expedition in 1980 at 73° S, constructed huts with walls of standard corrugated cardboard similar to those used in everyday packaging. This was light to move about, simple to erect, and provided excellent insulation (Fiennes, 1983). Problems of stations on ice shelves remain difficult. Buildings at Halley have been put below the ice surface to minimize accumulation of drift and disturbance to local conditions but these have gradually sunk and been crushed by ice pressure, becoming useless after only seven years. The latest, commissioned in 1983, was built within flexible plywood tubes, 9 m in diameter and 30–40 m long, constructed of interlocking insulated panels with sliding junctions to give more flexibility and steel tie bars to control horizontal stretch.[2] In 1988 these buildings were fast deteriorating and the first phase of a new construction using a different design was begun. This comprises buildings carried on platforms jacked to 3.7 m above the snow surface.[3] A third type of building is the geodesic dome at Amundsen-Scott Station.

5.8 The advent of satellites

The work habits of most Antarctic scientists have been changed by the advent of satellite communications. Since the early 1970s satellite imagery and satellite navigation have provided both displayed information and accurate positioning of the distribution of ice which have enabled scientists to be conveyed more quickly to their place of work. Satellites give synoptic views of most of the continent and its surrounding seas that can be used to produce maps very cheaply by a technique pioneered by the British Directorate of Overseas Surveys.[4] The production of topographical maps – a basic requirement not only for most kinds of environmental science but also for political purposes – had been a seriously limiting factor, field workers often having to produce their own maps before carrying out their scientific programmes. This limitation arose because adequate ground control was still lacking in many areas and because the cost of producing maps by conventional means was high.[5] Satellites can carry a variety of sensors, some of great potency. These developments will be discussed in their place in later chapters. Maslanik & Barry (1990) in a review of remote sensing over Antarctica and the Southern Ocean consider that the Earth Observing System (EOS) planned by NASA for launching in 1996 will be particularly valuable. A formidable array of sensors will be carried and most of these will have built-in flexibility to allow selection of channels and pointing angles. All of this, however, contributes to a general problem, that of the accumulation of data at a rate which outstrips the capacity to assimilate them. Even the collections and results of the classical expeditions took decades to work up and publish. Now the mass of information is becoming so enormous that, in spite of the help of computers in handling it, much remains unused. Another general point is that whereas before 1980, in spite of radio, the scientist was virtually cut off from advice and library resources for long periods, he now has them at his finger-tips. Again war was a stimulus, for following the 1982 conflict in the South Atlantic, BAS established satellite communication with its ships and bases. Before this, the use of a geostationary satellite link between the UK and Halley Station, for ionospheric investigations, had been shown to be practicable. Halley and Siple can communicate with each other and with institutions back home via geostationary satellites (INMARSAT for the former, ATS 3 for the latter).[6] A scientist in a laboratory at home in the northern hemisphere can thus carry out real-time experiments in the Antarctic. This advance in communications is perhaps not entirely to the good in every field of science; thrown on their own resources with a supervisor at the other end of the earth and no overwhelming mass of literature to cause confusion a really good young glaciologist or biologist, for example, had an ideal opportunity to strike out on an independent line of his own. The quality of the science done by the young, tenuously supervized, scientists of the British Antarctic Survey has in fact been remarkably high.

5.9 The impact of equality of the sexes

During the period we have been considering there has been a movement towards the equality of the sexes, a factor that can have considerable impacts both on the logistics and the morale of Antarctic scientists. Traditionally Antarctica has been a male preserve and although Norwegian whalers in the first half of the century frequently took their womenfolk south and the Soviets had women in their IGY expeditions from the beginning, the inclusion of women in Antarctic field parties has generally been a slow and reluctant process. This is not altogether chauvinistic; small all-male groups in the isolation of Antarctica have usually been found to adjust to give well integrated communities, and in a hostile and dangerous environment it is unwise to make any change unless one can be sure it will work equally well. It may be that 'exploration is a male compulsion generally beyond feminine comprehension' (Darlington,

Fig. 5.14 McMurdo Station, Ross Island. (US Air Force photograph courtesy of the National Science Foundation.)

1957) but the modern permanent scientific station is different from the base of a heroic age expedition (Figs 5.1 and 5.14). Women are as well able to withstand Antarctic conditions as men, and in choosing a person for a research task, ability and experience are the important desiderata, and sex should be irrelevant. Nevertheless, a pregnancy at an isolated station during the winter would certainly be disruptive to the science and might be dangerous. Whilst Argentina encourages births in Antarctica for political purposes (Chipman, 1986) West Germany has deemed it prudent when allowing women to overwinter at its Georg-von-Neumayer Station to insist that the team should be an all-female one (Dickman, 1989). The first women scientists to work with the US Antarctic Research Programme (USARP) were Mary McWhinnie (see p. 222) and her research assistant Phyllis Marciniak. They did their work aboard the *Eltanin* in the 1962–63 season. The first women to join the US and New Zealand research programmes on the continent stepped ashore in 1969. The books by Land (1981) and Chipman (1986) give accounts of work accomplished by these and other women scientists in Antarctica. By 1987 it had become normal practice, even with BAS, to include women in field work in the Antarctic, at least during the summer.

Endnotes

1 Brochure, *RV Polarstern*, issued by Alfred Wegener Institute, Bremerhaven, 1983.
2 *British Antarctic Survey Report* 1982–83, pp. 5–6.
3 Ibid 1988–89, p. 21.
4 BAS Scientific Advisory Committee Minutes *9*, 1975; *12*, 1976.
5 Ibid *6*, 1973.
6 Ibid *19*, 1981.

6

The modern period – the involvement with politics

6.1 The dependence of Antarctic science on public money

As the techniques of science became more diverse, precise and elaborate, and more dependent on technology, Antarctic research became increasingly expensive and had therefore to remain as it always had been, as big science, dependent less on private sources and more on public money. Thus the political influences which had affected science in the Antarctic in the past exerted an increasingly powerful control, braking or accelerating as imperialism, the possibilities of exploitation, or maintenance of national prestige, waxed and waned. There were also, of course, influences from within science itself affecting what was done in the Antarctic but these will be considered in the final chapter of this book, after the development of individual fields of science has been described.

6.2 Regulation of whaling and Antarctic research

The First World War left the nations which had been interested with few resources to spare for Antarctic expeditions but there was one problem which called for immediate attention. This was the regulation of whaling. The Norwegian, Captain C. A. Larsen, funded by Argentinian capital, had established a whaling station at Grytviken on South Georgia in 1904. This prospered and was quickly followed by others. Recognizing that the area, including South Georgia, the South Orkneys, the South Shetlands, the South Sandwiches and the tip of the Antarctic Peninsula was probably the most profitable whaling ground in the world, Great Britain constituted all the lands in this area Dependencies of the Falkland Islands by Letters Patent issued in 1908. The justification was that most of these lands had been discovered and claimed for the Crown by British nationals and no objection was raised at the time by other powers (Christie, 1951). This was enlightened imperialism in that the main outcome was that, on the initiative of W. L. Allardyce, Governor of the Falkland Islands from 1904–

15 and a resolute pioneer in conservation (Heyburn, 1980), regulations were introduced to restrain the slaughter of whales – perhaps the first ever legal instruments for conservation. The first steps were to restrict the number of shore-based whaling stations, factory ships and whale catchers and to protect female whales accompanied by calves. It occurred to a Norwegian whaling engineer, Andreas Morch, that the British government might go further and obtain useful information by requiring that the place, species, sex, pregnancy of females, and environmental data should be recorded for the whales caught. This suggestion was made in 1910 to the British Museum (Natural History), and the Keeper of Zoology, Dr Sidney F. Harmer, persuaded the Museum Trustees to make representations about the proposal to the government. An inter-departmental committee was set up in 1913 and a representative of the Museum was sent to South Georgia to obtain first-hand information but the outbreak of war put an end to this idea for the time being (Deacon, 1984). By 1917–18, six shore stations were operating on South Georgia, and Deception Island had become an important harbour for factory ships. The principal product was oil to be used in the manufacture of soap and margarine, with glycerol as a by-product, which was invaluable in war-time. Concern about the depletion of stocks continued and, although the war was at its height, another inter-departmental committee was set up through the vision of E. R. Darnley of the Colonial Office, to consider research and development in the Falkland Islands Dependencies. Evidence was taken from Dr W. S. Bruce, Captains C. A. Larsen and Th. Sorlle (the two Norwegian whalers), and Dr J. Hjort (the former Director of Fisheries in Norway); Dr Harmer from the British Museum (Natural History) was the scientist on the committee. The report was complete in August 1919 and published in April 1920. Although the whalers felt that stocks were not being seriously depleted, the committee concluded otherwise and recommended a programme of conservation measures and research to be financed by taxation of the whaling and sealing industries. The proposed research was far-reaching for it was clearly recognized that the behaviour and numbers of whales depended on plankton and that in turn on physical and chemical oceanography. Suggestions were also made for work on the meteorology, geology, mineralogy, zoology and botany of the Dependencies.

These recommendations were put into effect. As the committee had suggested, two ships were provided, one being Scott's old ship *Discovery*, from which the whole programme became known as the *Discovery* Investigations. *Discovery* sailed south again in July 1925 and was followed a year later by the smaller, purpose-built, *William Scoresby*. In 1929 *Discovery*, which had neither the power nor speed required for extended oceanographic work, was replaced by RRS *Discovery II*. The *Discovery* Investigations continued as such until 1951 when they were merged into the work of the newly established National Institute of Oceanography. Of course, these continued investigations provided the presence which was necessary to maintain the claim to sovereignty but they also produced a substantial body of science. Dr Stanley Kemp had been appointed director of research and leader of the expedition in the spring of 1924 and the great series of *Discovery* Reports eventually published are a fitting monument to his energetic planning and leadership (Hardy, 1967). The scientific results, which were as wide in scope as originally envisaged by the Inter-Departmental Committee, will be discussed in more detail in chapter 7. Here it is sufficient to note that a liberal interpretation of the responsibilities of an imperial power led to important advances in Antarctic science.

6.3 Nationalistic and imperialistic influences up to the Second World War

Another oceanographic vessel which set out for the Southern Ocean in 1925 was the *Meteor*. Under the Treaty of Versailles the German Navy was not permitted to send its vessels to foreign ports but in 1919 it was proposed to the German Admiralty that an unarmed research vessel would be another means of showing the flag. This idea, in accord with the Navy's tradition of research expeditions, met with favour but the financial state of Germany caused the project to be postponed repeatedly. Ultimately it was funded jointly by the Navy and private sources. The scientific programme was planned by Dr Alfred Merz, and, after his illness and death, was led by Captain F. A. Spiess with the scientific advice of Dr G. Wüst, who had been a student of Merz's, in a two-year survey of the Atlantic. The *Meteor* was well staffed with scientists and her equipment included two echo-sounders (Emery, 1980). She seems to have been the first oceanographic vessel to use the echo-sounder, which had been developed out of attempts at sonic submarine detection during the war, in extensive surveys (Schlee, 1973; Hackmann, 1984). Her cruise extended down as far as the ice-edge off the Antarctic continent. Important results were obtained which will be discussed in the next chapter. Again science had benefited from nationalistic assertion.

The *Meteor* expedition was without political significance for Antarctica but the *Discovery* Investigations were part of a wider scheme with ulterior motives. In 1919–20 the British government had decided for imperialist and strategic reasons to pursue a policy aimed at gradual acquisition of the whole Antarctic continent (Logan, 1979; Beck, 1986). To this end a secret committee, consisting mainly of civil servants with representatives from Australia and New Zealand was set up. After 1930 it became known as the Polar Committee and it continued until the outbreak of the Second World War, by which time it had met 130 times. In 1923 by Order in Council, sovereignty was asserted over the sector south of 60° S between 160° E and 150° W, henceforth to be called the Ross Dependency and to be administered by New Zealand. The French claim of control over the Adélie Land sector which followed in 1924 was a set-back to the grand design but although a powerful scientific lobby, headed by the intensely patriotic Sir Douglas Mawson, urged Australia to oppose this, Britain decided that it would be unwise. There were soon difficulties from other directions. Norway protested that Amundsen's claim had been ignored by Britain and her whalers began in 1926 an extensive series of surveying voyages around the continent (Riiser-Larsen, 1930; Christensen, 1935, 1939). Then there came news of an impending US expedition to the Ross Dependency to be led by Commander Richard Evelyn Byrd (Fig. 6.1). The British Ambassador in Washington was instructed to remind the US of British claims to this area but at the same time tactfully to offer assistance. The US had not made any territorial claims in Antarctica, did not intend to recognize any made by other nations unless they were substantiated by effective occupation, and delayed a reply until its expedition was firmly established in the Antarctic.

The first Byrd expedition to Antarctica was undertaken for the distinction of being the first to fly over the Pole, rather than for science. At a celebratory dinner after his successful flight to the North Pole, Byrd was asked by Amundsen 'What shall it be now?' and although his reply 'The South Pole' was afterwards reported as a joke, it seems that he had already given it some thought and that Amundsen gave him serious advice (Byrd, 1930; Bertrand, 1971; Carter, 1979). Although the great crash on Wall Street was some months off, money was still not easy to obtain. However, J. D. Rockefeller, Jr. and Edsel Ford contributed, and the expedition was able to set out with two ships and three aircraft in the late summer of 1928. The imperial establishment was worried but the people of New Zealand welcomed the Americans warmly *en route* (Quartermain, 1971). A base, called *Little America*, was established on the Ross Ice Shelf at the Bay of Whales not far from where Amundsen's *Framheim* had been. From there a flight to the Pole and back was successfully made on 28–29 November, 1929. Much exploration, making extensive use of aircraft, was carried out and important geological work was done. Of the 42 men who wintered at

Fig. 6.1 Richard Evelyn Byrd (right) and Floyd Bennett (? left). (Byrd Polar Archives)

Little America five were scientists, one of these being Byrd's second in command, Dr Laurence McKinley Gould (see Fig. 6.8), assistant professor of geology at the University of Michigan. The primary objects of the expedition were geographical exploration and the flight over the Pole, not the least function of the geological field parties being to provide ground control and rescue back-up for the aircraft (Gould, 1931b).

Radio enabled the world to know immediately of Byrd's success and his return to New York, as a newly-promoted Rear Admiral, was triumphant. From this time on Byrd's name was synonymous with Antarctica in the American mind. Apart from the firm establishment of modern technology in Antarctic exploration a major result of his first expedition was the revival of US interest in Antarctica, which extended to scientists as well as

Fig. 6.2 The Proclamation of King George V Land, 5 January 1931, by Sir Douglas Mawson during the British, Australian and New Zealand Antarctic Research Expedition. From Price, 1963, plate 39. (Courtesy of K. B. Price.)

the general public. Even before the Byrd expedition left, the American Geographical Society had arranged a symposium under the title *Problems of Polar Research* to which 31 Arctic and Antarctic experts, largely from outside the US, contributed (Joerg, 1928).

The geographical discoveries made by the Byrd expedition could have provided justification for a territorial claim by the US embracing the sector from the Ross Dependency at 150° W to the limit of the Chilean sector at 90° W. There was some talk of this but no action was taken. Nevertheless, the British government, together with those of Australia and New Zealand, was provoked into supporting an expedition to the sector marked out for an Australian claim. This British–Australian–New Zealand Antarctic Research Expedition (BANZARE) was led by Mawson and sailed in the *Discovery* in 1929. It carried out some excellent science but its primary motive was political and much was made of flag-raising ceremonies (Price, 1963; Fig. 6.2). The French claim was a complication that caused some delay but in 1933 an Order in Council affirmed sovereign rights over the sector south of latitude 60° S and between 45° E and 160° E with the exception of the sliver of Terre Adélie. The Bill of Acceptance by Australia was passed with little enthusiasm and not proclaimed until 1936. In New Zealand there had been concern about whaling, from which New Zealanders were getting little benefit, and there were calls for conservation, but there was little government interest in the Ross Dependency. The second Byrd expedition in 1933–35 and Lincoln Ellsworth's trans-Antarctic flight in 1935 gave the US further grounds for claims but still none were made and it has been doubted whether there was any real or coherent US policy in respect of Antarctica at this period except for the consistent theme of non-recognition of other nations' claims (Beck, 1986). The rescue of Ellsworth at the conclusion of his successful flight by RRS *Discovery II*, sent on Australian initiative, was first and foremost a gesture of goodwill but it also served to reinforce Imperial claims to the Ross Dependency. Nevertheless, in New Zealand, with the advent of a labour government in 1935, Antarctica receded even more into limbo (Logan, 1979). At the other side of the continent the British claim to the Peninsula was bolstered by the British Graham Land Expedition (1934–37). This small, privately organized, venture in an old French fishing schooner, renamed the *Penola*, was led by John Rymill and supported to the tune of £10,000 each by the Royal Geographical Society and the Colonial Office (Rymill, 1939). This was far from munificent; the expedition took a small plane but the scientists had to serve as the *Penola*'s crew. Nevertheless, extensive surveys, which showed that Graham Land was a peninsula attached to the continent, and valuable scientific work were accomplished. Thus, none of the powers concerned showed more than half-hearted interest in the Antarctic. This was to be changed by the outbreak of the Second World War.

Fig. 6.3 The first landing on the ice-shelf of New Schwabenland by members of the German Antarctic Expedition, 1938–39, from Herrmann, 1941, facing p. 157.

6.4 The Antarctic in the Second World War

The advent of the Second World War sharpened perceptions of strategic and political possibilities in the Antarctic. Even before hostilities began the German territorial claims resulting from the *Schwabenland* expedition, under the patronage of Reichsmarschall Hermann Göring (Fig. 6.3), had given Britain, Norway and the US food for thought. With this stimulus the popular interest in the US in Antarctica spread to government circles and whereas Byrd's first two expeditions had been sponsored privately, his third had substantial official support and the interest of President Franklin D. Roosevelt himself. Roosevelt summed up his thoughts in a letter to Byrd:

> The most important thing is to prove (a) that human beings can permanently occupy a portion of the Continent winter and summer; (b) that it is well worth a small annual appropriation to maintain such permanent bases because of their growing value for four purposes – national defense of the Western hemisphere, radio, meteorology and minerals. Each of these four is of approximately equal importance as far as we now know.
>
> (Roosevelt to Byrd 12. VII. 1939, quoted by Beck, 1986, p. 27.)

The US Antarctic Service Expedition 1939–41 had the co-operation of learned societies and took civilian scientists but was carried out by the Navy with Admiral Byrd as the commanding officer. However, it had behind it a civilian body, the US Antarctic Service, organized under the auspices of the Departments of State, Treasury, Navy and Interior. There seems to have been no scientist on its executive committee. Although the Service was set up in support of the one expedition, the President evidently intended it to be ongoing and the bases which it established should be permanent (Bertrand, 1971). The expedition, carried out by the two ships *Bear* and *North Star* and with four planes, sailed in November 1939 and set up its East Base on Stonington Island off the west coast of the Peninsula and its West Base, *Little America III* on the Ross Ice Shelf. This expedition, the best equipped and the most extensive in organization and objectives so far, accomplished much in exploration and research. There was an echo of the Wilkes expedition in Roosevelt's order of 25 November, 1939 to Byrd in which, among other things, he directed that all scientists should keep journals, that no journals, charts, specimens, photographs or other material should be passed to anyone not belonging to the Antarctic Service and that all such material should be surrendered at the end of the expedition (full text in Bertrand, 1971). Since the scientists were being paid at a derisory rate and were counting on publishing their results in the normal way, there was uproar when Byrd asked them to sign an undertaking to this effect. Mass resignation was averted only when Byrd promised to do his best to have the order rescinded (Siple, 1959, p. 66). In the end the substantial scientific results of the expedition were published in the usual scientific journals (see list in Bertrand, 1971). Roosevelt's order also reiterated that the US did not recognize any claims of sovereignty in the Antarctic and that no member of the expedition should compromise this position by word or deed. Appropriate actions were to be taken to provide a basis for US claims to sovereignty but no public announcements of this were to be made (Bertrand, 1971, p. 473). In the event no claims were made and in the light of subsequent events it seems fortunate that Roosevelt did not pursue his idea of permanent colonies. Siple (quoted by Carter, 1979, p. 216) considered that had he done so, the US might have been drawn into controversy with its wartime allies and the outcome for Antarctica might have been much less happy than it actually has been. With the worsening of the international situation the expedition was withdrawn after a full year in the field and the two bases were closed down.

Meanwhile the value of the lonely seas of the Antarctic as a hiding place for raiders had been realized by the Germans. The raider *Pinguin*, operating from Kerguelen, captured at a stroke an entire Norwegian whaling fleet as it lay at anchor off Queen Maud Land. The *Pinguin* was found and sunk by HMS *Cornwall* in May 1941, but its sister buccaneers *Komet* and *Atlantis* continued to cruise in the Southern Ocean and make forays northward to harass Allied shipping (Carter, 1979, p. 218). Argentine territorial claims in the Antarctic also presented problems for the British. In 1940, Chile, apparently stimulated to interest by the Norwegian claim made in 1939, had decreed the sector of the Antarctic between longitudes 53° W and 90° W to be Chilean territory (Christie, 1951). This overlapped the Falkland Islands Dependencies but does not seem to have perturbed the British government unduly. The Argentine claim which followed in 1943 was a different matter. Not only was the overlap with the British sector more substantial than was the Chilean one but the political complexion of the Argentinian government made it highly undesirable that that country should be in a position to control the southern side of the Drake Passage (Christie, 1951). The anxiety of the Foreign Office led the British War Cabinet to rethink its policy towards this hitherto largely disregarded region and in late 1943 a secret naval expedition, code named *Operation Tabarin*, was dispatched to establish a permanent British presence in the Peninsula.

Operation Tabarin was organized by the Navy but quickly assumed a scientific character. Search for young men with polar experience led to James Marr, who had been with Shackleton on his last expedition in the *Quest*, and who later had served as a biologist with the *Discovery* Investigations, being recalled from naval duties in the Far East to become the expedition's commander. The committee appointed to advise the expedition consisted of James Wordie (Fig. 6.4) – to whom Fuchs (1973) attributes the credit for turning a military operation to scientific account – Dr Brian Roberts (see Fig. 6.6), biologist on the British Graham Land Expedition, and Dr Neil Mackintosh, Director of *Discovery* Investigations. Before the war ended three stations (the first permanent bases to be set up in the Antarctic itself) had been established and programmes of surveying, meteorological observation and research in geology, glaciology and biology were begun. Political activity was limited to removing evidence of Argentine claims and erecting notices with *British Crown Lands* in large letters. The vessels which served the expedition, SS *Fitzroy*, SS *Eagle* and the *William Scoresby*, now HMS, were unarmed (Fuchs, 1982).

Fig. 6.4 Sir James Wordie. (Courtesy of the Scott Polar Research Institute.)

6.5 The Falkland Islands Dependencies Survey

With the ending of the war *Tabarin* ceased to be naval and became civilian, being renamed the Falkland Islands Dependencies Survey (FIDS). In 1947 V. E. (later Sir Vivian) Fuchs, a geologist who had worked in East Greenland with Wordie and who had led four expeditions in Africa, was appointed field commander, without, however, any specific directives on the scientific programme. The Survey retained a political function, indeed instructions to base commanders in 1947 stated that the primary object of the Survey was to strengthen the claim to sovereignty (Beck 1986). Ritual protest notes were exchanged with Argentinian, Chilean and US expeditions in the Dependencies but, apart from one or two occasions when Argentinians let off guns, Fids (as the personnel of FIDS came to be called) found this part of their duties irksome and childish, and established friendly and co-operative arrangements with their opposite numbers. At various times 18 bases were occupied. It was at this time that the ethos which still characterizes FIDS's successor, the British Antarctic Survey (BAS), was established. Through men such as Wordie and Priestley – of the heroic age – and Bingham and Roberts – of the British Graham Land Expedition – there was a direct connection with the traditions of the expeditions of the past and there was a sense of high adventure tempered with due respect for the hostile environment. Freedom from bureaucratic control gave the British genius for improvization free rein. Bases became quite cosy establishments in which everyone was expected to turn his hand to any task as need arose, as admirably described by Walton (1955). Survey and mapping were perhaps the predominant activities of FIDS and some magnificent sledge journeys were made (Fuchs, 1982).

6.6 The assertion of American interest

At the end of the war American interest in Antarctica had re-asserted itself with what still remains the largest individual expedition ever sent to these regions. This was *Operation Highjump* 1946–47 (Bertrand, 1971; Rose, 1980). Its objective was not primarily scientific: firstly, there was the need to continue exploration and occupation in case any territorial claims should be made and secondly, there was the value of the Antarctic for military training to prepare men and test equipment for possible deployment in the Arctic – a region that itself was too sensitive to use for such purposes (Beck, 1986). Byrd, who was designated Officer-in-Charge of the project but who actually had little control (Rose, 1980) evidently had some reservations about these objectives, for, flying over the Pole in February 1947, he dropped a cardboard box containing the flags of the United Nations as a symbol of his desire for international harmony in Antarctica. Siple (1959) later complained about the subordinate position of science:

> Many scientists had accompanied Highjump. Cruzen [Rear Admiral Richard H. Cruzen, USN Commander of the task force], however, considered them superfluous, choosing to emphasize the exploratory aspect of his orders at the expense of the scientific, and they were given so little opportunity to pursue their work that many of them vowed they would never return to the Antarctic with a Navy expedition. The result was that many of the best scientific opportunities were left to the plane pilots, most of whom unfortunately did not recognize them.
>
> (Siple, 1959, p. 79.)

He instanced the discovery of the Bunger Oasis at 66° 18′ S 100° 45′ E, inshore of the Shackleton Ice Shelf, an event which inspired the world's press to fantasy but at the time yielded scarcely any information of scientific significance. For an expedition which involved 4,700 men, 13 ships and nine aircraft, three scientific publications (Bertrand, 1971, p. 513) was a poor yield.

For the Navy *Operation Highjump* was a successful trial of their resources under conditions which sometimes were extremely adverse but for Byrd and his men it compromised their ideals of how exploration and science should be conducted. A final humiliation was the MetroGoldwyn Mayer film *The Secret Land* – a travesty of their activities in the Antarctic (Rose, 1980).

Operation Highjump was followed in 1947–48 by *Operation Windmill*, which again had the frankly military objectives of training and testing of the equipment but which followed up some of the discoveries made by *Operation Highjump*. A proposal by Byrd and his associates for *Operation Highjump II* to fill in the gaps left by *Operation Highjump* itself was killed by a combination of political, strategic, financial and diplomatic considerations among which was personal antipathy between President Truman and Byrd's brother, a power in Democratic party politics (Rose, 1980). In the same year as *Operation Windmill* there was another US expedition, supported both privately and from government sources and led by Finn Ronne, based on Stonington Island. This expedition was notable for, among other things, including the first two women, neither of whom were scientists, to overwinter within the Antarctic Circle (Darlington, 1957).

6.7 The growing problems arising from territorial claims

American activity was viewed with misgivings by other nations with interests in Antarctica. In Washington the British Embassy again intimated, to no effect, that notification was expected if US expeditions visited British Antarctic Territory (Rose, 1980) and on Stonington Island, where a FIDS base and Ronne's station were only a few hundred yards apart, there was reserve and mutual suspicion which fortunately gave way within a few months to profitable co-operation (Ronne, 1949; Walton, 1955; Darlington, 1957). Argentina and Chile saw *Operation Highjump* as a threat to their claims in the Peninsula area, and New Zealand and Australia also looked at it askance. The South American concern almost led the White House to cancel *Operation Highjump* but the Navy, which needed to boost its image, had its way (Rose, 1980). After *Highjump*, the US took the diplomatic initiative and in 1948 proposed to Argentina, Australia, Chile,

France, New Zealand, Norway and the UK that Antarctica should be territory under international trusteeship or under an eight-power condominium. Only New Zealand, whose government at that time showed little interest in Antarctica (Logan, 1979), among the nations approached was willing to surrender sovereignty and nothing came of the proposal. At the same time the State Department asked the National Academy of Sciences to consider a co-ordinated programme of Antarctic research but this, too, came to naught (Crary, 1982). US fears of developing Soviet interest, which had revived with the first venture of the *Slava* whaling fleet into the Antarctic in 1946–47, were realized in 1950 when the USSR forwarded a diplomatic note to those nations attempting to negotiate a condominium saying that it would refuse to recognize as lawful any decisions about the Antarctic regime taken without its participation (Beck, 1986). Other expeditions to the Antarctic at this time – the *Brategg* oceanographic expedition from Norway (1947–48), various expeditions from Chile between 1947 and 1955 and from Argentina between 1945 and 1955, the French expedition to Adélie Land (1949–53), and the Australian National Antarctic Research Expedition (1954–55) – all had territorial claims as objectives and some did only a minimum of science. The exception was the Norwegian–British–Swedish Antarctic Expedition (1949–52) which not only put specific scientific investigation before geographical discovery but was the first large-scale international expedition ever to be organized (Giaever, 1954). In 1949 Debenham suggested a low cost expedition to study ice mechanics and physics which was enthusiastically supported by the New Zealand Scientific Office in London but neither the UK nor the New Zealand governments were prepared to spend money on it (Logan, 1979). The Norwegian–British–Swedish expedition apart, the omens for Antarctica seemed gloomy, with quarrels over sovereignty, overriding of science by military considerations and, ever-present in the background, the 'cold war' between the western nations and the USSR.

6.8 The International Geophysical Year

At this juncture the only voice speaking clearly on Antarctica was that of science and in retrospect it seems almost miraculous that a fortuitous combination of political circumstances should have enabled it to prevail. The initiative which broke the log-jam was the International Geophysical Year (IGY) of 1957–58.[1]

The idea of the IGY seems to have been born in a conversation over a dinner on 5 April 1950 given by James van Allen at his Maryland home in honour of the visiting British geophysicist Sydney Chapman and attended by others interested in the physics of the upper atmosphere. The idea emerged that there should be another international polar year. This seems to have come up on the spur of the moment but no doubt crystallized thoughts that had been near the surface for sometime (Crary, 1982). The Second International Polar Year, which seems to have been proposed by the German meteorologist, Johannes Georgi, had taken place in 1932–33 and had included plans for a ring of stations around the Antarctic. However, it had been badly affected by the depression and these plans were abandoned, although the Argentines contributed observations from the South Orkneys. The Arctic programme nevertheless had gone ahead and been productive of results. Now, it was felt, scientific and technological advances justified a third polar year, which should concentrate on a solar maximum rather than a minimum as previously, even though it would be only 25 years since the last one. Chapman and Lloyd Berkner, the latter having risen to scientific eminence since serving as a radio engineer with the first Byrd expedition, put the idea to the Commission on the Ionosphere, which gave enthusiastic approval and forwarded it to the International Council of Scientific Unions (ICSU). The proposal was accepted in 1951 by the ICSU Executive Board and a special committee, Comité Speciale de l'Année Geophysique Internationale (CSAGI) was set up to plan a scientific programme and invite participation. In response to protests from meteorologists and magneticians that their studies, too, needed to extend to the polar regions the scope was widened and the title International Geophysical Year adopted. Chapman was elected President of CSAGI, Lloyd Berkner as Vice-President, and Marcel Nicolet of Belgium as Secretary-General. It was decided that IGY would run for 18 months, from 1 July 1957 to 31 December, 1958, to ensure adequate sampling of data and span an expected peak in sunspot activity. The general programme drawn up by CSAGI was elaborated by a council of national representatives and passed through national IGY committees to the organizations in each country responsible for carrying out the work. By May 1954, the deadline for submission of detailed programmes, over 20 nations had agreed to participate. A particularly massive contribution was promised from the US; President Eisenhower's endorsement of the IGY had ensured congressional support and the appointment of Laurence Gould, former chief scientist of the first Byrd expedition and now a powerful figure in the US academic establishment, as head of the Antarctic IGY Committee, provided both realistic advice on the Antarctic and access to the seats of

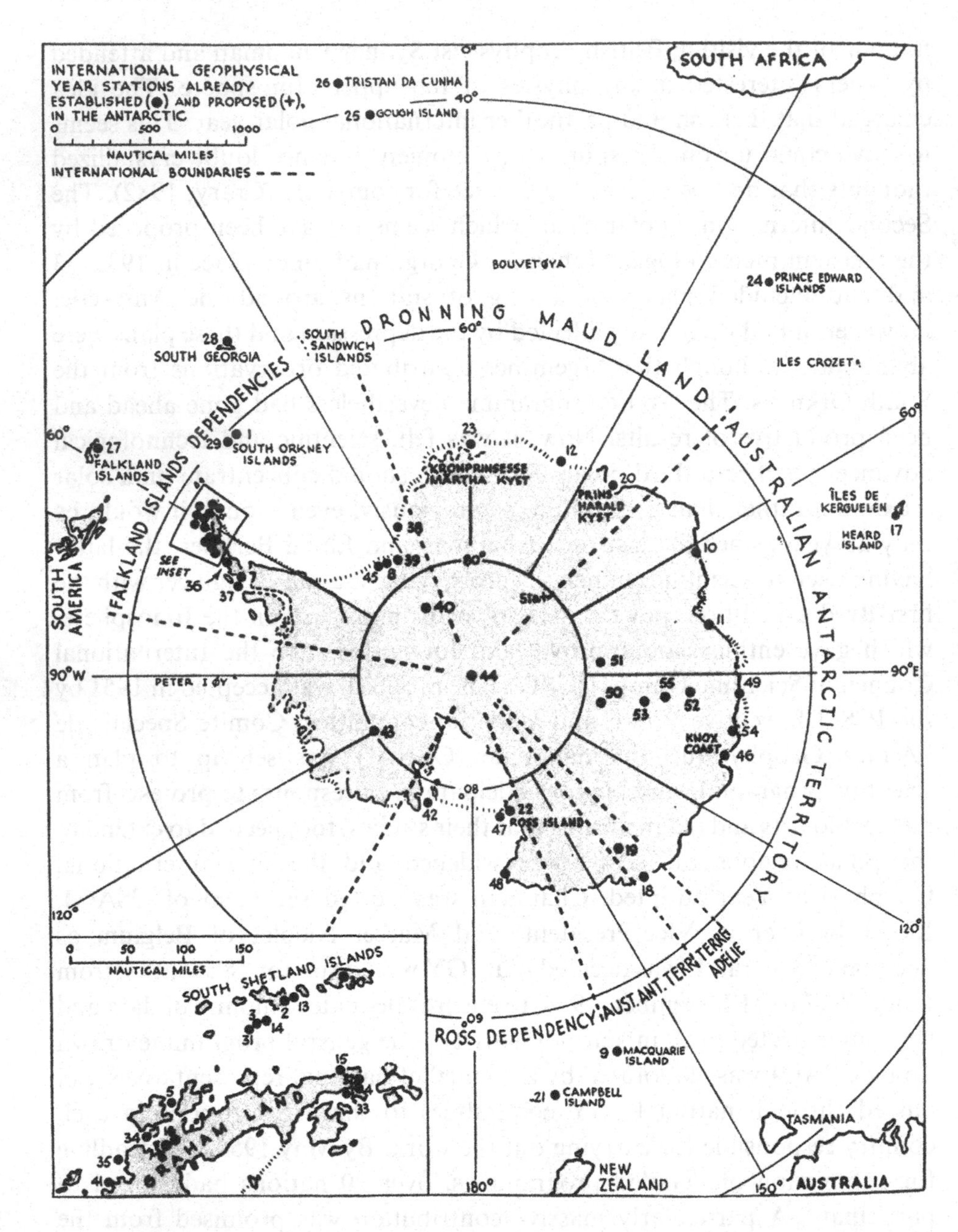

Fig. 6.5 Stations occupied in the Antarctic during the International Geophysical Year, 1957–58. Key: *Argentina*: 1 Orcadas, 2 Teniente Camara, 3 Decepción or Primero de Mayo, 4 Esperanza, 5 Melchior, 6 Almirante Brown, 7 General San Martín, 8 General Belgrano; *Australia*: 9 Macquarie Island, 10 Mawson, 11 Davis; *Belgium*: 12 Roi Baudouin; *Chile*: 13 Arturo Prat, 14 Pedro Aguirre Cerda, 15 General Bernardo O'Higgins, 16 Presidente Gabriel Gonzalez Videla; *France*: 17 Port-aux-Français, 18 Dumont d'Urville,

executive and legislative power (Crary 1982). However, nothing had come from the Soviet Union, evidently because at that time it did not belong to ICSU. Nevertheless, after the death of Stalin in 1953 Soviet scientists had begun to join more freely in international organizations and when CSAGI met in Rome in October 1954 it was notified by the Soviet Embassy there that the Soviet Academy of Sciences would participate. The following year the Soviets joined ICSU. Vladimir V. Beloussov, the chief Russian delegate at the Rome meeting of CSAGI, widened its scope not only by representing geotectonics in a gathering which was largely concerned with the atmosphere but by suggesting that it would be valuable to include in CSAGI countries of extensive area, such as China and India, as well as the Soviet Union. Both Chinas did eventually apply to join and thereby precipitated a crisis since neither would tolerate the presence of the other. CSAGI was firm in maintaining that its members represented their science and not their countries and would not exclude either party. The withdrawal of the Peking Chinese on the eve of IGY saved this intrusion of politics from wrecking the accord that had otherwise prevailed and, although not officially a member, mainland China did, in fact, carry out most of its agreed programme.

So, on 1 July, 1957 this great scientific venture began. It included scientists from 64 countries and involved some 4,000 stations (for those in Antarctica see Fig. 6.5). A programme of World Days, three per month, had been agreed for intensive observations in many related disciplines and there were also to be periodic 10-day Meteorological Intervals. There was to be free exchange of data and three World Data Centres, each to have a complete set of IGY data, had been set up, in the US, in the USSR and the third subdivided between western Europe, Australia and Japan. IGY was carried forward on a great wave of scientific enthusiasm and already hopes

Caption to Fig. 6.5 (*cont.*)

19 Charcot; *Japan*: 20 Syowa; *New Zealand*: 21 Campbell Island, 22 Scott; *Norway*: 23 Norway Station; *South Africa*: 24 Marion Island, 25 Gough Island, 26 Tristan de Cunha; *UK*: 27 Stanley, 28 Grytviken, 29 Signy Island, 30 Admiralty Bay, 31 Deception Island, 32 Hope Bay, 33 View Point, 34 Port Lockroy, 35 Argentine Islands, 36 Detaille Island, 37 Horseshoe Island, 38 Halley Bay, 39 Shackleton, 40 South Ice, 41 Prospect Point; *US*: 42 Little America V, 43 Byrd Station, 44 Amundsen-Scott Station, 45 Ellsworth Station, 46 Wilkes Station, 47 Williams Air Operations Facility; *US/New Zealand*: 48 Hallett Station; *USSR*: 49 Mirny, 50 Vostok, 51 Sovetskaya, 52 Pionerskaya, 53 Komsomol'skaya, 54 Oazis, 55 Vostok-I. From King, 1969, pp. 238–9. (Courtesy of Cassell PLC.)

were being expressed, on both sides of the iron curtain, that it would lead to greater understanding and co-operation in other spheres (Sullivan, 1961, p. 48).

The US programme had from early on included the dispatch of a large-scale expedition to Antarctica – the decision to do so having apparently been taken at the highest level by the National Security Council – but at the Rome conference it was agreed that special effort would be directed to two regions that were now accessible to technology – outer space and the Antarctic. With regard to the latter the conference report said that it represented:

> a region of almost unparalleled interest in the fields of geophysics and geography alike. In geophysics, Antarctica has many significant, unexplored aspects: for example, the influence of this huge ice mass on global weather; the influence of the ice mass on atmospheric and oceanographic dynamics; the nature and extent of aurora australis, for, although the aurora borealis has received considerable attention in recent years, the detailed characteristics of Antarctic aurora remain largely unknown; the possibility of conducting original ionospheric experiments northward from the South Polar Plateau during the long total-night season to determine the physical characteristics of the ionosphere during prolonged absence of sunlight. These and similar scientific considerations lead the CSAGI to recognize that Antarctica represents a most significant portion of the earth for intensive study during the International Geophysical Year. (Sullivan, 1961, p. 31.)

The first CSAGI Antarctic Conference, which was held in Paris in July 1955, was primarily concerned with the distribution of stations and co-ordination of logistics and scientific objectives. In addition to existing stations the establishment or reactivation of 14 more, including a US base at the South Pole itself, were contemplated. It was a historic meeting in that it not only determined the location of research stations, many of which have remained occupied ever since, but that it provided a framework of co-operation and free exchange of information that later developed into the Antarctic Treaty. A happy choice of chairman was made in the election of Colonel Georges Laclavere (see Fig. 6.8), Secretary-General of the International Union of Geodesy and Geophysics. His determination in putting science before politics combined with a forceful manner undoubtedly set the tone for all that was to follow (Sullivan, 1961; Gould, 1973).[2] The first critical situation that he had to face arose from the sensational announcement from the Soviets that they intended to send an expedition to the Antarctic. Hitherto, apart from whaling and intervention when the condominium was mooted, the USSR had shown no interest in Antarctica. Now the USSR proposed to establish three stations; one on the Princess Astrid Coast or the Knox Coast was upsetting to the Australians but otherwise reasonable enough as it completed the pole to pole chain through the centre of the USSR, but another was at the South Pole. This caused consternation; the Soviets, led by Beloussov, had arrived late but presumably knew of the US intentions. However, the threatened collision was averted when Laclavere pointed out that there was a vast gap in IGY coverage in the centre of East Antarctica and Beloussov amiably disavowed any insistence on the South Pole and agreed to take back to Moscow a recommendation for a station at the Pole of Inaccessibility. Another source of disquiet was that the Argentines and Chileans had sent, not scientists, but ambassadors as leaders of their delegations. Laclavere dealt with this when, in opening the first meeting, he ruled that political questions were not to be discussed, a decision to which the two ambassadors agreed. Another crucial agreement was that there should be exchange of personnel between the various national stations. This followed from the proposal that the staff of the *Antarctic Weather Central* at *Little America* should be international and agreement that this should include a Soviet meteorologist. The willingness of Moscow that in return an American could be stationed at the main Soviet base went a long way towards allaying fears and suspicions. A Soviet proposal for mapping of the Antarctic raised some difficulties but was deemed to be outside the remit of IGY. The US decided to curtail its ambitious cartographical programme but the Soviet expedition went ahead with its plans and an impressive atlas eventually resulted (Bakaev, 1966).

At a second conference in September 1955, recognizing that the establishment of so many stations in Antarctica would provide opportunities for useful biological and medical research, these fields were included beside the major physical programmes. Although this ancillary research was thus encouraged there was no attempt to co-ordinate it.

A venture which was not part of IGY but which interacted with it was the Commonwealth Trans-Antarctic Expedition which, under the leadership of Vivian Fuchs and Edmund Hillary, the mountaineer, traversed Antarctica from the Weddell Sea to the Ross Sea via the South Pole in 1957–58. The journey of 2,158 miles (3,473 km) was done in 99 days. A powerful motivation was to make Shackleton's dream of crossing the continent a reality at last – a politically attractive achievement for the UK – but it carried out a series of gravity determinations and seismic soundings of ice depth. It was also the means of bringing New Zealand into IGY. The second meeting of CSAGI had called on New Zealand to establish a station in the Ross Sea area but its government had not responded. Up to 1955 it had been understandably reluctant to get involved in the problems associated with sovereignty over the Ross Dependency and there had been little attempt to enforce even a paper control over the area (Logan, 1979). However, public opinion in New Zealand, focused by the Antarctic Society founded in 1933, was behind their hero Hillary and the Trans-Antarctic Expedition and the government eventually agreed to contribute towards this whilst remaining evasive about IGY. However, New Zealand scientists, for whom territorial aspirations were a minor consideration, realized that there was more future for them with IGY. When the US announced its intention of using a base in McMurdo Sound as a staging post and suggested a joint US–New Zealand base at Cape Adare, the government accepted with alacrity, realizing that here was a not-too-expensive and face-saving way both of making some assertion of its sovereignty and meeting its obligations both to the Trans-Antarctic Expedition and IGY (Quartermain, 1971; Logan, 1979). 'TAE was the swan song of the old imperial Antarctic dreams begun in 1920. However, it, and not the IGY, had given the initial impetus to get New Zealand into the Antarctic and provided the means to stay there' (Logan, 1979).

The other nations taking part in Antarctic IGY activities were more forthcoming with support and, of course, the resources poured into the Antarctic by the US and the USSR were massive. The total cost to all nations has been estimated at well over US $280 million (Law, 1959). The IGY expanded scientific activity in the Antarctic enormously and in many areas welded it firmly into the general corpus of science. This was particularly evident where physical investigations required data from multiple simultaneous observations made at many points on the globe, as in aurora and air-glow studies and ionospherics, which occupy 10 and nine volumes respectively of the IGY Annals (there is only one volume each for glaciology, oceanography and meteorology). A major conclusion was that East Antarctica is a continental shield whereas West Antarctica is an archipelago, and, at last, an estimate could be made of the amount of Antarctic ice. Biologists, envious of the success of IGY, were stimulated to enter into similar large-scale international co-operation and devised the International Biological Programme (IBP), again under the auspices of ICSU. This programme, which ran for 10 years from 1964, was concerned with the biological basis of productivity and human welfare, with basic research into how natural, man-modified and man-made ecosystems make productive use of solar energy, and into the physiological and genetical adaptations of man to different climates, altitudes and diets (Clapham,

1976). Antarctic research entered in to this to a marginal extent only, although the IBP Bipolar Botanical Project made a valuable comparison of plant productivity in the Arctic and sub-Antarctic (Callaghan *et al.*, 1976).

The scientific results were valuable but of even greater moment were the political consequences of IGY. There was no doubt that in its general aspects as well as in the Antarctic context it was a resounding success in the eyes of scientists and tremendously impressive for nonscientists. The question now arose of whether the research and international co-operation initiated by IGY should continue after 1958 – as, indeed, had already been envisaged by some before it had started (Crary, 1982). At an *ad hoc* meeting in September 1957, the US was in favour of continuation but there was vigorous opposition to this until the USSR delegate announced his country's intention of maintaining its stations in the Antarctic. The question was settled at a meeting of CSAGI held in Moscow in July 1958. There were first some tricky problems: the USSR refused to commit themselves to automatic dispatch of data about their satellites to the World Data Centres; unpleasantness arising from the apparent exclusion of the Taiwan Chinese from the meeting subsided only when it turned out that they had presumed that they would not be granted visas to enter the USSR and had not, in fact, applied for them; and there was censorship of news dispatches by visiting science reporters until Beloussov took action to stop it (Sullivan, 1961). However, it was the Soviets who proposed that IGY should be extended, the reasons given being that a number of tasks had not been completed, that the great investment in stations in Antarctica had as yet given only a small return, and that machinery was needed to provide for permanent international co-operation in geophysics. Newell & Townsend (1959) in commenting on this said:

> it appeared that the Soviet scientists need the IGY name and organization to help them maintain their position at home and their outside contacts and the freedom of intercourse that has been achieved so far. Apparently this need is a very urgent and demanding one in the opinion of the Soviet geophysicists and solar physicists.

This positive attitude of the Soviet scientists carried the day but some compromise in wording was needed to placate the governments of the US and other nations which had been given a definite understanding that IGY would terminate on the 31st July, 1958. A further year of IGY activity, to be known as the Year of International Geophysical Cooperation, was agreed upon and two months later ICSU endorsed this and decided to put IGY co-operation on a permanent basis. For Antarctic investigations this had the important effect of maintaining impetus and giving support to SCAR, the Special (later to be Scientific) Committee on Antarctic Research, which had been formed following a US suggestion at the fourth IGY Antarctic Conference, held in Paris in June 1957. SCAR had held its first meeting at The Hague in February 1958, its membership including representatives of the nations active in Antarctica, of ICSU and five of its scientific unions.

6.9 The Antarctic Treaty

Just before the Moscow meeting of CSAGI, in May 1958, the US government had circulated a note to the 11 other nations working in Antarctica proposing a treaty that would reserve the continent for scientific activities. For the US it was appearing that internationalization would serve its interests best, enabling continent-wide deployment of its technological superiority, whereas laying claim to a sector would restrict it (Berkowitz, 1986). Similar suggestions to avoid the persistent problems presented by territorial claims had been floated before but now there was the feeling that the moratorium on political argument that had prevailed during IGY might be maintained. The UK was now in favour of internationalization (although not under unpredictable United Nations auspices), having failed in its attempt to have its dispute with Argentina resolved by the International Court of Justice. This view had been elaborated by Prime Minister Harold Macmillan during a visit to Australia and New Zealand in February 1958. The US invitation was promptly and unanimously accepted and the first meeting of delegates was held in New York, in the board room of the National Science Foundation, on 13 June, 1958, under the chairmanship of the leader of the US delegation, Paul C. Daniels, a retired ambassador. It took 60 meetings spread over more than a year to arrive at an agreement acceptable to all parties. Meetings were chaired by the different nations in rotation. Because of the sensitiveness of the issues, great secrecy was maintained but memoranda on the meetings have been found among the papers of Admiral George Dufek (Beck, 1985), who commanded *Operation Deepfreeze*, and Daniels (1973) has written an account of the proceedings. Delegates came with varied ideas of what was to be achieved and for a time the Soviets adopted an intransigent attitude, maintaining that the purpose of the meetings was only to determine the date, place and procedures for a conference that was to follow. Eventually, Daniel's patience and conciliatory attitude prevailed, aided by much behind-the-scenes diplomatic activity – it was in this way, for example, that the ticklish problems existing between Argentina, Chile and the UK were kept from hampering the discussions – and finally a sudden transformation in the Soviet attitude (Quigg, 1983, pp. 146–7) enabled an agreed approach. The result was a draft for an Antarctic Treaty, the main object of which was to promote the peaceful use of Antarctica and particularly to facilitate scientific research in the area. It was proposed that territorial claims should be frozen for the period of the Treaty but that nothing in the Treaty should be interpreted as depriving any party of a claim or, on the other hand, as recognition of a claim. Military activity and testing of any kind of weapons were to be prohibited within Antarctica. On the positive side, information regarding scientific programmes, observations and results was to be exchanged between the contracting parties. Scientific personnel was also to be exchanged between expeditions and all areas of Antarctica and installations and equipment within them were to be open to inspection at all times. The possible involvement of the United Nations was felt to risk unwanted complications and, in the first place, the Treaty was to include only the 12 governments participating in the preparatory meetings.

It was fortunate that these preparatory meetings took place in a brief period of reduced East–West tension (Beck, 1985). Otherwise there would have been a real possibility that strategic considerations might have prevailed, as they already had in the Arctic, and the Antarctic become a testing ground for nuclear weapons. A major determinant for success was that neither the US nor the USSR, having established presences on the continent, was prepared to withdraw and leave the field to the other. Another factor was that the seemingly intractable British/Latin American problem was by-passed. The formal conference took place in Washington on 15 October, 1959 and the 12 governments concerned signed the Antarctic Treaty on 1 December, 1959. After ratification of the agreement by all the signatories the Treaty came into force on 23 June, 1961. Meanwhile the U-2 spy plane affair had precipitated the breakdown of the 1960 summit conference and the truce in the cold war was over.

The text of the Antarctic Treaty is given in many recent books (e.g. King, 1969; Quigg, 1983; Walton, 1987; Mickleburgh, 1987) and its operation and effects have been discussed in several others (Auburn, 1982; Beck, 1986; Polar Research Board, 1986; Triggs, 1987; Parsons, 1987). We need not give the politicians who framed it any great credit for altruistic desire to further science but they had the acumen to see that science offered a means of avoiding the problems which were besetting the Antarctic. The US government attitude was summed up by Henry Dater of the State Department in July 1959:

> Because of its position of leadership in the Free World, it is evident that the United States could not now withdraw from the Antarctic . . . national prestige has been committed. Our technical capabilities so frequently challenged in recent years are on trial in the Antarctic questions quite as much as in space . . . Our capacity for sustaining and leading an international endeavour there that will benefit all mankind is being watched not only by those nations with us in the Antarctic but also by noncommitted nations everywhere. Antarctica simply cannot be separated from the global matrix. Science is the shield behind which these activities are carried out. (Quoted by Beck, 1986, p. 64.)

In practice there need not be much difference between politics, the art of the possible, and science, the art of the soluble (Medawar, 1984). In that it made difficult problems soluble by presenting a way of tackling them, the Treaty is a thoroughly scientific document. The concept of sovereignty is meaningless and unenforceable, besides being scientifically irrelevant, in the vast inhospitable wastes of Antarctica and by implicitly recognizing this the Treaty circumvented the main obstacle in the way of progress. This 'imaginative juridical accommodation' (Scully in Vicuña, 1983) stopped the legal clock at 1961 and as with most stopped clocks it has become unnoticed. Those involved at the political level, however, are still double-tongued, emphasizing the strategic importance of Antarctica to justify funding but playing this down for Treaty purposes and there has been a distinct inverse correlation between the extent of military involvement in logistics and the scientific output of certain parties to the Treaty (Beck, 1986).

It is significant that a man who was perhaps more involved in the drafting of the Treaty than anyone else, Dr Brian B. Roberts (Fig. 6.6) of the British Foreign Office, was by training a scientist and had participated in the British Graham Land Expedition of 1934–37 (Law, 1979). A predilection of scientists is evident in the minimum of bureaucracy with which the Treaty has so far been administered. No provision was made for a permanent institution or secretariat and its business has been conducted by a series of Consultative Meetings held in turn in the different participating countries. In addition to regular general meetings there have been ones on special topics such as telecommunications and mineral exploitation. The recommendations of these meetings do not become effective until ratified by all the Consultative Parties and as one participant, Heap (quoted in Beck, 1986, p. 155) has observed, 'The ultimate power of an Antarctic Treaty State lies in refusing to take part in a consensus under which its freedom would be restricted. It cannot alter the *status quo* in its favour by refusing a consensus; the *status quo* can only be changed if all parties agree

Fig. 6.6 Dr Brian Birley Roberts, from *Polar Record,* **19**, p. 399. (Courtesy of the Scott Polar Research Institute.)

that it should be changed'. This framework has not only facilitated co-operation but has provided scope for evolutionary development to deal with new problems such as those of conservation and exploitation of resources. There has been growth in membership as more nations have begun scientific investigation in the Antarctic – Poland (1977), West Germany (1981), Brazil (1983), India (1983), China (1985), Uruguay (1985), East Germany (1987), Italy (1987), Spain (1988) and Sweden (1988) becoming Consultative Parties. The readiness of established Antarctic powers to pass on their experience to these newcomers – the BAS, for example, gave much advice to the West Germans – has undoubtedly helped to preserve unity and the 'Antarctic spirit' to which something of the success of the Treaty must be attributed (Beck, 1986, p. 177).

The Treaty organization has been criticized as an exclusive and secretive club but if science is to be its main business it is necessary that decisions should be taken by those with knowledge and practical experience and not by open debate in which ideology would inevitably prevail. Thus the idea that Antarctic affairs should be the responsibility of the United Nations has been alarming for the Treaty powers, especially when India, a non-Treaty state, began to be interested in the Antarctic in the early 1980s. However, the accession of India to the Treaty has given some assurance that non-parties will respect the principles of the Treaty, reducing the likelihood of a United Nations intervention in Antarctic affairs, and meetings of the Consultative Parties in the Treaty have been opened to non-consultative Acceding States, without extensive research commitments in Antarctica, of which there were 17 by 1989 (Headland, 1989). A different test came in 1982 when war broke out between two parties to the Treaty – Argentina and Britain. Both observed the provisions of the Treaty and hostilities did not extend south of 60° S, the limit of the Treaty area, and, indeed, delegates from the two countries encountered amicably in Hobart, Tasmania, while the war was on, to participate in a meeting of the Treaty Commission for the Conservation of Antarctic Marine Living Resources (CCAMLR). The BAS, a civilian organization, was placed in a difficult situation by the war. It provided the armed forces with intelligence and transport by sea outside the Treaty area but had to strike a delicate balance between patriotic inclination and not compromising its position *vis-à-vis* the Treaty (Perkins, 1986).

It has generally been thought that the ultimate test of the Treaty will come when economically exploitable mineral deposits or oil reserves are found in Antarctica. A strength of the Treaty organization has been that such problems can be considered in advance and in 1988 a mining convention (CRAMRA) was agreed. This would have allowed limited and tightly controlled exploitation of resources and the indications were that it would not have been necessary to put it into operation in the foreseeable future, there being no real evidence that economically useful reserves exist in Antarctica. Nevertheless, two Consultative Parties, France and Australia, under pressure from conservationists had second thoughts about agreeing to it. Scientists proved less adept at political lobbying than Greenpeace and similar organizations, other Consultative Parties followed suit and CRAMRA was not ratified. Instead at the Madrid Antarctic Treaty meeting in June 1991 it was agreed that there should be a moratorium for at least 50 years on all exploitation of minerals and oil in the Treaty area. The six years of delicate negotiation that went into CRAMRA were perhaps not altogether wasted since much of its substance was quarried to put together the Madrid Protocol in only three meetings. In itself the prohibition of mining must be welcomed – except perhaps by those Acceding States which have spent considerable funds in setting up Antarctic research facilities in the expectation of economic returns – and fears expressed by Laws (1989) and Blay & Tsamenyi (1990), that failure to agree on CRAMRA might endanger the Treaty through destruction of the

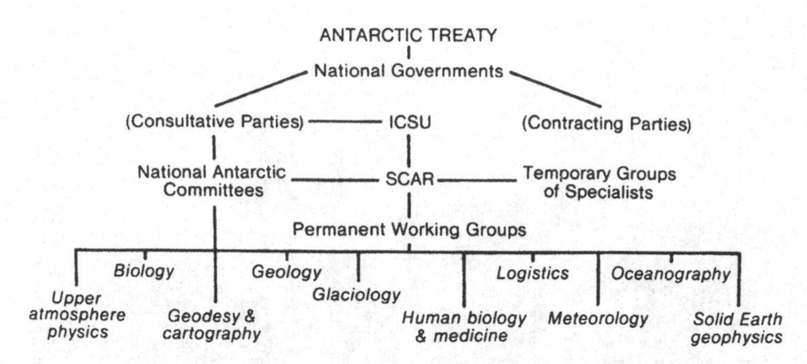

Fig. 6.7 The relationships of the Antarctic Treaty organization with the Scientific Committee on Antarctic Research and its various committees in 1987. (Courtesy of D. W. H. Walton, the British Antarctic Survey.)

consensus approach and leave Antarctica open to unregulated exploitation, have not been realized. The abandonment for political reasons of a carefully constructed and scientifically based agreement does not, however, bode well for the future.

6.10 The Scientific Committee for Antarctic Research

Although there is no formal connexion, the Treaty organization has been advised on scientific matters from the start by the Scientific Committee on Antarctic Research (SCAR), a non-governmental body organized by ICSU, and a mutually supportive relationship has developed. A useful oversight of the operation and achievements of SCAR has been given by Fifield (1987). As of 1987, eighteen nations were full members of SCAR, appointing their delegates through their national Antarctic committees, and there are also representatives of interested scientific unions (Fig. 6.7). SCAR's headquarters are in the Scott Polar Research Institute in Cambridge but the biennial meetings are held in rotation in the different member countries with the Presidency also representing different nations (Fig. 6.8). The bulk of the work has been carried out by permanent working groups dealing with topics such as biology, geodesy and cartography, geology, glaciology, human biology and medicine, logistics, solid earth geophysics and upper atmosphere physics (Fig. 6.7). Matters that did not fall easily within the remits of working groups, such as climate research, sea-ice, seals, Southern Ocean ecology, structure and evolution of the lithosphere, and Cenozoic palaeoenvironments, have been left to groups of specialists.[3] Antarctic oceanography has been incorporated into the remit of another ICSU committee, the Scientific Committee for Oceanic Research (SCOR). Routine synoptic meteorology is dealt with by the World

Fig. 6.8 The first four Presidents of the Scientific Committee on Antarctic Research; left to right – T. Gjelsvik (Norway) 1974–78; G. de Q. Robin (UK) 1970–74; L. M. Gould (US) 1963–70; G. R. Laclavère (France) 1958–63. (Courtesy of G. de Q. Robin.)

Meteorological Organization's working group on Antarctic meteorology and the scientific aspects by the International Commission for Polar Meteorology. This organization does not preclude the formation of specialist groups as the need arises. A notable example of how effective such groups can be is provided by the Group of Specialists on the Living Resources of the Southern Ocean. This was initiated under SCAR auspices but later was co-sponsored by SCOR and launched a major international programme under the title of Biological Investigations of Marine Antarctic Systems and Stocks, evidently chosen to give the handy acronym BIOMASS. BIOMASS has been largely centred on krill with the object of establishing the role of this exploitable crustacean in the ecosystem of the Southern Ocean. Its programmes of cruises, extending over 10 years, have involved as many as 16 vessels belonging to 12 nations, all contributing data to a central data base located in the BAS's headquarters (see also

p. 241). SCAR publishes a regular bulletin as well as reports on special topics.

Initially the countries subscribing to the Treaty and belonging to SCAR were the same, the custom of SCAR providing scientific advice was established early and the two organizations have evolved in parallel but with their respective meetings deliberately kept out of phase. The Treaty parties have thus been able to draw directly on the experience of the world's leading experts on Antarctic science and logistics who in turn have been able to exert considerable influence on the political decisions taken. All this has been accomplished by SCAR on a modest budget with the host countries paying the cost of its meetings and delegates providing their own travelling and subsistence expenses. There are undoubtedly shortcomings: Bonner (1988) has commented that priorities agreed by SCAR working groups are not invariably followed up, delegates are less able to commit their country's resources to particular programmes than they were when IGY was being planned, and the conflicting styles of institute-based research and programmes arising from peer-reviewed applications are difficult to reconcile.

6.11 National Antarctic research organizations and operations

At the time of writing (1990) there is planning of research programmes, exchange of scientific personnel and pooling of data, at the international level but nevertheless the actual scientific work carried out in Antarctica is still largely based on national organizations. With the exception of the Greenpeace base all Antarctic stations have been national (Fig. 6.9). To some extent international programmes may be contrived and artificial (Quigg, 1983, p. 217) and national facilities can give the greatest scope for innovative research. Different nations have widely different arrangements. We have seen something of the ways in which these evolved in the US and in the UK. The USSR, which launched itself abruptly into large scale Antarctic investigations, had a pre-existing institute for Arctic research with administrative and logistic support which had already sent nearly 300 expeditions north before 1945, on which to build. A British observer, Swithinbank,[4] who spent over a year with the Ninth Soviet Antarctic Expedition in 1963–65 and visited the Arctic and Antarctic Research Institute in Leningrad noted that there was almost complete standardization of equipment and that Novolazarevskaya, where he spent 11 months, was scarcely distinguishable from many Arctic stations. It should be noted, however, that the Soviet approaches to Arctic and Antarctic research have been different in one respect. The former has been

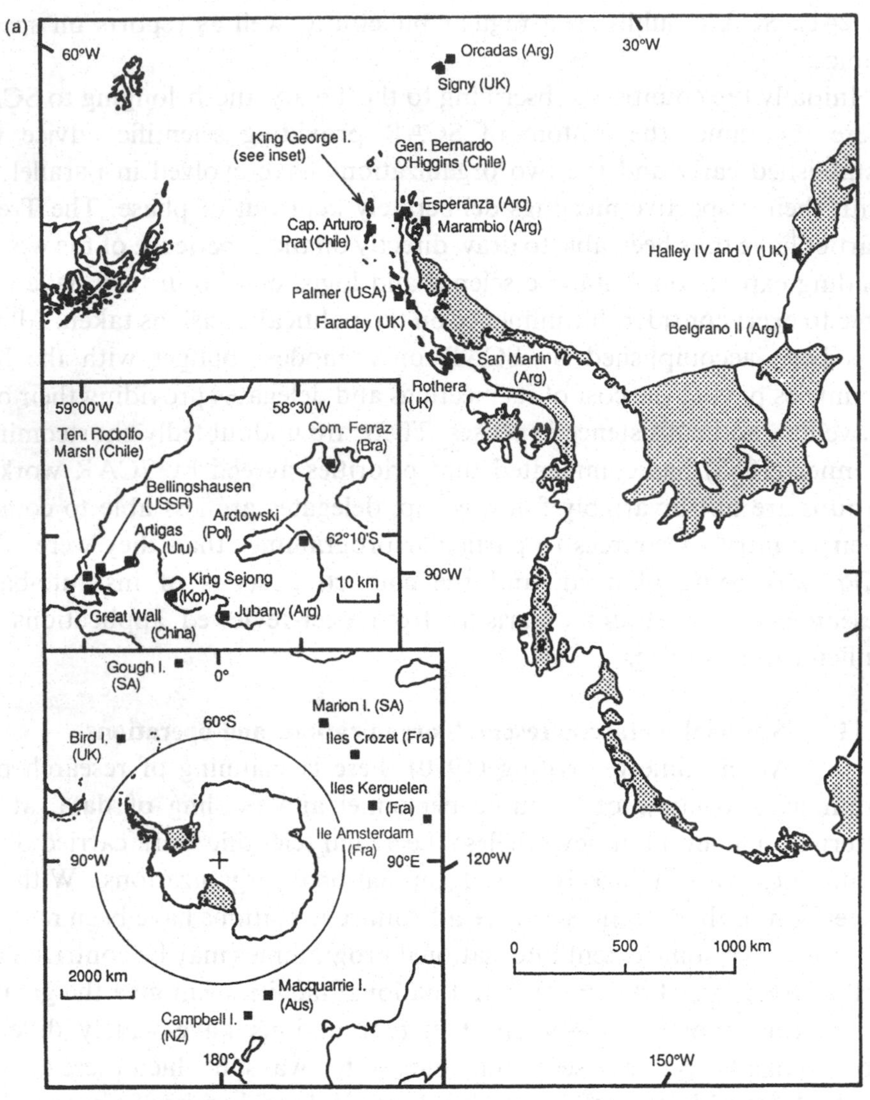

Fig. 6.9(a), (b) Stations operating in the Antarctic, winter 1991. (Courtesy of the Scientific Committee on Antarctic Research.)

heavily resource orientated and restricted by security considerations whereas the latter has been based on freedom of access to all parts of the continent and circumpolar waters, and regulation of resource exploitation by the Treaty organization has been accepted (Joyner, 1988).

It would not be profitable to discuss individual organizations in detail. Holdgate (1964a) and King (1969) have given general descriptions of the national agencies and Beck (1984, 1986) management diagrams for the contrasting organizations of the BAS and Brazil, but examples may be

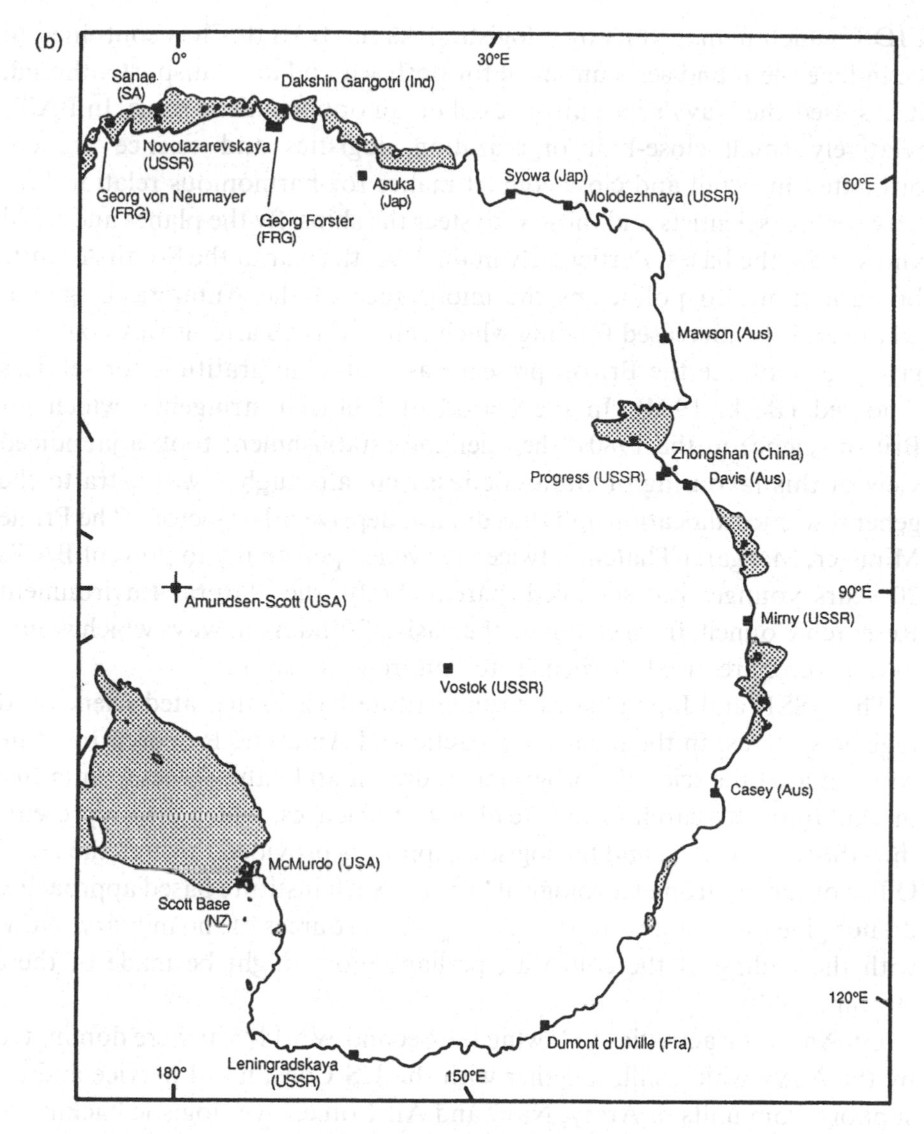

given of the different ways of dealing with two main problems, namely the provision of logistic support and the input of scientific ideas. Traditionally the former has usually been supplied by a navy and from the time of Halley onwards there have been frequent disagreements between scientists and professional sailors. This is illustrated in distressing detail in Mawson's BANZARE diary (Jacka & Jacka, 1988). We have seen how in the UK the Royal Navy regarded polar exploration as its special province until in the heroic age government parsimony left expeditions to private enterprise. The Royal Navy was responsible for *Operation Tabarin* but as soon as the Second World War was over this was succeeded by a civilian organization,

FIDS, which managed its own logistics. As the BAS this has continued to be independent and self-contained for both sea and air transport although it has used the Navy's ice patrol vessel on an opportunistic basis. In BAS's relatively small close-knit organization, logistics and science are co-ordinated in detail and close contact makes for harmonious relationships between the scientists and those who steer the ships, fly the planes and build and supply the bases. Particularly in the UK, the war in the South Atlantic brought home to politicians the importance of the Antarctic in global affairs and the increased funding which came to BAS after it was no doubt given to reinforce the British presence as well as in gratitude for services rendered (Beck, 1984). In the period of financial stringency which hit British science in the 1980s, the scientific establishment took a jaundiced view of this favouring of Antarctic research, although it was extra to the general science allocation and thus did not deprive other sectors. The Prime Minister, Margaret Thatcher, twice intervened personally to prevent BAS's 20 years younger but so-called 'parent' body, the Natural Environment Research Council, from changing the basis of funding in ways which would have severely reduced the benefit to Antarctic research.[5]

The USSR and Japan likewise run institute-based integrated science and logistic systems. In the former the Arctic and Antarctic Research Institute co-ordinated the scientific programmes drawn up by the Soviet Committee on Antarctic Research of the Academy of Sciences, which also represents the USSR on SCAR, and the logistic support is provided through the Head Office of the Hydrometeorological Service. Such institute based approaches do not give direct access to the vast logistic resources of the military. Now, with the ending of the cold war, perhaps more might be made of these (Robin, 1989).

US Antarctic activities following the Second World War were dominated by the Navy which still, together with the US Coastguard Service and air support from units of Army, Navy and Air Force, gives logistic backing to the US Antarctic Research Program, USARP. The US having such a multifarious collection of agencies dealing with the Antarctic, moves were made in the early 1960s to rationalize the position and create a unified agency but were twice defeated in Senate (Berkowitz, 1986). In 1979 the National Science Foundation gained full responsibility for funding and management and the right to hire private contractors to run the stations but not to organize logistics. This system has been much criticized by American scientists. Siple (1959, pp. 101–7) and Parfit (1988, p. 173) have mentioned some of the tensions and difficulties which have developed between scientists and the military. Military influence has declined but it seems that there is still an ultimate precedence of political presence over science (Parfit, 1988, p. 171). The involvement of three disparate organizations in running the operations inevitably produces bureaucratic complexity and inefficiency. Edward P. Todd, a Director of the National Science Foundation's Division of Polar Programs, is reported as saying 'Any graduate of a management school would turn green at this operation, not with envy but with nausea'. It works, he added, because many of the participants want to make it work (Parfit, 1988, p. 215). Berkowitz (1986) sees USARP as an entrenched bureaucracy, an organization of tremendous capacity struggling against internal squabbling, waste and inefficiency. He records a fall in the proportion of the US Antarctic budget spent on science from 100 per cent in 1953 to around 10 per cent in the mid-1980s, the comparable figure for other countries being about 20 per cent – although it should be remembered that many countries hide part of their expenditure on Antarctic research in military budgets so that cost analysis is difficult. At the same time the ratio of support staff to scientists in the summer season in US Bases rose to an unreasonable 20:1. Long before that a British visitor to McMurdo (Green, 1965) had commented on the separation between the scientists, who came mainly from the universities, and those who provide the means of getting to and existing in Antarctica. This results in an expensive way of doing things, with field workers having little in common with the base personnel, who tend to cocoon themselves against the alien environment. Llano (1988) has lamented the lack in the US of a national polar centre, an independent focus which would give continuity amid the continual turnover amongst USARP personnel. Such criticisms should not lead one to overlook the massive contribution which the US has made to Antarctic science and the great satisfaction which individual scientists get from working with USARP. A good picture of the serendipitous inception and daily operations of a programme on the comparative biochemistry of the proteins of Antarctic birds and fish has been given by Feeney (1974). This programme, like many others originated by individuals, developed in successive seasons and continued to be funded over a long period. A further point to be remembered is that the major projects sponsored by USARP have usually involved many different institutions and nationalities.

New Zealand is closely associated with the US in Antarctic matters and follows a similar pattern of organization. Research programmes are drawn up by the Ross Dependency Research Committee, which is appointed by the Minister-in-Charge of Scientific and Industrial Research. Detailed planning and logistic support are provided by the Antarctic Division of the

Department of Scientific and Industrial Research and transport by the Royal New Zealand Air Force and the Americans. Both Argentina and Chile rely on their armed forces for logistic support (see the comment by Parfitt, 1988, p. 270).

Similarly, there are great differences in the ways in which scientific programmes are generated. Until 1984 ideas came, to a large extent, from within the BAS, subject to approval and modification between 1971 and 1984 by a Scientific Advisory Committee, appointed by the Survey, of senior scientists, most of whom had Antarctic experience, and including representation from the Scott Polar Research Institute, Meteorological Office, Foreign Office and other bodies. This Committee presided over a period of expansion in which the output of research papers rose from 25 to 200 per year and preeminence was established in fields such as the atmospheric sciences. After 1984 the Natural Environment Research Council, the funding body, assumed more control, bringing in a wider range of expertise in formulating programmes but diluting the polar experience (a parallel instance is the Planning Committee of the Australian National Antarctic Research Expedition, a useful but not officially appointed body of experts, which was allowed to 'wither on the vine' by government bureaucrats (Law, 1983)). University scientists had always been able, through informal contacts, to participate in Antarctic research but already before 1984 more university participation had been invited and funded by BAS itself. Britain is represented on SCAR not by BAS but by the British National Committee on Antarctic Research, a separate body which has, however, some common membership with BAS committees, organized by the Royal Society. In the US the National Science Foundation is responsible for USARP. The general nature of the programme is overseen by the National Academy of Sciences through its Committee on Polar Research, which also represents the US on SCAR, and is advised by the Antarctic Policy Group, a government body including representatives from the Departments of State and Defense. The National Science Foundation, an independent Federal Government agency, works through its Office of Antarctic Programs which funds and administers research put forward by universities, government agencies and private corporations. Comparing the British and American systems a broad generalization would be that in the former the field work is mostly carried out by young graduates and is best suited to long-term programmes requiring year-round observations while the latter depends on a much greater proportion of senior scientists and is most appropriate for short-period intensive work. Other countries have more or less complex variations on these themes. France, the German

Federal Republic, Japan and Norway have organizations resembling that of Britain in so far as they have quasi-independence of the state whereas in Argentina, Australia, Chile, New Zealand, South Africa and the USSR there is more direct government control of scientific programmes.

The particularly close association of the US and New Zealand in the Ross Sea area calls for a little more comment. The essential basis is the provision by New Zealand of a staging post for the Americans in Christchurch in return for which the US provides some transport to and from the Antarctic and the ferrying by plane or helicopter of field parties within Antarctica. A reliable facility is thus traded against a service which is subject to the vagaries of the weather and therefore not altogether predictable and often in short supply. This arrangement will clearly only work well when good sense is shown by both parties. The leader of New Zealand's Scott Base has to be something of a diplomat. There may be occasional stirrings of patriotic resentment at the presence on territory, to which New Zealand lays formal claim, of an expanding US township. The attraction of the bright lights of that township, only two miles (3 km) away and served by a shuttle each afternoon and evening, has given rise to problems of another sort in Scott Base (Hayter, 1968).

An important recent development has been the establishment of a council of managers of national Antarctic programmes (MNAP).[6] To some extent this has taken over the function of the SCAR Working Group on Logistics, which has consequently been reconstituted with a modified remit as the Standing Committee on Antarctic Logistics and Operations (SCALOP). The remit of MNAP is wide, including finance and matters of general concern such as international programmes and tourism. A particularly valuable function is to establish personal contacts between managers which, among other things, will facilitate action in cases of emergency.

The various national organizations concerned with Antarctica are thus various and each complicated in its own way with different constraints in ideas, money and equipment. All this has to be taken into account when proposals for international programmes are drawn up.

6.12 Private expeditions

Private expeditions to the Antarctic have not ceased in spite of the dominance of the state-supported programmes, and many of those which have taken place have included scientific projects as a justification in their prospectuses. Among these we may note the British Joint Services expeditions led by Commander C. Furse to Elephant Island (1976–77) and Brabant Island (1983–84), the voyage of Dr David Lewis and his companions in an expedition mounted by the Oceanic Research Foundation to Macquarie Island, the Balleny Islands and Cape Adare (1977–78), the Transglobe Expedition's crossing of the continent under the leadership of Sir Ranulph Fiennes (1980–81), the *In the Footsteps of Scott* expedition (1984–85) and the 90° S expedition led by Dr Monica Christensen to retrace Amundsen's route to the Pole. Such expeditions have been made primarily for adventure but have nonetheless contributed sound and useful science. The 90° S expedition, for example, carried out studies on the flexural bending of the Ross Ice Shelf, microwave reflection and visible albedo of snow and ice and, incidentally, had the good sense to abandon adventure when it would have landed them in real difficulties (McIntyre, 1988). It is unlikely, however, that really innovative or substantial work can result when science is a secondary consideration, and beside the massive outputs of the state research organizations the results achieved seem meagre. Nevertheless, the establishment of the first commercial airline, Antarctic Airways, operating direct flights from South America to the continent means that there is the possibility that scientists without access to permanent bases can now do effective work in Antarctica (Swithinbank, 1988b).

6.13 The politics of conservation

The politics of conservation are increasingly having an impact on Antarctic science. In earlier chapters we noted flickers of concern about man's impact on the wildlife of Antarctica but even as late as the 1950s the general assumption amongst scientists was that the Antarctic was there to be 'developed' (e.g. Simpson, 1952). Nevertheless, at a time when the idea of conservation had not yet got its grip on the public imagination the Treaty powers were aware of the need for it although the code of conduct embodied in the Agreed Measures for the Conservation of Antarctic Fauna and Flora of 1964 now look rather naïve (Bonner, 1987a). In the same year a pilot sealing expedition from Norway brought realization of a potential threat to seal stocks and led to the 1972 Convention for the Conservation of Antarctic Seals. There was recognition of the need for general conservation measures and wide discussion amongst biologists of the basis on which they should be drawn up (Parker, 1972). The start of krill exploitation rang another alarm bell which resulted in an ambitious piece of Treaty legislation, the Convention for the Conservation of Antarctic Marine Living Resources (CCAMLR) signed in 1980. Based on a sound appreciation of the necessity of viewing the marine ecosystem as a whole this was scientifically unexceptionable but probably ineffectual in practice (Bonner 1987a). The agreement on a convention for regulation of mineral exploitation, already discussed above, would have completed the array of legislation for conservation if it had been signed. It is a merit of all this legislation that it has been drawn up before serious damage has been done. SCAR, of course, has played a major role in the forming of these conventions. The International Union for the Conservation of Nature (IUCN) has also been interested in Antarctica and in 1985 held a joint discussion meeting with SCAR. IUCN is largely a political body whereas SCAR, of course, is scientific and this meeting gave useful contact between international lawyers and those with Antarctic experience (Bonner, 1987a). Strain imposed on the Treaty system by conservation problems is exacerbated by popular concern. The misgivings which some conservationists have about the activities of Treaty powers have some justification; McMurdo Station is a source of appreciable pollution and Argentina's Esperanza and China's Great Wall stations are notably surrounded by litter (Berkowitz, 1986, see also p. 392). Such concern led to the setting up in 1987 of a quasi-permanent non-governmental base in Antarctica by the conservation organization Greenpeace. This was preceded by Greenpeace's Antarctic Declaration of 1984, which follows the Antarctic Treaty in many respects but would totally prohibit exploitation and has the aim of preserving Antarctica as a wilderness area (Mickleburgh, 1987). Dr David Bellamy, a botanist and conservationist, was proclaimed Greenpeace's first British 'Ambassador' to the Antarctica World Wildlife Park on 1st February 1986,[7] (Fig. 6.10). An intention that Greenpeace should carry out research in Antarctica was secondary to the political objective; the establishment of a research base in Antarctica might provide a basis for an application from Greenpeace to become a party to the Antarctic Treaty. However, although its aims have the sympathy of many Antarctic scientists its oversimplification of issues and confrontational tactics are looked on with less enthusiasm. The conservationist point of view has wide support amongst a variety of non-governmental organizations and their proposals, summarized by Bonner (1987a), tend to the conclusion that responsibility for the Antarctic should be transferred to the United Nations. As Laws (1985) has pointed out, the record of the United Nations agencies in conservation is in sorry contrast to the cost-effectiveness and success of management of Antarctic affairs under the Antarctic Treaty and Bonner (1987a) saw the greatest threat to the Antarctic coming not from the exploiters but from those claiming to be wholly dedicated to environmental conservation. Some rapprochement and a toning down of environmental protest was achieved by the Workshop on the Antarctic Treaty System sponsored by the US Polar Research Board which was held at the head of

Fig. 6.10 David Bellamy, botanist and conservationist, declared Greenpeace's first British 'Ambassador' to the Antarctica World Wildlife Park, 1 February 1986. (Courtesy of Times Newspapers.)

the Beardmore Glacier in January 1985 (Polar Research Board, 1986). As Schatz (1988) has commented, war dwarfs other environmental offences and the Antarctic Treaty is the best guarantee that the continent will never be used for military purposes. Such mutual forbearance between nations as the Treaty has brought about is not easily renegotiated and nurturing the Treaty system is far more likely to be environmentally protective than an idealistic but unrealistic substitute regime.

6.14 The problems of emergencies

The established Antarctic research organizations are more and more worried by a related problem – the moral responsibility that they have to assume for the safety of private expeditions, conservationist forays and increasing numbers of tourists that visit the Antarctic. The tradition of mutual help between expeditions from different countries is a strong one and emergencies have always called forth prompt and unstinted assistance in the field from those in a position to give it. The crash of the Air New Zealand DC10 tourist flight on Mount Erebus in November 1979, in which 257 persons were killed, illustrated in a terrible way how great the responsibility was becoming towards those who have no traditional or legal right to expect assistance. The discouraging attitude adopted by US Antarctic authorities, who have to bear the brunt of the responsibility, towards private expeditions is understandable if somewhat short-sighted in some cases (Mear & Swan, 1987).

Endnotes

1 The account which follows is largely based on those of Sullivan (1961) and Beck (1986).
2 BIOMASS *News Letter* 1987, **9** (1), p. 5.
3 Details of the composition and recommendations of working groups and groups of specialists are to be found in the SCAR *Bulletin* (Scott Polar Research Institute, Cambridge).
4 Swithinbank, C. 1966. *Technical notes on the Ninth Soviet Antarctic Expedition 1963–65*, typescript report, Scott Polar Research Institute, Cambridge.
5 *The Times* 3 March 1989 'Thatcher saved polar survey from fund cuts', also letter in the same issue from Lord Buxton of Alsa.
6 SCAR *Bulletin* No. 92, January 1989.
7 *The Times*, 1 February 1986.

7

The sciences of the Antarctic seas

7.1 The scope of the chapter

In considering particular branches of Antarctic science in the modern period it seems appropriate to begin with those concerned with the sea. Not only does one first have to cross the vast expanse of the Southern Ocean before reaching the continent itself but after the First World War oceanography in the Antarctic region developed in advance of other kinds of study. Physical and biological oceanography will be dealt with side by side since they have always been closely interrelated and particularly so in the Antarctic. The Polar Front provides a natural northern limit to the area to be considered. The southern limits will be taken as high tide mark and the inner edges of the ice shelves. Sea-birds and marine mammals will be regarded as belonging to marine or terrestrial habitats as suits our convenience.

7.2 Physical oceanography at the beginning of the twentieth century

We have already noted that oceanography in the nineteenth century was bedevilled by inordinately slow realization by those making observations at sea that unprotected thermometers give erroneous readings under pressure in deep water and that, unlike freshwater, seawater does not have a minimum in density at 4° C. As Deacon (1971) remarks 'it was extremely unfortunate for the future of marine sciences in the mid-nineteenth century that the distortion produced [in thermometers by pressure] was of the right size to give a convincing temperature if one accepted that sea water behaved like fresh water. The two errors combined lent each other a verisimilitude which neither could have maintained on its own'. These misconceptions were cleared out of the way by the time of the *Challenger* expedition (Fig. 7.1) but it was not until the close of the century that measurements of sufficient accuracy could be made to enable identification of water masses and theoretical calculation of their movements.

Fig. 7.1 Sea temperature measurements on HMS *Challenger*. Subsurface temperatures were measured with minimum thermometers, the most satisfactory instruments for this purpose at the time. From Thomson & Murray, 1885, p. 452. (Courtesy of the Royal Society.)

Then, Nansen and his associates developed techniques giving temperature and salinity measurements of an accuracy never before achieved at sea which enabled them to construct a three-dimensional model of the distribution of water masses in the Arctic Ocean and Norwegian Sea (Mill, 1900; McConnell, 1978). Scandinavia was, indeed, the centre of marine science at the end of the century. Otto Pettersson of Stockholm developed a theory of oceanic circulation in which he postulated that water from melting ice in polar regions provided the motive force (Pettersson, 1904). This was based on thermodynamic arguments and supported by elegant experiments (Fig. 7.2). Pettersson applied his theory in detail to the Arctic but as far as the Antarctic was concerned confined himself to pointing out that the ice effect would be considerably greater in the southern hemisphere and have climatic as well as hydrological consequences. His colleague, Vilhelm Bjerknes, put forward a mathematical theory of oceanic circulation which was applied by J. W. Sandstrom, B. Helland-Hansen and V. W. Ekman. A

Fig. 7.2 An experiment in a tank of water with additions of dye to show the effect of ice on water circulation. From Pettersson, 1904. Fig. 7. (Courtesy of the Scott Polar Research Institute.)

conference called on the initiative of G. Ekman and Petterson and held in 1899 led to the formation in 1902 of the International Council for the Exploration of the Sea, with Sweden, Norway, Denmark, Finland, Russia, Germany and the UK as its founding members. This organization, which was important in promoting the transition from the nineteenth century type of voyage of exploration to detailed long-term investigations, set up a laboratory, directed initially by Nansen, in Christiania (now Oslo, Norway) but later this was moved to the Council's headquarters in Copenhagen, Denmark. Several nations also established their own marine science laboratories around this time (Deacon, 1971).

Some of those involved in these developments, such as John Murray and W. S. Bruce, already had Antarctic experience and the new spirit quickly extended to investigations in the Southern Ocean. The picture of the distribution of water masses which had been constructed from the *Challenger* results was a crude one. The surface water in the region of 65° S was characterized as cold and of low salinity, both temperature and salinity increasing with depth (Fig. 7.3), and water of low salinity presumably from the Antarctic had been found not far below the surface at the equator. Large scale circulation was clearly taking place but the mechanisms involved were a matter of controversy (Deacon, 1971), although had the *Challenger* scientists had more confidence in their results they might have produced something resembling the modern theory of meridional circulation of the Atlantic Ocean (Deacon, 1968).

The *Belgica* (1897–99) and *Valdivia* (1898–99) expeditions, using up-to-

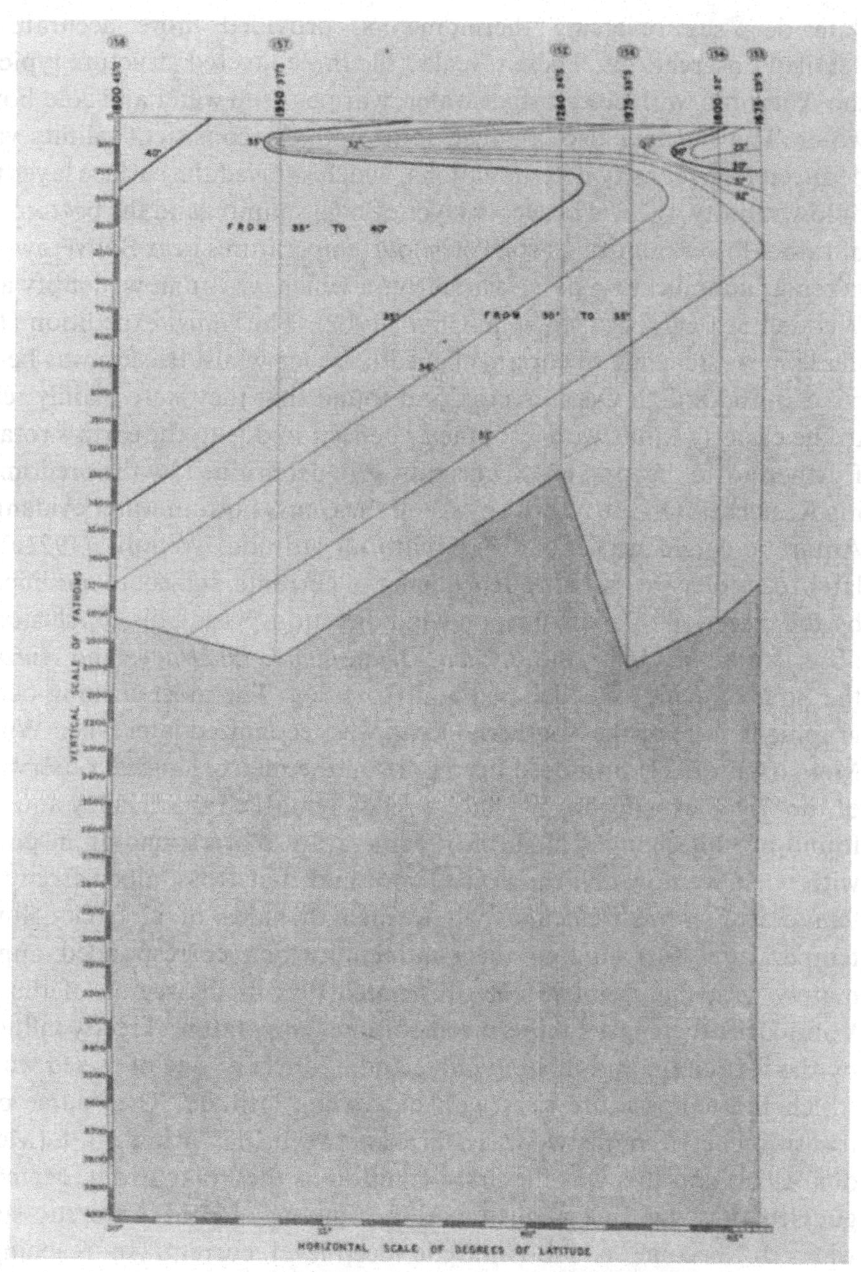

Fig. 7.3 Meridional temperature section between latitudes 50° and 65° south in the Indian Ocean obtained by HMS *Challenger*. From Thomson & Murray, 1885, p. 408. (Courtesy of the Royal Society.)

date deep-sea reversing thermometers, provided more accurate and detailed observations. These revealed the three-layered structure typical of the Antarctic, with cold surface water, warmer deep water and cold bottom water. The *Valdivia* also obtained some not very consistent salinity values deduced from density determinations, which showed the surface layer to be of low salinity, the warm deeper layer of high salinity, and the bottom layer of rather lower salinity. Particularly low temperatures near Bouvetøya were taken as an indication of a polar current, which we can now identify as the Weddell Sea current (Schott in Chun, 1902). The *Gauss* expedition (1901) made measurements of currents beneath the ice whilst its ship was beset at 66° S in the Indian Ocean sector, and found that they were mainly related to the easterly winds with a southerly deflection due to the earth's rotation. Further north, beyond 60° S, currents were determined by the predominating westerlies. On the return voyage of the *Gauss* bottom water evidently of Antarctic origin was found in equatorial latitudes. Wordie (1921a) was later to emphasize the same dependence of currents, subject to modification by the direction of coast-line, on wind direction, basing his conclusions on observations made by the *Belgica, Deutschland, Endurance* and *Aurora* in the South Pacific, Weddell Sea and Ross Sea. The most striking oceanographic feature of the Southern Ocean was recognized later when Wilhelm Meinardus (1923) produced his report on the meteorological observations of the *Gauss* expedition. We have already remarked that Halley and Cook found abrupt changes in surface temperatures corresponding in position with what we now call the Polar Front and that Ross, albeit from faulty temperature measurements, had formed the idea of a 'circle of mean temperature' surrounding the continent, which corresponded approximately with the Front. Meinardus noted that in the region of the West Wind Drift there is a southern zone of low temperature directly influenced by the formation and melting of ice and a northern one of mixed water in which the temperature rises with decreasing latitude. The course of the transition between the two was charted between 105° W and 80° E (Meinardus, 1923) and this was the first definition of the Polar Front. Meinardus guessed that this line might mark the sinking of cold Antarctic surface water to become a sub-Antarctic deep-level current. In reaching his conclusions he was able to take into account data obtained by Brennecke on the *Deutschland* expedition (1911–12), who had traced the sinking of low salinity Antarctic water as it travelled north beyond the Meinardus line (Brennecke, 1921).

Temperature and salinity or density measurements were made by the *Antarctic* (1901–3) and *Scotia* (1902–4) expeditions in the Atlantic sector.

The *Scotia* has been described by Bernstein (1985) as the most efficient oceanographic ship in Antarctic waters since *Challenger*. Her results were presented although not discussed in the expedition's reports (Bruce *et al.*, 1916–17) but Brennecke (1921) made use of them to show that the deeper water in the southern part of the Weddell Sea was warmer than that further north. This fitted in with the idea of a clockwise gyre in that sea, drawing in water from the south-west Indian ocean and expelling a tongue of cold, ice-laden, water from its northern margin. Wilhelm Brennecke, oceanographer on the *Deutschland*, had already had experience on the round-the-world voyage of the research vessel *Planet* (1906–7) and had the further advantage of the most modern equipment. This included both protected and unprotected reversing deep-sea thermometers, which can be used together to give an accurate estimate of depth from the difference in their readings. Brennecke also used the recently developed titration method for the direct determination of salinity. The careful observations which he made, together with a reappraisal of the results of previous expeditions gave a more definite picture of circulation in the Atlantic sector of the Southern Ocean. Antarctic bottom water was found to be formed by mixing of initially warm and saline water which had flowed south at a depth between 2,000 and 3,000 m and cooled during its circulation into the southern part of the Weddell Sea, with cold water raised to high salinity by ice formation over the continental shelf. Drygalski (1928) had a somewhat different explanation of bottom water formation, based on cooling without increase in salinity, but, as Mosby (1934) pointed out, the salinity determinations made on the *Gauss* were suspect and Brennecke's theory was more in accord with the facts. Brennecke's comparison of the carbon dioxide content of surface water with that of the air above it showed a disequilibrium evidently caused by the mixing-in of deep water rich in carbon dioxide (Deacon, 1984). E. W. Nelson (Fig. 7.4), one of the biologists on the *Terra Nova* expedition (1910–12), carried out a painstaking programme of temperature measurements and water sampling through the ice in McMurdo Sound. His results did not appear in the official report but an examination of them was published by Deacon (1975). He had found formation of ice crystals when the water had cooled steadily to the freezing point corresponding to its salinity and he had anticipated the later discovery of a striking increase in the salinity of Ross Sea water following ice formation.

All Antarctic expeditions at the beginning of the century reported on bottom sediments and their observations supported the conclusion drawn from *Challenger*'s results that the deposits are of a continental character.

Fig. 7.4 E. W. Nelson and D. G. Lillie using the Nansen-Pettersson water bottle to obtain samples on the *Terra Nova* expedition 1911–13. Photograph by H. Ponting. (Courtesy of the Scott Polar Research Institute.)

E. Philippi, geologist on the *Gauss*, designated all those collected from anywhere around the periphery of the ice in the far south as glacio-marine, since they consisted not only of the blue mud found around other continents but terrigenous glacial debris as well (Philippi, 1910). North of this zone, which was found to extend as far as the pack ice drifted, was diatomaceous ooze and north of that, beneath warmer waters, globigerina ooze. Philippi considered that these bottom sediments related to present and former currents and, finding diatom ooze below globigerina ooze at 55° S 83° E (Philippi, 1910, p. 592), deduced that there must have been a retreat in Antarctic influence since an earlier ice age. Earland (1935) reviewed the records of foraminiferan remains in marine deposits reported by expeditions of this period together with later ones obtained by *Discovery* Investigations. The Weddell Sea fauna was not the same as those of the Scotia – Bellingshausen Sea and the Ross Sea, which resembled each other. Differences in temperatures and salinities did not seem sufficient to account for this and he concluded that it must have resulted from a former separation of the two sea areas by a land bridge, now represented by the Scotia Arc.

7.3 Marine biology and biological oceanography in the early twentieth century

Marine biologists in the period just before the First World War were mainly concerned with the description of flora and fauna – a very proper and necessary preliminary to other kinds of studies – but biological oceanography developed outside the Darwinian framework (Mills, 1989). Victor Hensen, who coined the word 'plankton', had been the first to adopt a physiological approach in biological oceanography and to attempt quantitative estimates of plankton abundance and its relation to physical and chemical conditions. He led Germany's first major oceanographic venture, the Plankton Expedition in the *National* which cruised the North Atlantic in 1889. It was found that, contrary to expectation, cold and temperate waters supported a much greater mass of planktonic life than did tropical waters. Hensen's protégé, Karl Brandt, extending the ideas of contemporary agricultural chemistry to the sea, explained this in terms of the availability of nutrients such as nitrate and phosphate (Schlee, 1973; Mills, 1989). Hensen founded an influential school but none of his immediate disciples worked in the Antarctic although one of them, H. Lohmann, accompanied the *Deutschland* as far as Buenos Aires and this, like the previous *Gauss* expedition, made a point of obtaining data on chemical characteristics of the seawater which were likely to be related to phytoplankton growth. Their finding that Antarctic surface and intermediate waters both had a high nitrate content established the Southern Ocean as, potentially, the most productive sea in the world. From the ice-bound *Gauss* Vanhöffen sampled the plankton and found a marked decline, especially in diatoms, during the austral winter. His colleague Gazert (1901, 1927), studied the bacteriology of the seawater by means of plate counts, concluding that both nitrifying and denitrifying bacteria were sparse. A central feature of Brandt's conception of the control of plankton production by the nitrogen cycle was that denitrification was responsible for maintaining the balance. Polar waters were thought to be richer in nitrate than those in the tropics because denitrifying bacteria were inhibited by low temperatures (Mills, 1989). Harvey Pirie (in Bruce, 1907–1915) on the *Scotia* expedition also found no nitrifying activity in seawater although he did demonstrate denitrification. It is interesting to note that autotrophic nitrifying bacteria still remain difficult to find in Antarctic habitats (Smith, 1985). These failures, which were perhaps due to inactivation of the nitrifiers by light, made it hard to account for the abundance of nitrate in Antarctic waters and Gazert surmised that there must be fixation of atmospheric nitrogen at the ice-edge. Nathansohn (1906) pointed out that Nansen's explanation of the enrichment of Arctic waters by inflow of continental rivers would not work for the Antarctic. He interpreted the high productivity, correctly as we now know, in terms of his general theory of nutrient supply by vertical circulation. Gazert supported Brandt's theory of nitrate-limitation of phytoplankton growth and suggested that pelagic animal life was abundant because organic matter was not being broken down to any extent by bacteria.

Work on Antarctic phytoplankton at this time was almost entirely of a systematic nature: Karsten (1905) described material collected by the *Valdivia*, Mangin (1915, 1922) that collected by the *Pourquoi Pas?* and by the *Scotia*, and Van Heurk (1909) that collected by the *Belgica*. Amongst the zooplankton, krill (*Euphausia superba*) striking in its abundance and interesting as the food of whales, was a frequent object of opportunistic observation and comment (see Marr, 1962), and Zimmer (1913) gave an account of its morphology but, apart from this, as in zooplankton studies generally, the approach was mainly systematic. Indeed, the voluminous expedition reports of this time are largely composed of systematic accounts of the various animal groups. For example 12 of the 20 thick volumes of the *Gauss* expedition (Drygalski, 1911–31) are devoted to this kind of zoology. This applies to reports on benthic as well as pelagic animals. E. Racovitza made dredgings from the *Belgica* on the continental shelf off the Peninsula and found a fauna similar to that of oceanic depths (in Cook, 1900). Dredging was carried out from the *Gauss*, an emperor penguin being employed to thread a line between two holes in the ice surrounding the ship. Amongst other things, sessile larval forms of crinoids, isopods, amphipods, heliozoa, bryozoa, pycnogonids and foraminifera were brought up from 385 m. Bottom dwelling fish, *Notothenia* and *Lykodes* were also caught. The *Scotia* made a collection of marine invertebrates that for a long time was unmatched by any other Antarctic expedition

Fig. 7.5 D. G. Lillie with siliceous sponges from the Ross Sea; *Terra Nova* expedition, 1911–13. From Huxley, 1913, vol. II, p. 475.

(Bernstein, 1985). Bottom trawling from the *Terra Nova* in McMurdo Sound brought up animals, such as enormous sponges (Fig. 7.5), that except for those requiring calcium for shells or skeletons, were much larger than their counterparts in warmer waters, (British Museum, 1914–35).

7.4 The inter-war period and the *Discovery* Investigations

The First World War convinced the naval establishment of the value of science. However, both in Britain and the US, there was friction between the scientists, who wanted to establish a theoretical background, and the naval authorities, who wanted equipment for immediate use (Hackmann, 1984). Underwater acoustics and echo-sounding devices – which, incidentally, had been envisaged by Maury in 1854 – but not applied successfully – were needed for submarine detection, for example, and this had its spin-off in the rapid development of echo-sounding. However, when peace returned basic research was still largely dependent on private funding (Schlee, 1973). The *Meteor* expedition of 1925–26 continued the German tradition of professionalism and innovation and this cruise is often taken as the beginning of a new phase in oceanography. The basic programme was a survey of temperature, salinity and oxygen at different depths in the Southern Atlantic Ocean, enabling the description of surface, sub-surface, intermediate, deep and bottom layers and linking them with Antarctic water types by an extension of the cruise into the Atlantic sector of the Southern Ocean. A great advance was to relate this precise and detailed information about water masses to the topography of the basins which contained them, as revealed by extensive use of the echo-sounder, to deduce their origins and circulation patterns (Wüst, 1928). These results showed that the Atlantic must be regarded as a single entity rather than as two independently circulating masses, a point emphasized by the penetration of recognizable Antarctic bottom water as far as 17° N. This general picture of ocean circulation put forward by Wüst (1928) is much as we know it today. It was at this point that the term *Convergence* (Polar Front) replaced *Meinardus Line*. Details, such as the extent to which the intermediate layer water moved to the surface and cooled were, however, speculative. A full explanation of this had to wait for another 30 years (Gill, 1973).

The *Discovery* Investigations were, as we saw in the previous chapter, paid for by a levy exacted by the British Colonial Government from the whaling industry. Enlightened in conception, they were designed for the long-term benefit of whaling and it was recognized from the beginning that basic studies in oceanography – physical, chemical and biological – were essential. This was an approach that was being strongly developed at this time under the direction of E. J. Allen at the Marine Biological Association in Plymouth, UK (Mills 1989). It was recommended that 'The remuneration of the civilian scientific staff should be provided for at considerably higher rates than have been given in the case of previous expeditions' (Inter-departmental Committee 1920) and the Investigations got staff of the highest quality. None of them had much experience in oceanography on joining but several of them eventually achieved international eminence

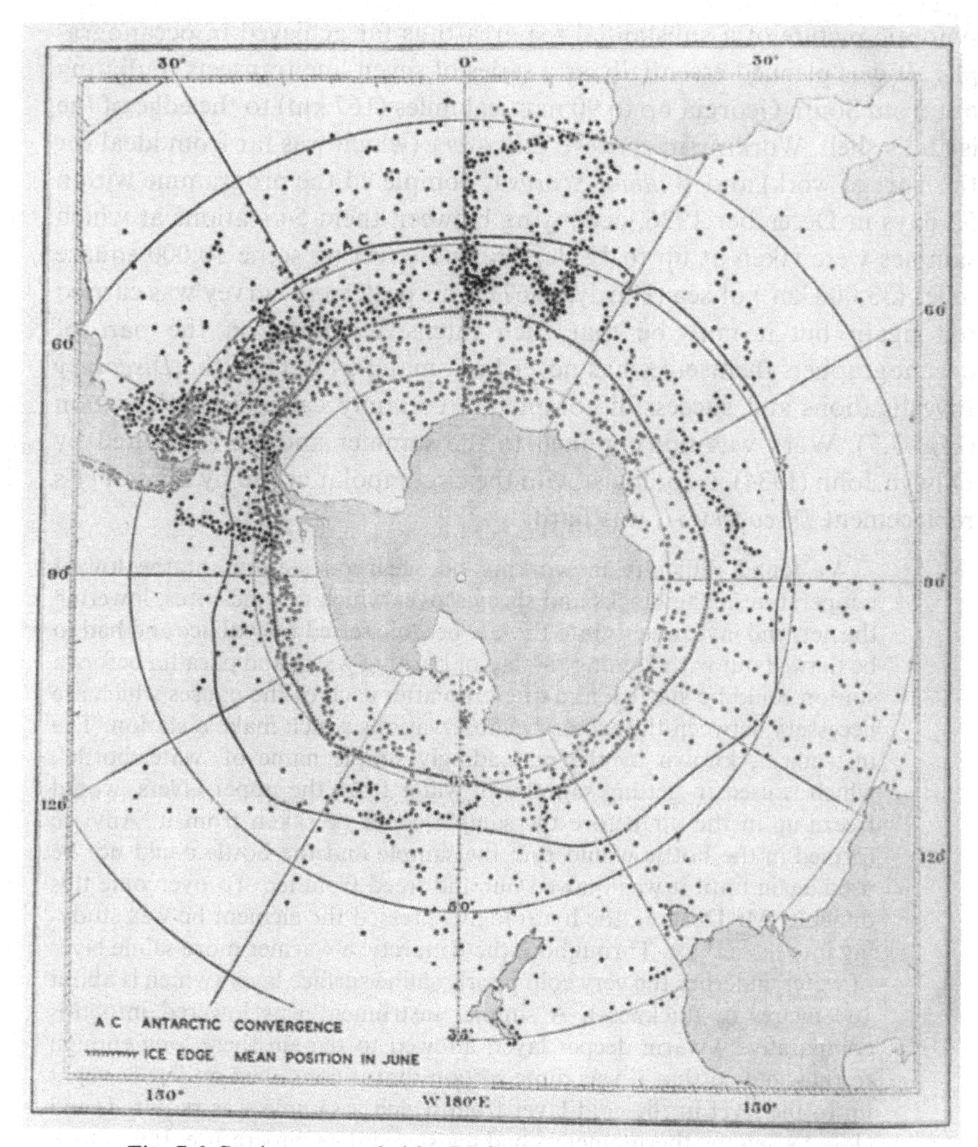

Fig. 7.6 Stations, occupied by RRSs *Discovery*, *Discovery II* and *William Scoresby* between 1926 and 1939, for plankton observations. From Marr, 1962, Fig. 1. (Courtesy of *Discovery* Investigations.)

in marine science. The chart showing stations occupied for krill observations during the 15 years of the Investigations (Fig. 7.6) gives an idea of the ultimate extent of the coverage; it was soon realized that for an understanding of the oceanography of the Southern Ocean it was necessary to study the whole of it (John, 1934). There was also intensive investigation around South Georgia which gave the nearest approach to a detailed synoptic picture of a substantial sea area thus far achieved in oceanography. It was planned essentially as a series of seven line transects, radiating out from South Georgia up to 90 nautical miles (167 km) to the edge of the island's shelf. Working in concert, *Discovery* (which was far from ideal for this sort of work) and *William Scoresby* completed the programme within 52 days in December 1926, occupying between them 54 stations at which samples were taken at up to 17 depths, and covering some 10,000 square miles (33,000 km^2) of sea (Hardy, 1967). This pattern of survey was carried out again but it may be that such intensive efforts on the part of oceanographers themselves has never been matched outside the *Discovery* Investigations and successful completion certainly called for celebration (Fig. 7.7). Work was not confined to the summer and, as recounted by Dilwyn John (1934), a zoologist with the circumpolar cruise by *Discovery*'s replacement *Discovery II*, was hard:

> We found difficulty in working our gear on stations at the lowest temperatures. The blocks and sheaves over which ran the wires, lowering the nets and instruments into the sea, became seized up with ice and had to be thawed out with flaming torches of burning waste and paraffin before a station could be started; and often too after each of the pauses which are necessary between the successive observations which make a station. The instrument, known by the misleadingly simple name of 'water-bottle', which is used in getting samples of water from the upper layers, would freeze up in the air before the sample could be taken from it. Any ice formed in the bottle would ruin the sample and the bottle could not be used again until it was thawed out and freed from ice. To overcome this difficulty Mr Deacon, the hydrologist, pressed the element he was studying into his service. Throughout the Antarctic a warmer more saline layer of water underlies the very cold poorly saline surface layer (which is about 100 metres in thickness). A simpler instrument was lowered into this comparatively warm deeper layer; allowed to remain there long enough to thaw out so that it was quite certain that it contained no ice; brought up to the level in the cold layer from which a sample was required; and closed when the thermometers had registered the temperature. It was then hauled to the surface, taken quickly off the wire, and rushed to the laboratory.

As a result of these investigations, which were largely carried out by already established methods aided by such newly available devices as the echo-sounder, a more complete picture emerged of the water masses, currents and ice of the Southern Ocean. N. A. Mackintosh, one of the zoologists, published charts showing the entire Polar Front, the position of which at any given longitude he found to vary only by about 50 nautical miles (93 km) at any time of year (Mackintosh, 1946). Deacon (1984) in

Fig. 7.7 Caricature of Alister Hardy singing 'Yip-i-addy-i-ay', drawn at South Georgia, Christmas 1926, by J. W. Ridley, Chief Engineer of RRS *William Scoresby*. He is depicted with a tow-net in one hand and waving his plankton recorder (in miniature) with the other, with a cloud of krill above his head and a small winch for working nets in the background. From Hardy, 1967, p. 291. (Courtesy of M. Hardy.)

referring to this work, remarked on 'the wonder of such a relatively stable feature in the middle of a wide ocean'. H. F. P. Herdman, who was hydrologist from 1924 onwards, had been joined by G. E. R. Deacon in 1927. The latter, a 21-year-old graduate in chemistry totally without experience of the sea, later became Sir George Deacon, FRS, Director of the National Institute of Oceanography (Charnock, 1985). He summed up the hydrological results in several reports (Deacon, 1933, 1934 and 1937). A major finding was that nearly all the Antarctic bottom water originated in the Weddell Sea. Deacon suggested that the stability of the Polar Front was the combined result of the character of deep-water circulation in the

Southern Ocean and its bottom topography. About the same time Sverdrup (1933, 1934) argued that the Front was purely a feature of the drift current and dependent on the meridional weakening of the wind. Clowes (1933) showed that, contrary to Wüst's idea, there was a considerable input of Pacific water through the Drake passage into the south-west Atlantic where it formed a warm, more saline, layer between 1000 and 2500 m. He estimated the flow through the Drake Passage as $110 \times 10^6 \text{ m}^3 \text{ s}^{-1}$, a figure of the same order as a more recent one (p. 230). A variation in temperature and salinity in the Antarctic Intermediate Current as it flowed to the north in the South Atlantic was supposed to originate in seasonal variations in the sinking Antarctic surface water which contributed to it and, following a suggestion made by Brennecke in 1909, was used to estimate a speed of between $1\frac{1}{2}$ and $2\frac{1}{2}$ miles (2.8 and 4.6 km) per day. Later, Deacon (1934) felt that although this estimate happened to be more or less correct the method itself was questionable in view of what is now known about the variability of water movements in the ocean.

During the period 1927–30 oceanographic work was also done by Norwegian expeditions financed by the whaling magnate Lars Christensen. Mosby (1934), from his own observations made from the *Norvegia* together with those of previous workers, plotted the topography of isobaric surfaces in the Atlantic sector of the Southern Ocean. This showed a cyclonic movement around a centre of elongated shape stretching from the Weddell Sea to the north-east, in which surface water was estimated to take about 11 years to circulate. The same pattern was seen in the winds of this region and these, with some influence from the Circumpolar Current, were seen as the principal cause of the movement.

The biologists recorded quantitatively, from counts of net samples, the distribution of plankton and its seasonal variation. To complement the account of the tribulations of the physical oceanographer just given, we may quote from F. D. Ommanney (1938), a biologist with the *Discovery* Investigations from 1929–39.

> Every station, or series of scientific observations at sea, was conducted in the same way. The ship was brought round with her head in to the wind and continued to steam slowly into the wind at a speed sufficient to keep the wires to which we attached our instruments hanging vertically in the water. It required skill on the part of the Officer of the Watch to keep her thus for if she steamed too fast, or if her head fell away from the wind, the wires trailed away astern and there was a danger that the wires carrying the water-sampling instruments would become entangled with those carrying the biological nets, which were being hauled vertically from the

> after part of the ship at the same time. There was a further danger that the wires carrying the nets would get round the screw. If, however, she steamed ahead too slowly the wires trailed forward and there was again a danger that they would become foul of one another. All these things happened from time to time, providing a change from the monotony of routine and something to talk about next day. However, we never lost our tempers over these mishaps and sometimes, when the scientists and a group of seamen were struggling to unravel a tangled skein of steel wire brought slowly to the surface after some submarine calamity of this sort, there would be heard Dick's [the second mate's] gusty laughter from the bridge. Why worry?
>
> (from *South Latitude* by F. D. Ommanney, reproduced by permission of Curtis Brown Ltd., London)

Hart (1934) found that diatoms dominated the phytoplankton. Because the Southern Ocean is an uninterrupted circumpolar belt it would be expected that the distribution of plankton species would be circumpolar too and Baker (1954) showed that this is true for the more important phytoplankton species and larger zooplankton species. Striking correlations were found between the amount of phytoplankton and depletion of phosphates in the water and between the distributions of krill and baleen whales but not between krill and the phytoplankton on which it feeds. This led A. C. Hardy (Hardy & Gunther, 1935) to put forward his animal exclusion hypothesis which supposed that the frequently observed inverse relation between numbers of phytoplankton and zooplankton arose not just from grazing but also from active avoidance by the animals of dense concentrations of the plants. Whether this is so or not remains doubtful but the device invented by Hardy to record the patchy distribution of phytoplankton known as the *continuous plankton recorder* has become a standard and widely used piece of oceanographic equipment. Towed behind a vessel cruising at normal speed it sampled the plankton and preserved it on a continuous, moving strip of gauze so that afterwards the numbers of the different species along the track of the vessel could be counted in the laboratory. Hardy's continuous plankton recorder was devised by him for the *Discovery* Investigations and first used successfully in 1926 (Fig. 7.8) to give a continuous record of the plankton right across the Drake Passage (Hardy, 1926, 1936a,b). Since 1948 it has been used in regular surveys of the North Atlantic and the North Sea, providing a unique record of long-term changes in plankton (Colebrook, 1979). The continuous plankton recorder has recently been improved by the incorporation of an undulating mechanism to enable it to sample over a range of depths (Aiken 1981).

The only other country to investigate the plankton of the Southern

Fig. 7.8 Shooting the continuous plankton recorder from RRS *Discovery*, 1926. The person in the broad-brimmed hat is presumably Alister Hardy. (Courtesy of the Institute of Oceanographic Sciences, Deacon Laboratory.)

Ocean at this stage was Norway. In the austral summer of 1929–30 J. T. Ruud sailed on the whaling factory ship SS *Vikingen* along the ice-edge of the Weddell Sea, collecting plankton and water samples for chemical analysis (Ruud, 1930). His results were discussed by his teacher, H. H. Gran (1931) who used them to support his new hypothesis – which was to become a corner stone of biological oceanography – that phytoplankton development was retarded by vertical mixing and that only after some stabilization of the water column could a bloom develop and deplete nitrate and phosphate.

Apart from investigations of the distribution and development of plankton, including, of course, *Euphausia superba* (Marr, 1962), there were anatomical and systematic studies of both benthic and planktonic animals. There was no shortage of material. Hardy (1967) wrote of collecting from the *Discovery* on a cruise from South Georgia to the Peninsula in February 1927.

> We made up to the lee of Clarence [Island, South Shetlands], for the weather, none too good, was freshening, and here we took a cast with the

> dredge in 343 metres depth. The life below appears to get richer and richer; I am only recording a few of the hauls we made. On this occasion our dredge was again on the bottom for just five minutes: yet such was the varied collection of animals it brought up that, with all of us working hard, we have only to-night, twenty-four hours later, got all the material sorted, labelled and put away.

The primary object of the *Discovery* Investigations – the whales – were studied mainly at the whaling station at Grytviken, South Georgia, from a shore-based laboratory at the nearby King Edward Point. Anatomical and statistical data were obtained on the whales brought in and a marking programme provided information on migrations. Early whale-marking cruises by the *William Scoresby* were unsuccessful, evidently because, although the marks fired into the whale were found to hold firm in preliminary tests, they had little penetration and were soon rejected. A simple 25 cm bullet of stainless steel with a weighted head was found to be more satisfactory and was brought into use in the 1932–33 season (Fig. 7.9). By 1939 some 5,219 whales had been marked out of which 33 (4.9% of 668) had been returned from blue whales, 118 (3% of 3915) from fin whales and 36 (6.6% of 548) from humpbacks (Rayner, 1940). This showed, for example, that the humpback makes extensive migrations, northwards into tropical waters in the austral autumn for breeding and southwards to feed in the spring. A vast amount of information on the distribution of the commercially important whales was thus obtained. N. A. Mackintosh and J. F. G. Wheeler (1929) laid the foundations of our knowledge of the anatomy, growth and reproduction of whales but their work, which depended on the dissection of carcases, could not have been done in the absence of commercial whaling or without the full co-operation of the whalers. Whilst these researches were being done the character of whaling was changing. The first pelagic whaling factory was introduced in 1925. By 1930 the number of these vessels had risen to 41 and with this the possibility of individual government control of whaling quotas had gone. Between 1925 and 1931 the number of whales killed per year rose from 14,219 to 40,201. The first significant step towards international control was the Convention for the Regulation of Whaling signed in Geneva in 1931 and which came into force, after ratification by a sufficient number of nations, in 1935. Thereafter all whaling statistics were collected in Oslo. Germany and Japan refused to adhere to the Convention and only the UK and Norway among the pelagic whaling nations observed it. Another attempt at regulation was made with the signing in London in 1973 of an International agreement for the Regulation of Whaling. The Council for

DISCOVERY COMMITTEE,
COLONIAL OFFICE,
LONDON, ENGLAND.

"DISCOVERY" INVESTIGATIONS.

WHALE-MARKING.

The Discovery Committee is making renewed experiments in whale-marking in the Antarctic.

The mark used, consists of a metal tube approximately 10 inches in length and 5/8 of an inch in diameter, with a leaden head. It is designed to penetrate the blubber and enter the muscle for a few inches and should be found with little difficulty when the blubber is stripped. The tube has an inscription and a serial number at each end. A reward of £1 will be paid by the Discovery Committee, Colonial Office, London, England, for the return of each mark accompanied by the following particulars:—

Number of Mark 6308

Date of capture 12th February 1938

Position of capture S. 63°20' - W. 40°00'

Species of whale Finwhale

Sex Female

Length from tip of upper jaw to notch in the middle of tail (measured in a straight line) 67 feet

Was the wound completely healed and without suppuration? Yes.

Was the whale in good general condition? Yes.

Name of finder Herman Jacobsen

Address Viksfjord, Norway. per Larvik.

If the finder wishes to retain the whale mark as a memento it will be returned to him after examination.

Please return whale-mark.

Please use a separate form for each mark.

From: Fl. /f. N. T. Nielsen-Alonso
c/o: Messrs. Melsom & Melsom
Nanset per Larvik, Norway.

Fig. 7.9 Whale-mark return form issued by the *Discovery* Committee. (Courtesy of *Discovery* Investigations.)

the Exploration of the Sea, concerned about the number of blue whales that were being killed, called another conference in 1938, in which, again, Japan and Germany refused to participate. The establishment of a whale sanctuary in the area between 70° and 160° W and south of 40° S and complete protection for the humpback were agreed upon (Small, 1971). Much of the information upon which such regulations were based came from the *Discovery* Investigations. Thanks to the far-sighted action of the Colonial Office in accumulating funds for research, investigations at sea were able to continue in spite of dwindling yields of tax from the land-based whaling operations but they had perforce to stop with the outbreak of the Second World War.

7.5 The impact of the Second World War on oceanography

War again gave a fillip to oceanography, a matter which Schlee (1973) and Hackmann (1984) have considered in detail. In the US, expenditure at the Naval Research Laboratory grew from $1.7 million in 1940 to $13.7 million in 1945 and the budget of the Woods Hole Oceanographic Institution rose so dramatically that the annual treasurer's report in 1942 was not published for fear of revealing the magnitude of government interest in certain lines of research. Increased funding, of course, meant increased government direction and with the return of peace the private oceanographic institutions found that return to their pre-war independence was impossible without unacceptable contraction, so that they must henceforth continue to be involved with government. Oceanography was thereby changed in character as well as scale, the most obvious effect being that physical studies gained importance and marine biology lost its former pre-eminence. In Britain the idea of a National Institute of Oceanography, growing out of the *Discovery* Investigations, was considered before the war ended but it did not come into existence until 1949. It was at first largely an Admiralty responsibility but Deacon, who was appointed as its first director, organized it on a broad and liberal basis to include biological as well as physical oceanography – a feature that can be directly related to his Antarctic experience. At the same time the *Discovery* Committee was dissolved, its funds going to support the Falkland Islands Dependencies Survey and its ships, the *William Scoresby* and the *Discovery II*, being transferred to the new institute with their first task under the new regime being to round off their pre-war work (Charnock, 1985). The *Discovery* Investigations staff were effective advocates for the continuation of work in the Antarctic at this time (Coleman-Cook, 1963).

Oceanography in *Operation Highjump* in 1946–47 reflected the US

Navy's preoccupation with wartime problems of submarine detection by sonar – a name which has replaced the earlier British one, Asdic. Extensive use was made of the bathythermograph, an instrument developed in the US during the war to enable a depth-profile of temperature to be obtained from a ship under way, so revealing the location of thermal discontinuities which affect underwater acoustics (Hackmann, 1984). The closely spaced vertically continuous profiles which it gave showed the temperature structure of Antarctic waters to be much more complicated than was evident from the widely spaced sampling that had been carried out hitherto. The bathythermograph has continued to be used routinely on most Antarctic supply ships and has provided data essential for the study of frontal zonation structure and heat storage (Gordon, 1986). The 'deep scattering layers' which had perplexed the sonar experts and which seemed attributable to zooplankton, were found to disappear entirely in the region of the Polar Front and to be present in Antarctic waters only when there were some hours of darkness (Dietz, 1948). A post-war expedition in the classical style was that of the *Brategg*, planned by the Norwegian Geographical Society's Committee for Oceanographic Investigation and sponsored by the Federation of Norwegian Whaling Companies. In 1947–48 it worked stations in the 90° W to 174° 41′ W sector and took the opportunity to leave a copy of the Norwegian occupation statute of 1929 on Peter I Island (Mosby, 1956).

There followed a 10-year hiatus when, although there was much oceanographic activity elsewhere, little more than occasional observations from ships in passage were made in Antarctic waters. This situation remained essentially the same during the IGY, when oceanography in the Southern Ocean, with the exception of the cruises of the Soviet vessel *Ob* in 1956–58 (Maksimov, 1958; Lebedev, 1959; Kort, 1964), was mostly carried out opportunistically from supply ships without any specific orientation towards Antarctic problems. The result was rather uneven coverage (Fig. 7.10). However, the permanent interest of the US in the Antarctic established by the IGY led to the National Science Foundation (NSF) acquiring the *Eltanin* (Fig. 7.11) as an Antarctic research vessel in 1961. She had been built in 1957 as an ice-strengthened cargo ship for the US Navy and after conversion could accommodate up to 39 scientists and a technical support party. She served as a national facility for institutions carrying out research with NSF support and as such was able to cater for short-term projects covering a wide range of marine science. In addition, during her $10\frac{1}{2}$-year service with the NSF, during which she made 55 cruises and remained at sea for 75 per cent of the time, she made long-term oceanographic surveys

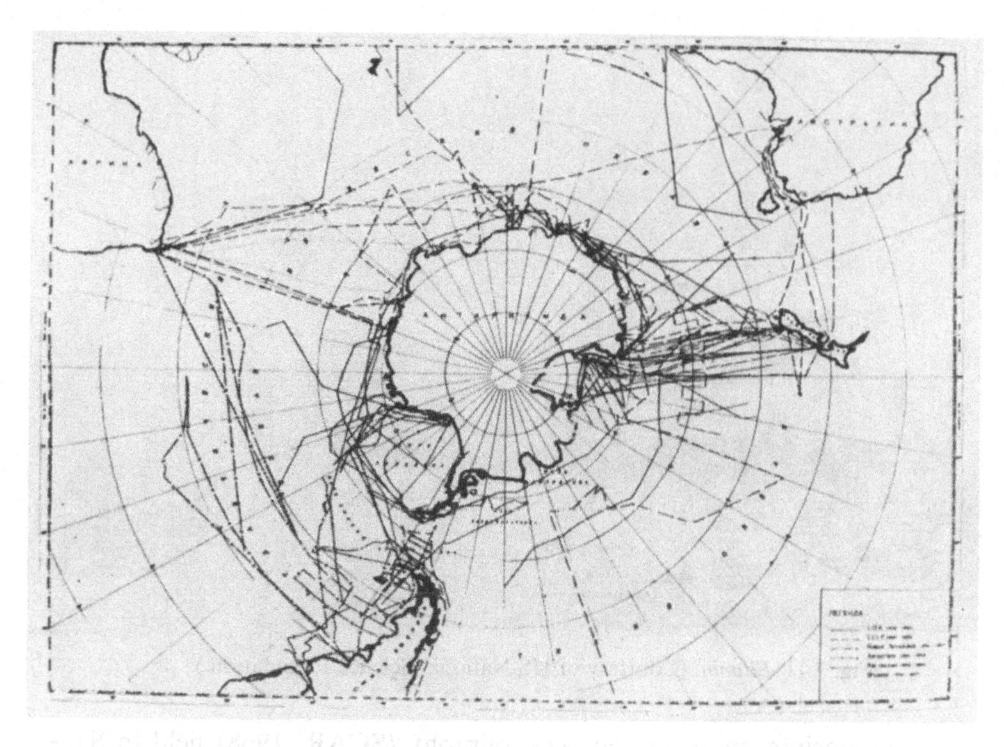

Fig. 7.10 Tracks of Antarctic marine expeditions during IGY. From Kort, 1964, Fig. 1. (Courtesy of MIT Press.)

which contributed massively to our knowledge of the Southern Ocean (Fig. 7.12). During this period oceanographic techniques continued to evolve. In particular, observations at serial hydrographic stations began to be replaced with continuous *in situ* measurements of temperature and conductivity which gave far better resolution of thermohaline stratification. Tracer geochemistry of stable isotopes of oxygen and of tritium, carbon-14 and freon, began to come into use for studying exchange processes. The *Southern Ocean Atlas* (Gordon *et al.*, 1982), made substantial use of the data on temperature, salinity, oxygen content and plant nutrient concentrations which the *Eltanin* gathered. In the same period Soviet oceanographic surveys accumulated data on general circulation patterns, heat exchanges, basic water masses and distribution of sea-ice (Kort, 1964) much of which were incorporated in the *Atlas of the Antarctic* (Bakaev, 1966). In 1972 *Eltanin*'s programme was discontinued because of shortage of funds and after a period in which, under the name of *Islas Orcadas*, she was operated jointly by the US and Argentina, she reverted to the US Navy in 1979 and was decommissioned.

Fig. 7.11 *Eltanin*. (Courtesy of US National Science Foundation.)

A symposium on Antarctic oceanography (SCAR, 1968) held in Santiago, Chile, in September 1966 gives a good cross-section of the state of knowledge at this stage. In introducing it R. J. Currie wrote as follows:

> Little is known, however, of the dynamics of the whole system of the Southern Ocean, physically and biologically. We have, as yet, no clear understanding of the transference of energy between the atmosphere and the ocean. The processes which bring about the circulation are still the subjects of controversy, and little can be said about the transport of energy within the ocean. In biology, there is an equivalent lack of information concerning the quantitative and qualitative transference of energy in the organic regime, and an inadequacy in our understanding of the interrelationships of even the major elements of the flora and fauna.

The symposium was truly international in nature with most of the Treaty nations, in particular the US and USSR, contributing substantially. Sponsored by SCAR, SCOR, IAPO and IUPS (see 'A note for the reader'), it had a correspondingly catholic coverage, the principal sections dealing with: (1) surface and upper layers, (2) deep water, (3) ocean floor, (4) coastal waters, (5) pack ice regime, and (6) biological productivity. This brought about some curious juxtapositions, as of mathematical analysis of the effect of the Drake Passage on the Circumpolar Current and the protozoa associated with a relict crustacean, but the point was made that

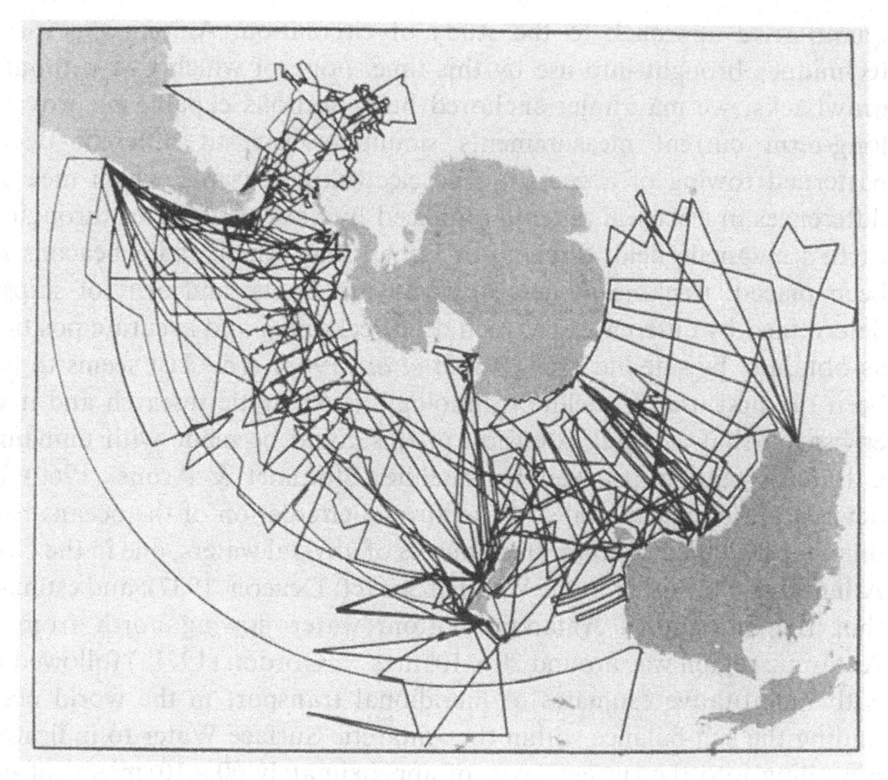

Fig. 7.12 Tracks of *Eltanin*'s 52 cruises, 1962–72. From Bushnell, 1975, plate 15.

different disciplines must co-ordinate their studies if best use was to be made of the limited logistic resources available.

The surface features of the Southern Ocean were appearing more complex. It had become evident that the Polar Front was determined by distribution of winds and atmospheric pressure as well as by sub-surface circulation and bottom topography; both Deacon and Sverdrup were right (see p. 208) and air–sea interaction had to be recognized as important (Kort, 1964, 1968). Equally, the involvement of the Polar Front in the biology of the pelagic organisms and geochemistry of the bottom sediments was evidently profound. The Antarctic Divergence also had its biological significance, playing a part similar to that of upwelling in coastal regions, in providing nutrients to maintain phytoplankton productivity. It was more than ever clear that biological oceanography is dependent on physical oceanography for essential background.

The accumulation of basic hydrographic data and the development of new methods of observing currents were allowing a more precise and quantitative approach to the study of circulation. Among the various techniques brought into use by this time, none of which was without its drawbacks, we may note: anchored buoy stations capable of providing long-term current measurements simultaneously, at different depths; patterned towing of a geomagnetic electrokinetograph, which measures differences in electrical potential induced by flow of seawater through the earth's magnetic field; tracking of icebergs on which radio beacons had been placed; tracking of neutrally buoyant floats; and drift of ships as determined by differences between dead reckoning and accurate positions, as obtained by satellite fixes (Rubin *et al.*, 1968). This last seems to have been the first use of satellite technology in Antarctic research and it was envisaged that a useful extension of this could be made with unmanned automatic stations tracked by satellite. Stommel & Arons (1960) had devised a theoretical model of the abyssal circulation of the oceans based on two postulated concentrated sources of abyssal waters, one in the North Atlantic and the other in the Weddell Sea (cf. Deacon, 1937), and estimated that the amount of Antarctic bottom water flowing north from the Antarctic region was around $20 \times 10^6 \, m^3 \, s^{-1}$. Gordon (1971a) followed this with quantitative estimates of meridional transport in the world ocean, finding the salt balance within the Antarctic Surface Water to indicate an upwelling into the surface layer of approximately $60 \times 10^6 \, m^3 \, s^{-1}$ of deep water during the summer. During the winter about one half of this amount sinks and entrains with deep water to give a total southward transport of $10^8 \, m^3 \, s^{-1}$ at this period. This recirculation between the superficial Antarctic water masses and the Circumpolar deep water was large, exceeding by several-fold the inflow of North Atlantic deep water, which was necessary to maintain the salinity of Antarctic waters in a steady state. The axis of the Antarctic Circumpolar Current lay slightly south of the Polar Front and almost directly below the axis, or jet stream, of the westerly winds and Gordon (1971a) speculated on the transfer of energy from the westerlies into the Circumpolar Current. Circumpolarity and absence of meridional obstructions provide a unique environment for development of waves. Some investigations of waves and swell in the Southern Ocean were made during IGY but in 1966 there were still technical difficulties in using buoy stations in such tempestuous seas and working from shore stations was always a handicap.

The crucial involvement of ice in the dynamics of the Southern Ocean was only taken into account by Gordon (1971a) in the most general terms. Indeed, knowledge of sea-ice had advanced little beyond the descriptive state reached, for example, by Wordie in 1921. It could scarcely be otherwise at this date; airborne remote sensing had not yet been used to provide synoptic information on the distribution and character of ice nor had *in situ* observations on the physics of ice formation and behaviour been made within the pack and fast ice at all seasons of the year. Only tentative guesses could be made of the effects of shelf-ice, by way of calving of icebergs and melting, on the energy budget and salt balance of the ocean. Much more information was available about Arctic sea-ice – a review on the physical properties of sea-ice published by Pounder in 1962 was entirely based on Arctic work and it was an expert on this who contributed to the 1966 Antarctic symposium (Weeks, 1968) – but while, of course, it is true that the physics is basically the same at the two poles, some important distinctions were realized (Lewis & Weeks, 1971).

Study of bottom deposits had extended in several new directions. Measurement of the thickness and stratification of sediments by seismic reflection profiling (Fig. 7.13) in a major project to cover the world oceans by the Lamont-Doherty Geological Observatory, in the US, was extended into Antarctic waters (Ewing *et al.*, 1966). Deep sea floor photography carried out by scientists from the same laboratory revealed evidence of deep sea currents (Heezen *et al.*, 1966) although techniques needed to be improved before they could be of real value. Extension of palaeomagnetic dating methods for deep sea sediments further back than five to 10 million years and longer cores to penetrate into pre-glacial sediments were needed before the beginning of the Antarctic glaciation could be dated. The tectonics of the floor of the Southern Ocean will be best considered in relation to earth sciences (see chapter 8).

7.6 Marine biology in the immediate post-Second World War years

Marine biologists continued to study the kinds and distributions of planktonic organisms, although hampered by a shortage of taxonomic and systematic experts and difficulties of publication. A new approach to the study of plankton had, however, come with the development of the radiocarbon technique for determining primary productivity by the Danish oceanographer E. Steemann Nielsen in 1952. The only direct method available previously, that depending on measurement of oxygen production introduced by Gaarder and Gran in 1927 (Mills, 1989), was insufficiently sensitive for use in any but the most productive waters. Determination of chlorophyll concentrations, which measured standing crop of phytoplankton, not its photosynthesis, carried out in the Gerlache and Bransfield Straits in 1959 by the Americans Burkholder & Sieburth (1961) working from Argentine vessels, had indicated that although some areas

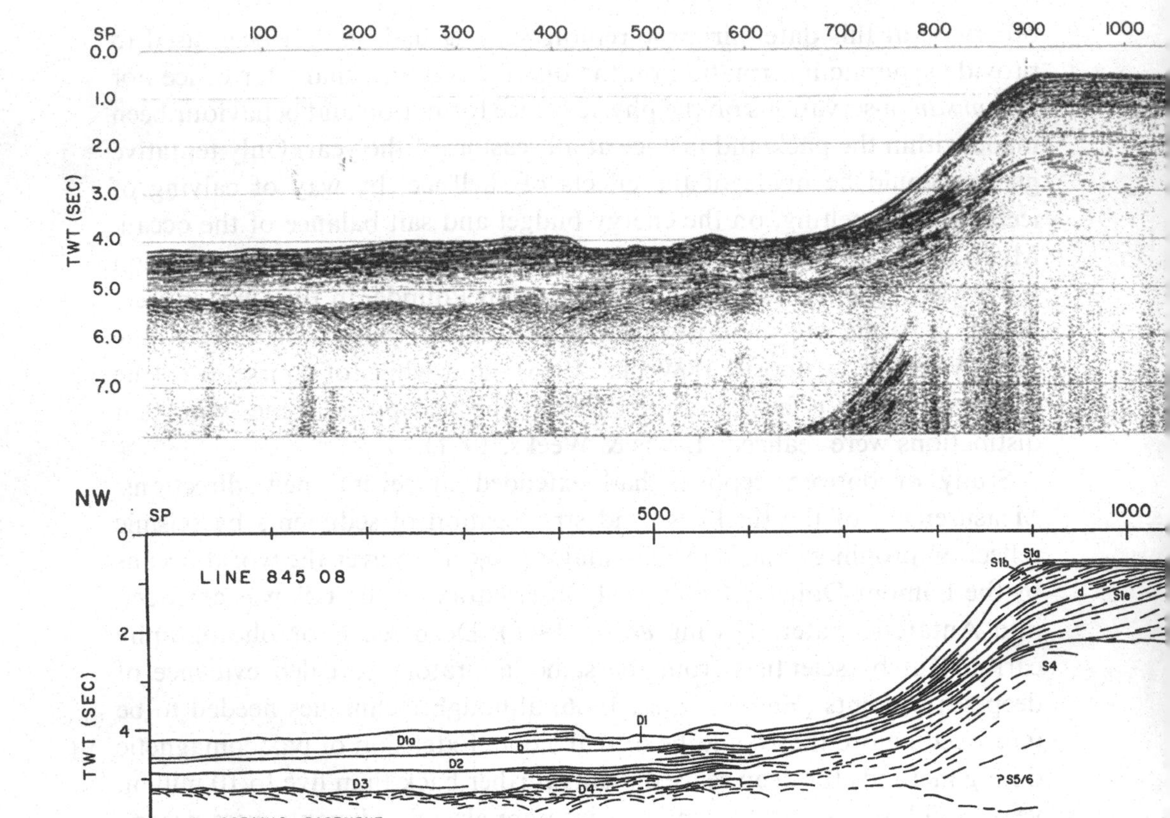

Fig. 7.13 Multichannel seismic reflection profile across the Pacific margin of the Antarctic Peninsula at *c.* 64° S, with interpretive line drawing. (R. D. Larter, British Antarctic Survey.)

were unproductive others might have rates of primary productivity equalling those of fertile temperate waters. The radiocarbon technique could be easily adapted to any desired sensitivity and gave gratifyingly precise results. It was only after many years that it was realized that it did not necessarily give a straightforward measure of the production of organic matter by photosynthesis (Fogg, 1975). Nevertheless, it gave an immediate means of comparison and of examining the basis for the apparently great biological productivity of the Antarctic seas. The first radiocarbon determinations in these waters were made by Dyson in 1959[1] from the Australian-chartered IGY supply vessel *Magga Dan*. This was followed by others from

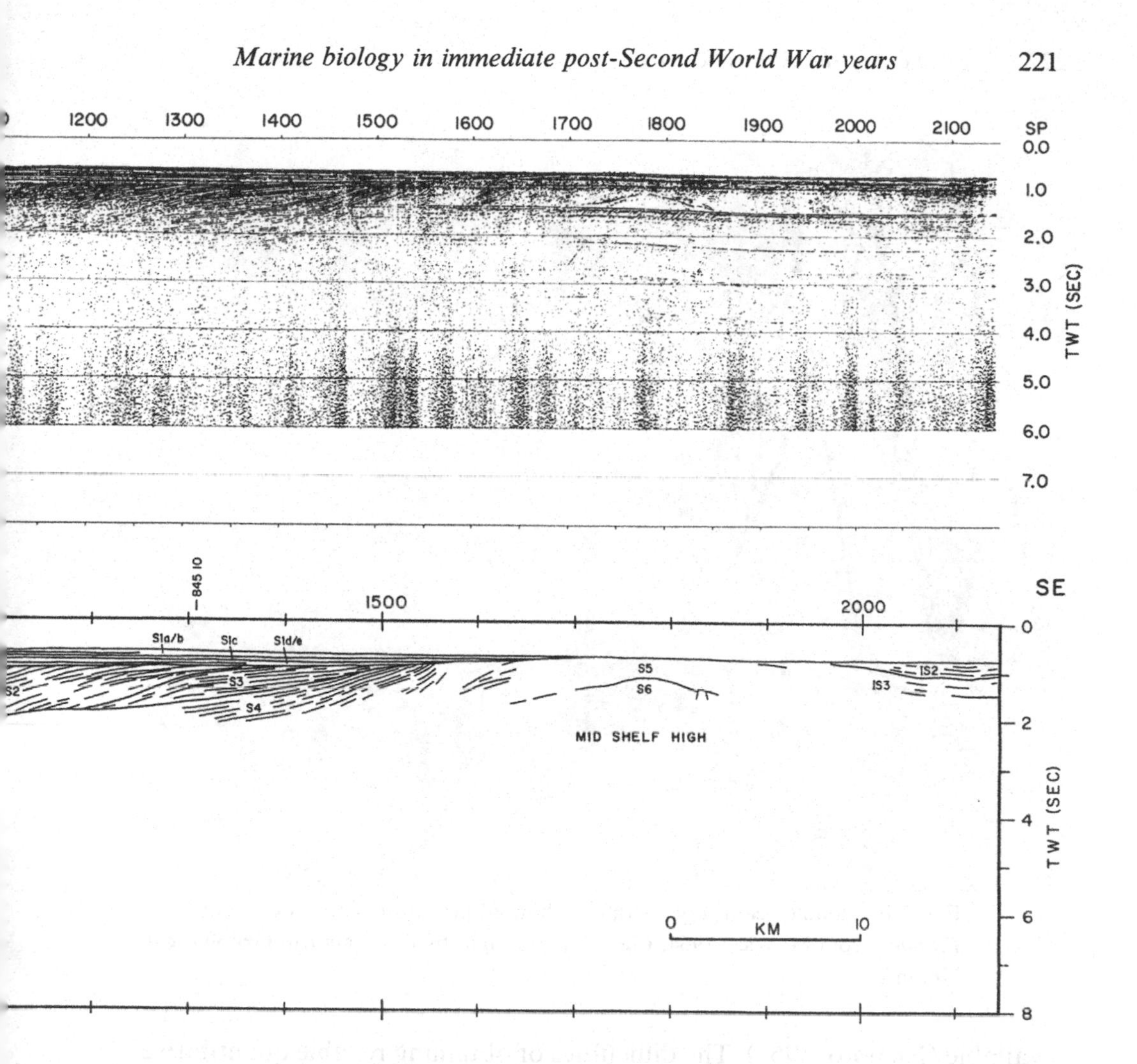

the Soviet ice-breaker *Ob* in 1960 (Klyashtorin, 1961) and the *Umitaka-maru*, the research vessel of the Tokyo University of Fisheries in 1961–62 (Saijo & Kawashima, 1964). In 1966, S. Z. El-Sayed, based at Texas A. & M. University, published the first of an extensive series of radiocarbon studies on the Southern Ocean. All these determinations were made from ships in passage and thus variations in time and space were confounded. However, it became clear that besides areas of high primary production, which as Hart (1934) had realized were mostly areas of comparative stability of the water column, there were vast areas of unproductive ocean (El-Sayed, 1966).

The life-histories and distribution patterns of several zooplankton organisms had by now been studied (David, 1966). Krill, *Euphausia superba*, remained a major object of attention, and details of the mechanisms by which whales harvest it and its own grazing on diatoms were becoming

Fig. 7.14 Double beam trawl with cage hoisted preparatory to shooting on *Eltanin*. From Menzies, 1964, Fig. 13. (Copyright by the American Geophysical Union.)

available (Nemoto, 1966). The difficulties of obtaining reliable quantitative data on krill, which arise from its patchy distribution and swarming habits, had not yet been overcome. Side scan sonar had entered service as an anti-submarine device in 1948 and was developed for study of the topography of the sea floor in the late 1960s (Hackmann, 1984). The idea of using such acoustic scanning for investigations of krill distribution and density was put forward by Mackintosh in 1966. The need for this information was becoming urgent because of the possibility that krill might be harvested on a large scale as human food or animal feed (Nemoto, 1966). Physiological studies were initiated by Mary A. McWhinnie's work (1964) on cold adaptation in *Euphausia superba* and the lobster krill, *Munida gregaria*. She suggested that the former is more adapted to lower temperatures, perhaps through enhanced oxidative metabolism, than the latter, which is sub-Antarctic in distribution.

The sampling of benthic faunas had been confined to particular areas and much remained to be learned of distribution patterns. However, had the wishes of the zoogeographers for more sampling been granted the material collected might well have remained unidentified since the increased pace of collecting in general after the war was overwhelming the small number of specialist systematists. The ridges which appear to connect the Antarctic continent with the major land masses to the north seemed the obvious routes along which fauna might have migrated but as detailed bathymetric information became available it was evident that they lay more deeply than zoogeographers liked. Slight evidence was available to indicate that some molluscs had indeed invaded Antarctica by such routes, notably the Scotia Ridge, but for other groups there was no such evidence (Dell, 1966). For deep water faunas dredges, trawls and grabs remained the only means of sampling, the major improvements since the days of the *Challenger* being the ancillary use of echo-sounders, steel trawl wires, double beam trawls (Fig. 7.14), bottom contact indicators and protection of the trawl bag from entanglement with the trawl wire (Menzies, 1964). However, photography of the sea bottom began to promise useful information. Within the limits of the pack ice, littoral plants and animals are few because of ice-scour, and information came largely from the sub-Antarctic islands. Murray & Barton (1895) had compared Arctic and Antarctic seaweed floras, concluding that there is a general similarity and some identity of species between them, although the largest seaweeds, belonging to the Fucaceae and Laminariaceae, are either Arctic or Antarctic but not both. In the sublittoral, sampling had depended on the random action of a dredge, often bouncing from rock to rock, and Skottsberg (1941), who with Gain (1912), was the first to record the abundance of inshore plants and animals in the Antarctic, considered that a bathysphere or diving suit would be the only means of getting an accurate idea of what was there. Drygalski in 1901 took diving equipment of the type invented by Siebe in 1830 on the *Gauss* and it was used in Antarctic waters to carry out repairs to the ship but, apart from cursory examination of the underface of the ice, was not put to scientific use. The aqualung developed by J.-Y. Cousteau and E. Gagnan in 1943 opened new vistas. Neushul (1961, 1964), who with Argentine colleagues R. O. Dains and A. Carosella was the first to apply this technique (SCUBA) for survey and collection in the Antarctic, described some of the problems.[2] These were not so severe as they might seem, it being possible to spend as much as an hour in the water in a suitably designed wet-type foam-neoprene suit (Fig. 7.15). Neushul's deepest dives, made around the South Shetlands and off the west coast of the Peninsula, were to 20 m. Below the scour of the ice the seaweeds were found to be surprisingly luxuriant. The French began to use diving in studies on infra-littoral ecology around Kerguelen in 1962 (Delépine, 1964). Although SCUBA diving was not to be applied extensively for some years yet, many papers on the descriptive ecology of sub-Antarctic and Antarctic intertidal and sublittoral zones were available for review by Knox in 1966. Studies such as those of Pearse (1965) on the reproductive periodicities of the common Antarctic starfish, *Odontaster validus*, began to show how benthic invertebrates are adapted to continuous low temperature and seasonal pulses of primary production. Physiological studies had scarcely begun.

By now the inventory of the species-poor fish fauna of the Antarctic seemed relatively complete (Marshall, 1964). Most of the hundred or so species known were bottom dwellers belonging to the order Notothenii-formes. Some of them were remarkable in being 'bloodless' (i.e., lacking in haemoglobin) a fact well known to Norwegian whalers but first referred to in print by L. H. Matthews (1931, p. 36). However, studies of physiology, growth and reproduction were so far rather tentative although Wohlschlag (1964) had obtained evidence seeming to show that nototheniids such as *Trematomus bernacchii* were cold-adapted, with elevated rates of metabolism, and that there was good agreement between direct ecological observations and data obtained in the laboratory. One of the most intriguing problems presented by these fish – how they manage to escape the hazard of freezing of their body fluids – was beginning to attract attention. Although Antarctic fish had been caught for the pan since the early days there was no definite suggestion as yet of a commercial fishery.

Around this time biologists were beginning to realize that sea-ice presents a variety of different niches and contains distinct communities of plants and animals. The presence of diatoms in pack ice had been known since the time of Hooker but had been investigated mainly in the Arctic (Horner in Dunbar, 1977) – the discovery of the same species of diatoms in sea-ice from near Greenland, Spitsbergen and Jan Mayen, was one of the facts used by Nansen in formulating his theory of Arctic Ocean currents. Diatoms in ice were briefly mentioned in the report of the *Gauss* expedition but otherwise they were little noticed, except as contaminants in plankton tows (Horner in Dunbar, 1977). The idea – now abandoned – that polymers of the water molecule, which would be concentrated around thawing ice, are particularly favourable for living organisms was put forward as an explanation for the abundance of algae in the ice (Lebedev, 1959). Experimental work began when Bunt (1963) used SCUBA diving to examine algal sea-ice communities *in situ*. From determinations of chloro-

Fig. 7.15 The first SCUBA diving for survey and collection in the Antarctic, 1958. A. Carosella (left) and M. Neushul (right) wearing wet-type foam-neoprene suits. (Courtesy of M. Neushul.)

phyll concentrations and of photosynthesis by the radiocarbon method he deduced that the enormous area of sea-ice might add appreciably to the primary production of the Southern Ocean. Bunt (1967) later obtained sea-ice algae in pure culture and examined their physiological characteristics, finding that they were obligately psychrophilic (i.e., able to grow at low temperatures only) and possessed an extreme capacity for shade adaptation. This was the first physiological study on Antarctic marine microalgae.

7.7 Physical oceanography in the modern period: the advent of remote sensing

Thus far, oceanographic research in Antarctic waters had been to some extent opportunistic and largely organized on a national rather than international basis. With increasing international pressure to exploit the resources of the oceans this has changed. For many years there had been US–Argentine collaboration and this extended in 1967 with the first of a series of International Weddell Sea Oceanographic Expeditions, the US Coast Guard oceanographic contribution to the US Antarctic Research Programme. It involved participants from Argentina, Canada, Norway and West Germany as well as from the US. The icebreakers *Glacier* and *General San Martin*, from the US and Argentina respectively, were used to place current meter arrays far south in the Weddell Sea to measure deep currents associated with the formation of Antarctic bottom water. Severe ice conditions and damage to the *General San Martin* hampered operations in the second and third expeditions but in the fourth, in 1972–73, two of the four arrays placed in 1968 were recovered and were found to have recorded the first winter data from the Weddell Sea (Dater in Bushnell, 1975). From 1968 Soviet oceanographic investigations in the Southern Ocean were extended and included POLEX-South (South Polar Experiment), a long-term, large-scale, study of air–sea interactions around the continent. For the first time mesoscale and seasonal oscillations of the Antarctic Circumpolar Current were measured (Brigham, 1988). In 1981 a joint US–USSR expedition of the US-chartered Norwegian vessel *Polar Duke* and the Soviet vessel *Mikhail Somov* penetrated for the first time into the northern area of the Weddell Sea for short periods in the winter. The West German *Polarstern*, in 1986, was the first vessel to move, more or less freely, in the ice of the Weddell Sea for the entire winter (Hempel, 1988). Sea-ice conditions were recorded along transects and intensive observations made of ice structure and physics.

By this time oceanographic techniques had taken another step forward in sophistication. Measurements on water samples were routinely replaced by those made *in situ* by an electronic package, the CTD (conductivity-temperature-depth, Fig. 7.16), which measured temperature with a platinum resistance thermometer, conductivity by an induction method, and depth *via* pressure measured by a strain gauge technique, all the information being sent up to the ship through the supporting cable for immediate display. The package included reversing thermometers and sample bottles which could be operated to obtain further information as required. Much of the agony of making temperature and salinity obser-

Fig. 7.16 Conductivity/temperature/depth recorder (CTD) with water-bottle rosette and profiling current meter suspended beneath, being used on RRS *John Biscoe*, 1985. (Courtesy of the British Antarctic Survey.)

vations had gone. Apart from using current meters and other standard techniques, eddies were studied from satellite images of the sea surface. These methods were used in another co-operative project organized by the US, the International Southern Ocean Studies, which aimed at improving understanding of the Antarctic Circumpolar Current. This was initiated at the request of the Office of Polar Programs of the US National Science Foundation (USNSF) in 1974 and funded by the Office for the International Decade of Ocean Exploration of the NSF. Operations were principally in the Drake Passage and an area south-west of New Zealand. There was a considerable Soviet contribution and, in all, scientists from 12 countries participated (Neal & Nowlin, 1979). In the final experiment an array of 91 instruments on 24 moorings was deployed in the Drake passage over a year. This programme revealed some of the fine structure of the current, notably showing that it was not a single broad flow but concentrated into two or more relatively narrow jets marked at the surface by the sub-Antarctic and Polar Fronts. Numerical models of wind-driven ocean circulation failed when applied to the Antarctic Circumpolar Current, predicting currents far stronger than that observed (Whitworth, 1988).

Remote sensing, of particular value in oceanography, has been included in the space programmes of the US, USSR, Canada, the European Space Agency, France and Japan (Robinson, 1985). International co-operation is, of course, highly important in the use of satellites, not only in design and launching but in calibration, ground control of instruments, and standardization and transmission of data. A vast amount of data has been accumulated, not all of which has necessarily been properly considered by oceanographers or those interested in the Antarctic. One of the most dramatic advances was that the distribution of sea-ice was mapped, with a completeness never achieved before, by the NIMBUS-5 satellite, launched in December 1972 with scanning microwave radiometer which could detect radiation even in darkness and the presence of cloud. About three days were needed to accumulate data to give adequate coverage of the polar regions, the resolution being about 30 km. The microwave emissivity of ice varies with its salinity and thus satellite radiometry can detect different types of ice but the ground control necessary for this has come mostly from the Arctic. The false colour images produced from NIMBUS-5 data were spectacular and extremely informative about seasonal change. A major feature discovered by this means was the existence in mid-winter of large (*c.* 100 km), ice-free areas, *polynyas* (a word coined by the Soviets). These seem to be maintained by the upward flux of massive amounts of deep warm water balancing a downward flux of cold surface water in convective overturning and are not necessarily recurrent features; the Weddell Polynya appeared in 1974 and disappeared in 1977. Besides being of importance in deep ocean overturning, polynyas greatly alter the extent of exchange of heat and fresh water and must have important climatic effects (Gordon, 1988). Colour scanners on the NIMBUS satellites showed the distribution of phytoplankton but the signal came from the top few metres of water only and there are difficulties in making satisfactory corrections for atmospheric absorption so that no reliable index of primary production could be obtained from them (Walton, 1987; Fifield, 1987). Remote sensing of chlorophyll is more feasible from low-flying aircraft but direct use of remote sensing for detecting krill swarms did not seem possible in 1974 (El-Sayed & Green, 1974) and still does not at the present.

Remote sensing techniques of great promise were used on SEASAT, the first satellite to be dedicated to oceanography, which returned 100 days of data in 1978 before a power failure terminated the mission. The synthetic aperture radar which it carried not only gave images of high resolution but provided altimetric data with a potential resolution of 10 cm. It was thus capable of giving detailed information about sea-ice but was specifically designed for studying sea surfaces, the mean 'slope' of which is related to ocean currents and features of bottom topography. This gives a method of monitoring oceanic circulation which is far superior to the traditional methods, which give data too hopelessly confounded in space and time to permit proper assessment on climatic time scales (Cartwright, 1983). The Global Atmospheric Research Programme (GARP), which extended over a year from December 1978, incidentally gave information about surface currents since the drifts of the 300 or more buoys released into the oceans of the southern hemisphere were tracked by satellites. Also received were the temperature and pressure data which the buoys relayed, which were processed in near-real time by the French ARGOS system (IOC/WMO, 1978; Walton, 1987; Fifield, 1987). Such observations are of particular value in studying a sea area so vast, remote and inhospitable as the Southern Ocean. However, the straightforward application of mid-latitude techniques is not always possible because of complications which arise from the presence of sea-ice and ice-sheets. Thus at present sea-surface temperatures and wind speeds cannot be reliably mapped within pack ice (Maslanik & Barry, 1990).

The welter of information produced in recent years by these various programmes needs time to settle before one can select from it with confidence the salient and substantial items. Meanwhile one can only make a subjective choice of some interesting points.

Recent research has reinforced the idea that the Southern Ocean is a sea area with its own distinct characteristics and a dominant role in determining the circulation of the world's oceans. Features of the circumpolar circulation to which it owes its uniformity, and of the meridional exchanges through which it exerts its effects on other sea areas, are now well established. The suggestion, publicized by Humboldt although he perhaps did not originate it (Deacon, 1971, p. 210), of a density-driven circulation causing a cold bottom current to flow from poles to equator, has been substantiated although the sources of cold water are more circumscribed than Humboldt ever imagined. As we have already seen, the principal source of Antarctic Bottom Water has been thought to be the Weddell Sea. Here, below a depth of about 4000 m the water is cold (*c.* −0.4 °C) and saline (*c.* 34.66 ‰). Mosby (1934) from data obtained on Norwegian expeditions showed that this was formed by mixing in approximately equal proportions of water flowing down from the continental shelf (−1.9 °C and salinity 34.62 ‰) and Antarctic circumpolar water (0.5 °C and salinity 34.68 ‰). Neither the Ross Sea, from the continental shelf of which sinking is restricted by a submarine ridge, nor the Adélie Coast have been found to generate bottom water in significant amounts (Gordon, 1972; Deacon, 1984). To understand properly how Antarctic bottom water is produced it is necessary to take into account the equations of state for water at low temperature when considering the observational data. Approximate methods which work well for warmer waters sometimes give erroneous results for water near freezing point. In an elegant study on this more rigorous basis, Gill (1973) produced a clear picture of the circulation in the Weddell Sea. He concluded that the dense water on the continental shelf acquires its salinity from net brine release during ice formation with the desalinated pack ice continually being removed by being blown offshore. The circulation which results was estimated by Gill to exchange the water on the shelf once every seven years and to remove an amount of salt equivalent to the formation of 0.5 m of ice per year.

The Circumpolar Current is deep reaching and restricted only at the Drake Passage and south of the Australian land mass where the deep channel connecting the Indian and Pacific Oceans is 60° of longitude in length and varies from 20° to 35° of latitude in width. Most of the flow of nearly $235 \times 10^6\ m^3\ s^{-1}$ crosses the Pacific sector with only a relatively small amount turning north (Gordon, 1972).

In one respect the Southern Ocean has not proved to have the crucial influence on the world's oceans that was at one time supposed. John William Lubbock in 1832 envisaged that in the absence of continents the tidal wave would proceed from east to west and that the co-tidal lines would be meridians (see Deacon, 1971, p. 255). However, there are continents in the way and the only clear track right round the earth is at 62° S, and it is the tides in these latitudes that might be expected to have exceptionally large amplitudes. As far back as 1836, Captain Robert Fitzroy of the HMS *Beagle* questioned, on the basis of his own observations, whether the idea which follows from this of 'Atlantic tides being

(a)

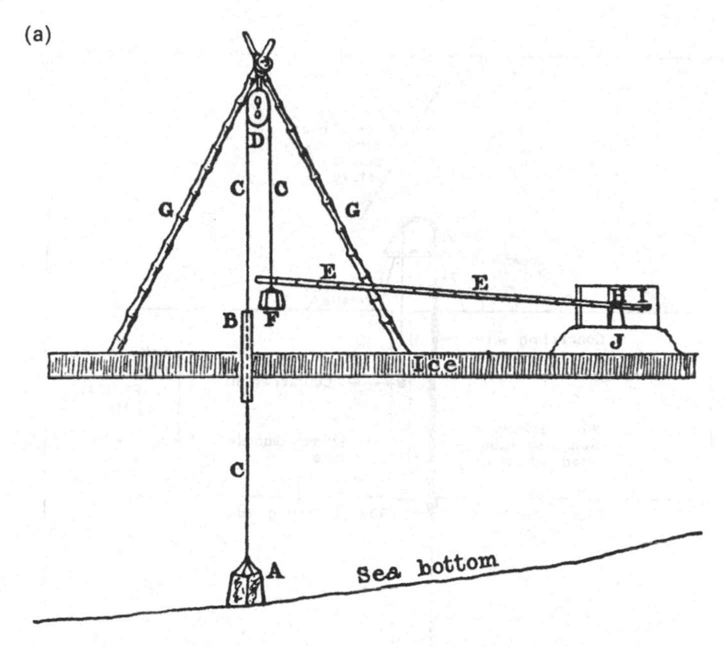

Fig. 7.17(a) Tide-gauge used on the *Nimrod* expedition, 1907–09. A, weight; B, metal tube fixed through the ice and filled with oil; C, wire passing over pulley, D, which is supported by tripod, G, and kept taut by weight, F; E, lever moving on axle, H, and its short end, I, connected with the pen of a barograph, which is the recording part of the apparatus. From Murray, 1911, Fig. 1.

principally caused by a great tide wave coming from the Southern Ocean' was correct. Tidal observations made from shore stations, such as those of the German Transit of Venus expedition on Kerguelen in 1874, can be misleading on such a point. Tidal and associated current measurements, conveniently made from a carefully chosen area of ice adjacent to land, were made by the *Gauss, Discovery, Nimrod* and *Terra Nova* expeditions (Drygalski, 1904; Shackleton, 1909; Murray, 1911; Darwin, 1910; Figs 7.17a and b). Pratt (1960c) studied tides at Shackleton Station on the Filchner Ice Shelf, where there were no fixed points from which conventional measurements might be made, by means of a gravimeter. He found the amplitude of the principal lunar component of the semi-diurnal tide to be about 60 cm, but, of course, this too was an unrepresentative position – tidal dissipation by flexing of ice shelves may be considerable. The question was settled by direct observation with deep-sea pressure sensing capsules. These registered no large tidal waves in the Southern Ocean. In any case such waves seem unlikely since resonance occurs along a single latitude whereas the tide wave extends over 2000 km (Irish & Snodgrass, 1972).

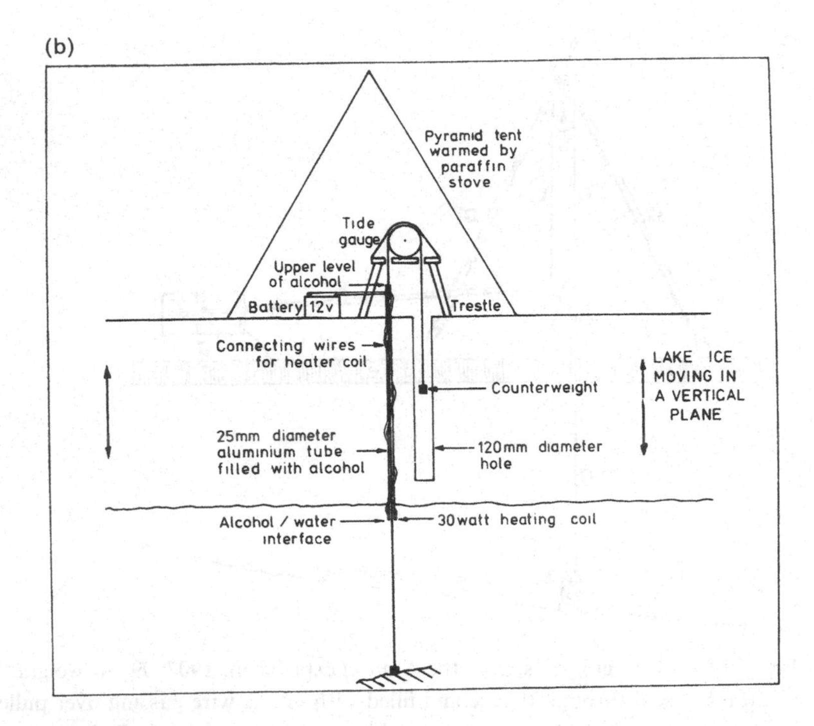

Fig. 7.17(b) Tide-gauging equipment used under extremely cold conditions on George VI Ice-Shelf. It was installed on the ice of a lake which was connected with the sea. The tide-gauge itself was a Russian 'Mashpriborintorg GR-38'. From Bishop & Walton, 1977, *Polar Record*, **18**, 502–5. Fig. 1.

Plots of the surface topography of the ocean obtained from satellite radar altimeters finally disposed of Lubbock's idea. Sea level measurements from SEASAT have shown temporal variations but these extend over months and are related to variation, which may amount to as much as 30 per cent of the mean transport, in the Antarctic Circumpolar Current (Fu & Chelton, 1984).

7.8 Studies on sea-ice and icebergs

Knowledge of sea-ice, so necessary for logistic, climatic and biological purposes as well as being of intrinsic interest, has advanced unevenly in recent years. On the one hand, satellite surveillance, which has just been touched on and which began with automatic picture transmission from ESSA 2 in 1967 (Potocsky & Kniskern, 1970), gave marvellously comprehensive and detailed pictures of its distribution. On the other hand, the complex variability in distribution has eluded precise prediction. There is as yet no widely used model to describe persistent areas of pack ice even in the much frequented vicinity of the Ross Sea (Keys, 1990).

The situation is similar for the most conspicuous features of the Southern Ocean, the icebergs. Their distribution, of course, has always been of concern to navigators. Horsburgh (1830) and Russell (1895) discussed aspects of their distribution and seasonality. More recently, whalers collected data of the same sort. Mawson, on his 1914 expedition, noted down a careful description of the break-up of a large berg (Jacka & Jacka, 1988, p. 241). Wright & Priestley (1922) classified the different types and discussed their formation. Wordie & Kemp (1933) considered the nature of the morainic and bottle-green bergs, which appear to be characteristic of the Weddell Sea area but to be absent from the Ross Sea. They commented on the difficulty of obtaining samples for analysis and on the nature of bottle-green ice, which can have a sharp plane of demarcation from white ice in the same berg and still remains without satisfactory explanation (Dieckmann *et al.*, 1987). Wordie & Kemp also catalogued unusually large tabular bergs seen in the Weddell area. Now, such bergs can be tracked in satellite images. The *Discovery* expedition of 1937 recorded 4,300 icebergs but only 20 could be measured.

The literature on icebergs remains largely descriptive (Keys, 1990) with little about their physics. N. N. Zubov considered the disintegration of icebergs on a theoretical basis, concluding that thermal deformation plays an important role (Lebedev, 1959) but *in situ* observation of their physical properties has been scanty. However, landing by helicopter has enabled the placement of accelerometers, tiltometers and strain meters on bergs and these, together with simultaneous recording of waves and thickness measurements by airborne echo-sounder showed how they respond mechanically to ocean waves (Wadhams & Kristensen, 1983). Numerical models have been constructed to predict the wave height and period likely to cause break-up. Ablation has been found to be most rapid in a narrow zone around the water-line but below this, although the rate of melting per unit area is of an order of magnitude slower, the loss is greater since it takes place over the greater surface (Keys & Williams, 1984). The heat transfer across the ice–water interface is of obvious interest here (Johnson & Mollendorf, 1984). On the *Slava* whaling expedition of 1954–55 to the Antarctic, V. Lebedev and I. F. Kirillov investigated temperature regimes in an attempt to understand the complicated hydrography in the vicinity of icebergs (Lebedev, 1959). The plume of mixed, slightly cooled and diluted, seawater behind a drifting berg is enriched with nitrate from the ice and may entrain nutrient-rich deeper waters but Arctic icebergs do not appear to have any pronounced effects on the phytoplankton in the adjacent water although the situation may be different with the larger Antarctic bergs (Shulenberger, 1983). Neshyba (1977) proposed that icebergs in the Weddell Sea may cause upwelling but the recent work on meltwater patterns around icebergs have been carried out only in the Arctic (Hirst, 1989).

7.9 Biological oceanography: productivity and the pelagic ecosystem

The classic view of the wax and wane of phytoplankton populations is that they are controlled by the availability of light, by the concentrations of limiting nutrients such as nitrate and phosphate, and by zooplankton grazing. In the view of Cushing (1959) the single peak in phytoplankton population found around midsummer in polar waters was the result of exponential growth in the presence of ample nutrients with increasing availability of light and a long lag in the development of herbivores, low temperatures delaying their peak until a time when phytoplankton is already declining as a result of exhaustion of nutrients. The availability of light is a function of the stability of the water column as well as the flux of solar radiation since mixing may carry the phytoplankton down to a depth where there is insufficient light for photosynthesis and thus prevent growth (Sverdrup, 1953). Vitamins, such as B_{12}, thiamin and biotin, have been implicated in the control of phytoplankton growth and may affect its floristic composition in Antarctic waters (Carlucci & Cuhel, 1977) but not total standing crop or productivity (Hayes *et al.*, 1984). Whitaker (1982), who was the first to follow the sequence of events over a significant period at one station in Antarctic waters, found that the decline in phytoplankton biomass in Borge Bay, Signy Island, South Orkneys, began before nitrate was exhausted from the water, when light intensities were still high, and when grazing by zooplankton was not appreciable. Maximum concentrations of nitrate and phosphate in the Southern Ocean are usually some five times higher than those of productive inshore waters in temperate regions and it has been generally observed that these nutrients are never limiting for phytoplankton in Antarctic seas. There is, however, evidence that silica, a specific requirement for the growth of diatoms which are the main component of Antarctic phytoplankton, may sometimes be in short supply in surface waters (Le Jehan & Treguer, 1983). Whitaker (1982) concluded as Gran (1931) and Hart (1934) had done before him, that stability of the water column was the decisive factor controlling phytoplankton growth. He found that phytoplankton began to increase as fast ice disappeared in the spring and continued to develop rapidly so long as melting ice was stabilizing the water column by diluting the surface layers with freshwater. However, disappearance of the ice removed protection from wind and in each of two years over which observations were made the decline of the phytoplankton began with a period of high winds. Nevertheless, it is too simplistic to think that primary production is strictly correlated with stability. A study carried out by the BAS off King George Island, South Shetlands, showed that a patch of high phytoplankton biomass was retained within an eddy but water column stability was not any different within the eddy from elsewhere in the Bransfield Strait and growth did not appear to be enhanced. The patch seemed to have been transported into the area and then caught by the eddy, which, however, was not 'water-tight' and leaked high phytoplankton concentrations down current (Heywood & Priddle, 1987). Elsewhere, in temperate and tropical seas, the physical structure of the water mass has been found to be a dominant factor in determining phytoplankton abundance and it is clear that in the Antarctic close collaboration between physical and biological oceanographers is essential for progress in this field. The patchiness of phytoplankton in the Southern Ocean is only one factor among several which makes estimation of its total contribution to world primary production a difficult matter. Although some areas have rates of primary production among the highest reported anywhere, others are less productive and Holm-Hansen *et al.* (1977) estimated the total as 0.65×10^9 t of carbon fixed per annum in an area of 3.2×10^7 km^2, which suggested that as a whole, Antarctic waters are less than half as efficient as other parts of the world's oceans.

The concept of primary production has, however, become nebulous in recent years. This is because of the growing realization that much of the carbon fixed by photosynthesis is retained only transiently within the cells of phytoplankton. Apart from products of photosynthesis which are immediately respired, some are released into the surrounding water as extracellular products of healthy cells, and, of course, much is eaten by predators which may liberate a good deal of soluble material from them by sloppy feeding. Dissolved organic matter from whatever source may be taken up by bacteria. A direct coupling of bacterial and algal growth in a marine habitat seems to have been first demonstrated *in situ* by Grossi *et al.* (1984) in a study on sea-ice microbial communities in McMurdo Sound. Since the beginning of the 1980s it has been recognized that a considerable proportion of the photosynthetic product passes quickly into this 'microbial loop' and at what point it should cease to be regarded as part of primary production is difficult to decide (Williams, 1984). Kriss *et al.* (1967) were the first to make a world-wide survey of the distribution of marine bacteria. In cruises in the Indian Ocean in 1956–57 which went as far south as 69° 46′ S they made counts by the classical plate method. The bacteria which they isolated were classified into various biochemical groups and the conclusion drawn that the population of heterotrophic bacteria in Antarctic, as well as in Arctic, waters was very low in comparison with that in warmer seas. The proportion of strains decomposing protein and fermenting sugars, however, was several-fold higher than for the equatorial regions. Plate counts are now known seriously to underestimate the number of viable bacteria in seawater and a further source of error in Kriss's work was that cultures were grown at temperatures far above those in the sea. However, measurements of rate of microbial uptake of various organic substrates by Morita *et al.* (1977) showed that although these were low north of the Polar Front they were moderate to high south of it. Studies of the effect of temperature on uptake indicated the presence in waters from the Ross Sea of microbial populations adapted to low temperatures, with activities at 0° C comparable to those of the microflora in more temperate waters. High microbial activity was also found by Hodson *et al.* (1981) in McMurdo Sound, and by Hanson *et al.* (1983) in the waters of the Drake Passage – who found, by direct counts, between 10^4 and 10^5 bacterial cells per ml as compared with the 10^5 per ml which is usual for temperate oceanic waters. These and other results show the presence in the Antarctic of a microbial loop just as active in breaking down dissolved organic matter, regenerating mineral nutrients and producing new cell material as that in warmer waters. However, dissolved organic matter seems to assume a particular significance in the Antarctic. The Ross Ice Shelf Project (see p. 274) found life below the Shelf more than 400 km away from the open sea. Low bacterial densities and sparse crustacea were found. Their basal metabolism was low, comparable with rates found in the abyssal ocean, and there were no benthic organisms in the sediments (Azam *et al.* 1979). Fish had previously been found below shelf-ice in George VI Sound, Alexander Island, 100 km from the nearest open sea but the evidence that these subsisted on organisms that had grown under the shelf is not conclusive (Heywood & Light, 1975). Under the Ross Ice Shelf there could have been no photosynthesis, and particulate organic matter from the open sea is scarcely likely to have been carried so far underneath the shelf. The only possible source of energy for life appears to have been dissolved organic matter made available to the larger forms via bacterial assimilation. Radioisotope studies indicate that some of the water beneath the shelf is exchanged with the open sea in a period of less than six years so that fresh, dissolved organic matter could be brought in (Michel *et al.*

1979). Beneath the inner part of a shelf dissolved organic matter is excluded from ice crystals. This process must occur too, in pack-ice and there may contribute, together with excretion of dissolved organic matter by photosynthesizing algae, to the considerable heterotrophic activity of the biota which it contains (Sullivan & Palmisano, 1984; Grossi *et al.* 1984). The abundance of choanoflagelletes at the ice-edge in the Weddell Sea (Buck & Garrison, 1983) is one indication of intense heterotrophic activity in this zone.

For studies on Antarctic zooplankton the greatest desideratum has been a reliable method for estimating population densities of that highly mobile and intermittently gregarious organism, *Euphausia superba*. The traditional plankton net as used in the *Discovery* Investigations had the disadvantage, realized at the time, that the bridle and release gear gave sufficient warning of the approach of the net to enable krill to take avoiding action. The electronically controlled rectangular midwater trawl, devised by Clarke (1969) for other purposes, being much larger and without attachments directly in front of the net, was found to be more effective. An improved version of the Hardy plankton recorder, equipped with an acoustic opening and closing device, gave invaluable information on krill swarm structure (Bone, 1986[3]). Echo-sounding as a means of estimating krill biomass has been used following preliminary studies carried out in collaboration between BAS and O. A. Mathisen of the College of Fisheries, Washington, USA. Integrated data from an echo-sounder operating at an appropriate frequency can be converted to the acoustic parameter – mean volume backscattering strength – from which krill density can be obtained assuming a target strength determined experimentally. Calibration of the rectangular midwater trawl against measurements made by echo-sounder has shown that there is still effective avoidance of the trawl during the day but not during the night, suggesting that the avoidance depends on vision (Everson & Bone, 1986). The acoustic method has enabled estimates of krill biomass to be made in particular sea areas. Thus the area of about 140,000 km^2 around South Georgia was estimated to have a late summer standing stock of eight million tonnes of krill, an amount which is only about one-twentieth of that estimated to be consumed by predators such as whales, seals and penguins during the summer. This suggests an improbably high production rate but redetermination of the target strength of krill, to which the acoustic method is very sensitive, has given a lower value which leads to much higher values of krill abundance in line with those determined indirectly from predator consumption rates and life-span (Everson *et al.*, 1990). A further problem was that krill virtually disappeared from South Georgian waters in the 1983 winter (Atkinson & Peck, 1988). The *Polarstern* found dense aggregations of krill feeding on ice algae in the pack of the marginal zone between 59° and 76° S of the Weddell Sea in 1986 – previous evidence on this has been conflicting (Marr, 1962). Such a remarkable ability to change from pelagic life to a different habitat and food source in winter may explain the krill's success (Marschall, 1988). Knowledge of the biology and ecology of krill, which was greatly extended by BIOMASS (see p. 242), was reviewed by Miller & Hampton (1989). There is still much to learn about krill but other zooplankton forms cannot be ignored and, whilst it seems to be a key organism in some areas, it is an over-simplification to regard the Southern Ocean as unique among pelagic ecosystems in being a very simple one centring on this one organism.

The potentialities of krill as a source of protein for human food or cattle feed began to attract attention in 1962, when the USSR started investigations. Two years later the *Akademik Knipovich* was catching krill in Antarctic waters and in 1970 various krill-based foods were marketed in Russia. The Japanese sent their first vessel to try out the krill fishery in 1972. The popular press found this interesting,[4] it often being assumed that krill that was not being eaten by the depleted whale stocks was there for the taking so that amounts of the order of 10^9 tonnes might be harvested. The Food and Agricultural Organization (FAO) of the United Nations was not slow to seize on the possibilities and, initiating a Southern Ocean Fisheries Survey Programme, published reviews on the living resources in general (Everson, 1977), and the harvesting (Eddie, 1977) and utilization (Grantham, 1977) of krill. Soon it was being found possible to achieve catch rates averaging 40 tonnes per hour and the total catch approached perhaps 600,000 tonnes per annum (Laws, 1985). There were, however, difficulties in processing; autolytic enzymes break down the body proteins rapidly so that a catch destined for human consumption must be dealt with inside three hours. Also the exoskeleton contains toxic amounts of fluoride. Nevertheless, krill is sold as a delicacy in the USSR and the Far East and there are possibilities of using the chitin from its exoskeleton in medicine and for various industrial purposes (Nicol, 1989). The need for conservation of krill and other marine resources had not been considered properly when the Antarctic Treaty was drawn up and to remedy this, the 1972 Convention for the Conservation of Antarctic Seals and then the 1980 Convention on the Conservation of Antarctic Marine Living Resources (CCAMLR) were signed. There were objections that what were essentially fisheries agreements had no legality if negotiated in the absence of the existing exploiting parties by a group of nations only some of which had (doubtful) sovereignty in the region involved and that the interests of the Third World ought to have been taken into account (Bonner, 1981). One reply to this was that no state outside the Treaty had made a significant contribution to Antarctic marine research. Research has made it abundantly clear that the situation is far from simple and that unlimited harvesting of krill might well be ecologically disastrous. It should be noted that unrestricted fishing of fin-fish soon reduced stocks to a level at which stable recruitment was doubtfully possible. Around South Georgia there was a decline during the period 1977 to 1981 in fish biomass of 46 per cent, with one species, *Pseudochaenichthys georgianus* being reduced to 2.3 per cent of its original abundance (Laws, 1985). CCAMLR is based on the ecosystem concept and specifies that harvesting should not decrease a population below levels which allow stable replenishment, that ecological relationships among species must be maintained and that no changes in the marine ecosystem should be allowed to occur that are not potentially reversible over two or three decades.[5] So far, the convention has not been applied effectively.

This ecosystem approach was an unusual one with a fishery problem and it has produced two major new advances in marine ecology. One was a simple general model for a multispecies fishery. Management of established fisheries had hitherto depended on the concept of maximum sustainable yield, which aims at regulating the catch of one particular species at a level sufficient to ensure that future yield is not reduced. It was designed to deal with fish at the top of the food chain and assumed, among other things, a steady state in the stock to be exploited, a constant carrying capacity in the environment and a density-dependent mechanism producing a compensatory rise in recruitment to the breeding stock when numbers in the stock are reduced. This concept has been of doubtful value in regulating the exploitation of fin-fish and its assumptions certainly do not apply to krill, which occupies a middle position in the food chain. A model produced by May *et al.* (1979) took into account the sort of interactions with other species which affect krill numbers and showed clearly that the maximum sustainable yield concept in its conventional form cannot apply at this trophic level, thereby pointing the way to a new approach to fisheries problems. The other major advance has been to quantify the activities of sea-birds so that they can be included in marine ecosystem models. The breeding habits and lack of fear of man shown by Antarctic sea-birds make it relatively easy to monitor their foraging, food intake and growth, a matter which will be discussed further in chapter 10. Such work, which has mainly been carried out by BAS at Bird Island off the northern tip of South

Georgia, has shown that sea-birds are a substantial component in the pelagic marine ecosystem (Croxall & Prince, 1980).

The feeding and foraging behaviour of seals have been studied and some idea gained of their importance in the marine ecosystem (Laws, 1984). Again, much of our knowledge has come from land-based studies and will be dealt with later (chapter 10) but the three pelagic species – the crabeater, the Ross and the leopard seals – together with the other species in their non-breeding phases, must be observed at sea. During the IGY, seal counts were made from US ice-breakers in the pack ice of the Ross Sea and the Vincennes Bay area of the Indian Ocean sector (Eklund, 1964) but the census methods used were open to criticism (discussion in Carrick *et al.* 1964, pp. 455–7). A census in the Pacific sector was carried out by Gilbert & Erickson (1977) using aerial photography, ice-breakers and helicopters to sample strips orientated north–south to penetrate the ice to a maximum distance from the outer edge. The counts for each species were adjusted to correct for differences in proportions of seals on the ice according to the time of day. Information was also obtained on habits and associations between species. Such work has led to reasonably reliable estimates of total populations (Laws, 1984). The crabeater seal, with a population size of between 15 and 30 million, proved to be the world's most abundant seal.

It may be noted that another animal group with an undoubtedly great impact in the food web of the Southern Ocean, the squids, is still largely unknown. Their abundance has been deduced for a long time from their numerous remains in sperm whale stomachs but the fact that the species so identified are different from those which can be caught in research trawls shows the extent of our ignorance (Clarke, 1985).

The subject of whales has become highly emotive with at the same time an insufficiency of information on which to form firm conclusions as to what is happening to the populations of different species. It was unfortunate that the *Discovery* Investigations, admirable though they were in other respects, did not include quantitative studies on the population dynamics of whale stocks and their reactions to exploitation. From a general survey, in which he was mainly dependent on statistics from the whaling industry itself, Harmer (1931) concluded that stocks were rapidly deteriorating following the introduction of pelagic whaling factories. The Second World War gave some respite, and the establishment by the United Nations in 1946 of the International Whaling Commission, a pioneer organization in the field of conservation, gave some grounds for hope. This hope has scarcely been realized. To some extent this was because of the lack of good quantitative data on which to base decisions but also it was because the

Commission could not effectively enforce decisions once they had been taken. It was only after the Commission appointed a small committee of scientists in the early 1960s to advise it that necessary information was assembled. Even with clear evidence of catastrophic decline of stocks there was difficulty and delay in taking action because of the commercial and political interests involved but now the pendulum has swung in the other direction and somewhat extreme conservationist views predominate (Gulland, 1987). Paradoxically, with the cessation of large scale whaling, an important source of information necessary for management of whale stocks has been extinguished.

However, it has been found possible to apply molecular genetics techniques to 200–300 mg skin samples collected from free-ranging whales by means of a dart-tipped arrow, fired from a cross-bow and tethered to a fly-casting line. DNA fingerprints usually allow certain identification of individuals, and the technology to determine fundamental information about populations without killing whales is thus at hand (Hoelzel & Amos, 1988).

In general, it is true to say that our concept of the Antarctic marine food web is qualitatively exact but woefully inadequate quantitatively (Walton, 1987).

7.10 BIOMASS

For many years SCAR played little direct part in oceanographic investigation, considering that with so many international bodies interested it had no need to maintain a permanent working group on Southern Ocean research and that it would be more appropriate for SCOR to provide a focus for physical and chemical oceanography in this area.[6] Nevertheless there has been one major oceanographic project organized by SCAR itself – Biological Investigations of Marine Antarctic Systems and Stocks (BIOMASS). In the 1970s the exploitation of the living resources other than whales was already under way in the Antarctic, indeed the commercially valuable fish, *Notothenia rossii* was already fished out around South Georgia.[7] BAS began to think of switching effort into a programme of off-shore research.[8] Likewise recognizing the need for a scientific background if exploitation was to be managed wisely, SCAR set up a subcommittee on the Marine Living Resources of the Southern Ocean in August 1972. Later, upgraded under the title SCAR Group of Specialists on Living Resources of the Southern Ocean and co-sponsored by SCOR as its Working Group 54, it organized, at a meeting in Cambridge in 1975, an international conference which was held in Woods Hole, Massachusetts, USA, in August

1976. At this conference background review papers were presented (El Sayed, 1981) and the development of an international programme of study was discussed. The biology, ecology, population dynamics and distribution of krill were to be the primary objects of investigation, using up-to-date methods, but potentially important stocks of pelagic and demersal fish and cephalopods were also to be looked at. The necessity of a solid background of physical and biological oceanography for the success of the project was recognized and the elaboration of models of the quantitative interaction between the different components of the ecosystem was envisaged. The existing Group of Specialists was entrusted with the scientific planning. SCAR, of course, could not fund the programme and the function of the Group was to co-ordinate methods and techniques, programmes for field work, and analysis of data in the different national contributions. To this end, working parties were arranged and a comprehensive series of handbooks of methods and reports of meetings issued. Operations were to concentrate on the Atlantic sector but to include Drake Passage and its western approaches. Preparatory work was to begin in 1977, and data analysis and synthesis were envisaged as being completed by 1986. Within this there were to be two main international multidisciplinary multiship experiments, the First International BIOMASS Experiment (FIBEX) in 1980–81, and the Second International BIOMASS Experiment (SIBEX) in 1983–84, the latter being perhaps the largest multiship exercise so far in biological oceanography. Research vessels from Argentina, Australia, Brazil, Chile, Federal Republic of Germany, France, Japan, Poland, South Africa, UK, US and USSR participated in SIBEX, which was focused in the Atlantic sector but included cruises in the Indian and Pacific sectors also.[9] The object of SIBEX was to build on the results of krill biology and population dynamics obtained by FIBEX and to extend them, particularly in the direction of trophodynamics, to include the roles of the higher predators – fish, birds and mammals.[10] Whilst working to a co-ordinated plan, there was some overlap of cruise tracks – not at all a bad thing – and, as will have been realized from perusal of the section on biological oceanography above, many contributions formed part of a nation's own ongoing programme of research. Thus the UK cruise was part of the BAS's Offshore Biological Programme which had been started in 1977 and still (1990) continues. Being mainly in the same area as that covered in detail in the *Discovery* Investigation BAS's contribution thus gave the possibility of studying changes with time. An innovation in Antarctic oceanography introduced by BAS into the BIOMASS programme was to make a Lagrangian rather than a Eulerian approach in the study of krill populations by following individual parcels of water, using icebergs as markers.[11] Later, another innovation was the use of a microprocessor to display temperature and salinity data in real time so that water masses could be identified on the spot. Offers to provide facilities for the database needed to deal with the vast amount of information generated by BIOMASS were made by the Alfred-Wegener Institut at Bremerhaven in Germany and by the BAS at Cambridge, UK, the latter being chosen because of the more advanced state of its computer facilities.[12] The host institute, however, has to carry the major part of the cost of running the database. The data, which are subject to a straightforward validation test before being loaded on to the computer, can be subjected to various statistical treatments and used to produce maps, diagrams and graphs. The availability of the database enables the development of mathematical models, for example, of krill population dynamics. Although the US has contributed to BIOMASS, notably by providing the chairman of its executive committee, S. Z. El-Sayed of Texas A. & M. University, its interest has not been commensurate with its resources. This may be partly because US vessels have never fished for krill but the National Science Foundation was sceptical of the programme's value, considering it to be diffuse (Quigg, 1983, p. 83). BIOMASS is planned (1990) to end with a final analysis and evaluation meeting in Bremerhaven, Germany, in 1991.

7.11 Inshore marine biology

Much of the early work on the marine biology of the Antarctic was done from ships and, as already mentioned (p. 144) the first shore-base specifically for such studies was the *Discovery* Committee's laboratory on South Georgia, completed in 1925. It was over 30 years before others were established, mostly at already existing stations: at McMurdo (US) in 1958, Mirny (USSR) *c.* 1958 on Haswell Islands, Signy (UK) in 1962, Palmer (US) in 1965, Kerguelen (France) in 1968, Syowa (Japan) in 1965 and Henryk Arçtowski (Poland) in 1977. It may be noted that 1964 was the first year of the field programme of the International Biological Programme but except at Syowa (Headland, 1989) there was no direct connexion between this and the increase of biological work in the Antarctic. The establishment of these permanent shore-based laboratories allowed more sophisticated investigations on the general biology, ecology, physiology and biochemistry of marine organisms. Whereas, hitherto, most work has been carried out on preserved material by specialists without experience of Antarctic conditions, year-round direct observations could now be made with the result that many confusions were cleared up and significant advances in understanding made. To a large extent the pattern of these researches in the Antarctic has followed that of similar work elsewhere. Mostly the projects have been modest, carried out by one or two individuals and liable to suffer in competition with 'big' science. Thus BAS diverted resources from inshore marine biology in order to support its offshore programme.[13] Nevertheless, shore-based research has contributed much important knowledge of the biology of Southern Ocean marine invertebrates and fish. The bases have also been the source of living material to be transported to home laboratories for further studies. Bunt (1967) successfully cultured several species of sea-ice algae from McMurdo Sound during the 1962–63 austral summer, some of which were later obtained in axenic culture and used in physiological studies in the US. Special life support systems enabled living fish and invertebrates to be conveyed first by ship and then by road from the Antarctic to BAS's laboratories in England for the first time in 1974.

Work since IGY has confirmed the impressions formed before then that Antarctic benthic communities are characterized by high biomass, gigantism, high levels of endemism, an incomplete range of invertebrate groups and the relative absence of pelagic larval stages. The Polar Front emerged as an effective biogeographical barrier for benthos and for plankton, and species diversity appeared as remarkably high (White, 1984).

Cold adaptation and cryobiology are subjects of concern to biologists generally, albeit of particular interest in the polar situation. Because of its locally high productivity it was supposed that phytoplankton in the Southern Ocean must have special physiological features suiting it to low temperatures. Steemann Nielsen & Hansen (1959) supposed that cold water phytoplankton might offset the slowing of enzymic activity by low temperature by having higher concentrations of enzymes in their cells. However, evidence seeming to support this idea was obtained with a faulty experimental approach and Jacques (1983) concluded that Antarctic phytoplankton have no features of photosynthesis that might be considered as adaptations to low temperature. Their growth rates are what would be expected of temperate water phytoplankton at near freezing point. They are, nevertheless *psychrophilic*, that is they are only capable of growth at low temperatures, a feature which appears to depend on their respiration increasing relatively more rapidly than photosynthesis when the temperature is raised so that net photosynthesis decreases to zero at about 10° C (Bunt *et al.*, 1966). Studies of the relation of metabolic rate to temperature in Antarctic fish was initiated by Wohlschlag (1964) at McMurdo Station in 1958, some five years after similar work on Arctic species. However, the relatively high rate of metabolism which Antarctic invertebrates, as well as fish, have been supposed to show is an artefact of experimental procedure and routine basal rates are much the same as would be expected of warm-water species at the same temperatures (Clarke, 1980). In this respect, therefore, these animals are not cold-adapted but many Antarctic species have been found to contain enzymes which are more active at temperatures around zero than are their counterparts in organisms from temperate or tropical waters (Walton, 1987). There is also a definite adaptation in respect to freezing which has been particularly studied in Antarctic fish. The ambient temperature can be as much as a degree below the freezing point of a fish's body fluids. Supercooling without freezing is possible and a fish can survive providing it avoids contact with ice crystals or other freezing nuclei. A species that lives in deep water away from ice can do this. Fish frequenting shallow waters need some other way of avoiding freezing and do it by producing an 'antifreeze'. These were first discovered in the blood of Arctic fish by Per Scholander and colleagues and then extensively studied by Arthur L. DeVries at McMurdo Station in Antarctic fishes (DeVries & Somero, 1971; see also Feeney, 1974). The body fluids of fish likely to be exposed to ice crystals were found to contain unique high molecular weight glycoproteins or peptides which produce, together with the inorganic ions and small molecular weight organic compounds present, a depression of freezing point of about one degree Celsius. It seems likely that these antifreeze glycopeptides, the comparative biochemistry of which has received much attention, exert their effects by direct adsorption to the ice lattice, inhibiting its growth (Ahlgren & DeVries, 1984). Such studies on cold adaptation and energetics in Antarctic fish have radically altered our view of life in cold seas.

Endnotes

1 *IGY World Data Center A, Oceanography*. Production measurements in the world ocean part II. *IGY Oceanography Report* 1961, No. 4, 517–19.

2 Personal communication from Dr M. Neushul.

3 British Antarctic Survey Report 1979–80, p. 48; 1982–83, p. 39–40.

4 Hagan, P., 'Daily Krill and Chips? Could the Krill, a small Antarctic crustacean alleviate our desperate world food shortage?' Daily Telegraph Magazine No. 552, 27 June 1975, p. 9.

5 *SCAR Bulletin* January 1981. No. 67.

6 *SCAR Bulletin* No. 73, p. 39.

7 BAS Scientific Advisory Committee Minutes *1*, 1971.

8 Ibid., *3*, 1972.
9 *BIOMASS Newsletter* 1984, **6**(1).
10 Ibid., (1982) **4**(2).
11 *British Antarctic Survey Report* 1978–79, p. 46.
12 *BIOMASS Newsletter* 1984 **6**(1).
13 *British Antarctic Survey Report* 1977–78, pp. 40–52.

8

The earth sciences

8.1 The geological outlook at the beginning of the twentieth century

In the early part of the nineteenth century there had been controversy, fired by philosophical and theological prejudice, between those with 'uniformitarian' and those with 'catastrophic' views on earth history. This passed most geologists by and in the less contentious fields of stratigraphy and palaeontology there was steady advance. By the end of the century there was some approach towards agreement on a general framework for geological thought (Greene, 1982). Many British and American geologists were still followers of Lyell, for whom geology was the application to the understanding of the history of the earth of chemistry, physics, mineralogy and biology, the laws of which he supposed to have remained always the same. That much was not disputed but the other idea included in Lyell's uniformitarianism, namely that the magnitude of geological forces has never been greatly different in the past from what it is at present, made it difficult to give a convincing explanation of mountain formation and was unacceptable to geologists on the continent of Europe. There, Léonce Elie de Beaumont (1798–1874), who held that mountain ranges are formed by uplifting arising from contraction of the earth as it cooled, had been particularly influential. He was followed by Eduard Suess (1831–1914) whose *Die Entstehung der Alpen*, published in 1875, marked the beginning of a new movement in Europe towards structural-dynamic studies of the earth's crust. Uniformitarian and catastrophic views were combined by Suess into an acceptable theory visualizing earth history as divided by episodes of violent disturbance into periods of relatively quiet readjustment. He thought mountain ranges were formed by lateral movements of whole sections of crust over hundreds of kilometers (Suess 1904–9). At the time of Suess's death in 1914 the contraction theory was manifestly inadequate but no other suggestion as to what might be the motive power for mountain formation was forthcoming. The continental glaciation theory of Louis Agassiz (1840), based on field work in his native Switzerland and later in Scotland and North America, gave a satisfactory explanation of many other features of physical geology, including the occurrence of erratic boulders which British geologists, following Charles Darwin's comments on observations in the Southern Ocean (p. 50), believed must have been transported by icebergs.

There was little hard information available specific to the Antarctic but those who were thinking about expeditions to those parts at the turn of the century saw that there could be many interesting problems to be tackled. Georg Von Neumayer, addressing the International Geographical Congress in 1895, quoted Hanns Reiter's[1] attempt to fit Antarctica into the picture of the development of the crustal structure of the earth put forward by Suess. Reiter had concluded that it should be regarded as a continent but saw structural links between its coastal features and islands with the mountain chains of South America and Australasia. Murray (1898) emphasized the importance of searching for more fossils if the geological history of the continent was to be elucidated. W. T. Blandford, ex-Geological Survey of India and author of the section on geology in *Antarctic Manual* (Murray, 1901), was more specific, pointing out that the *Glossopteris* flora characteristic of the Permo-Carboniferous of the southern hemisphere seemed to have had its origin in the Antarctic region and that if it were to be found there this would suggest that the continent must once have been linked with others *via* land bridges, an idea that went back as far as Hooker in 1847 (see p. 89). The Duke of Argyll, an amateur geologist of firm catastrophic persuasion, considered that Quaternary geology would benefit most from further exploration in the Antarctic which, as the region where glacial activities are at their maximum, would have most to show of how glacial agencies operated in the ice age. The importance attached to Antarctic geology at this time is evident in the choice of leaders for the three expeditions which sailed in 1902; Nordenskjöld was a geologist, Drygalski, a physical geographer, and the Royal Society's nominee for the position eventually filled by Scott was the geologist Gregory. All the expeditions up to 1914 included one or more geologists.

8.2 Geological reconnaissance

The first task was obviously to undertake geological survey of the sort that was by then standard in all other continents (the Geological Survey of Great Britain had been instituted in 1835). Blandford (in Murray, 1901) gave admirably clear instructions that would enable ex-

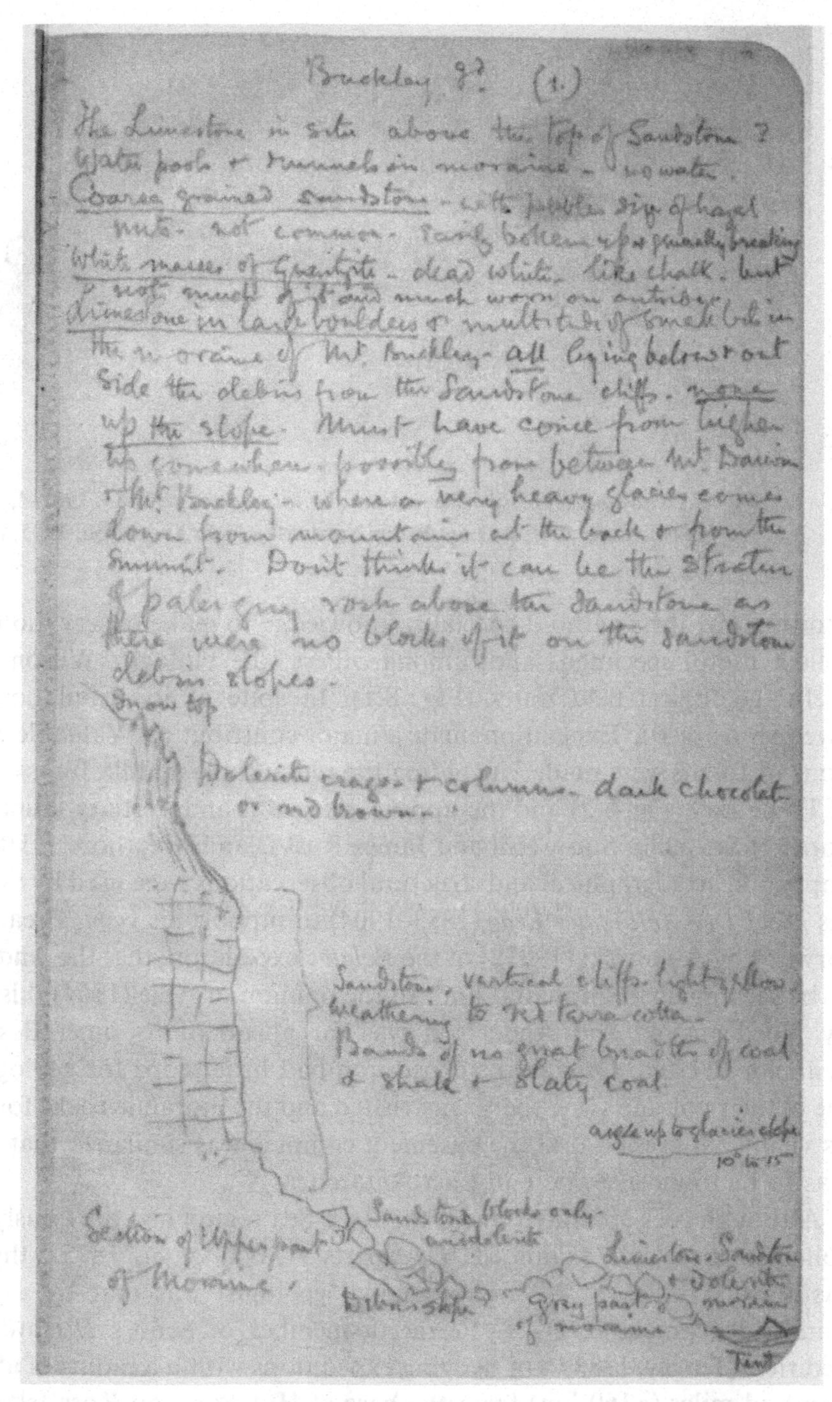

Buckley 9th (1.)

The Limestone in situ above the top of Sandstone ?
Water pools & runnels in moraine – no water.
Coarse grained sandstone – with pebbles size of hazel nuts – not common. Easily broken up & gradually breaking
White masses of Quartzite – dead white like chalk. but not much of it and much worn on outside.
Limestone in large boulders & multitude of small bits in the moraine of Mt Buckley – all lying below & outside the debris from the Sandstone cliffs. none up the slope. Must have come from higher up somewhere – possibly from between Mt Darwin & Mt Buckley – where a very heavy glacier comes down from mountains at the back & from the summit. Don't think it can be the stratum of pale grey rock above the sandstone as there were no blocks of it on the sandstone debris slopes.

Fig. 8.1 Page from E. A. Wilson's sketchbook with notes on the geology of Mount Buckley, 9 February 1912. (Courtesy of the Scott Polar Research Institute.)

Fig. 8.2 *Otozamites*; fossil collected on the *Antarctic* expedition, 1901–4, from the Jurassic flora at Hope Bay. From Nordenskjöld & Andersson, 1905, p. 447.

pedition members without specialist knowledge to make observations and collect useful specimens and, among others, the biologist Wilson made useful geological field notes (Fig. 8.1). In spite of its tribulations the Swedish Antarctic Expedition made a major contribution. Valuable collections of fossils were made,[2] including the remarkable middle Jurassic flora of Hope Bay (Fig. 8.2) and the upper Cretaceous and Tertiary faunas and floras of Seymour, Snow Hill and James Ross islands (Nathorst, 1904). Its important stratigraphical and structural observations were used by Suess in his great *Das Antlitz der Erde* (1883–1904) to support his view, already put forward by Arçtowski (1901b) of the *Belgica* expedition, that the Andes are to be seen again in Graham Land. In the opinion of Adie (1964) this work by Nordenskjöld and Andersson was far ahead of its time. Beset off Wilhelm II Land, Drygalski's expedition had little scope for geology but the extinct volcano Gaussberg was visited and the morainic rocks found in its vicinity suggested that the basement complex was similar to that which was to be found elsewhere in East Antarctica.

Although rock specimens from the Ross Sea region had previously been collected and later examined by experts (David *et al.* 1895), the first sustained work there by a professional geologist was that by a young Cambridge graduate, H. T. Ferrar, a member of Scott's *Discovery* expedition (Tingey, 1983). On sledging expeditions within a radius of about a hundred miles (*c.*160 km) from the base at Hut Point on Ross Island, he established the basic stratigraphy of the mountains of Victoria Land as an igneous and metamorphic basement complex overlain by mainly horizontal sedimentary strata, the Beacon Sandstone, intruded by dolerite sills and dykes (Ferrar, 1907). The Beacon sandstone was sparsely fossiliferous and the few plant fossils found did not suffice at the time to determine its age more precisely than late Palaeozoic or early Mesozoic. Only later did further work on his specimens reveal that they contained *Glossopteris* remains and were therefore late Palaeozoic in age (Edwards, 1928). The distinguished Australian geologist, Professor T. W. Edgeworth David FRS (see Fig. 8.10), accompanied Shackleton to Ross Island and was then persuaded to stay instead of returning to Australia on the *Nimrod.* Together with David's former student, Douglas Mawson, lecturer in mineralogy and petrology at the University of Adelaide, and Raymond Priestley, fresh from the University College of Bristol, Shackleton had a powerful geological team. There was a spectacular first-ever journey by David and Mawson to the South Magnetic Pole and the first inspection of the crater of Mount Erebus. The work begun by Ferrar was extended, particularly by Priestley's further exploration of the Ferrar Glacier area. Shackleton did not take any of his geologists with him on his furthest south journey but himself brought back valuable information and specimens. The Beacon Sandstone formation was found to extend to the Beardmore Glacier and fossils of *Archaeocytha* found in erratics from that area suggested that somewhere there were sedimentary rocks of Cambrian age between the basement and the Beacon Sandstone. Outcrops of coal-bearing strata were found near the head of the glacier. The monograph by David & Priestley (1914) on the geology of southern Victoria Land remained for many years the authoritative account. The major structural feature which was recognized was named the Antarctic horst of South Victoria Land, a block rising to heights ranging from 8000–15,000 feet (2,440–4,570 m), 1,000 miles (1,610 km) long and 50–100 miles (80 to 160 km) wide. On its eastern edge the displacement appeared to be from 5,000–6,000 feet (1,520–1,830 m) but the occurrence of down-faulted blocks on its western flank could only be inferred.

Scott has been criticized for taking his second expedition to the same area as his first, when there was obviously so much of the Antarctic still to be explored, and Mawson, who felt that scientific discovery was worth more than getting to the Pole (Mawson, 1964; Jacka & Jacka, 1988, p. 53), declined Scott's invitation to join him. The geologists who did go with Scott were Priestley (see Fig. 8.10) and two of David's former students, Frank Debenham (see Fig. 5.10) and Thomas Griffith Taylor (see Figs. 5.10 & 8.10). They extended the work of their predecessors in the area. Taylor and Debenham worked south along the west coast of McMurdo Sound and the

Ross Ice Shelf, one of their notable discoveries being of fossil fish scales of Devonian age in the lower Beacon Sandstone. Priestley worked along the Victoria Land coast from Cape Adare southwards. His party was forced to overwinter in the field in two successive seasons because the relief ship was unable to reach it and the privations it endured naturally limited the amount of science which could be done. On their return from the Pole, Scott's party spent a few hours geologizing at Mount Buckley (now Buckley Island) where Shackleton had collected previously (Fig. 8.1). Scott recorded in his diary:

> 'Wilson with his sharp eyes has picked several impressions the last a piece of coal with beautifully traced leaves in layers also some excellently preserved impressions of thick stems showing cellular structure.
>
> (Scott Diaries, 8th Feb. 1912.)

At Wilson's request and without demur from Scott, 35 lbs (16 kg) of rock samples from there were carried with them for the remaining 50 days of their journey and this extra weight may well have contributed to their deaths. To dismiss these specimens as worth almost nothing and the carrying of them as 'a pathetic little gesture to salvage something from defeat at the Pole' as Huntford (1979) does, is the judgement of a non-scientist bent on creating sensation by debunking a legend. The 'beautifully traced leaves' were, in fact, the first *Glossopteris* material to be recorded from Antarctica, establishing the age of the coal-bearing rocks of Victoria Land as late Palaeozoic and linking them with similar formations in other land masses in the southern hemisphere. Wilson and Scott did make a significant contribution by not jettisoning them (Seward, 1914; Priestley & Tilley, 1928; Young, 1980; Tingey 1983) and more ought to have been made of them later in the context of continental drift. Publication of the geological results of Scott's last expedition was piecemeal and slow, the final collection of papers not appearing until 1964 (British Museum, 1964).

Breaking new ground, the expedition of 1911–14 led by Douglas Mawson found evidence that the basement complex and Beacon Sandstone formation of Victoria Land extended into King George V Land and the Adélie Coast. Shackleton's plans for his trans-Antarctic expedition included ambitious geological forays at the head of the Weddell Sea but in the event nothing could be done, although Wordie (1921b) did his best with what was to hand on Elephant Island. Survey was greatly extended by the Byrd expedition of 1928–30, on which Dr Laurence M. Gould (see Fig. 6.8) was geologist. Aircraft were used, to little effect, in taking geological parties out into the field. A visit to the Rockefeller Mountains was hampered by bad weather and curtailed by the wreck of a plane, but sufficed to show these low mountains as erosional remnants of coarse-grained granite with intrusions of pegmatite dykes and younger granite. A dog-sledge expedition, covering 1,500 miles (2,400 km), to the Queen Maud Mountains was important in showing that the Beacon Group extended along the Trans-Antarctic Range, 300 miles (480 km) further across the continent than had previously been known (Gould, 1931a; for bibliography see Bertrand, 1971). Gould (1931b) described the moment of discovery as follows:

> We had still to climb about an eighth of a mile to our right along a steep slope with a great yawning chasm only 200 or 300 feet below us. A serious slip and we might all have ended up in it without leaving any trace at all. But we carefully picked our way along the slope and in due time arrived at the coveted rocks.
>
> I had had the unhappy and almost dismal reaction of finding that the higher we climbed, the lower my spirits became, for the nearer we got to the long sought flat-lying rocks, the less they looked like the sedimentary layers I wanted so much to find, and the more they looked like a great series of volcanic flows . . .
>
> I had climbed and clawed my way up over several thousand feet of glacier to the rocks and had actually to hold a piece in my hand before my very eyes, before I realized that it was after all not volcanic rock, but sandstone. That little piece of rock which I first picked up and which was not half as large as the palm of my hand had repaid me for the whole trip. Had it been necessary for me to have turned around that moment and started back to Little America, I should still have felt that it had been a most profitable journey. (Courtesy of the author.)

Brief landings by Mawson during the BANZARE expedition of 1929–31 indicated that the rocks of Mac. Robertson Land were of typical basement complex type. The second Byrd expedition of 1933–35 had two geologists and made great use of mechanical transport. Further examination of the Queen Maud Mountains revealed the most southerly fossiliferous coal-bearing formations then known and the Ford Range, discovered to the east of the Rockefeller Mountains, was supposed, erroneously as it now appears, to be petrographically and structurally similar to the Antarctic Peninsula (Byrd, 1935; for bibliography see Bertrand, 1971). On the US Antarctic Service Expedition 1939–41, the geology of southern Palmer Land was explored from East Base (sited at Stonington Island in West Antarctica), and from West Base (sited on the Ross Ice Shelf off East Antarctica) there were further forays into the Rockefeller Mountains and Ford Range (Bertrand, 1971). The British Graham Land Expedition had also investigated the southern part of the Peninsula and eastern Alexander Island, which is geologically very different from the mainland, with a richly fossiliferous sedimentary succession on its east coast later identified as upper Jurassic – Lower Cretaceous (Adie, 1964).

Following the Second World War the Falkland Islands Dependencies Survey initiated a new phase, under the leadership of V. E. Fuchs and R. J. Adie, with a systematic series of long sledging journeys which eventually covered nearly all the coastal stretches of Graham Land and its archipelago (Adie, 1964). These surveys showed that West Antarctica has had a more varied tectonic history than East Antarctica and the idea, first put forward by Nordenskjöld in 1913, that East and West Antarctica are separate land masses was confirmed (Adie, 1964). The Trinity Peninsula Series sediments were thought to have been contorted and overfolded by earth movements associated with the formation of the Scotia arc in late Palaeozoic to early Mesozoic times. The post-Triassic rocks which overlaid them were more widely distributed. In late Cretaceous times there was further folding, rocks of the widespread and important Andean Intrusive Suite were formed, and block faulting responsible for much of the present topography occurred. There was further volcanism in the Tertiary (Adie, 1964; Warren, 1965). The US operations *Highjump* of 1946–47 and *Windmill* of 1947–48 collected rock specimens but geological studies were incidental to their main objectives of aerial survey. The Norwegian-British-Swedish Expedition of 1949–52, in addition to its geophysical and glaciological work, carried out geological surveys in Queen Maud Land. Walton (1987) has tabulated geological investigations in Antarctica up to the IGY, giving names and nationalities of the geologists and the areas in which they worked.

8.3 Geology during and after the IGY: the dry valleys

The IGY was primarily concerned with synoptic geophysical observation and therefore not with geology of the sort we have been discussing but nevertheless it provided a great stimulus to it. Most IGY stations had their geologists and some of these were able to take part in the extensive traverses undertaken as part of glaciological and crustal geophysical investigations. There was renewed investigation of the geology of the dry valleys of Victoria Land discovered by Scott (1905). Shackleton had recognized the geological interest of the extensive areas free from ice-cover which they presented and, under his aegis, David and Priestley (1914) carried out surveys. Further surveys were made by Griffith Taylor (1922) on Scott's second expedition. After this there had been no further investigation until in 1947 aerial photography was carried out by *Operation*

Highjump. These photographs aroused the interest of the biologist-meteorologist with the New Zealand section of the Trans-Antarctic Expedition, R. W. Balham, and he inspired two young geology students, B. C. McKelvey and P. N. Webb, to the extent that, after much pestering, they were permitted to infiltrate the 1957–58 New Zealand expedition as temporary general assistants. They lived a hand-to-mouth existence but managed to get included in visits to the dry valley area. The geological work that they did was published promptly (Webb & McKelvey, 1959) and they made such a good impression that, following their return to New Zealand, they were sponsored by the Victoria University of Wellington, with the approval of the Ross Dependency Research Committee and logistic help from the US Antarctic Support Force to investigate the dry valley system more thoroughly (Hatherton, 1967; Quartermain, 1971). Since then the dry valleys have been the subject of intensive work, culminating in the Dry Valley Drilling Project (Fig. 8.3). When the US NSF came to consider its post-IGY programme it was obvious that the dry valleys presented excellent opportunity for stratigraphical investigation and that with laboratories at McMurdo Station and helicopters available, logistics would be comparatively easy. The proximity of the New Zealanders at Scott Base made it logical that they should be involved and then, through the Antarctic Treaty organization and ICSU, the project became international (Smith, 1981). The third major participant was Japan, mainly, perhaps, because of scientific contacts established by Harry Kelley, Assistant Director of the NSF, who had previously been in Japan as a member of McArthur's staff. Each of the three countries became responsible financially for a particular part of the programme. Because of the high costs of drilling, a preliminary aeromagnetic survey was carried out to give accurate base data and field trials of drilling were done on Ross Island. Protocols were drawn up to minimize environmental damage. In this, the first rock drilling on the continent, 2 km of cores were obtained from 15 sites with 93 per cent recovery, although difficulties were encountered in penetrating the deeper glacial-marine strata (McGinnis, 1981). These gave material for study of the geological history of the area over the past 10 million years. Investigation of the geology of another 'oasis' area, in the Bunger Hills was begun by the Soviets in 1956. Their work showed that the absence of ice and snow was due to topographical and meteorological factors rather than to underground heat sources (Lebedev, 1959, Solopov, 1967).

Elsewhere on the continent geological studies continued intensively after IGY and field mapping extended to nearly all areas where there were rock exposures, for example in Ellsworth Land and the southernmost part of the

Fig. 8.3 Dry Valley Drilling Project, drilling rig, 1971–75. (Courtesy of the US National Science Foundation.)

Antarctic Peninsula (Rowley *et al.*, 1983). Geological reconnaissance of the Peninsula was virtually complete by 1974,[3] about 95 per cent of the work having been done by BAS.[4] Absolute dating by radioisotope ratios, made particularly by the Soviet workers M. G. Ravich, I. Ye. Starik and G. E. Grikurov, began to provide a framework for the geological history of the continent (for references see Warren, 1965; Halpern, 1971). There was also great advance in Antarctic palaeontology. However, these developments are best considered in the context of the revolution in geological thought brought about in the 1950s and 1960s by the general acceptance of the theory of continental drift.

8.4 The continental drift theory and the tectonic structure of Antarctica

The antecedents of the theory of continental drift may be seen as far back as 1620 in Francis Bacon's comment on the similarity of the coast-lines of Africa and South America. Antonio Snider-Pellegrini in his book *La Création et ses mystères dévoilés* (1858) included maps showing the original conjunction and subsequent drifting apart of the Atlantic continents but this idea was considered too outrageous to be taken seriously. Suess (1904–9) recognized connections between continents and postulated the supercontinent of Gondwanaland, embracing central and southern Africa, Madagascar and peninsular India, but it was F. B. Taylor in 1910 who gave the first logically constructed hypothesis of continental drift. His prime object was to explain mountain formation and he supposed that the moon was not captured by the earth until Cretaceous times and that the tidal disturbances resulting from its capture had displaced continents, crumpling up their leading edges as they moved. He envisaged the fragments of Gondwanaland in the southern hemisphere spreading out from around the South Pole. He did not support this hypothesis with independent evidence and, again, the geological community was unimpressed. It was Alfred Wegener, born in 1880, trained as an astronomer but later becoming a meteorologist and Arctic explorer, who first assembled evidence in support of the idea that the continents were once united into a single land mass, which he called Pangaea. He was evidently ignorant of previous suggestions that continents may have drifted (LeGrand, 1988) and it is said, plausibly although none of his writings confirm it, that the idea that such a land mass might have split up and the bits drifted apart came to him as he watched a floe of Arctic ice breaking up. He aired this idea and discussed the palaeontological evidence which supported it in a lecture to the Geologische Vereinigung in Frankfurt-am-Main in 1912 (see Shea, 1985) and developed it further in his book *Die Entstehund der Kontinente und Ozeane* published in 1915. The strength of his case lay in his demonstration that, beyond the fit of outlines with each other, there was a match of geological and palaeontological patterns across the continental boundaries when they were fitted together. Wegener published revised versions of his book in 1920, 1922 and 1929, the 1922 edition being translated into English and the last including evidence from palaeoclimatology. His hypothesis did not at first attract much interest but in 1922 the British Association and in 1928 the American Association of Petroleum Geologists held discussion meetings on the subject. At these meetings there was general hostility and, analysing this, Frankel (1976) concluded that the common factor was that Wegener's theory had no distinct advantage over the competition within any specific area, the corroborating facts being so diverse that their great variety and cumulative force was not appreciated by specialists in restricted fields. LeGrand (1988) sees the situation as more complicated. The idea of continents moving horizontally was not in itself a difficulty – after all the accepted concept of isostasy assumes that land masses move vertically in a viscous substratum so it would be illogical to deny that they might move laterally – but the motive power was a stumbling block. Wegener could supply no more than the tentative suggestion that gravitational forces were responsible but it was easily shown that these would have been far too weak. The mathematical geophysicist, Sir Harold Jeffreys, whose major treatise *The Earth* appeared in 1924, was contemptuous of Wegener's theory, dismissing his geological and biological evidence scornfully without any proper consideration (LeGrand, 1988). Among Antarctic geologists, Mawson rejected the theory, holding it absurd that solid rocks could be thought to flow like pitch (Oliphant, 1983). Wright entertained the idea of drift without using it for explanatory purposes (Wright & Priestley, 1922) and David towards the end of his life seems to have been well disposed towards it (LeGrand, 1988). Opposed by distinguished specialists the theory remained in limbo, although, building on the suggestion made by John Joly in 1909 that radioactivity provided a virtually eternal source of heat within the earth, Arthur Holmes in 1929 put forward a plausible mechanism for horizontal displacement of continents by sub-crustal convection currents. Wegener died in 1930 on the Greenland ice-cap but the South African geologist, A. L. du Toit, in his book *Our Wandering Continents* (1937) carried the torch forward, amassing further telling evidence, but perhaps spoiled the effect by showing himself over-prejudiced in favour of continental drift. In all this, Antarctica, apart from being fitted neatly into place in the continental jigsaw puzzle did not figure largely in discussion of Wegener's hypothesis (Holmes, 1944; Hallam, 1973; Bishop, 1981) although du Toit described it as the key piece of Gondwanaland. In 1928, Priestley & Tilley considered that the palaeoclimatological problems presented by Antarctica were an insuperable obstacle to the acceptance of what they otherwise found to be a 'fascinating and all-embracing theory'.

It was not until some time after the Second World War that the accumulation of evidence for continental drift began to become irresistible. Before the war the Dutch geophysicist, F. A. Vening Meinesz, by making gravity measurements in a submarine showed that ocean floor sediments were thin, usually less than 1 km, and covered basaltic rocks. Furthermore, the Mohorovicic discontinuity, the point taken as the boundary between crust and mantle, was only about 10 km below the ocean bed as compared with 30 km below the continents. Later, the Deep Sea Drilling Project, which began in 1975, was to demonstrate that ocean sediments are relatively young, being nowhere older than Jurassic. The Atlantic and Indian Oceans appeared not to have existed 200 million years ago and the ocean which had surrounded the single land mass of Pangaea seemed to be the precursor of the modern Pacific. Compelling evidence for continental drift had meanwhile come from studies of the magnetism of rocks. Igneous rocks containing magnetically susceptible minerals have permanent, although weak, magnetization in the direction of the earth's field as it was at the time and location at which they cooled below a particular temperature, that is, they are fossil compasses. By 1952, P. M. S. Blackett had devised an astatic magnetometer sufficiently sensitive to determine such magnetization and realized that it could be used to extend the history of the earth's magnetic field back from 400 years to 500 million. With others, he found that some rocks are magnetized along axes which indicate a significant shift in position of the magnetic poles since the time that they were laid down and that there have been periodic reversals of the earth's magnetic field. These variations form a pattern with time which can be used as a means of dating of not only igneous rocks but certain types of sedimentary rocks and sediments containing magnetic particles. S. K. Runcorn (1956) was the first to show that the track of magnetic polar wandering was different for Europe and North America and it soon became evident that the tracks varied from continent to continent in a way that could best be explained by continental drifting. H. H. Hess put forward a theory, which Dietz (1961) was first to express in print, that spreading of the ocean floor occurs as the result of upward convection of molten crustal material at the mid-oceanic ridges and it is this which provides the motive power for continental drift. This differed from Holmes' model which was centred on the dry land, in that the action was seen as taking place in the ocean floor (LeGrand, 1988). Study of magnetic anomalies in sediments along transects across ridges produced remarkable evidence that this actually happens. A regular pattern of stripes of reversed polarity, parallel to the ridges, was found. The Cambridge geophysicists, Vine and Matthews (1963) – the latter having become interested in the drift hypothesis through his own experience of geological mapping in the Antarctic Peninsula area – explained this pattern as being produced by the basaltic crust forming at the ridge becoming magnetized in the direction of the prevailing field and acting very much as a tape recorder. Dating of the stripes suggested that the Atlantic Ocean has widened at a rate of about 2 cm a year.

In reviews of Antarctic geology Adie (1964) and Warren (1965) were cautious and brief in referring to continental drift. In North America geologists were, in the main, skeptical and it was not until the annual meeting of the Geological Society of America in 1966 that they began to accept the theory (LeGrand, 1988). Magnetic anomaly profiles collected by *Eltanin* and Opdyke's (1968) demonstration that two separate measures of the palaeomagnetic reversal age scale, from Antarctic sediment cores and continental rocks, were mutually consistent, were particularly persuasive evidence for seafloor spreading (LeGrand, 1988). From a survey, by questionnaire of 300 members of the Geological Society of America selected at random, Lemke *et al.* (1980) concluded that acceptance of the idea of continental drift was the result of a sort of 'chain reaction' or general shift in opinion which uniformly altered the attitude of the majority, rather than of individual judgements of the evidence for and against the theory. LeGrand (1988), however, saw the situation as far more complicated than a 'bandwaggon effect'. Perhaps geologists particularly concerned with the southern hemisphere were more ready to accept the theory but soon it was appropriate to devote a section comprising five chapters to the subject in a symposium on Antarctic research (Quam, 1971). Thereafter, studies on continental drift relating to Antarctica burgeoned. Antarctic geology from this time on, like world-wide geology, became less historical and descriptive and more explanatory, a trend begun with the IGY. Nevertheless, some Antarctic geologists, notably V. V. Beloussov (see p. 171; Beloussov, 1964), remained resistant to the idea of drift (LeGrand, 1988). In any case, detailed descriptive work, such as continued to occupy the majority and occupied the greater part of the 1982 symposium on Antarctic earth sciences (Oliver *et al.*, 1983), remained largely independent of high level theory.

By 1970 it was accepted that the primary geological subdivision of the earth's surface should be into plates of crustal material bounded by zones of tectonic activity rather than into continents and oceans. The delineation of the tectonic plate or plates carrying the Antarctic land masses was work for the sea-going geophysicist rather than the Antarctic geologist. Pitman & Heirtzler (1966) on a cruise in the *Eltanin* found symmetrical linear patterns of magnetic anomalies on either side of the Pacific–Antarctic Ridge which indicated a spreading rate of 4.5 cm on a year averaged over the last 10 million years. From a consideration of bathymetric features, magnetic anomalies, the distribution of earthquake epicentres and palaeomagnetism, Heirtzler (1971) was able to produce a chart of the Antarctic Plate and a series of maps showing the break-up of Gondwana and subsequent movement of the continental fragments which was still broadly acceptable in 1987. The movement of continents is deduced from palaeomagnetism, which gives an estimate of latitude with respect to the magnetic pole which one may assume to be statistically the same as the pole of rotation over a period (Frakes & Crowell, 1971). In 1971 there were very few palaeomagnetic determinations from Antarctica and these suggested that the continent had remained in nearly its present position for the last 200 million years (Frakes & Crowell, 1971). By the same token the other land masses of the southern hemisphere have moved in a general northerly direction and the palaeomagnetic evidence, as far as it went, agreed with the geographical and geological evidence in showing their original configuration around Antarctica in Gondwana. Correspondences between the Gondwana fragments were found in distinctive beds of tillite left by glaciations in the Carboniferous and Permian in southern parts of South America, Africa, Australia, peninsular India and Madagascar as well as in Antarctica. Geosynclines were also matched up between some fragments. A geosyncline is a structural concept, now fallen from favour, visualized as starting as an asymmetric, linear, subsiding trough, bounded on one side by a stable continental block and on the other by a tectonically active land mass. Characteristically, predominantly calcareous deposits accumulated on the former side and thicker argillaceous deposits on the other. Griffith Taylor had realized early on (Sanderson, 1988, p. 80), that the late Precambrian/early Cambrian geosyncline in Antarctica was comparable with the Adelaide geosyncline, both containing *Archaeocyatha* (fossils of limestone reef forming organisms indicative of warm marine conditions). There is similar correspondence between the Antarctic Palaeozoic geosyncline and the Tasman geosyncline (Hurley, 1968). Examination of radiometric ages obtained in the previous five years enabled Halpern (1971) to demonstrate a geochronological province lying tangentially to the older continental shield and corresponding with the major 'Samfrau' geosyncline postulated by du Toit (1937). Frakes & Crowell (1971) found good matching of Antarctica to Australia and Africa in respects of alignments of geochronological and orogenic belts and other structural features. There were, however, difficulties for Heirtzler (1971) in fitting the New Zealand plateau and the Antarctic Peninsula into the pattern. The Scotia Ridge also presented problems and these became a particularly British preoccupation. The studies of a group at the University of Birmingham showed it to consist of blocks mainly of continental origin, assembled in their present position during Tertiary times (Barker & Griffiths, 1972). One of the most practical reconstructions of Gondwana resulted from the work of this group (Barker & Griffiths, 1977) whose collaboration with BAS was formalized in 1975.[5] West Antarctica has proved to have a complex structure. Clarkson (1983) proposed that it is composed of five main tectonic plates, some of these built from microplates with other microcontinental fragments perhaps present in the Weddell embayment. A project initiated in 1982 in which there was British/US collaboration in studies including gravity measurements and aeromagnetic survey substantiated this.[6] There is still much discussion about the precise fits of Gondwana fragments.

Palaeobotanical evidence for the continental drift theory had existed since examination of the samples collected by Scott's party had shown the presence in the Beardmore Glacier area of the *Glossopteris* flora characteristic of the southern continents (Seward, 1914). Of course, some plants may be disseminated by airborne spores or water-borne propagules, so that evidence from fossils of the larger terrestrial vertebrates, which cannot be carried over wide expanses of sea by wind or currents, or of freshwater fish, which cannot tolerate salt water, is more telling. Fishes of Jurassic age were discovered near the Beardmore Glacier in 1966–67 in deposits which seem to have been laid down in small ponds or lakes, and in the following season a fragment of an amphibian lower jaw was discovered in a pebbly lens in a sandstone bluff in the same area as a follow-up to a geological report made in the 1961–62 season by V. McGregor (Barrett, 1968; Colbert, 1971). This fragment proved to have come from a labyrinthodont amphibian of Triassic age and in 1969–70 more complete remains, together with those of therapsid and thecodont reptiles, were obtained from the same formation. Among the therapsids was one virtually identical with *Lystrosaurus murrayi*, which is especially characteristic of the Lower Triassic of South Africa, thus establishing a close palaeozoological link between Antarctica and southern Africa. Moreover, this particular association of fossil amphibians and reptiles is one that occurs not only in Antarctica and South Africa but in India, western China and Russia. Contiguity to Africa gave plausible explanation for its presence in Antarctica (Colbert, 1971) and would also explain numerous, otherwise disparate occurrences of marine and non-marine invertebrate fossils (Tasch, 1971). Thus, these palaeontological discoveries provided strong evidence for the existence of Gondwana, clinching the argument about continental drift as far as most biologists were concerned. In recent years the continental drift hypothesis has been generally accepted and Antarctica has featured prominently in reconstructions of Gondwanaland (Oliver *et al.*, 1983).

8.5 The ice-cap and the land underneath it

By the beginning of the twentieth century the general outline of the Antarctic continent had been delineated and it was apparent that within it there was little to be seen at the surface except snow and ice. How deep the ice was and what the disposition of the supporting land mass might be were questions that took a long time to answer. Croll (1897) had made an estimate of the thickness of the cap on the supposition that ice does not flow if the slope of its surface is less than one degree. Projected over the 1,400 miles (2,253 km) from the edge of the continent to the Pole this gave a thickness in the centre of 24 miles (39 km). Thinking this rather excessive

Croll guessed at a thickness of four miles (6 km) on the basis that some of the larger Antarctic icebergs were estimated to be more than a mile (1.6 km) thick. The *Gauss* and *Endurance* expeditions found the continental shelf to be at about twice the average depth of the shelves of other continents but Philippi's (1910) explanation that this was the result of planing down by ice at a time when the ice-cap was of greater extent was not accepted (Wordie, 1918). Nordenskjöld and others pointed out that the chief factor causing this abnormal submergence must be the weight of the continental ice (Priestley & Tilley, 1928). The implication here that over much of its extent the ice must be of considerable thickness was not, however, generally accepted; the widespread protrusion of mountains above the ice in West Antarctica suggested that the ice cover was thin in this region and it was widely believed that a high rocky plateau with gentle relief lay beneath the ice of East Antarctica (Hayes, 1928; Gould, 1940; Bentley, 1965). Direct determination of ice thickness was beyond the techniques available early in the century. Hand operated ice-drills become hard to use beyond 15 m and the greatest depth drilled by Drygalski (1904) was 30 m (Fig. 8.4). Before any definite idea of ice thickness could be obtained exploration had given a general picture of its horizontal disposition. The expeditions of Scott, Shackleton and Amundsen had shown that beyond the range of mountains which formed the western coastline of the Ross Sea and Ice Shelf was a more or less level plateau at an altitude of two to three thousand metres. The Mawson expedition of 1911–14 found a more gentle rise to a plateau on a journey towards the South Magnetic Pole in the Adélie Land sector. Air reconnaissance during the Byrd expedition of 1928–30 showed a similar plateau in Marie Byrd Land. The only major features relieving the monotony of this vast expanse were the Trans-Antarctic Mountains, stretching in an unbroken chain for 2000 km from Cape Adare to the Queen Maud Mountains (discovered and named by Amundsen and surveyed by the Byrd expedition), and the mountains of the Antarctic Peninsula. From aerial inspection in 1928 Wilkins thought that the Peninsula was separated from the main continent by a strait but on his flight across the continent in 1935 Ellsworth concluded that this was an error and the continuity was proved conclusively on the ground by the British Graham Land expedition of 1934–37. The second Byrd expedition, of 1933–35, established the outline of the east coast of the Ross Ice Shelf and showed that there is no obvious strait leading from the Ross Sea to the Weddell Sea. Remote sensing by satellite is capable of determining surface elevations of ice as well as horizontal distribution. The SEASAT satellite recorded altimeter data only to 72° S in 1978 and subsequent satellite

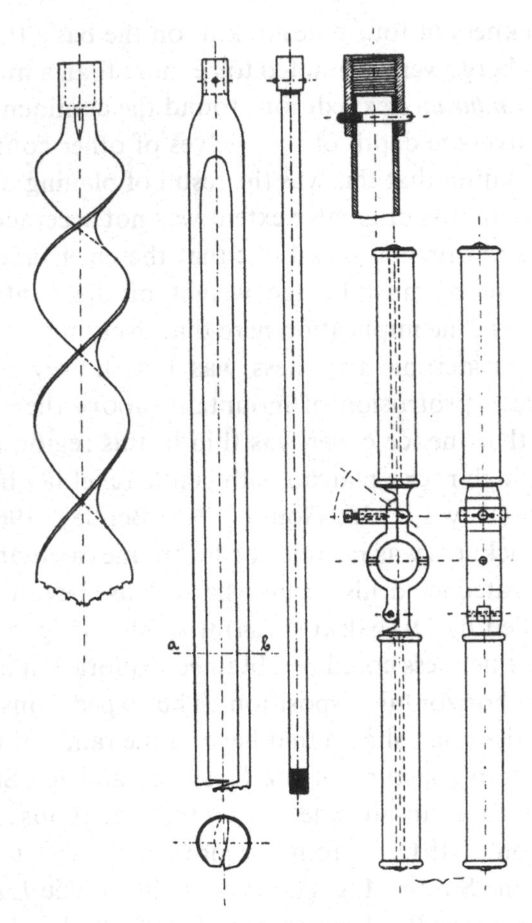

Fig. 8.4 Ice drills used on the *Gauss* expedition, 1901–3. From Drygalski, 1904, p. 282.

missions went no further south than 82° S (Maslanik & Barry, 1990), leaving a hole in which are areas where non-steady-state behaviour of the ice sheet is most likely to be happening. Computer models have been devised to describe the configuration and behaviour of the whole ice-sheet (eg. Drewry *et al.*, 1985). Such models, which have a good general fit to the real ice-sheet but show discrepancies in detail and in certain areas, provide an important input into the large general circulation models constructed to predict changes in global climate.

In 1919 Mawson had proposed that since there is sufficient difference in elasticity between rock and ice, a modified echo-sounder might be used to measure ice depth (Mawson, 1928). This was not followed up and it was at the suggestion of Dr R. A. Daly and the Harvard Committee on Geophysical Research that Byrd used seismic sounding on his second expedition. This technique of studying subsurface structures by means of the reflection and refraction of shock waves from explosions detonated near the surface had been in routine use for geophysical surveying in the petroleum industry since around 1930. It had already been used in the 1920s in a rudimentary form to determine valley glacier thickness in the Alps (Clarke, 1987a). Its use in the Antarctic was only possible when tractor transport capable of carrying the heavy equipment that was needed became available. However, in the first instance it was only used on the ice shelf in the vicinity of the Bay of Whales and on the northern tip of Roosevelt Island and provided no indication of the thickness of the inland ice sheet (Poulter, 1950). No seismic soundings were made on the US Antarctic Service Expedition of 1939–41. The first detailed soundings in the interior were made on a traverse by members of the Norwegian–British–Swedish expedition in 1951–52 which took them 650 km through the mountains of Queen Maud Land onto the central plateau (Robin, 1953, 1958). The ice was found to be up to 2,400 m thick with the underlying rock surface, on average, only a few hundred metres above sea level and, in places, well below it (Fig. 8.5). Taken in conjunction with observations by Expéditions Polaires Françaises, made about the same time, that the rock surface in central Greenland lies near sea level, this led to an immediate upward revision in estimates of the thickness of the Antarctic ice-cap (Bentley, 1965). However, it was not until the IGY that seismic sounding became general. Seismic soundings were begun by the First Continental Expedition of the USSR in April 1956. By 1962 soundings had been made by parties from Australia, Belgium, the British Commonwealth, France, Japan, US as well as the USSR, along 25,000 km of traverse giving adequate reconnaissance all but the 15° W to 45° E sector of the continent (Bentley, 1965). The traverse made by the Commonwealth Trans-Antarctic Expedition of 1955–58 using reflection shooting passed through the South Pole where the ice was determined as being 1990 m thick (Pratt 1960b) over a rock surface 810 m above sea level. However, high surface noise level following detonation when the temperature of the surface snow is below −25 °C leads to poor quality results for most of Antarctica over much of the year, and Woollard (1962) found it necessary to recalculate both these depths and those of the Soviet traverse to the Pole of Inaccessibility on the basis of gravity data. A consideration of these by Bentley (1964) and other errors which may arise in seismic sounding of ice led to the conclusion that few of the soundings reported up to 1963 are valid.

The equipment for seismic sounding was cumbersome and a single

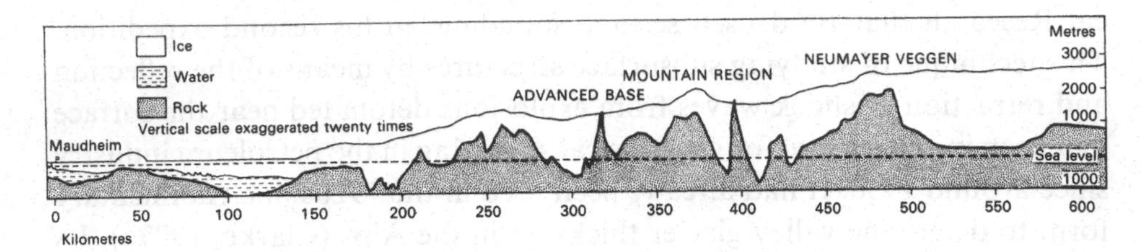

Fig. 8.5 Detailed profile of under-ice topography in Queen Maud Land based on seismic survey by the Norwegian–British–Swedish Expedition in 1949–52. (Courtesy of G. de Q. Robin, 1958.)

measurement took from two to three hours. It was a great advance on drilling, which with improvements in techniques had become more efficient but at the same time very expensive, but itself was inferior to radar sounding, which can produce an individual measurement in 50 microseconds and can give a continuous profile of ice thickness from a surface vehicle or an aircraft. That cold ice is reasonably transparent to radio waves in the VHF band became evident when accidents to aircraft were traced to erroneous readings of radio-altimeters over ice and the first successful use of this property was by A. H. Waite of the US Signal Corps in 1958 who obtained spot soundings on ice up to 600 m thick near Wilkes Station (now Casey Station) on the Budd Coast. Further development to produce a continuously working instrument specifically for glaciological use was carried out at the Scott Polar Research Institute (Evans, 1969; Robin, 1972). The system was tested first in Greenland in 1964 and then in the Antarctic in a joint programme between the Scott Polar Research Institute and the US NSF in December 1967. Since then the technique has been used extensively in the Antarctic by the USSR, France, Australia, Belgium, South Africa and Japan, as well as in continued collaboration between the UK and the US. One of its earliest achievements was to confirm the existence of the Gamburtsev Mountains, a group rising to some 2,600 m above sea-level but completely covered by ice, in the region of the Pole of Inaccessibility. These mountains had been detected on the Soviet traverse of 1958–59 under conditions extremely unfavourable for seismic shooting. The aircraft making the radar soundings followed nearly the same track as the surface traverse and found a strong alpine relief although the ice depths measured were greater than those indicated by the seismic method (Robin *et al.*, 1968). Another finding was that in places the signal changes from one characteristic of an ice–rock interface to one indicative of an ice–water interface, suggesting the existence of lakes trapped between the bedrock and the ice (Robin *et al.*, 1968). Geothermal heat warms the bottom of the ice and calculations confirmed that the thermal conductivity of ice is such that at depth the temperature can be at melting point although the surface temperature may be as low as −60 °C. The largest of these lakes is one of about 8,000 km² area near Vostok. By 1975 it was possible to produce a map showing the ice thickness and bedrock topography for the sector *c.* 90° E to 180° (Drewry, 1975). By 1982 more than half of continental Antarctica had been sounded over a 50–100 km grid, entailing more than half a million km of flight track (Drewry *et al.*, 1982). To be useful, of course, such soundings and other geophysical measurements made from aircraft need to be backed by navigation of a high degree of accuracy (Smith, 1972) and a by-product has been the correction of maps. Thus, it was by this means that Charcot Island was found to be 30 km out of position on the then existing map (Doake, 1976). On the small scale, radio echo-sounding traverses made on the ground across glaciers at intervals of a few days can give accurate measures of rates of movement (Swithinbank, 1972).

Because ice is only about one-third as dense as the rocks immediately below it, changes in its thickness may also be estimated from the magnitude of gravitational anomalies. This is most useful when there is a nearby rock outcrop for reference or when it is necessary to interpolate between points at which ice-depth has been established by other means. The gravity traverse made by the Trans-Antarctic Expedition of 1955–58 was intended to provide such interpolation but needed accurate determinations of elevation before it could be used for that purpose (Pratt, 1960a). The problems of determining elevations accurately have been discussed by Bentley (1964) who estimated that with barometric altimetry, the usual method, there might be error of ±50 m over most of the continent. Radar altimeters flown on GEOS-C and SEASAT satellites have yielded reconnaissance data for Antarctica north of 72° S and a numerical model for handling them has been proposed (Drewry, *et al.*, 1984).

By 1965, Bentley could give a reasonably accurate overview of the land beneath the ice except for large areas in Queen Maud Land and Wilkes Land which were still unexplored. The subglacial floor of East Antarctica had been found to be mostly above sea-level adjusted to allow for isostatic recovery were the ice to disappear, this being the best estimate of the level of the rock surface prior to glaciation. A basin in the northern part of Victoria Land appeared to be the only substantial area below this level. In striking contrast, most of the rock surface of West Antarctica was found to be well below adjusted sea-level, a great subglacial depression, the Byrd

Basin, extending from the Ross Sea across to the eastern Amundsen Sea and nearly to the Bellingshausen Sea and apparently joining with the Weddell Sea beneath the Filchner Ice Shelf. However, a direct connection of the Ross and Weddell Seas as suggested in 1914 by Taylor appeared to be blocked by a ridge between the Ellsworth and Thiel Mountains. The Trans-Antarctic Mountains seemed to be breached only by a gap to the east of the Thiel Mountains, forming a barrier which produces an ice divide noticed by both Amundsen and Scott, and an asymmetrical drainage pattern with the major flow towards the Weddell Sea. Evidence of the crustal structure of the continent was provided by seismic refraction profiles and gravity determinations supplemented the meagre information from exposed outcrops (Woollard, 1959). The striking depression in West Antarctica appeared to divide a volcanic province in Marie Byrd Land from a metasedimentary province in the Sentinel Mountains and the region to the south-west, a conclusion supported by magnetic data. On the Victoria Land plateau the prominent sedimentary section including the Beacon Formation, as found in the Trans-Antarctic Mountains, is absent. The crustal thickness below East and West Antarctica had been found to be strikingly different, estimates being about 40 and 30 km respectively, with an abrupt transition across the Trans-Antarctica Mountains. Nearly 20 years later the picture of the general structure of the continent was much the same (Bentley, 1983). Palaeomagnetic data were still insufficient to pin down the past relative positions of the two parts of the continent and it appeared that in the Cretaceous, West Antarctica must have been separated into at least three tectonic units – Ellsworth Land with the Antarctic Peninsula, Marie Byrd Land, and the Ellsworth-Whitmore block, which seemed once to have been a section of the Trans-Antarctic Mountains. The amount of ice held on the continent, a matter of interest to the meteorologist as well as to the geologist, was estimated by Gow (1965), using a value of 2000 m for the average thickness including the ice shelves, as 27×10^6 km^3 but Bardin & Suyetova (1967) gave a lower figure of 24.031×10^6 based on less simplified morphometric data. Drewry *et al.* (1982) brought together data on the surface configuration, ice thickness and bedrock characteristics of the Antarctic ice-sheet. With the benefit of data from airborne radio echo-sounding and satellite surveillance its volume, including shelves, was calculated to be $30.11 \pm 2.5 \times 10^6$ km^3, a value not too different from these previous ones. Their estimate of the mean sub-glacial bedrock elevation of the continent was, however, -160 m as against Bardin & Suyetova's (1967) $+410$ m.

8.6 Glaciology

Mawson (Wordie, 1918, discussion p. 232) emphasized strongly that ice should be regarded as a rock in the geological sense and be treated petrologically as such. No doubt most geologists acquiesced in this and the significance of ice in physical geology had been recognized since the mid-nineteenth century, but it was actually a long time before glaciology and ice physics came together. What Clarke (1987a) has described as a 'compelling mixture of the classical and romantic' drew many distinguished scientists of the nineteenth century – Agassiz, Forbes, Huxley and Tyndall – to the study of the glaciers of the European Alps. Agassiz (1840) had favoured a theory of glacier motion in which dilation on freezing advanced the ice down the slope with a sort of rachet action. Forbes (1846) opposed this with his viscous flow theory based on experiments with models made of plaster of Paris and glue as well as on measurements made on glaciers themselves. Tyndall & Huxley (1857), without rejecting Forbes' theory, concluded that the 'simulated fluidity' of ice arises from 'regelation', that is, melting under pressure and refreezing. If this were correct then most glaciers in Antarctica would not flow, being too cold. The viscous flow theory, coupled with recognition that sliding also occurs, prevailed and at the turn of the century stimulated a number of studies connecting glacier flow to fluid mechanics as well as experiments on the deformation properties of ice (Clarke, 1987a). The International Commission of Glaciers was established in 1894 to encourage the monitoring of variations in glaciers. However, the glaciology carried out in the Antarctic at this stage, mainly on the *Nimrod* and *Terra Nova* expeditions, was naturally more descriptive than quantitative or theoretical although it covered details of crystal structure as well as the grosser features (David & Priestley, 1914; Mawson, 1916). Mawson (in Shackleton, 1909) described the horizontal stratification seen in the ice shelf face and noted that even at some distance below the horizontal surface the ice was recognizably compacted snow and different from glacier ice. The *Terra Nova* expedition report on glaciology (Wright & Priestley, 1922) has become a classic in its field. Towards the end of his life Sir Charles Wright (in Neider, 1974) recalled that glaciology in the Antarctic was an open field, those who had worked on it in Europe having little conception of the totality with practically all the ice in the world in Greenland and the Antarctic. There was much evidence of the recession of ice in recent times, and the need for further work on palaeoclimatology was emphasized. Wright (in Huxley, 1913), using values for snow accumulation on the Ross Ice Shelf obtained by the *Discovery* and

Nimrod expeditions, calculated the total amount of ice added to the shelf in a year and derived from this an estimate of the rate of advance of the edge, finding a reasonable agreement with the observed rate of about 500 yards (457 m). This unique feature of Antarctic glaciology, the ice shelf, continued to provide expedition members with material for speculation for many years to come (Debenham, 1948) but the techniques to answer some of the most crucial questions were not yet available.

The International Glaciological Society was founded in 1936 and following the Second World War there was a resurgence of interest, recalling, Weertman (1987) thought, the golden age in the nineteenth century. Snow-pit sampling was first used in the Antarctic Service Expedition 1939–41 (Fig. 8.6), following a programme based largely on suggestions from British glaciologists (Wade, 1945b). The emerging theoretical glaciology strongly influenced the Norwegian–British–Swedish expedition of 1949–52 and the modern phase of Antarctic glaciology began. The improved logistics and new techniques that came with IGY provided further impetus and 11 nations combined their efforts in measuring net mass budgets in the Antarctic. Apart from radar sounding and its various applications there was Döppler-satellite positioning which enabled positions to be fixed to within a few metres so giving more accurate measurements of ice movements (Anon, 1974), and various kinds of drill, including the electrothermal type which had been used for the first time in 1948 to drill a 137 m hole in the Jungfrau Glacier (Clarke, 1987a). One of the principle aims in this period was to determine the mass balance of the ice-cap. Accumulation of snow is most simply measured against stakes planted in the surface but this requires observations over an extended period and to obtain estimates along a traverse resort must be had to measurements of the distance between annual layers exposed in the walls of a pit, a method that can give records of snow accumulation back to 200 years or so before present (Bentley, 1964). In these ways the general accumulation rates over the entire continent were estimated soon after IGY (Hoinkes, 1964) giving the net accretion of snow as 1.7×10^{15} kg water per year (Gow, 1965). Where layering was difficult to interpret, as happens through recrystallization at depth, the employment of dating techniques, such as that based on ^{18}O:^{16}O ratios, could be envisaged but had not been applied successfully in Antarctica by the early 1960s. Ablation did not appear to cause a major loss of ice but the large masses of snow blown into the sea by persistent strong offshore winds seemed to be significant, perhaps amounting to $1\text{–}2 \times 10^{14}$ kg per year (Loewe, 1956; Gow, 1965). What was happening under the ice shelves was uncertain. Debenham (1948) had argued that

(a)

(b)

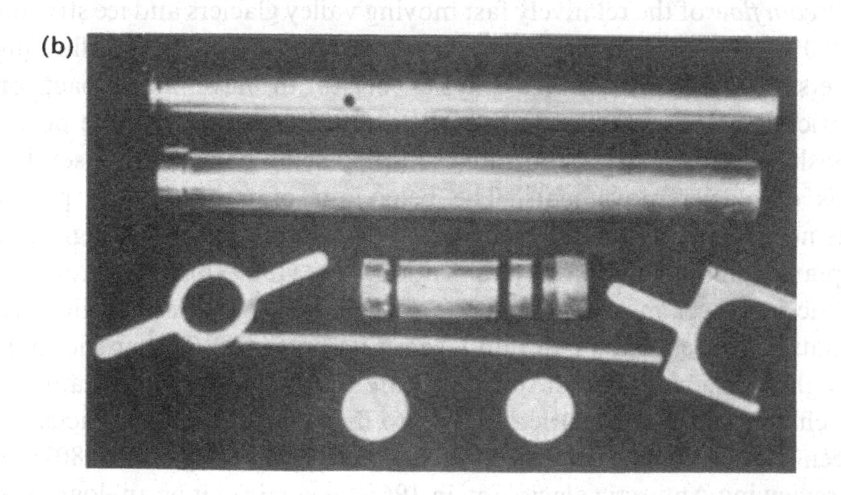

Fig. 8.6 (a) Snow samples being taken in a pit at West Base during the US Antarctic Service Expedition, 1939–41. (b) The snow sampler used, made of steel and brass. (Official photograph US Antarctic Service, 1945.)

accretion of sea-ice on the bottom of shelves occurs but temperature profiles in deep drill holes and the absence of sea-ice from the bottom of cores suggested that melting is the predominant process. The main wastage is by calving of icebergs and this could be estimated from rates of flow and data on thickness of ice edges as around 5.7×10^{14} kg per year. Wexler (1961) concluded from an examination of eight different ice budgets that there was an excess of accumulation over wastage of 10^{15} kg (10^{12} tonnes) of water equivalent per year, a reversal of the previously held idea (Odell, 1952) that the Antarctic ice sheet was diminishing rapidly.

Numerous studies during IGY confirmed the useful fact that the temperature of the ice at about 10 m depth remains almost constant over the year and is close to the mean annual temperature at the surface, a relationship proposed in 1876 by P. A. Kropotkin (Lebedev, 1959). Such measurements put the 'cold pole' of the continent, with a mean annual temperature of -57 °C, on the high part of the East Antarctic plateau, between the Pole of Inaccessibility and Vostok. Temperature was found to fall steadily with depth and the expected rise due to geothermal heating was not seen, evidently because drilling could not be taken sufficiently deeply. The all-important measurements of ice movement were difficult to make because of the paucity of fixed reference points and the slow rates, necessitating observations over several years, if astronomical or satellite fixes were to be used. However, Mellor (1959) was able to distinguish between *sheet flow*, the general outward movement of the ice sheet (1–30 m a^{-1}), *stream flow* of the relatively fast moving valley glaciers and ice streams (100–760 m a^{-1}), and *ice shelf movement*, the lateral spreading of floating ice-sheets (300–500 m a^{-1}). Ice physics began to have an impact on Antarctic glaciology, the flow law of Glen (1955), stating a simple power relationship between stress and rate of strain, being particularly useful in analysis of glacier movements. The behaviour of ice is neither purely viscous nor purely plastic but involves slippage along the crystallographic basal plane combined with recrystallization. Nye (1959) proposed that the Antarctic ice sheet moves more or less as a sliding block with internal deformation concentrated in a thin layer at its bottom. Whether frictional heat might cause melting and thus lubricate the sliding was uncertain.

The change of a newly settled snow into firn and thence into glacier ice had been studied in the Alps for 100 years before IGY (Seligman, 1980) and Gow, reviewing Antarctic glaciology in 1965, recognized it as analogous to the conversion of a sediment into rock then, ultimately, metamorphosed rock. During IGY, investigations of Antarctic snow and ice covered density, permeability, porosity, crystal structure, bubble structure, viscoelasticity, ultrasonic wave propagation, radio wave transmission, seismic wave velocity in relation to temperature and density, strength and deformation under a variety of stress and temperature conditions. Deep drilling at Byrd Station to a depth of 309 m produced a depth-density curve in which the transition from 'old snow' into 'permeable firn' occurred at 10 m and from that to 'impermeable ice' at around 65 m, corresponding to an age of about 300 years. For what is to follow it should be noted that in this second transition the air spaces, which make up about 10 per cent of the volume of firn, are sealed off and compressed until they eventually become spherical with contents at between 200–300 atmospheres. The presence of traces of impurities and solid particles was remarked on and the presence of radioactivity from fission products showed that the Antarctic atmosphere was not cut off from that of the rest of the world (Rubin, 1964) but the exploitation of these for study of conditions in the past was yet to come. Ice crystal size showed a marked increase with depth (Gow, 1965). Bore holes gradually close with time at a rate conforming with the power law of flow derived from laboratory studies.

Glaciology began to assume importance in a wider context after 1975 with the realization that sudden retreat of Alaskan glaciers might result in calving of icebergs which would interfere with shipping. Identifying the sources of flow instability became a preoccupation which has persisted and the first major seminar on the causes and mechanics of glacier surges was held in Quebec in 1968 (Clarke, 1987b). The switching of flow between slow and fast suggested that different physical processes were involved. Attention was then drawn to the possible catastrophic consequences of collapse of the West Antarctic ice-sheet should global warming by the 'greenhouse effect' provide a trigger (Mercer, 1978). This was perhaps alarmist but Weertman (1976) could fairly describe the behaviour of this ice-sheet as 'glaciology's grand unsolved problem'. There is actually a complex of interrelated problems among which one of the most intriguing is that of the ice streams which drain from the sheet into the Ross and Filchner-Ronne Ice Shelves. These streams, which have velocities, about 1 km yr^{-1}, similar to those achieved temporarily by surging glaciers, are clearly demarcated flows through the relatively static ice surrounding them and are not obviously related to channels in the bedrock. The fast flow results from fast sliding, not fast creep, and various theories have been put forward to explain it (Clarke, 1987b). Perhaps the most striking fact to emerge has been that ice stream B on the Siple Coast of the Ross Ice Shelf is underlain by a layer of highly porous, water-saturated, and easily deformed material, presumably till, about 8 m in thickness (Alley *et al.*, 1986). The lubrication provided by this layer seems to explain how fast sliding can be produced by the very low driving stresses typical of ice streams. A basis for assessing general changes in the extent and behaviour of Antarctic glaciers in the long-term was provided in Swithinbank's magnificent satellite image atlas (1988a).

The ice shelves are a unique feature of Antarctica through which is exerted much of the effect of the continent's ice on the global environment. The Ross Ice Shelf Geophysical and Glaciological Survey (RIGGS), carried out during the period 1973–78, made a great contribution to knowledge of these structures (Bentley, 1984). Air-lifts enabled measurements to be made at 200 stations on the Shelf. These included accumulation rate, strain rate, ice thickness, subglacial water depth, gravity, temperatures, rate of movement, seismic and radio wave velocities, electrical resistivities and radar polarization. As access is more difficult the Filchner-Ronne Ice Shelf has received less attention but, as we have seen (chapter 7), it is of especial importance for ocean circulation as the main source of Antarctic bottom water. A major feature of the Ronne Ice Shelf is a thin central region which arises because of shear between thick fast-moving ice streams along the eastern and western margins. Cyclonic circulation raises water from beneath the thicker to the thinner areas and, because the decrease in pressure raises the freezing point of the water, saline ice is formed. However, bottom melting predominates under most of the Shelf (Robin *et al.*, 1983).

In the world of science, glaciology is but a minor branch of physics but in Antarctica in recent years it has developed a dominant role. Not only is Antarctica the natural place to study basic glaciological phenomena and processes but the recognition of the crucial part played by its ice-cap in global heat exchanges and circulations of atmosphere and ocean make these studies of general importance. The mathematical models used to investigate global warming will be of questionable value for prediction so long as accurate and complete data on interactions with the ocean through melting and freezing of the ice shelves are lacking.

8.7 Climatic history and the records in ice cores

A variety of indications – erratic boulders and striated rock surfaces above the present ice level, ice-free but glaciated valleys, and submarine moraines some distance seaward of the present ice edge – suggest that the Antarctic ice-sheet was formerly of greater extent than it is now. This retreat of the ice had been obvious to the geologists of the early expeditions, whose work was confined to coastal regions (Arçtowski,

1901b; Drygalski, 1904; Ferrar, 1907; Wright & Priestley, 1922), but during the IGY similar indications were found in the interior, although whether the thinning of the inland ice was contemporaneous with the recent retreat of the ice edge was not clearly established (Gow, 1965). Some thought that the recession was correlated with the widespread retreat of glaciers in other parts of the world but the estimate of 10,000 years before present for the onset of marginal retreat in Antarctica was unsupported by absolute dating. The palaeontological evidence makes it clear that Antarctica was not always glaciated to the extent that it is now and that it had experienced climatic amelioration, for example, in Triassic times. Before that, in the severe Permo-Carboniferous glaciation, Antarctica was evidently the centre from which ice flowed outwards to leave tillites and other evidences of glacial action in other parts of Gondwana. The time of the onset of the most recent glaciation was speculative when Adie (1964) wrote his review but the finding of fossil penguins in the late Oligocene to lower Miocene of northern Graham Land suggested that the climate was getting cooler then. It seemed that the present ice-sheet had begun to form in the mid-Pliocene contemporaneously with that in the northern hemisphere. By the early 1960s there was considerable evidence from work on raised beaches, wave-cut platforms, submerged sea caves revealed by SCUBA diving and morainic deposits interbedded with lava flows, of late Tertiary fluctuations in sea-level in the Peninsula and Scotia Arc area but accurate dating had not been carried out (Odell, 1952; Adie, 1964). These fluctuations seemed to have been greater in West Antarctica, where they were related to world-wide sea-level changes which caused the major outlet glaciers to rise in level as their outlets dammed up. The ice-sheet in East Antarctica seemed to have been more stable (Hendy *et al.*, 1979). Drilling during the McMurdo Sound sedimentary and tectonic study showed that glaciomarine conditions in that area go back to late Palaeocene, 60 million years ago (Webb, 1983). A global temperature drop was inferred from isotope records to have occurred at the end of the Eocene, 40 million years ago, but there was no evidence from the Antarctic to support this. Ice-sheet formation presumably began about 25 million years ago when the continents comprising Gondwanaland had separated sufficiently to allow the Circumpolar Current to become established and isolate Antarctica. The major build-up of the East Antarctic ice-sheet appeared from oxygen isotope determinations on benthic foraminifera recovered from Deep Sea Drilling Project cores obtained by the *Glomar Challenger* to have started in middle Miocene, 12 million years ago (Savin, 1977; Frakes, 1983). The apparent agreement between studies on oceanic sediments and continental drift in delineating the glacial history of Antarctica was reassuring but Frakes (1983) warned that the assumption by geologists that present knowledge of the hydrography of the Southern Ocean can be extrapolated back to define past climates was based on insecure foundations.

Glaciologists, during the IGY period, noted the presence in ice of included particles of various sorts, bubbles and traces of certain chemical substances. The idea that these contaminants, preserved in chronological order of their deposition down the length of an ice core, might be used to give a picture of climatic conditions and global events in the past, developed later. By shifting one's attention from the macroscopic to the microscopic and from the qualitative to the quantitative, one could learn much of the past by applying the palaeontological approach to recent deposits. This has been demonstrated by the success of pollen analysis in Europe and the US. Introduced in Sweden by Lennart von Post in 1916, it had become an effective instrument for investigating long-term vegetational changes and providing a time scale against which climatic and geological events could be viewed. Following the invention of the Jenkin corer in 1938, lake deposits were studied by W. H. Pearsall, W. Pennington and F. J. H. Mackereth (who, incidentally, had served as an inspector on Antarctic whalers) in England and by E. S. Deevey and G. E. Hutchinson in the US, not only by palynological techniques but also chemically and palaeomagnetically, to reveal much about general changes as well as those in lakes themselves. E. Philippi (1910), as we have seen (p. 200), deduced changes in Antarctic water masses from the sequence of different kinds of superficial sea bed deposits but it was not until B. Kullenberg developed his piston corer in 1947 that such investigations could be carried much further. Ice, in many ways, provides a much more satisfactory deposit for this type of study. The first significant ice-core to be recovered from the Antarctic was 100 m long obtained by the Norwegian–British–Swedish expedition in 1949–52 from the Maudheim ice shelf using modified rock-drilling equipment. The innovative programme of deep-core drilling in ice started in Greenland by the US Army Cold Regions Research and Engineering Laboratory was extended to Antarctica during IGY and continued there afterwards. Dating was done by observations of visible stratigraphical features such as wind crusts and grain size variations and by seasonal density variations (Langway *et al.*, 1971). The use of isotope studies, particularly measurements of the oxygen isotope ($^{18}O:^{16}O$) ratio, to determine seasonal changes was mainly developed in the Greenland work as was a new down-borehole gas extracting device tested in the deep borehole at Byrd Station in 1968–69. Cores from Byrd Station showed layers of what seemed to be ash from nearby volcanoes and preliminary studies of oxygen and hydrogen isotope profiles indicated that major climatic changes in Antarctica during the Wisconsin period, which began about 25,000 years ago, were essentially synchronous with those in the northern hemisphere (Langway *et al.*, 1971). Conventional palynological analysis of post-glacial organic deposits on South Georgia indicated that the climate there had ameliorated by 10,000 years BP and that soon afterwards many elements of the present flora were flourishing. Since few obvious peaks in the pollen profiles were found, no clear inferences about subsequent climatic changes could be drawn (Barrow, 1983).

Attention has been focused on chemical studies on ice-cores to provide a baseline against which the extent of global atmospheric pollution may be judged (Fig. 8.7). The Antarctic ice-fields afford a unique situation for such work since they are remote from sources of anthropogenic contamination as well as storing chemical impurities in chronological order of deposition with minimal thermal and bacterial breakdown and diffusion. The history of this line of research has been one of developing realization of the need for extraordinary precautions against contamination if reliable results are to be obtained. The organochlorine insecticide, DDT, eventually proved to be present in snow samples in concentrations from a fortieth to a hundredth of those first reported from central Antarctica, indicating that the global airborne load of this pollutant was smaller than previously supposed (Peel, 1975). Similarly when heavy metal analyses were done on ice-cores collected and sampled with meticulous precautions to prevent and detect contamination, values were considerably lower than the earlier ones reported from Antarctica and than those from Arctic ice (Boutron & Patterson, 1983). When sampling resolution was for periods much shorter than a year it was evident that there is great variability which is controlled, not so much by volcanism as had been supposed, but by meteorological conditions (Landy & Peel, 1981). Air entrapped in bubbles in an ice core from Siple Station was found to have increased in methane (another 'greenhouse' gas) concentration almost two-fold over the last 200 years (Stauffer *et al.*, 1985). The idea that bubbles might yield information about ancient atmosphere seems to have originated with N. W. Rakestraw (Sisler, 1961).

The outstanding achievement in this field has been the remarkable record of the earth's climatic history obtained by Soviet–French collaborative work on a 2,200 m core from Vostok. The engineering behind this was itself impressive. The Soviets used a thermal drilling technique in which the 10 cm diameter tube was electrically heated so that it penetrated the ice

Fig. 8.7 Snow sampling by British Antarctic Survey, 1985–86. Samples for analysis for trace constituents are handled with clinical cleanliness to avoid contamination. (Courtesy of the British Antarctic Survey.)

without damaging the sample. The melt water had to be recovered quickly to prevent refreezing and resultant distortion which would make retrieval impossible. At depth the glaciostatic pressure had to be balanced by fluid (alcohol) not liable to freeze at the temperatures encountered at the surface. The core extracted represented a period of 160,000 years, thus going back to the ice age before the one that ended about 8,000 years ago. Both ^{18}O and deuterium contents were determined and the latter taken as the better indicator of atmospheric temperatures. A total temperature amplitude of about 11 °C was found and the warmest period in the last interglacial was about 2 °C warmer than any since the last ice-age. The Vostok record is in good agreement with the marine records as far back as 110,000 years, which encourages one to think that the results are of global significance but before this there are discrepancies (Jouzel *et al.*, 1987). The core also yielded direct evidence of past changes in the carbon dioxide content of the atmosphere with major variation over 100,000 years between glacial and interglacial periods and lesser cyclic changes with a period of some 21,000 years (Barnola *et al.*, 1987). These results are, of course, of particular interest in relation to the 'greenhouse effect', showing as they do a close correlation between carbon dioxide concentration and mean atmospheric temperature. Thus, carbon dioxide concentration emerges as the major factor forcing, or following, temperature changes but it appears that these are triggered by and amplify insolation changes brought about in both the northern and southern hemispheres by orbital forces (Genthon *et al.*, 1987). A commentator (Campbell, 1987) expressed surprise that after initiating ice-core studies so brilliantly the US should have fallen behind in this field and attributed this to the diffuse nature of the US programme on the one hand and the successful informal collaboration within multidisciplinary glaciological laboratories achieved by the Europeans on the other.

8.8 Meteorites on the ice-sheet

Ice-cores contain micrometeorites and presumably there is a remote chance that one day a corer will hit a large meteorite. However, the first meteorite to be found on the Antarctic continent was discovered on the surface in December 1912 in Adélie Land by an Australasian Antarctic Expedition party led by F. L. Stillwell:

> It measured approximately five inches by three inches by three and a half inches and was covered with a black scale which in places had blistered; three or four pieces of this scale were lying within three inches of the main piece. Most of the surface was rounded, except one face which looked as if it had been fractured. It was lying on the snow, in a slight depression, about two and a half inches below the mean surface, and there was nothing to indicate that there had been any violent impact.
>
> (Mawson, 1915, II, p. 11.)

The early IGY expeditions discovered three more, then in 1969 Japanese scientists came across hundreds of fragments from dozens of meteorites

Fig. 8.8 US team collecting meteorite. (Courtesy of Dr P. Wasilewski, NASA Space Flight Center.)

scattered on the surface of blue ice in the Yamoto Mountains, East Queen Maud Land (Yoshida *et al.*, 1971). A similar find by Americans on the Allan Hills, to the west of the Dry Valleys in Victoria Land, soon followed and by 1985 more than 7,000 fragments representing between 1,200 and 3,500 separate falls, a number of the same order as known non-Antarctic meteorites, had been collected (Lipschutz, 1985). The reason for this extraordinary harvest is that under the right conditions the ice sheet acts as a collecting mechanism. In areas of blue ice, with high ablation, strong wind and slow horizontal movement, meteorites remain on the surface (Fig. 8.8) and accumulate against any suitably placed obstruction blocking the ice flow. Antarctic meteorites are of great interest not only because of their quantity but for their variety. Many are of types common in non-Antarctic falls but some are of rare or unique type. Among Antarctic meteorites, for example, are the only known naturally transported lunar samples on earth (Eugster, 1989). Some carbonaceous chrondrites contain inclusions of an anomalous composition suggesting an origin outside the solar system. It seems from statistical studies on their composition that this variety is not simply a matter of preservation but reflects some major difference in meteorite flux between Antarctica and elsewhere so that Antarctica may be a potentially unique source of information on extraterrestrial objects (Lipschutz, 1985; but see Wright & Grady, 1989).

8.9 Denudation processes

In 1898, the Duke of Argyll thought that Antarctic research would dispel differences among geologists as to what glacial agencies did during the ice age (Murray, 1898). For the expeditions which followed, Drygalski's work on the Greenland ice, which had been published in 1897, served as a useful model. Ferrar (1907), Wright & Priestley (1922) and Taylor (1922) described glaciers and periglacial phenomena but recognized that most of the Antarctic glaciers which they saw were incapable of causing much erosion. In striking contrast to the glaciers of Switzerland and New Zealand, supraglacial rock debris was found to be scarce on those of Antarctica. Many of the existing topographical features, Taylor concluded, must have been shaped in warmer times, in the initial stages of glaciation, to be preserved under the ice-sheet until revealed by its recession and then, perhaps, modified by fluvial or marine processes or subaerial erosion. Priestley & Wright (1928) summed up by pointing out that nearly everywhere on the continent the ice-sheets extend over the sea so that glacial debris is deposited below water and consequently the Antarctic explorer tends to adopt a very conservative estimate of the power of ice to erode. Holmes (1944) in his important book *Principles of Physical Geology* took no examples of glacial erosion from the Antarctic literature. Among other agents of denudation, frost action was recognized as active in the South Shetlands by Eights (1846) and by Bernacchi (1901) at Cape Adare. However, Ferrar (1907) working 6° further south than Bernacchi considered that, except for dolerite which split easily, frost action was negligible because the prevailing low temperatures ensured a general absence of liquid water. He put wind at the head of his list of agents of denudation. This not only removes and transports rock debris but, charged with ice or sand particles, has considerable abrasive power. Boulders cut with facets by this agency, *ventefacts*, were described by members of the *Challenger* expedition on Heard Island (Thomson & Murray, 1895, Fig. 8.9). When Scott discovered the Taylor Valley in 1903 he noticed that:

Fig. 8.9 Ventefact from Heard Island. From a sketch made by J. Y. Buchanan on the *Challenger* expedition, 1874; 'The side towards the west, with the high light in the woodcut, is being rapidly worn down by the sharp sand blown against it, which has cut an irregularly fluted pattern in it.' From Thomson & Murray, 1885, p. 372. (Courtesy of the Royal Society.)

> the hillsides were covered with a coarse granitic sand strewn with numerous boulders, and it was curious to observe that these boulders, from being rounded and sub-angular below, gradually grew to be sharper in outline as they rose in level. (Scott, 1905.)

Ferrar (1907) paid little attention to ventefacts but was more concerned with the hollowed granite boulders which he encountered and which he attributed to wind action, although he noted an internal incrustation of calcium carbonate in one. Water action he considered of limited consequence although he recognized that in the summer, melt-water redistributes mud and sand from moraines over the ice. Chemical action was obvious and he described spectacular accumulations of sodium sulphate near some moraines between White Island and Black Island.

More detailed study of denudation processes had to wait until 1957 when attention was again focused on the dry valleys; other areas, South Georgia for example, where frost shattering is spectacular, attracting little notice. The first long-term study in the peri-Antarctic islands of the daily changes in a tablet of indigenous rock appears, in fact, to have been that by Hall (1988) carried out on Signy Island. The recent work on the dry valleys has been summarized by Campbell & Claridge (1987). There is general agreement that the main elements of present day topography were formed before the continental glaciation, perhaps as long ago as 15 million years, during a period of alpine glaciation when the ice was mainly wet-based. The later dry-based glaciation has preserved, rather than significantly modified, these earlier-cut features. In ice-free areas frost action now appears to be more important than Ferrar thought; rocks exposed to the sun, besides being subjected to purely thermal strains, may undergo as many as 100 freeze-thaw cycles per year (Black, 1973). Additionally, salt weathering, in which fragmentation is brought about by pressures exerted by salts crystallizing within pores and fine cracks, has been found to occur particularly in the cavernous weathering of coarse-grained rocks noticed by Ferrar (1907) and Priestley (1923). These boulders usually have case-hardened polished outer and upper surfaces and the salt weathering proceeds on protected sides, wind removing the fragments, until holes may extend right through the rock (Calkin, 1964). Now that general survey of Victoria Land is virtually complete, it seems from the relative absence of aeolian deposits that salt weathering cannot be as vigorous as supposed, otherwise there would be extensive sand deposits either in the valleys or inshore. Patterned ground, well known from the Arctic and sub-Arctic – it was noticed in Spitsbergen by Scoresby (1820) – is also a prominent feature of the maritime Antarctic and the dry valleys of the continent. Drygalski (1904, p. 102) described stone stripes on both Kerguelen and Heard Islands but thought they were produced by flowing water. The mechanism of the formation of polygons, stripes and other patterns and the sorting of material that occurs, which is essentially based on the expansion and contraction of ice, was studied on Signy Island (Chambers, 1970) and in Victoria Land (Campbell & Claridge, 1987). Apart from its interest in relation to weathering it is suggested that patterned ground may be of use in dating changes in environmental conditions (Berg & Black, 1966). Chemical weathering is actually slow under Antarctic conditions but it is significant. The striking accumulations of salts – sodium sulphate, calcium chloride, calcium sulphate – that are sometimes found have several different origins. Near the coast, marine influences dominate as is shown by isotopic abundances as well as by distribution (Keys & Williams, 1981). The situation is complex and controversy continues but remarkable insights are being obtained into geochemical processes (see Heywood, 1984).

8.10 Soil

Whether or not they are dignified by the name of soils, the surficial deposits of ice-free areas are worthy of study. The presence of peat in the McMurdo Sound area was noted by David and Priestley and they figured a profile of a deposit in a small lake, formed from the remains of 'a large fungoid plant' (Shackleton, 1909) which was presumably a cyanobacterial mat. However, analyses carried out by Jensen (1916) on samples of typical soils of this area, collected by Mawson on the *Nimrod* expedition, showed them to be of characteristic desert type with coarse textures and having soluble salts and calcium carbonate at or near the surface. No further studies were done until the 1950s when the Soviets, who have a tradition of pre-eminence in soil science, carried out reconnaissance work near Mirny. New Zealanders became interested through a casual sample scooped up during the construction of Scott Base. The New Zealand Soil Bureau dispatched J. D. McCraw and G. G. C. Claridge in 1959 to prepare a soil map of the Ross Dependency but in the event their work had to be largely confined to part of the Taylor Valley (Campbell & Claridge, 1987). US soil scientists followed with work in the same region and by 1966 their studies on Antarctic soils were sufficiently advanced to have a volume of the *Antarctic Research Series* devoted to them (Tedrow, 1966). The whole field was reviewed by Campbell and Claridge in 1987. The chief feature is that organic component in soil formation is largely missing and the weakness of the biotic factor makes these soils not only interesting in providing a key to the understanding of soil processes but useful as test materials, in some respects similar to Martian soils, for ideas and techniques in space exploration (Cameron, 1963). It has also made them difficult to fit into traditional schemes of soil classification. In the maritime Antarctic this difficulty diminishes. Systematic investigations of soils in this region were carried out by Holdgate *et al.* (1967) on Signy Island, South Orkney group. Breakdown of rock by frost action was found to be rapid but solifluction processes extensively influenced the upper layers of debris and there was no clear soil stratification on many slopes. In stable areas supporting bryophyte communities there was accumulation of peat, a radiocarbon date of about 1840 years BP from the base of a bank being considered a minimum (Collins *et al.*, 1975). The most developed soils approximated to brown earths (Holdgate *et al.* 1967). Further north, on South Georgia, peat deposits are more extensive with the oldest basal layers dated at 9,000 years BP. Brown soils occur on well-drained sites and podzols predominate as a result of the high precipitation and low temperatures (Headland, 1984) but detailed investigations of these have not been carried out.

8.11 Physical limnology

The ponds and lakes of the McMurdo Sound area chiefly interested the biologists (see chapter 10) of the heroic age of exploration and little was done on their physical limnology except that Mawson reported on the formation of vertical prismatic ice and crystallization of salts and Murray reported on seasonal changes in temperature at various depths in lake ice (Shackleton, 1909). Heywood (1967) gave a description of the catchment areas and morphometry of the small lakes which occupy the more low-lying parts of the glacially overdeepened valleys on Signy Island. An account of the physics and chemistry of these waters (Heywood, 1968) completed his survey which was of value more in providing a background for biological studies than for revealing any remarkable physical features. However, the peculiar hydrology of the dry valleys has created situations of great interest to the physical limnologist. The lakes between the Wright and Mackay Glaciers in the McMurdo area were the first to be examined, by a party from the Victoria University of Wellington, New Zealand, in 1957 (Hatherton, 1967). Lakes having fresh and saline waters were found, an extreme and unique example being Don Juan Pond, which contains a nearly saturated solution of calcium chloride. It never freezes, even at temperatures as low as −51 °C and was found by Torii & Yamagate (1981) to have crystals of a mineral new to science, antarcticite ($CaCl_2 \cdot 6H_2O$), in and around it. Dry valley lakes are without outlets and such salt concentrations are reflections of the accumulation which has already been touched on (p. 283). Lake Vanda, on the other hand, a permanently ice-covered lake fed by continental Antarctica's longest river, the Onyx, has surface waters which are fresh. In the austral summer of 1960–61 a biological expedition from the University of Kansas, USA, found that the bottom water of this lake was saline, that is the lake was meromictic, and had a temperature of about 25 °C, 46 °C above the mean temperature of the region. It was at first suggested that this was caused by geothermal heating but during the Dry Valley Drilling Project in 1971–75 a hole was drilled through the 12 m of sediment at the bottom of this lake into the bedrock and this showed a heat flux from the water into the sediments rather than the other way round. It is now generally accepted that the hypothesis of Wellman & Wilson (1962), that it arises by transmission of solar radiation through the exceptionally clear prismatic ice and water, with storage of heat in the bottom layer, which is stabilized by its high density, is correct (Vincent, 1987). A numerical model with solar radiation as the forcing function accurately describes the thermal structure. The origins of the meromictic condition of Lake Vanda are not altogether clear but it is evident that the lake has undergone large fluctuations in area and depth during the last few thousand years and possibly, after a period of salt concentration by ablation of the ice-cover during a colder period, the river Onyx resumed its flow and overlaid the brine with fresh water (Vincent, 1987). Lake Vanda is Antarctica's most famous lake and a permanent station on its shore was established by New Zealand in 1967.

8.12 The wider role of geologists in Antarctica

This chapter should not be concluded without some reference to the major role which geologists have played in general Antarctic affairs. It was mentioned at the beginning that geologists were included on the staffs of all the main early expeditions but several of these men thereafter became devoted to promoting Antarctic exploration and research in a wide sense and transmitted their enthusiasm to disciples. Ferrar, after taking part in Scott's first expedition, continued to work as a geologist but had no further important contact with the Antarctic. Edgeworth David not only took a deep interest in Antarctic matters both before and after his participation in Shackleton's *Nimrod* expedition but trained students who were to become key Antarctic figures (Fig. 8.10) – Mawson, the driving force behind Australian ventures south; Griffith Taylor, geologist on Scott's second expedition who afterwards became one of the foremost physical geographers of his time; and Debenham, another geologist on that expedition, who became Director of the Scott Polar Research Institute in its formative years. Priestley, who served as a geologist with both Shackleton and Scott, was not one of David's students but was closely associated with him on the *Nimrod* expedition. As recounted in chapter 5, Debenham, Priestley and Wordie, geologist with Shackleton on the *Endurance* expedition, were the founders of the Scott Polar Research Institute and Wordie was influential in getting science incorporated into *Operation Tabarin* and in the setting up of its successor, the Falkland Islands Dependencies Survey. Priestley, who became the first Director of the Survey, established its Geological Section and appointed Adie, one of Wordie's students, to lead it. Another of Wordie's students, Fuchs, followed Priestley as Director of the Survey and of the British Antarctic Survey which it became, and led the Commonwealth Trans-Antarctic Expedition. This may savour of an 'old boy network' but its outcome has been wholly admirable. Among Americans, Gould, who went as geologist on Byrd's first expedition, directed the Antarctic programme of the US National Committee for the IGY, was chairman of the Polar Research Committee of the National Academy of Sciences, and represented his country on SCAR (see Fig. 6.8).

Fig. 8.10 Sir Edgeworth David, FRS, and his scientific progeny; (a) Sir Edgeworth in 1927, (b) Sir Douglas Mawson, FRS, in 1956, (c) Sir Raymond Priestley in 1913, (d) Professor Griffith Taylor circa 1956. From Taylor, 1958, facing p. 32. (Courtesy of Robert Hale Ltd.)

Endnotes

1 Later published in *Die Südpolarfrage und ihre Bedeutung für die genetische Gliederung der Erdoberflache*, 1896.

2 Fossils had been found before this in Antarctica (p. 46). What he thought, erroneously, to be the first fossils brought back from the Antarctic were described by Weller (1903). These were collected from a talus slope on Trinity Peninsula by F. W. Stokes, artist on the *Antarctic* expedition – not the *Belgica* as Weller states – who did not overwinter and returned north that year. The 12 specimens were mainly molluscs and related to forms found in the Middle or Upper Cretaceous of southern India.

3 British Antarctic Survey Scientific Advisory Committee Minutes, 7, 1974.

4 Ibid, 20, 1982.

5 Ibid, 9, 1975; 10, 1975.

6 British Antarctic Survey, Annual Report 1985–86, p. 50.

9

The sciences of atmosphere and geospace

9.1 The atmospheric sciences at the end of the nineteenth century

In the nineteenth century the main developments in meteorology were in standardization of observational techniques and in international collaboration, these together enabling the establishment of meteorological services based on synoptic observations from networks of weather stations. This was promoted, on the one hand, by the introduction of public electrical telegraph systems in 1845 and, on the other, by international congresses of meteorologists such as that held in Vienna in 1873. Observations were confined to land – or sea-level – although as long ago as 1749, Alexander Wilson of Glasgow had used kites to raise thermometers. Joseph Gay-Lussac in 1804 had made balloon ascents to as much as 7,000 m during which he made measurements of temperature, humidity and magnetic field and took air samples for analysis. The first station for the investigation of the upper air was established near Berlin in 1899. The study of atmospheric electricity, first seriously begun by Benjamin Franklin in 1746–52, was given impetus in 1850 by Sir William Thomson (later Lord Kelvin), who presented an interpretation of all the then known facts in terms of electrostatic theory, developed new instruments, and inaugurated programmes of continuous recording. The aurora, which had early on been linked with magnetic disturbances by Halley, Humboldt and others (e.g. Farquharson, 1830), had been located as occurring at no great height above the earth – John Dalton (1828), for example, estimated 100 miles (161 km) – and was therefore presumably an atmospheric phenomenon, albeit still completely mysterious. Heinrich Hertz in 1886–88 had demonstrated the propagation in space of electromagnetic energy, as foreseen by James Clerk Maxwell, but at this stage there was no thought of using radio waves in the investigation of the atmosphere.

Orthodox thinking about global circulation had become set in a mould based on what was known of tropical regions, the poles being looked on as inert areas. Hermann von Helmholtz (1888) had provided the basic hydrodynamical theory needed for the study of atmospheric circulation and focused attention on the role of radiation losses at the poles, which by increasing the density of the surface layer of air would bring in more from above and force a radial outward surface distribution. However, the mathematics proved intractable and no comprehensive theory had emerged on this basis. It was assumed in theories put forward by authorities such as Maury, Ferrel and James Thomson, that there was symmetrical cyclonic circulation around the South Pole. Thus in the book *Elementary Meteorology* by Professor Davis of Harvard (1894), it was stated that air flowing southwards along the surface ascends at the Pole and thence proceeds northwards in currents of enormous intensity and volume to give high pressures at sea-level in the tropical region. The indications, found by explorers from Cook onwards to d'Urville, Wilkes and Ross, of higher pressures at high latitudes were ignored (Mill, 1926) as was von Helmholtz's theoretical work. However, by the end of the nineteenth century there were those who were prepared to take another look at such data as were available from the Antarctic. Apart from the detailed observations made on South Georgia by the German International Polar Year expedition in 1882–83, the only information came from ships. Summing up what was known at the Royal Society discussion meeting in 1898, John Murray concluded that Ross's observations of southerly to south-easterly winds bringing clear weather in the extreme south indicated a vast permanent anticyclonic area over the Pole, inside the belt of low atmospheric pressure with westerly or north-westerly wind south of 49° S. This view was amplified by J. Y. Buchanan, who considered the observations made from the *Challenger* supported the idea of upper air currents flowing towards the Pole to make good the drain caused by the outflowing south-easterlies at the surface. This was diametrically opposed to the orthodox view and presented the difficulty that it implied a drier atmosphere over the Pole with consequently little precipitation, whereas the cyclonic situation would, of course, provide ample precipitation to maintain the ice-cap. This problem was to dominate Antarctic meteorology for the next 50 years.

Georg von Neumayer at the same meeting, besides emphasizing the need for observation of winter temperatures on the continent for general climatology, also had perceptive words (which probably sounded better in the original German) to say about magnetism and atmospheric electricity:

> a discussion of the magnetic state of our earth for the epoch 1885 has yielded a curious fact (perhaps merely a coincidence), namely, that Dr Schmidt, of Gotha, has calculated that part of the magnetic action of our

earth which lies outside it, above the earth's atmosphere probably, and has arrived at the conclusion that this part amounts to about 1/50th part of the entire potential. The curves which he constructed, based on this calculation, show likewise a close coincidence with the frequency of the southern lights. This requires to be further investigated, for it is perhaps a mere coincidence. The question of atmospheric electricity, yet under the shadow of a hypothesis of more or less probability, may yield, in connection with the matters touched upon, some results more definite than science has hitherto been able to divine. (See Murray, 1898.)

9.2 Heroic age meteorology

While this discussion was going on the ice was gripping the *Belgica* and enforcing direct experience of the Antarctic winter. Meteorological observations were carried out throughout the year for which the ship was beset; the lowest temperature, −43 °C, was recorded on the 8th September, 1898 and the lowest average for a month, −12 °C, in July. The barometric pressures were higher than expected but the variations were such as to make it uncertain whether this was characteristic of latitudes around 70° S or not. These data were, of course, for a sea-ice area but as the *Belgica* was being released the *Southern Cross* expedition led by Borchgrevink was settling down for a sojourn that was to provide a year's data for the continent itself, at Cape Adare. The physicist Louis Bernacchi was responsible and it is worth listing the equipment at his disposal and the scope of his observations (Bernacchi in Borchgrevink, 1901, and Murray, 1901) to give an idea of the state of the art at that time. Observations were obtained as far as possible as at a meteorological station of the first order, with two-hourly readings between 9 a.m. and 9 p.m. for nine months of the year and two-hourly readings day and night during the three winter months. These readings were supplemented with continuous recordings from barograph and thermograph (Fig. 9.1), records from a Campbell-Stokes sunshine recorder modified to cope with the low elevations of the sun, maximum and minimum temperatures, and measurements of solar radiation by black bulb thermometer *in vacuo*. Spirit thermometers had to be used because mercury froze at winter temperatures but were somewhat variable in their readings, a point reiterated by Murray (in Shackleton, 1909) who found that they could show errors of several degrees because of displacement of the spirit. The Robinson anemometers were demolished by a wind which exceeded 90 miles (145 km) per hour and even before this happened were probably giving inaccurate readings because of wear and tear. Bernacchi recommended that future expeditions should take pressure anemometers in addition to those of the cup type.

The three major expeditions which departed in 1901 thus had some experience on which to draw. In the Antarctic manual prepared for the *Discovery* expedition (Murray, 1901) J. Y. Buchanan contributed a section on meteorological instruments and R. H. Scott introduced a multi-author account of the climate of Antarctica in which data from the *Balaena* of the Dundee whaling expedition, from the *Belgica*, from the *Valdivia*, and from the voyage of the *Antarctic* in 1894–95, were discussed. Also in the manual, Dr Julius Hann, author of the definitive work *Klimatologie* was equivocal about the existence of a continental anticyclone, maintaining that if it existed it must be very shallow, but both Dr A. Supan and Louis Bernacchi concluded that there was such a thing. By now the controversy was beginning to resolve itself into a matter of definition. The change in prevailing winds seen from the *Belgica*, from westerly in winter to easterly in summer, was taken as an indication that this anticyclone shifted, moving further towards the eastern hemisphere in winter presumably because a cold centre develops on that side – an argument in favour of the view which we now know to be correct that the major part of the continental land mass is situated there.

These expeditions contributed a mass of data on Antarctic meteorology. Nordenskjöld at 64° S on the peninsula was able to make limited observations only (Fig. 9.2) but the weather he endured may be summed up as continuing south-westerlies and depressions. He found the temperatures on the east coast of the Peninsula to be much lower than those on the west. The *Gauss* expedition at 66° S experienced easterly winds, often extremely strong, for most of the 11 months of her stay in the ice and Drygalski (1904) wrote a heart-felt account of the difficulties of making meteorological observations under such conditions. The *Discovery* expedition at 77° 30′ S found relatively high pressures and prevailing south-easterly winds but Shaw (1908) was not convinced that these were anything more than local conditions. His doubts arose because on some journeys away from base across the ice shelf there was uncertainty in the records as to whether the wind directions were magnetic or true and because there was no reliable information about the slope of the ice shelf surface to enable barometric pressures to be adjusted to sea-level so there was no real evidence of the predicted pressure rise towards the Pole. Hobbs (1926) was later to point out the danger of ignoring, as so often such commentators did, the effects of local topography and of assuming observed wind directions to be prevalent over extensive regions. He cited David's discussion of wind circulation in terms of the geographical pole rather than of ice topography in spite of the experiences of Shackleton and himself on their journeys, with

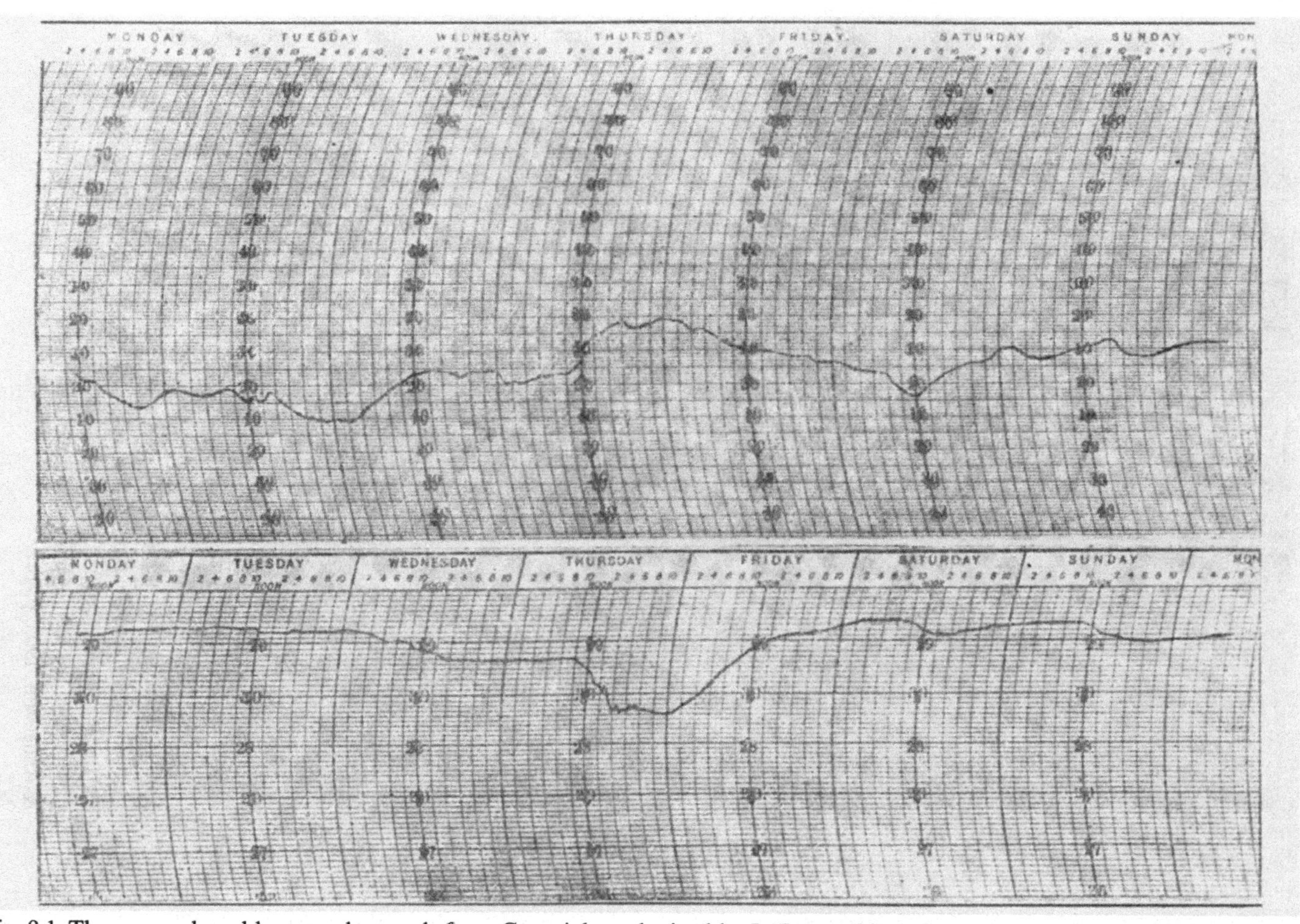

Fig. 9.1 Thermograph and barograph records from Cape Adare obtained by L. Bernacchi on the *Southern Cross* expedition, 1898–1900, showing rises in temperature with falls in pressure; Borchgrevink, 1901, p. 309.

Fig. 9.2 Meteorology at Snow Hill on the *Antarctic* expedition, 1901–4. On the wall can be seen the barograph, an evaporimeter, the recording equipment for the anemometer, paper for the sunshine recorder and an anaeroid. G. Bodman, the meteorologist, is at work at the table. From Nordenskjöld & Andersson, 1905, p. 147.

blizzards blowing down slope. Nevertheless, Hepworth (1908) and Meinardus (1909) in appraising the information available agreed that a constant succession of cyclonic depressions passing from west to east encircled a permanent anticyclone over the continent. Meinardus continued to find difficulty in postulating a normal anticyclone over Antarctica since this would imply dry conditions and leave a problem about the replenishment of the enormous quantities of ice which were obviously being lost from the periphery of the continent. He supposed that there must be a cyclone above the anticyclone to provide the necessary inflow of moist air. Observations of clouds and smoke from Erebus showed that there was indeed a poleward flow of the upper air and that the cold surface layer of air was probably no more than some 1500 m thick (Shaw, 1908). Neither Drygalski (1904), who went up to 500 m, nor Scott who attained less than half that height, made any serious meterological observations during their balloon ascents. On the *Scotia* expedition to the Weddell Sea area in 1902–4 a large kite was taken to carry aloft an instrument package to record pressure, temperature and humidity. Unfortunately it could not be flown successfully from the ship nor on land (Brown *et al.*, 1906). Drygalski (1904) likewise had no success with kites.

Dr William Bruce had trained himself in meteorology under arduous conditions by setting up a station on the summit of Ben Nevis in Scotland. The meteorological and magnetic station which he established on Laurie Island in the South Orkney group was given to Argentina after the UK government had declined to accept responsibility for it. He handed over the buildings and stores free of cost, took four Argentinians from Buenos Aires down to Laurie Island to staff it, and left R. C. Mossman to guide them through the first year (Brown, 1923). Since 1904, this station, which was renamed Orcadas, has been kept in continuous operation by Argentina, although it was manned by Britons, Norwegians and Danes until around 1920. It has provided a unique long-term record of maritime Antarctic weather of great importance as a base line for studying climatic change.

In the next 10 years Shackleton's *Nimrod* expedition, Charcot's voyage in the *Pourquoi-Pas?*, Scott's second expedition, Filchner's *Deutschland* voyage and Mawson's expedition to the George V Coast all added substantially, in various ways, to the store of information. Shackleton gave a clear description of conditions on the plateau for the period he was there but there was no inkling in 1912 that its annual temperature variations were unlike those anywhere else on earth. When the US station was established at the Pole in 1956 it was found that it had a 'coreless' winter, in which temperatures were not as low as expected and remained at around the same

Fig. 9.3 Sending up a meteorological balloon on the *Terra Nova* expedition, 1910–13. (H. Ponting.)

level for about six months, and a 'pointed' summer of short duration. Amundsen reached the Pole at the height of summer but Scott was there when temperatures had already begun their plunge to the winter level.

The detailed meteorology on this expedition of Scott's was confined to the base on Ross Island where the meteorologist was George Clarke Simpson (1878–1965, Fig. 9.9), a scientist of distinction (Gold, 1965). When he was invited to go to Antarctica, Simpson was working in India where in the time he could spare from routine inspection of meteorological stations he was investigating the generation and distribution of electricity in thunderstorms. Before that he had studied atmospheric radioactivity in Lappland and in 1905 had been the first lecturer in meteorology at a British University. In the Antarctic, Simpson used balloons for determining the temperature of the upper air (Fig. 9.3). Ascents were only possible in calm weather but heights of up to two or three miles were reached. The recording instrument was attached to the balloon by a release mechanism and was recovered by searching in the direction in which it was seen to fall. In summer a steady decrease in air temperature of about 6 °C per 1000 m up to the greatest height attained was observed. In winter the instrument carried a mile-long light silk thread by means of which it could, sometimes, be traced to where it fell. It was found that in winter the temperature rose from the surface upwards to about 1000 m but then fell and continued to do so up to 2500 m. This was the first observation of the characteristic stable temperature inversion which occurs over the Antarctic in winter. Another line of investigation, suggested by Dr W. N. Shaw, Director of the Meteorological Office, was on the question of at how low a temperature could supercooled water droplets exist in the atmosphere. On one occasion Simpson saw a fog-bow, which could only be formed by droplets of liquid water, when the temperature was −29 °C. His observations led him to doubt whether ice crystals could produce diffraction effects; ice needles could do so but as they would tend to float with their axes horizontally they could not produce symmetrical coronas such as were seen. The iridescent clouds frequently seen in Antarctic skies he thought, erroneously as it now appears, must also consist of water droplets although, being of cirrus type, they would be at an altitude where the temperature was well below the ice-point. Examining the records of barometric pressure which he collected in the Antarctic he became interested in periodic pressure variations. As in other parts of the world there was a semi-diurnal variation. In middle and low latitudes the phase of this wave of variation is nearly constant for places near sea-level; when time is measured from local midnight the maxima occur around 10 a.m. and 10 p.m. However, in polar regions, as first noticed by Greely in the Arctic in 1888, the maxima occur at nearly the same Greenwich time, regardless of longitude (see p. 333 for parallel phenomena in the thermosphere). Simpson confirmed this by examination of the available data for stations north of 70° N, spread over 280° of longitude, and discussed the anomaly in terms of two hypothetical waves, one travelling from Pole to Pole and the other from east to west – the first mathematical model in meteorology. He also described pressure oscillations of longer period crossing the Ross Sea area. The mean period of these waves was 150 hours, the mean variation amounted to 14.5 mm, and they travelled at a speed of about 40 miles (64 km) per hour. Simpson believed that such pressure waves were of constant importance throughout Antarctica but he was unable to explain their origin. His conviction that depressions were not their cause was proved wrong during the IGY. He synthesized his observations into a theory of atmospheric circulation over Antarctica which combined Meinardus's idea of a superimposed cyclone with emphasis on the role of the dome of the ice-cap promoting drainage of cold air. The continental mass of Antarctica, he insisted, would establish its own regime in the atmosphere above it and not merely protrude inertly into the circulation. He wrote:

> Over the snow-covered surface of the Antarctic, whether at sea-level or at the height of the plateau, radiation is so strong that the air is abnormally cooled, especially in the layers immediately above the surface. This cooled air is heavier than the surrounding air, and therefore the pressure increases from the exterior to the interior of the polar area; in other words the pressure distribution is anticyclonic and the air movement in general outwards. Above each anticyclone a cyclone forms on account of the relatively rapid vertical pressure change caused by the cold dense air. The descending air is warmed giving clear cloudless skies . . . as one penetrates the Antarctic.
>
> The clear skies in their turn facilitate radiation, as does the small absolute humidity of the air. In consequence the air and the snow surface became abnormally cold and there is a great tendency to the formation of temperature inversion, especially in the lower atmosphere. On these normal fine weather conditions are superposed a series of *pressure waves* which travel more or less radially outwards from the centre of the continent. These waves alter the surface pressure conditions, and cause air motion which is frequently accompanied by forced ascending currents. The abnormally cold surface air is forced upwards in these currents, rapidly cooled in the ascent, and the water contained is precipitated as snow, which when combined with the high surface winds produces the typical Antarctic blizzard. (Simpson, 1919.)

It was W. H. Hobbs, a geologist whose glaciological studies impelled him towards meteorology, who brought the various ideas together into a concept of the Antarctic anticyclone much as it is accepted today. His first essay on this subject was in 1910 and he elaborated it in a book published in 1926. His principal point, already foreshadowed by Helmholtz in 1888, was that increase in density of air, resting on a downward slope, by withdrawal of heat is as potent a means of maintaining circulation as a decrease in density of air, resting on an upward sloping land surface, by continued addition of heat. His innovation was to apply this to wide regions where refrigeration is continued over long period. He used the name *glacial anticyclone* to denote a permanent and complete circulatory system in three dimensions in a mass of air extending from the ground to the top of the troposphere and including an indraught of air from a distance as well as the

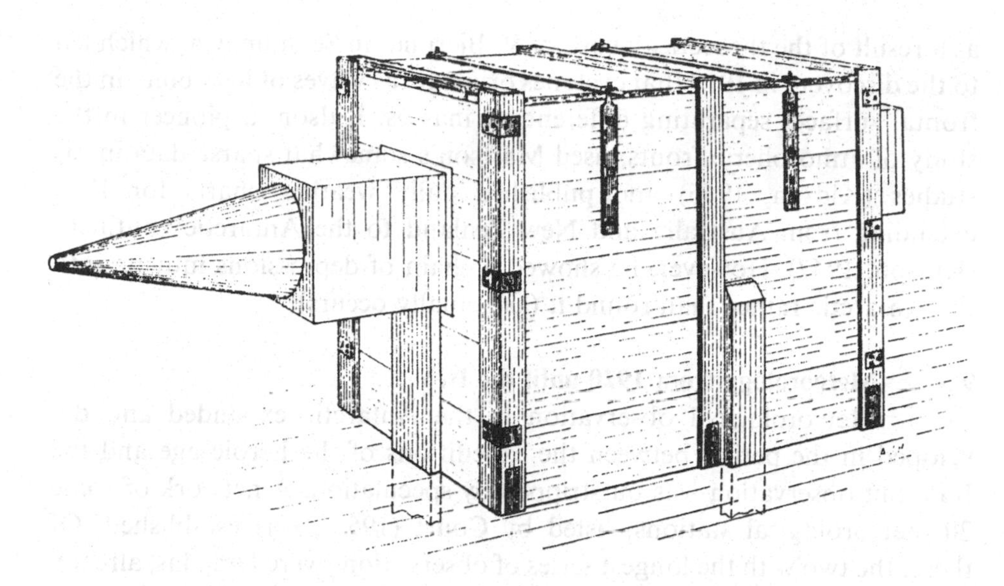

Fig. 9.4 The drift gauge used on the Australasian Antarctic Expedition, 1911–14. From Mawson 1915, vol. I, p. 123.

downward and centrifugal flow of air on the surface. He saw cirrus clouds as an adequate source of moisture to maintain the ice-cap at the very low temperatures prevailing.

Mawson in selecting Adélie Land for his 1911–14 expedition had inadvertently picked one of the most relentlessly windy spots on the continent. In March and April the wind often blew for days on end at speeds between 96 and 128 km h^{-1}, gusting to 160 km h^{-1} and occasionally touching 320 km h^{-1}. Over the first year the wind averaged nearly 80 km h^{-1}. So extreme were these records that the meteorological establishment was sceptical and Mawson's instruments had to be tested and recalibrated independently in London before the observations were accepted (Walton, 1987, p. 198). Sometimes wind velocity was best measured by the displacement of a weight suspended on a wire. Drift of snow was measured by a gauge consisting of a large drift-tight box with a long metal cone tapering to an aperture on the windward side (Fig. 9.4); the drift-laden air entering the aperture had its speed checked in the capacious interior of the gauge and the snow which dropped to the bottom could be collected and measured. Meteorological observations were obtained over two years at the Adélie Land base and in addition there were those from the base 48° to the west over one year, from Macquarie Island for two years, and those made on the ship and on sledging expeditions.

During this time weather models and forecasting underwent a revolution as a result of the theoretical work of V. Bjerknes in Scandinavia, which led to the discovery in 1919 that cyclones originate as waves of instability in the frontal surfaces separating different air masses. Kidson, a pioneer in the study of atmospheric fronts, used Mawson's somewhat sparse data in his studies (Kidson, 1946) and published daily weather charts for 1912 extending from Australia and New Zealand to the Antarctic continent (Kidson, 1947). However, he showed a chain of depressions moving into the Antarctic rather than round it (as actually occurs).

9.3 Meteorology from 1920 until the IGY

Meteorological observation in the Antarctic expanded and developed in the period between the expeditions of the heroic age and the IGY but observation was outstripped by speculation. A network of some 20 meteorological stations, listed by Court (1951), was established. Of these, the two with the longest series of observations were Orcadas, already mentioned, and South Georgia, established in 1904. As a result of Argentine and British rivalry in the Antarctic during the Second World War, more or less permanent stations were established in the Peninsula area. This was of great importance meteorologically since the Peninsula, the two sides of which present tremendous contrasts in weather with isotherms running lengthwise rather than latitudinally, is a climatic divide of the first order (Schwerdtfeger, 1970). The French station which operated at Port Martin, near Mawson's old base at Cape Denison, during 1951–52 was also well placed meteorologically and able to corroborate the extraordinary wind records from this area. Much information came from whaling fleets but, whalers being reluctant to reveal their positions to competitors, it was necessary to provide ciphers which only the Weather Bureau at the collecting centre in South Africa could decode, before this information was forthcoming in useful quantity (Walton, 1987, p. 199). Even at best the data were insufficient to give convincing synoptic charts of southern hemisphere weather. However, Koopmann (1953) was able to put together from this source a chart of atmospheric pressure for the area between 25° and 75° S which enabled him to deduce wind velocities and wind-driven movements of water and to plot positions of divergences. The meteorological observations of the Byrd expeditions provided information on the upper air but less complete surface observations than had been made at the beginning of the century. Barkow had made kite meteorograph flights from the *Deutschland* in 1912 and both kites and aircraft were used at Little America in 1928 and 1934 (Grimminger & Haines, 1939; Grimminger, 1941; Court, 1951). Between 1927 and 1934 improvements in radio techniques enabled the balloon-borne radiosonde to be developed for meteorological purposes but when first used in the Antarctic, by Dr Jörgen Holmboe on Ellsworth's 1934 expedition, its performance was so erratic that the data obtained were never used. The first successful radiosonde ascents in the Antarctic were from the *Schwabenland* in 1939 but the records from these were mostly destroyed during the Second World War (Court, 1951). During 1940–41 there were 190 successful ascents made from Little America III with the expenditure of only 200 radiosondes. These were the first to go above the tropopause (the boundary separating the stratosphere from the atmosphere below) in the Antarctic and it was found that this discontinuity disappeared towards the end of the winter and that the stratosphere became exceptionally cold; these are features which are not parallelled in the Arctic (Court, 1942). *Operation Highjump* in 1947 provided for the first time a network of simultaneous observation sufficient to provide twice-daily synoptic weather maps of the Antarctic. An advance in technique was to use the rawinsonde – radiosonde with the addition of the direction-finding equipment at the ground station to give direction and speed of winds at various levels. Although obtained primarily to enable forecasting for flying, these observations shed new light on the characteristics and movements of air masses and frontal systems. However, at this time there was still virtually no information, other than isolated data obtained opportunistically, for inland on the continent.

Meinardus (1938) published an extensive summary and discussion of the climatic data available up to the mid-1930s but still no consensus had been reached on the Antarctic anticyclone. There were obviously seasonal variations, possibly there were several anticyclones rather than one large one, and the upper air observations made on the Byrd expeditions showed the circulation to be much more complex than had been thought (Grimminger, 1941). Summarizing on the problems of Antarctica's anticyclone, Court (1951) wrote that it:

> exists only around the edges according to Meinardus and Shaw, is a thin layer according to Simpson, Barkow, Kidson and Grimminger, is broken up into several cells according to Lamb and the Navy aerologists (Operation Highjump), shrinks markedly from winter to summer according to Serra and Ratisvonna, Coyle and Gentilli, and is a major feature fed by an upper cyclone according to Hobbs, with Palmer considering the upper cyclone to be an abstraction derived from averages of widely varying conditions.

There was similar uncertainty and argument about other features, such as whether or not there is a continuous Antarctic front and how the cyclones around the continent originate (Court, 1951). Answering such questions was made more difficult by the lack of knowledge of the topography of the interior of the continent. The nature of the waves of pressure change – more correctly designated surges according to Court – discovered by Simpson, remained obscure. The region from which Simpson supposed they originated had been visited in the spring of 1935 by Ellsworth in the course of his transcontinental flight. He encountered 12 days of variable bad weather but whether this corresponded with a surge minimum – which might be checked from records made over the same period by the British Graham Land Expedition – does not seem to have been established. Court (1951) concluded that surges are caused by breaking out of anticyclones from the interior of the continent, which may be associated with the poleward motion along other longitudes of anticyclones from lower latitudes. Although radiosonde observations were made by the Australians, South Africans and French in the maritime Antarctic at this time, the 1940–41 series from Little America III remained the only ones from the entire troposphere and lower stratosphere throughout the year. In summer the tropopause was found at roughly 9 km with a temperature of −50 °C and in winter somewhat higher, 10 km, and colder, −70 °C. The extreme cold of the winter stratosphere over Little America had suggested to Court (1942) that:

> lack of circulation between Antarctica and lower latitudes would permit air there to cool unmolested during the winter until it attained lower temperatures than anywhere else on earth

and that the sub-Antarctic westerlies

> may largely confine the antarctic air.

Willet (1949) took up this idea of a circumpolar cyclonic vortex manifest in zonal westerlies repressing transport of heat and kinetic energy so that, as Bjerknes and other had suggested earlier, meridional flow in high southern latitudes should be much less than in the Arctic. Like its ocean, Antarctica's atmosphere seemed to be marked off distinctly from that of the rest of the world.

Just before IGY there were two substantial programmes of field work on Antarctic meteorology. Observations were made by the Norwegian–British–Swedish Expedition to Queen Maud Land in 1949–52 on radiation balance and turbulent heat transfer. Sverdrup, 30 years earlier, had made far-reaching studies on energy exchange in the Arctic but these were the first to be carried out in the Antarctic. They provided a classical study of energy exchange for a climatically vulnerable ice shelf together with detailed analysis of the boundary-layer processes involved (Liljequist, 1956–57). The meteorological results from this expedition were of particular value in planning operations in the Antarctic during IGY.[1] On a wider scale Loewe (1956), on the French expedition to Port Martin (66° 49′ S 141° 24′ E) in 1951–52 obtained data from which he was able to produce the first attempt at a detailed account of the energy, mass and water budgets of the continent.

9.4 Meteorology during IGY

The well-distributed network of IGY stations reporting to the Weather Central at Little America provided regular synoptic data for Antarctica for the first time. It was manned mainly by American meteorologists but other countries, notably the USSR and Argentina were represented. Sixteen of the manned stations sent up weather balloons which were tracked by radar, and the movements automatically recorded and reported to the Weather Central. Launching of these balloons in winds of often hurricane force was extremely difficult but surprisingly few scheduled launchings were missed. During the IGY, the first automatic weather stations (AWSs) in the Antarctic were set up by the Australians on Lewis Island off the Wilkes Coast and by the Japanese at Syowa on the Prince Olaf Coast. Such stations had been in use for some time in the northern hemisphere but had not been called upon before to withstand such rigorous conditions. Some of the difficulties in operating them under Antarctic conditions were described later by Butler (1988):

> [the] AWS site was at a position called AO28, at 68° 24.4′ S, 112° 12.6′ E. We arrived there on 26 September (1984) after poor weather had cost us two days and soft conditions had forced us to double-haul. The weather then held up erecting the AWS mast for two days. When we were able to put it up the conditions were still not pleasant – the temperature was below −30 °C and a twenty-knot (37 km/h) wind blew drift over us. We made frequent trips to a nearby living van to warm up, especially when only thin gloves could be worn for the fiddly jobs. With the instruments on the mast, I connected the cabling to the electronics drum which we buried with the battery box in a two-metre-deep trench downwind of the mast.
>
> The next day I stood by the antenna and checked that the AWS was transmitting. It was not. Disheartened, I tracked down the fault to the battery box connector. It had been manufactured with its locating keyway for the plug on the opposite side to where it should have been and the supply voltage had been applied in reverse. There were dozens of this kind of connector used in the weather stations and the only defective one had found its way into this critical spot.

> I replaced the charred printed circuit board, exchanged the bad connector on the battery box and put the AWS on the air. My overnight hopes that it was now working were dashed when I received the news from head office: 'data confused'. The best course of action was to take the electronics drum with us, try to fix it and deploy it on the way home.
>
> (Butler 1988, p. 89.)

In the first year both the Australian and Japanese AWSs ran for four months of the 12 expected of them. The meteorological data obtained in IGY were used in South Africa until the mid-1960s to prepare a series of daily charts, which gave the first detailed pictures of the entire situation in the Antarctic, and these formed the base material for much subsequent research. However, radio communications did not function as well as expected and the feeding of information to the Weather Central and the broadcasting of weather analyses to other continents was not altogether effective. For this reason an International Antarctic Analysis Centre was established in Melbourne, Australia, early in 1959 to take over the work of the Weather Central. As an aid to meteorology in the IGY the South African Weather Bureau produced a summary of its Antarctic data. In reviewing it Deacon expressed the hope that more fundamental studies would emerge and emphasized the importance of ocean-atmosphere interactions (Deacon, 1958).

For the first time winter weather in the interior of the continent had been observed and, as expected, records were broken. In September 1957, Amundsen-Scott Station at the South Pole recorded −74.9 °C then −87.4 °C was reported from Vostok in July 1958. Remarkable temperature inversions were found; at the South Pole and Sovetskaya at 427 m and 917 m, respectively, above the ice surface temperatures were remarkably constant at between −40 °C and −45.5 °C. The temperature regime of continental Antarctica appeared as distinctly different from that of the Arctic. The coreless winter and pointed summer (referred to on p. 295) are characteristics of the former but not the latter and average temperatures are lower in the Antarctic. There are several causes for these differences, including the different distributions of sea and land masses at the two poles, the penetration of warm ocean currents into the Arctic and the inhibition of heat transfer from equatorial regions into the Antarctic by the west wind drift. The annual variation in temperature over the Antarctic was found to be greatest in the stratosphere, which warms up extremely rapidly on the return of the sun in the spring, another different situation from that in the Arctic.

The measurement of precipitation has always been a problem in the

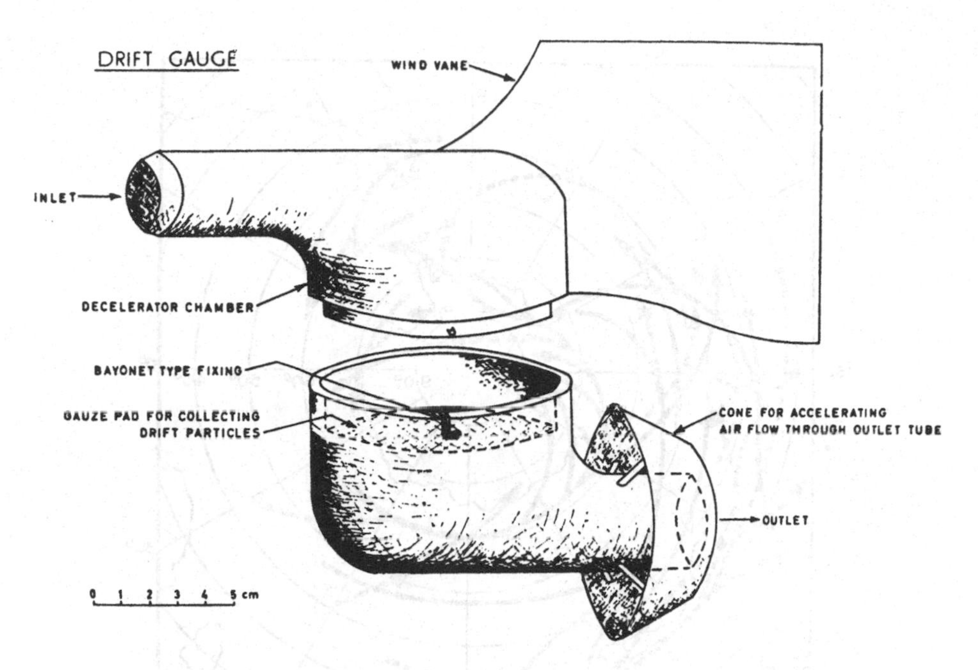

Fig. 9.5 The drift gauge used on the Commonwealth Trans-Antarctic Expedition, 1955–58. From Lister, 1960, Fig. 3. (Courtesy of the T.A.E. Committee.)

Antarctic and Lister (1960), who observed precipitation and drift at South Ice during the Commonwealth Trans-Antarctic Expedition, wrote as follows:

> The measurement of snow precipitation is very difficult, frequently inaccurate, and sometimes impossible. Because of the low specific gravity of snow it is subject to the slightest air movements. Hence any instrument to measure the snowfall almost inevitably interferes with the existing natural conditions.
>
> The path of a falling snow particle varies from near vertical when there is no wind, to near horizontal when the wind speed is moderate or high. A snow particle travelling near to the horizontal passes across the collecting can, instead of into it.

He measured drift using a gauge which permitted the flow of air but trapped the particles (Fig. 9.5) and formulated an empirical relation of drift to wind speed and height above the ground. Others made similar measurements and a tentative conclusion was that drifting snow is not of great importance for the mass budget (Hoinkes, 1964). Very little wind was encountered at Sovetskaya on the central part of the ice-sheet and the snow was so fluffy and loosely packed that tractors became bogged down. In

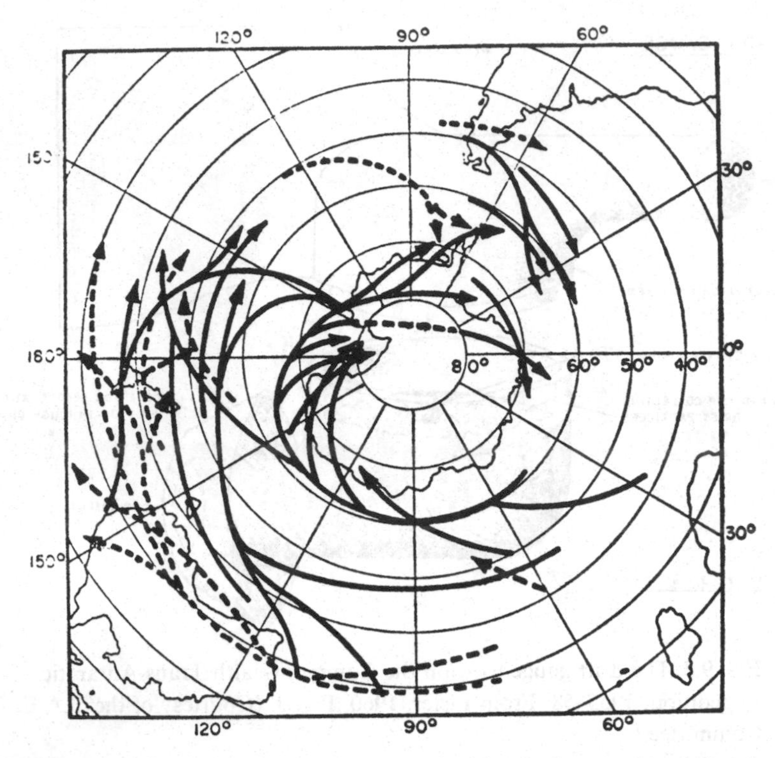

Fig. 9.6 Antarctic storm tracks as compiled by the International Geophysical Year Weather Central at Little America for the austral winter of 1957. The solid lines represent cyclonic systems, the dashed lines are anticyclone tracks. From Sullivan, 1961, p. 235.

contrast at Oazis, the Soviet out-station on the Queen Mary Coast, there was no snow cover and the meteorological hut was torn apart by the wind, pieces of it being carried for almost a mile (1.6 km). Such observations were systematized by Bugayev (1963) who recognized five distinct climatic zones in Antarctica. Dalrymple (1966) proposed delimitation of four climatic regions on the Antarctic plateau on the basis of data collected from nine stations between 1957 and 1965.

A dominant feature in the general picture of Antarctic weather which emerged was the succession of cyclones in latitudes around 60° S, these rarely penetrating into the high central plateau of the continent but often swinging into the Ross Sea taking warm air with them (Fig. 9.6). Nearly stationary continental anticyclones were found over Marie Byrd Land and the Pole of Inaccessibility. The Soviets made the interesting suggestion that the shape of the continent is determined in part by storm tracks, pointing out that the ice does not conform in distribution with what little rock appears at the surface, its thickness and extent seeming more dependent on the pattern of storm movement (Hatherton, 1958; Sullivan, 1961). However, the gravity-impelled katabatic winds were found to be related to the general large-scale slope of the surface and, usually, to over-ride the cyclonic circulation. These strong winds record their direction in the orientation of sastrugi on the snow surface and aerial photographs were found useful by meterologists in plotting wind direction (Hoinkes, 1964).

From such studies it became possible to make tentative quantitative estimates of meridional mass transport, the mass of air moving to and from high latitudes. Various investigations were made of heat budgets and, reviewing them, Hoinkes (1964) considered that the thorough analysis of data obtained at the South Pole by Dalrymple and his colleagues (1966) was the most remarkable contribution to glacial micro-meteorology during the IGY period. Because of instrumental difficulties and different methods of computing, comparison of results from the various IGY stations had to be confined to orders of magnitude (Hoinkes, 1964). Nevertheless, Rubin (1962) found a satisfying balance between the influx of sensible and latent heat on the one hand and the heat used in melting and evaporation plus the radiative heat loss from the top of the atmosphere, on the other. Mass budget quantities, however, could not be easily related to heat budget quantities, as had been found, too, even for intensively studied glaciers in Europe and North America (Hoinkes, 1964).

The IGY meteorologists met in Melbourne, Australia, for a symposium on their way home (Antarctic Meteorology, 1960). This brought together an enormous amount of useful material but, as Lamb (1961) commented, it was too soon for proper digestion. A general conclusion on long-term trends was that Antarctica was 'warming up'. The records from Little America showed that over a period of 45 years there had been a secular rise in temperature of 2.6 °F, in accord with glaciologial evidence of a slight recession of the ice-sheet.

9.5 Post-IGY meteorology

After the IGY, meteorological observation continued in the Antarctic but the number of manned weather stations did not increase. In fact, with the dwindling activities of whalers in the 1960s the input of information from the Southern Ocean drastically decreased. Up to 1986 there were only 25 meteorological stations operating the year round south of 60° S; Weller (1986) pointed out that per unit area the US itself had 50 times as many. Furthermore, of these 25 stations only two were on the interior plateau and many of the newly established peripheral ones had been sited for political reasons or ease of access rather than for meteorological usefulness. It was not until 1980 that Russkaya was installed at 74° 48′ S 136° 5′ W, filling in a gap between 60° W and 162° W (Schwerdtfeger, 1984). These deficiencies in number and distribution of manned stations were, however, offset to a large extent by improved observational systems. Automatic weather stations became more reliable, with improved power supplies and reporting via polar orbiting satellites. One at Byrd Station operated continuously for over three years. Others were deployed in the Ross Sea and Ice Shelf area; inland from the French station of Dumont d'Urville (66° 40′ S 140° 1′ E) up to Dome C (74° 50′ S 120° E) specifically for the study of katabatic winds; inland from Casey Station (66° 16′ S 110° 32′ E); and on the Antarctic Peninsula. Similar equipment can be mounted on a buoy for use at sea but with the exception of 1978–79 during the first GARP (see pp. 229 and 315) experiment when ocean buoys were released into the oceans of the southern hemisphere, reporting of data from the open Southern Ocean and its pack ice has been sparse. Information on both ice and cloud distribution had been meagre for the Antarctic as compared with the Arctic (Weller, 1974). After TIROS-1 took the first television pictures of the atmosphere in 1960 remote sensing by satellite provided a practical means of obtaining repetitive large-scale coverage of Antarctica over long periods of time and was used from the early 1970s (Fig. 9.7). Information on cloud formations – and hence atmospheric circulation – temperatures and sea-ice could be obtained daily. Surface winds in the vicinity of mountains, of vital importance for aircraft operations, however, could not be monitored in this way. The SCAR group of Specialists on Antarctic Climate Research (Allison, 1983) made the comment that there were data gaps in the multifrequency passive microwave imagery available for Antarctica, lack of accurate ice-sheet elevations obtainable by altimetry from a polar-orbiting satellite, and poor accessibility, processing and compilation of such accurate climate-related satellite information as did exist. A major change in organization was the incorporation of Antarctic meteorology as an integral part of the World Weather Watch in the mid-1970s. Interest in Antarctic weather had by then spread beyond the SCAR specialists into the major meteorological numerical analysis and forecast centres, including the European Centre for Medium Range Weather Forecasting. This called for improved communications between Antarctica and the Global Telecommunications System which included geostationary communications satellites to ensure prompt delivery of data (Weller, 1986; Fifield, 1987; Walton, 1987). A reliable cloud

Fig. 9.7 Mosaic of cloud pictures taken in the infrared from a NOAA satellite.

climatology spanning several years is nevertheless still lacking, important though it is for modelling of the global climate (Maslanik & Barry, 1990).

Fairly complete accounts of the weather and climate of Antarctica could be given by the mid-1960s (Bakaev, 1966; Weyant, 1966, 1967; Rubin, 1946, 1966; Schwerdtfeger, 1970, 1984; Weller, 1986). The general picture which had been emerging was consolidated with the main features appearing as a kabatic boundary-layer regime over the continent which is only weakly coupled to the air flow above it, and surrounding this is a strongly coupled and highly energetic wind regime over the Southern Ocean. Using data from ESSA satellites a classification of cloud vortices was used by Streten & Troup (1973) to determine the general features of behaviour of southern hemisphere depressions. Persistent maxima of early vortex development were found in the western South Atlantic and the central South Pacific, cyclones drifting gradually southwards as they tracked round the globe and usually terminating as they reached high latitudes near the

Antarctic coast. The circulation in the stratosphere was seen as following two basic patterns during the year. In summer there is continuous absorption of solar ultraviolet radiation by ozone, raising the temperature and resulting in a weak easterly flow. During the rest of the year, heating in this way is minimal and the stratosphere over the pole cools. The circulation then becomes westerly and intensifies to produce a stratospheric circumpolar vortex. This is more intense and more stable than the similar vortex over the Arctic because perturbations by surface topography and variations of albedo are less (Farman, 1977) and, unlike that, is not subject to disruption by sudden incursions of warm air lower latitudes. The beginning of summer in the Antarctic is marked by a single warming event. The Antarctic polar vortex effectively seals off the stratospheric region from that in the surrounding lower latitudes. This was demonstrated dramatically by the finding that the fine dust projected into the stratosphere by the eruption on Mount Agung in Bali in March 1963 did not spread over the Antarctic until the winter polar vortex broke up in late spring (Farman, 1977).

9.6 Atmospheric chemistry: ozone

This brings us to a finding of Antarctic research which has attracted much public attention in recent years – the depletion of stratospheric ozone. Interest in atmospheric ozone began with G. M. B. Dobson, a British meteorologist. Studies of meteor trails led to the discovery of a region in the stratosphere at a height of about 50 km which owed its high temperature to absorption of ultraviolet radiation by ozone. Dobson invented a spectrophotometric method of measuring its total amount in the air column and in the 1930s set up a chain of ozone measuring stations (Lloyd, 1989). Systematic measurements using the Dobson instruments at Argentine Islands (Faraday) and Halley Bay (Halley), were begun in 1957. Measurements were also made at other IGY Antarctic stations including Little America. The interest was that ozone, being produced photochemically at heights of between 20 and 50 km, mostly at low latitudes or, in the summer only, at high latitudes, could be used as a tracer of atmospheric circulation at high levels. It was found that a major increase in total ozone occurs in the course of breakdown of the Antarctic winter stratospheric vortex. Long-term trends, however, seemed to be small, less than those at lower latitudes (Farman, 1977).

Meanwhile concern had been growing about effects of human activities on the ozone layer which might result in penetration of damaging amounts of ultraviolet radiation to ground level. Lovelock (Lovelock *et al.*, 1973), on a cruise on RRS *Shackleton*, which took him into Antarctic waters, found that chlorofluorocarbons (Freons or CFCs), much used in industry as aerosol propellants and refrigerants, and entirely anthropogenic, were accumulating in the lower atmosphere. Rather incautiously, although there were no indications to the contrary at the time, Lovelock remarked that these substances constituted no conceivable hazard. When it was realized that chlorine can be an agent for the catalytic destruction of ozone, Molina & Rowland (1974) pointed out that CFCs transported into the stratosphere would be photolysed to yield reactive chlorine, which in turn would destroy ozone. This possibility was taken seriously and in the UK in 1979 the Department of the Environment in its second report on the subject, concluded from theoretical studies that in the steady state for release of CFCs at the current level, stratospheric ozone would be reduced by between 11 and 16 per cent. Around this time the use of CFCs as aerosol propellants was banned in the US and reduced in the UK. The need for better understanding of processes in the stratosphere, more extensive and accurate monitoring of stratospheric ozone, and intergovernmental action to reduce global production of CFCs was recognized.[2] Detecting the effects of man-made chemicals on stratospheric ozone is difficult because ozone fluctuates considerably in concentration through natural causes. In addition to a ground-based network of Dobson spectrophotometers supplemented by balloon-borne ozone sondes, the US launched an ozone measuring system on the near-polar orbiting satellite Nimbus 7 in 1979. This was the total ozone mapping spectrometer (TOMS) which uses backscattered solar ultraviolet radiation to produce daily maps of vertically integrated (total) ozone over the entire sunlit portion of the globe. The calibration of this instrument has been a problem, there being uncertainty as to whether a drift in time relative to ground-based measurements of ozone values is an artefact due to calibration drift or real drift due to global or local increases in tropospheric ozone, which would affect the ground-based instruments more strongly. However, it seems reasonably certain from a critical consideration of the satellite data over the period 1979–86 the actual decrease in global-mean total ozone was about five per cent (Bowman, 1988), an appreciable but not catastrophic fall. It was therefore sensational when BAS observations by conventional spectrophotometric measurements from the ground at Halley Station showed a deep minimum in spring with the 1984 values down by about a third from the 1957–77 values (Farman *et al.*, 1985). The US satellite ozone measuring system had evidently missed this because it was programmed to discard low values, which it was presumed would be due to instrumental error, but reexamination of the data confirmed the dramatic depletion in the Antarctic spring (Stolarski, 1986).

These variations in concentration of stratospheric ozone are clearly related to the behaviour of the polar vortex, which during the winter, when there is little or no ultraviolet radiation to stimulate local photochemical production, cuts off the stratosphere over Antarctica from replenishment with ozone by transport from lower latitudes. When the vortex breaks down in the spring, transport can take place and the ozone concentration rises sharply. During the rest of the year processes destroying ozone predominate and Farman *et al.* (1985) found that the declines in total ozone content over Halley and the southern hemisphere were correlated with increases in CFC concentrations. A correlation does not, of course, prove a causal relationship and the possible chemical processes which might be responsible were a matter for argument. Farman's hypothesis was that the very low temperatures prevailing from midwinter until several weeks after the spring equinox make the Antarctic stratosphere uniquely sensitive to increase in inorganic chlorine concentration, primarily by the effect on the ratio of the nitrogen oxides which are also involved in processes affecting ozone. There was no doubt, however, about this local disappearance of ozone and its potentially serious implications. The UK's response to Farman's lead was inadequate but that of the US was immediate, massive and effective. A series of National Ozone Expeditions, beginning in 1987, was initiated by John Lynch, the Atmospheric Sciences Programme manager of the Department of Polar Programmes of the US NSF. The US National Aeronautics and Space Administration (NASA) enlisted the co-operation of the governments of Argentina, Chile, France, New Zealand and UK and help from other organizations such as the Chemical Manufacturers Association, in a combined airborne and ground-based investigation. Observations were made from specially instrumented high-flying aircraft, two DC-8s and an ER-2 (a modified US spy-plane), flying from Punta Arenas in Chile and capable of reaching the centre of the ozone hole at 18 km height. It involved 150 scientists and flights totalling about 175,000 km (Kerr, 1987). Concentrations of chlorine and nitrogen compounds and the composition of stratospheric ice crystals were determined as well as ozone concentrations. The results were consistent with the suspected trend of ozone loss modified by a two-year cycle. The general ideas of chemists about the formation of the hole were confirmed but it was found that extreme cold plays an essential role by locking up water in ice crystals that absorb nitrogen oxides which normally prevent chlorine from destroying ozone (Kerr, 1987; Stolarski, 1988). These ice crystals are visible

Fig. 9.8 Joseph Farman of the British Antarctic Survey addressing the 'Saving the Ozone Layer' conference held in London, 5–7 March 1989 by the Department of the Environment in association with the United Nations Environment Program. (Courtesy of Times Newspapers.)

as the nacreous clouds which are characteristic of the polar stratosphere. Satellite ozone data from 1979 to 1986 have shown that decreases of total ozone have not been confined to the Antarctic spring season but are global in extent (Bowman, 1988). However, this is not the place to review the spate of publications on stratospheric ozone and related matters and the public discussions (Fig. 9.8) that have followed Farman's discovery. It gave impetus to growing concern about the environment and led to the first global agreement, signed at Montreal in September 1987 after discussion between representatives of 35 countries, to limit pollution, i.e. by limiting the production of CFCs (Johnston, 1987).

Other aspects of atmospheric chemistry, although overshadowed, have received attention (Weller, 1974). The chemical composition of atmospheric particulate matter at the South Pole, which was designated as one of the US National Oceanic and Atmospheric Administration (NOAA) four base-line stations for monitoring climatically important variables at localities remote from anthropogenic influences, was investigated with great thoroughness by Zoller *et al.*, (1974) using atomic absorption analysis and neutron activation techniques. One group of trace metals in these particulates, including among others aluminium, manganese, chromium, sodium and calcium, was derived from either crustal weathering or the ocean. The other was of relatively volatile elements such as zinc, copper, selenium and lead, and apparently derived from other sources but no particular natural or anthropogenic processes which might be responsible could be identified.

9.7 Energy balance and modelling

In many respects Antarctic situations are relatively simple and lend themselves to mathematical modelling. Lettau (1971) picked out three great advantages: (1) the exceptional uniformity of the physical structure of the Antarctic snow surface, (2) the large horizontal scale of the topographical gradients of the continental ice-dome, and (3) the relative paucity of short-time disturbances of the dominant long-period (seasonal) variation of insolation. These factors he considered to justify comparison of atmospheric structure with laboratory results and he demonstrated this with reference to snow, micrometeorological stratification of air temperature and wind speed, temperature profile structure and optical refraction phenomena, wind systems, and the calculated climate assuming that the South Pole region was snowless. Rusin (1961) had produced the first important summary of the meteorological and radiational regime of Antarctica, based largely on the Soviet IGY data. He emphasized the role of turbulent heat exchange, finding this to increase nearly four-fold from the interior of the continent to coasts subjected to katabatic winds. Subsequent measurements at several plateau stations confirmed the largely negative radiation balances which he found. It is not the ice-sheet which is the main global heat sink but the higher layers of the Antarctic atmosphere. Carrol (1982) made energy balance measurements at the South Pole using modern equipment over a period of three years and, in agreement with previous work, found a net radiative loss from the surface over the year. Even during the polar night there was much short-term variability arising from the influx of maritime air. The strong radiational cooling of the surface produces the katabatic regime and massive drift of snow. This was studied by Budd *et al.* (1986) at Byrd station. Radiational cooling produces remarkably stable stratifications, wind shears, and associated optical phenomena. The vertical profiles through the inversion layer of temperature and wind were found by Kuhn *et al.* (1977) to be highly complex. The acoustic sounder is a research tool that has been used successfully in the Antarctic since 1976, for example, to add detail in studies of this layer. The origin of the precipitation, which had so perplexed the earlier meteorologists, was studied directly by Miller & Schwerdtfeger (1972). They found ice crystal formation in the warmer air above the temperature inversion at inland stations under clear sky conditions and showed that this contributes a significant part of the annual accumulation.

In the mid-1970s Antarctic meteorologists showed increased interest in climatic studies – it was at about this time that there was some concern about the onset of another ice age. BAS, for example, began a collaboration with the Climatic Research Unit of the University of East Anglia, in England.[3] An analysis of temperature records for a number of Antarctic stations showed consistent upward trends of around 0.2 °C per decade in agreement with predictions of the effect of rising carbon dioxide concentrations made from contemporary models.[4] Nevertheless, major problems remained to be solved in relation to the Antarctic heat sink and its variations. The First GARP Global Experiment (FGGE), organized jointly by ICSU and WMO in 1978–79, made good use of the special advantages which the Southern Ocean offers for studying the interactions between oceans and atmosphere. Information on this was required for better understanding of heat transfer processes and heat budgets for incorporation in general circulation and climate models. Buoys relaying information via polar-orbiting satellites provided data on sea-level atmospheric pressure and sea-surface temperature for vast areas which previously had not been covered even by ships' reports. Synoptic analyses benefited greatly and the pressures in cyclones over the oceans were found to be even lower than had been thought (Guymer & Le Marshall, 1980). Model calculations of global circulation have many short-comings, errors tending to be particularly large in high latitudes. 'Spectral' models, which came into use in the late 1970s and adopt a different mathematical approach to solve the governing equations, tend to give more realistic high latitude simulations than 'grid-point' models (Mullan & Hickman, 1990). A coupled atmosphere–ocean–sea–ice model was developed at the Meteorological Office in the UK as a means of investigating climate response to such factors as changing carbon dioxide concentration (Foreman, 1989). It did not adequately predict the strong annual cycles in sea-ice in the Antarctic and it was concluded that heat fluxes at the surface of the Southern Ocean were still too poorly understood.

9.8 The beginnings of study of the upper atmosphere

At the beginning of the century, as we have seen, there was awareness of things of an electrical and magnetic nature going on aloft and particularly in the polar regions. It was traditional to make magnetic measurements on

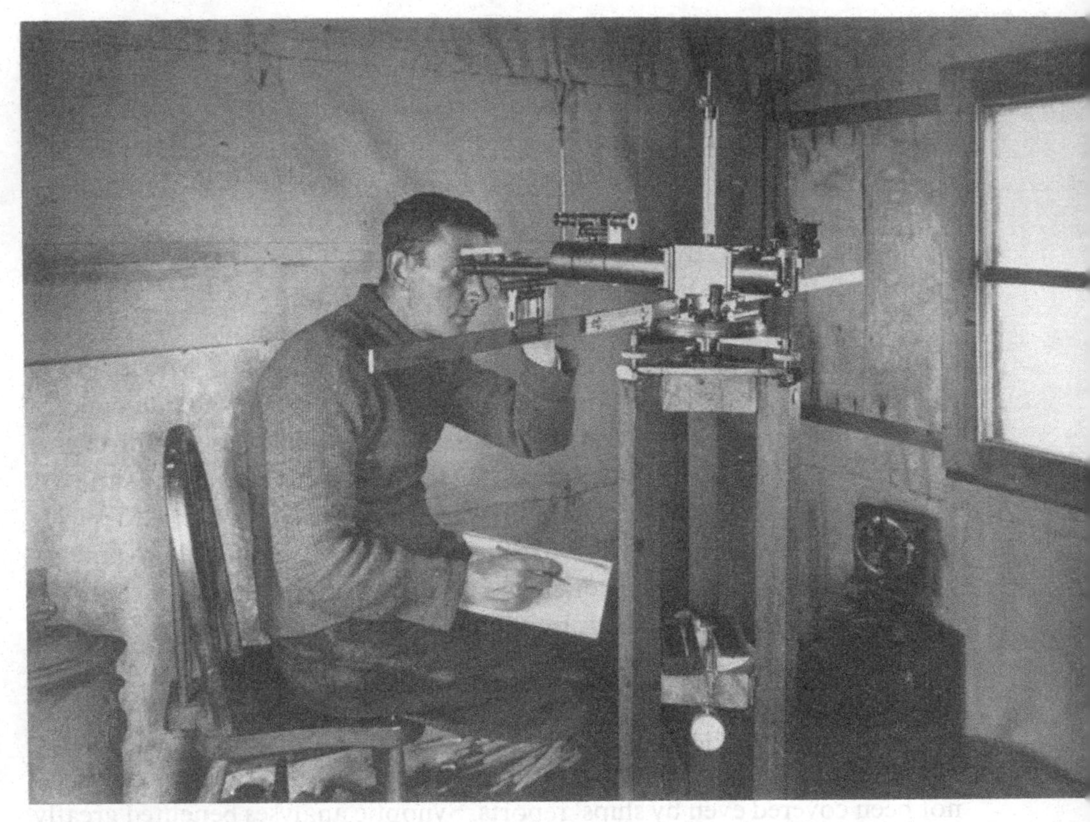

Fig. 9.9 G. C. Simpson using the unifilar magnetometer at Cape Evans on the *Terra Nova* expedition, 1910–13. Photograph by H. Ponting. (Courtesy of the Scott Polar Research Institute.)

polar expeditions (Fig. 9.9) but now, in addition, observations were made on other phenomena some of which subsequently proved to be interrelated with magnetism. In the heroic age such observations were associated with meteorology, if not actually made by the meteorologist himself. This was logical because the upper atmosphere, with which these phenomena seemed to be associated, is a continuum with the lower atmosphere which is the concern of meteorologists. However, the new branch of science which developed has largely remained apart from meteorology and is better dealt with separately.

Bernacchi, the physicist and meteorologist on Borchgrevink's 1898–1900 expedition made magnetic observations but had difficulties. The rocks of the Cape Adare region were highly magnetic and the delicate and heavy unifilar magnetometer with which he was provided was in his opinion 'the worst instrument that could possibly be taken to the Antarctic regions for the determination of intensity'. Disturbances associated with the aurora, which was seen almost every night, were great. A consistent connection between the brilliance and rapidity of displays and large displacements of the magnetic field was noted.

Measurements of electric potentials were made but on the *Discovery* expedition it was found that Kelvin's portable electrometer was unpleasant to handle in subzero temperatures and, possibly because of an accident to the instrument, there was some doubt about the absolute accuracy of the readings. The mean potential difference found between the ground and air at 1.22 m was rather lower than that at Kew in England but became abnormally high in drifting snow, but not, it seemed, because of electrification by friction. Auroral displays, except possibly exceptionally intense ones, which seemed to be associated with low values, had no obvious effect on the electrical field, as Wilkes (1845, vol. II, pp. 35 and 328) had found 60 years before. Observations on the aurora consisted mainly of records of its occurrence and appearance, Bernacchi's account being accompanied by some beautiful drawings by Wilson (Fig. 9.10). The magnetic observations made on the UK, German and Swedish expeditions were carried out according to internationally agreed protocol but the extensive data obtained do not call for any particular discussion here. Professor Kr. Birkeland (in Chree, 1909) in an examination of the correspondence between Arctic and Antarctic magnetic disturbances, which he found to be on the whole simultaneous, again made the point that the magnetometers taken to the south were over-elaborate. He considered that instruments of less sensitivity but with capacity to yield continuous records over extended periods of time could have produced more important results.

On Scott's second expedition, Simpson together with C. S. Wright made measurements on the ionization of the air on the voyage down to Antarctica. They found that a significant proportion of this was due to something other than local radioactivity and presumably related to what was then called penetrating radiation which Wright, a physicist (but the expedition's chemist nevertheless), had been studying in Cambridge. This was an important contribution to the growing knowledge of cosmic rays. The crucial observation that the residual ionization of the air was due to radiation coming from above was to be made on a balloon ascent by Victor Hess in the following year, 1911.

In geomagnetic studies a dramatic achievement of the *Nimrod* expedition was to plant the British flag on the South Magnetic Pole. A magnetic meridian was followed as closely as possible inland from the coast of Victoria Land until the dip readings were vertical. This was at 72° 25′ S

Fig. 9.10 Auroral curtain, 5 July 1902, 1 a.m. to 2 a.m., seen in north; drawing by E. A. Wilson. From *National Antarctic Expedition, 1901–4, Physical Observations* IV, Plate CLXI (1908).

155°16′ E on 16 January, 1909. The party was nominally led by Professor Edgeworth David but from Mawson's diary (Jacka & Jacka, 1988) it is clear that at the age of 50, tough as he was, he found the strain of this, the longest unsupported man-hauling sledge journey in history, too much and Mawson was the *de facto* leader. The attainment of the magnetic pole was only approximate and proved nothing of great scientific importance. It was achieved at the cost of foregoing an examination of the dry valleys, which would have been geologically highly rewarding, but it was the decision of David – a distinguished geologist – himself to go for the more glamorous objective. As physicist to the *Nimrod* expedition, Mawson otherwise concerned himself with a variety of topics. From his record of auroral occurrences he noted that brilliant displays coincided with interruption of telegraph services by earth currents in Australia and speculated that the aurora might be related to electric currents flowing high in the atmosphere (he evidently did not know of Birkeland's experiments, see p. 323). Such coincidence, as well as those between auroral displays and magnetic disturbances were difficult to study further without some means of continuous registration of aurorae unaffected by sunlight or cloud (Gillmor, 1978). Although Carl Störmer (1955) already in 1910 had had some success in photographing the aurora borealis, Mawson's attempts in 1908 were of little scientific value. Mawson commented on the alignment of faint auroral effects in the azimuth of the sun whereas bright ones seemed to be arbitrary in direction, following the coast line more than anything else (actually geomagnetic east to west on average; Jacka & Jacka, 1988, p. 196). As on the *Discovery* expedition, no connection was found between electrical potential in the air and auroral displays.

As we have already noted (p. 135) Mawson on his 1911–14 expedition was the first to use radio in the Antarctic and it is clear from his diaries that he realized its potential value for investigation of the physics of the earth's atmosphere. He had 500 kHz spark transmitters, giving a nominal 1.5 kw of power, but the size of antenna he was able to build was limiting. By establishing communication with Tasmania via a relay station on Macquarie Island, over a distance too great for receiving the ground ray, he effectively demonstrated the existence of the reflecting layer now known as the E-region. He noted that radio reception was poor when the aurora was good but at the same time found an absence of 'static' and asked the question 'Is there anything in static when no aurora?' (Jacka & Jacka, 1988). Indeed there is! Unfortunately, he was frustrated in his attempts to carry such observations further by having a neurotic wireless operator, recording in his diary:

> I sadly now give up all hope of getting much of scientific value from the wireless as advancing the subject of aether waves. Jeffryes [the wireless operator] is certainly not the man for such work, he appears to have no conception of scientific analysis.
>
> (Jacka & Jacka, 1988, p. 191, entry for 26 May 1913.)

> I can't get him or Bickerton to take the subject up scientifically. This is another argument in favour of true scientists for everything. If I were choosing another staff I would get specialists for each branch, true scientists capable of assisting with sledging.
>
> (Ibid, p. 192, entry for June, 1913.)

Jeffreys had been rejected by Mawson when he first applied to join the expedition but had been taken on the following year when the *Aurora* sailed south again, apparently in ignorance of Mawson's decision. The radio observations that were obtained remained unexamined because of the First World War and it was not until 1929 that they were written up by Charles Wright, of Scott's second expedition, and not until 1940 that they were published (Wright, 1940) so that they were not known, for example, to the Byrd expeditions. One of the more important findings was that reception became bad in the spring and seemed inversely related to illumination of the upper conducting layer of atmosphere between Macquarie Island and the base at Cape Denison. There was a remarkable asymmetry in reception between these two stations at the equinoxes, that from Cape Denison to Macquarie being better than in the reverse direction. Wright pointed out the desirability of carrying out further investigations with modern equipment and extending over a sufficient length of time to obtain measurements in years of high and low solar magnetic activity.

Mawson's magnetometrician, Eric N. Webb, was a trained scientist and a good man in the field. He was a New Zealander and a graduate in civil engineering. Before going south he received intensive training for four months in magnetic observation under experts from the Carnegie Institution and the Melbourne Observatory (Lugg, 1984). His equipment, a set of Eschenhagen quick-run photographic self-recording instruments, had been chosen for him, on the advice of Dr C. Chree of the Kew Observatory in the UK, but he was left to design and erect the housing for it himself. This he did with meticulous care to achieve the most even temperatures possible and exclude draughts. The remarkably tight building which was the outcome is now a historic monument (Webb, 1975; see Fig. 5.9). His observations were made with similar care. Soon after his return from the Antarctic he was swept into the First World War but the working up of his data was completed, in their spare time, by 12 young women students of the Physics Department of Canterbury University, who became known as 'the magnetic ladies of Canterbury College' or 'the Mawson club' (Anon., 1978). An exceptionally arduous sledge journey into the magnetic pole area yielded a wealth of high quality field data which placed the Pole at 71° 10′ S 150° 45′ E (Fig. 9.11). This position in December 1912 meant that the probable location in 1909 was some 130 km off David and Mawson's extreme station. Webb (1977) complained that it was the position reached by that 1909 field party which had got into the scientific literature rather than his much better supported and more accurate fix. Between April and October 1912, observations by Scott's second expedition and those by Webb were synchronous and Chree (1923) was able to make detailed comparisons. One of his conclusions was that the South Magnetic Pole must have a daily oscillation in position and that during violent magnetic storms it might wander as much as a hundred miles (161 km) from its mean position. The first satisfactory photographs, with an exposure of only 10 s, of the aurora australis were another achievement of Mawson's expedition.

9.9 The concept of geospace

The baffling and disconnected observations just described can now be seen to fall into place in one of the most magnificent concepts of

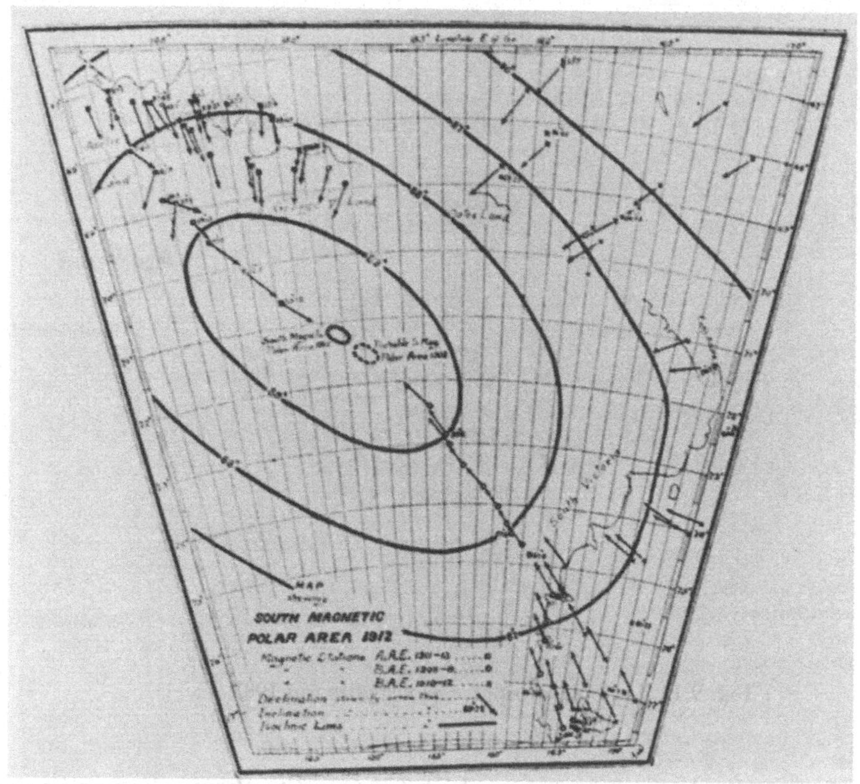

Fig. 9.11 E. N. Webb's map of the South Magnetic Pole area in 1912, showing declinations and dips. From Webb, 1977, Fig. 1. (Courtesy of the Australian Antarctic Division.)

contemporary science, that of geospace.[5] The science underlying this is highly specialized and intricate so that the contribution of Antarctic studies, even though it has been seminal and substantial, cannot be separated clearly from the rest. To provide some orientation for the reader it seems best to abandon a strictly chronological treatment at this point and to preface the account of the development of upper atmosphere research in Antarctica since the heroic age with a sketch of the structure of geospace as now visualized. The main features of this structure were worked out by the 1970s (Lanzerotti & Park, 1978) and have remained virtually unaltered since (Walton, 1987). It emerged as a consensus and is not associated with the names of any particular scientists.

Geospace encompasses the region surrounding the earth in which the sun's atmosphere, the interplanetary medium with its magnetic field, and the earth's atmosphere and magnetic field interact together (Fig. 9.12). The sun emits at supersonic speeds a stream of plasma – atoms from which electrons have been stripped, leaving them with a positive charge, and the

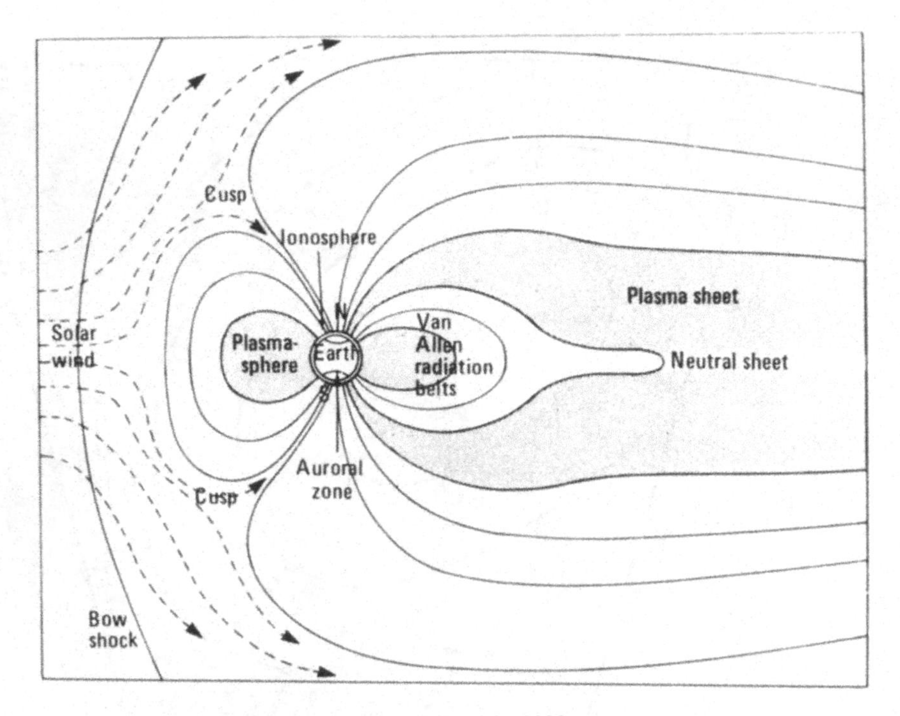

Fig. 9.12 The magnetosphere as visualized in 1982.

negatively charged electrons which belonged to them – known as the solar wind. The concept of this wind originated in 1951 with Biermann's observations on the behaviour of the tails of comets (Biermann, 1985). Its strength varies on time scales between seconds and decades in relation to irregular phenomena, such as solar flares and sunspots, as well as more regularly with the solar rotation period and the 11-year cycle of solar activity. The flow of the solar wind around the earth generates the magnetosphere. On the sunward side of the earth the geomagnetic field is compressed whereas on the leeward side it is stretched out into a comet-like tail extending thousands of earth radii into space – 'magnetosphere' is therefore something of a misnomer. Just to the earthward of the bow wave produced in the solar wind is a boundary layer, the magnetosheath, in which deaccelerated solar plasma streams round the earth at about 10 earth radii distance. Beneath this is another boundary layer, the magnetopause, through which solar plasma does not penetrate. However, cusps in the magnetosphere, one over each magnetic pole, allow plasma from the solar wind to penetrate deep into the polar atmosphere. Solar plasma can also enter the magnetosphere, as the plasma sheet, through its tail. Plasma consists of equal numbers of positive and negative ions and is therefore electrically neutral but capable of conducting electric currents. An electric field, aligned from dawn to dusk, is generated by the flow of the solar wind past the magnetosphere and causes a corresponding sunward flow of plasma, setting up convection cells. This provides the plasma which forms the inner and outer Van Allen radiation belts which in turn support an electric current, the ring current, which girdles the earth above the equator. The orientation of the magnetosphere is determined by the sun but within it is the rotating magnetic dipole of the earth. The transition between the inner, rotating, magnetic field and the outer, fixed, magnetic field is known as the plasmapause. The doughnut-shaped volume within the plasmapause contains plasma derived from the ionosphere. Going up through the plasmapause one passes abruptly from dense low energy plasma to low density high energy plasma. The boundaries between the various regions are in a constant state of disturbance because of the gustiness of the solar wind, major eruptions on the sun causing magnetic storms which in turn affect the ionosphere. The ionosphere also responds strongly to wind, temperature and composition changes in the neutral thermosphere which lies between it and the magnetosphere. The magnetic field lines of the magnetosphere, being also approximate equipotential lines, delineate the flow of plasma or the transmission of an electric field. The field lines which go through the boundaries and other features just described – the tail, the cusps, the plasmapause – all come down into the earth's atmosphere in the polar regions. The value of this knowledge of geospace, both for science and technology is immense (Rycroft, 1989). Geospace contains the only natural cosmic plasma available for carrying out *in situ* physical investigations and space technology, remote sensing, radio communications and electrical power transmission are all to a greater or lesser extent affected by conditions and events in it.

9.10 Ionospherics up to the IGY

The long-established geomagnetic work and descriptive recording of the aurora provided the soil in which the seed could flourish but prior to 1957 very little ionospheric investigation had been carried out in the far south. The Arctic had been more productive of work in this field. In 1881, Hermann Fritz from a systematic study of the distribution and frequency of the aurora borealis noted that activity waxed and waned with the sunspot cycle and further concluded that displays were most often seen in an oval band centred, neither on the magnetic pole nor on the geographic pole, but on a geomagnetic pole marking the axis of the earth's magnetic field in space. In the 1890s, Kristian Birkeland in Norway found that this pattern could be mimicked in the luminous rings which surrounded a small magnetized sphere exposed to a discharge of electrons in a near vacuum. From this followed the idea that the aurora is produced by interaction of charged particles emitted by the sun with the rarefied gases of the upper atmosphere. Trajectories of charged particles entering a two-poled magnetic field such as that of the earth were calculated by Carl Störmer and published in 1907 in a paper, the significance of which was appreciated only much later. Particles entering the field at an angle would follow a spiral course, the turns of which would get tighter until a point of reversal was reached. In this way the particle would be trapped, flying backwards and forwards within a crescent-shaped circuit.

Another phase of research began when with increasing experience of radio communication it became clear that long distance transmission must depend on an ionized reflecting layer in the outer atmosphere. Balfour Stewart in an article which appeared in the *Encylopaedia Britannica* in 1882 had already postulated such a conducting region in the earth's outer atmosphere to explain systematic changes in the earth's magnetic field. The ionosonde, an instrument for study of the ionosphere with vertically directed pulses of radio waves, was devised by Edward Appleton, then at the University of London, in 1924 (Appleton, 1963). Three years later it was used by M. P. Hanson, chief radio officer on Byrd's first expedition to the Antarctic to determine seasonal variation in the height of the Kennelly-Heaviside Layer (E-layer), the lowest reflecting layer in the ionosphere, on which long distance medium-wave radio communication depends. His equipment was not automatic; he used the base's main transmitter and a portable receiver sited up to 15 km from base and observations were sometimes made under appallingly difficult conditions. He had difficulty in operating the oscillograph and photographic recording equipment and in keeping the receiver batteries from freezing in a small tent with temperatures as low as −55 °C. The oscillograph time-base was obtained by blowing a calibrated musical pitch pipe into a telephone receiver and recording the audio voltage output on the second element of the oscillograph but the pipe had to be modified to be sucked, not blown, otherwise its reed froze up with the moisture from his breath (Gillmore, 1978). His outstanding discovery was that the upper F-layer of the ionosphere, that reflecting high frequency radio waves, persisted during the polar night and showed a consistent pattern of diurnal variation in height, being lowest during the local day (Fig. 9.13). On the other hand, he found no evidence of an E-layer in winter. There was an increase in the virtual heights of reflecting layers with the frequency used. These results were reported promptly at the spring meeting of the American Geophysical Union in 1932 but although relevant to the work of the Second Polar Year they had little impact, possibly because the publishing plans of Byrd's first expedition were never completed. A fuller discussion of Hanson's work is given by Gillmor (1978). On Byrd's second expedition the chief radio officer, John N. Dyer, began studies of very low frequency (VLF) phenomena in the little time that could be spared from operating a programme of commercial broadcasts as well as general communications. Although he had virtually no electronic gear he made good use of phonographic recording equipment and many of his disc recordings of whistlers – whistle-like sounds of descending pitch which had been noticed during the First World War – and other atmospherics still exist. He attempted to correlate these with the aurora but his results were never published, probably, Gillmor (1978) surmised, because he was a broadcasting engineer with many responsibilities and few academic contacts, and there was little scientific interest in whistlers and the like in 1935. Gillmor made a study of the data which Dyer left, and with the reservation that it is senseless to try to find things in the observations which were only discovered years later, summed up as follows:

> Dyer heard whistlers at a high-geomagnetic latitude station inside the southern polar zone, and he described them well, noting something of their frequency and amplitude characteristics. He found that they occurred frequently but irregularly in winter and not at all after the coming of spring. He noted that atmospherics were heard much more frequently in summer than in winter. He observed that there was a rather clear association between audio hiss and the general occurrence of active aurorae, though there was no association between an individual atmospheric and a discrete auroral form. He noted an association between aurorae and 'rising tones' and 'jingling' sounds. He measured the terminating low frequency of tweaks. Finally, he made observations of the frequency of occurrence of whistlers and atmospherics as a function of latitude during a sea voyage which covered almost 70° of latitude, from inside the polar circle to the equator. Certainly, some of his observations would have been of use to later expeditions and to those who revived the study of VLF after World War II. (Gillmor, 1978.)

Auroral studies advanced little during this period. It was not until the US Antarctic Service Expedition of 1940 that the position of the southern auroral oval was roughly sketched out and two-station simultaneous photography used to determine auroral height (Wiener, 1945).

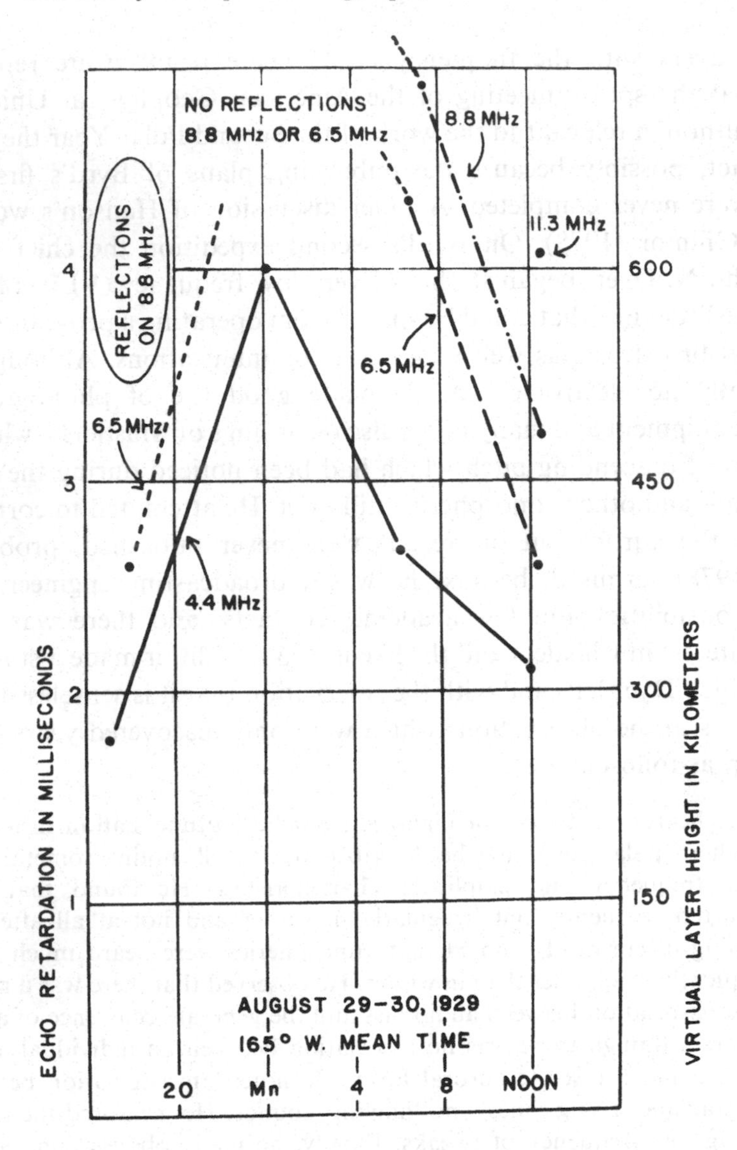

Fig. 9.13 Virtual layer heights in the ionosphere measured by M. P. Hanson at Little America, 29–30 August 1929. From Gillmor, 1978, Fig. 5.2. (Copyright by the American Geophysical Union.)

The Second International Polar Year ionosphere investigations had not extended to the Antarctic but Appleton and a team of investigators made ionosphere soundings at Tromsö in Norway, near the zone of maximum auroral activity, and were the first to report properly the occurrence of a polar radio blackout (Ratcliffe, 1966) although Mawson had encountered this phenomenon nearly 20 years earlier (Jacka & Jacka, 1988). Appleton was unaware that some of his other findings had been anticipated by Hanson on Byrd's first expedition. Ionosonde techniques as an adjunct to radio communication developed rapidly during the Second World War and an automatic sweep-frequency ionosonde was used shipboard on *Operation Highjump*, giving the first detailed data on diurnal variation of critical or penetration frequencies of the ionospheric layers. Ionospheric soundings from Terre Adélie in 1951 provided the first extended observations from the continent itself (Beagley & King, 1965).

One of those responsible for the development of the ionosonde, Lloyd V. Berkner, was a guest at the dinner, already mentioned in chapter 6, at which the idea of a Third International Polar Year crystallized. The principal guest, Sydney Chapman of the University of Oxford, had recently proposed a model of the earth's atmosphere and the space surrounding it in which tidal movements of the atmosphere, which he had demonstrated as occurring, generated electric currents by causing movements of electrically charged particles through the earth's magnetic field. The simple view, that the earth's magnetic field extended indefinitely out into space and that its atmosphere remained qualitatively much the same as it decreased in density with altitude until it blended with the near vacuum of interplanetary space, was being questioned and there were those present who had the means to provide answers. The host, James Van Allen, a protégé of the Antarctic physicist Thomas C. Poulter, was working on the use of rockets for sounding the upper atmosphere (Sullivan, 1961). This had begun in the US in 1945 with WAC Corporal, a small liquid-fuelled research rocket, and captured German rockets. Van Allen had used such rockets to investigate weak cosmic radiation over the North Magnetic Pole. For such a group of enthusiasts the polar regions, with their unique dispositions of magnetic and geographical co-ordinates, were focal points of interest. Not only do the extreme geophysical conditions in the neighbourhood of the poles allow critical observations not possible elsewhere but the absence of dense centres of population and overflying commercial aircraft reduce radio interference and allow greater freedom in the use of rockets. As we have already seen the idea of a concentrated effort in atmospheric sciences in these regions immediately struck a responsive chord in the international community of geophysicists and expanded into a world-wide scheme for comprehensive geophysical studies to coincide with a maximum of sunspot activity.

9.11 Ionospherics during the IGY

The plans for the Antarctic IGY programme were drawn up at four conferences held under the auspices of CSAGI (see 'A note for the reader') and resulted in more than 50 observatories being set up on the Antarctic continent and peri-Antarctic islands. The co-ordinates and other information relating to these stations have been tabulated by Law (1959). Not all of them, of course, carried out the full range of observations in the atmospheric sciences. All made surface meteorological observations with greater or lesser thoroughness. Twenty-three stations on the continent observed the aurora and air-glow, many of them using all-sky cameras for the first time in an intensive way to obtain records of the morphology of these displays and their changes in time and space (Paton & Evans, 1964). Spectrographs and photometers gave further information, one unexpected result being the detection of a transient lithium line in the auroral spectrum seen at several Antarctic stations which later was related to US hydrogen bomb tests at high altitudes. Radar of 40–100 MHz was used to detect highly ionized regions at auroral levels and to relate these to visible manifestations. Measurements of the velocity of ionospheric winds were made at the Australian base, Mawson, using a Döppler-radar instrument to track the trails of ions left by meteors passing through the 90 km level. Fifteen stations carried out ionospheric investigations using ionosondes of standard patterns to measure the virtual heights of ionized layers, the variations in these heights, and degrees of penetration by radio waves of various frequencies – observations of considerable practical value for communications. The ionosonde used by the BAS, the Union Radio Mark II designed and built just after the Second World War, incorporated what was then the latest technology such as one of the first solid state rectifiers, besides more tried and trusted devices such as thermionic valves and bicycle chains (Rodger, 1982; Fig. 9.14). This type of instrument was first used in Antarctica in 1953 and gave sterling service, the last being replaced by a microchip version, weighing only 40 kg as against over 500 kg, in 1982. The observations obtained yielded estimates of electron densities in the ionized layers and information about the processes involved in ionization by ultraviolet radiation from the sun and particles impinging from outer space. US workers made the perplexing discovery that ionization in the E- and F-layers persists during the darkness of the Antarctic winter in spite of the absence of ultraviolet radiation. This is now known to be caused by the anti-sunward flow of plasma into the area of the polar night produced by the convection pattern mentioned on p. 323.

The study of whistlers, particularly by Americans, also provided valuable information. These originate in lightning flashes, the impulses from which are ducted by the lines of force of the earth's magnetic field, arching far out into space before coming back to earth on the side of the equator

(a)

(b)

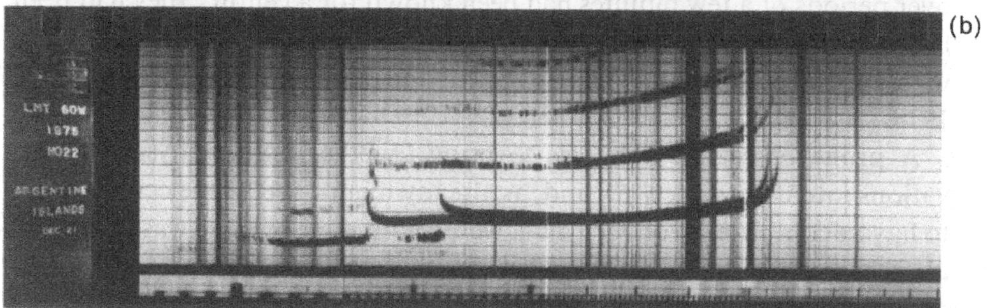

Fig. 9.14 (a) The Union Radio Mark II ionosonde at Argentine Islands (Faraday Station). The decorations on the instrument are of more psychological than ionospherical interest. From A. Rodger, 1982, Fig. 2. (b) Ionogram (photographic record of the returned signals displayed on a cathode-ray oscilloscope) for 02.13 Local Time on 21 December 1975, from Argentine Islands, showing echoes from the E- and F-regions of the ionosphere. From A. Rodger, 1982, Fig. 1.

opposite to their starting point, from whence they can be reflected back. They provided a means of investigating geospace beyond the reach of the ionosonde. The information which they gave about electron densities was confirmed by rocket and satellite observations. At the beginning of IGY, Van Allen had taken part in an Antarctic cruise on which a number of rockoons – experimental rockets launched from balloons at high altitudes – had been used. One of them detected a region of extremely high radiation count in the auroral zone similar to that which he had already found in the Arctic. These were the first indications of the existence of what was later to be called the Van Allen Belt. Altogether, during the IGY and its follow-up, the year of International Geophysical Cooperation, the US sent up 300 research rockets, the USSR 175, and Australia, Canada, France, Japan and the UK a few others, but nearly all these were launched in the northern hemisphere. They provided information on temperatures, density, winds, chemical composition, electron densities, auroral particles, radiation and magnetism. The US satellite, *Explorer* IV, had an orbit passing near enough to the poles to include the edge of the auroral zones and showed an upward curve of the contours of equal radiation intensity in this region, suggesting that this might be the bottom of an inner radiation zone.

Accurate geomagnetic observation was needed to support much of the work on ionospherics. In addition, these provided measurements of declination south of 45° S which called for drastic revision of the previously accepted values for high latitudes which had been mostly extrapolations from those further north. Although the existence of magnetic variations over periods of a few minutes had been known for a century they had been little studied until the IGY, when it was found that they were associated with auroral phenomena such as optical and X-ray emissions (Lanzerotti, 1978).

A prime object of IGY was to bring together all geophysical observations into a synoptic view and, as already mentioned, three centres, each to have a complete set of IGY data, were set up. One of these was in the US, another in the USSR and the third was divided between western Europe, Australia and Japan. Vice-Admiral Sir Archibald Day, recently retired from the position of Hydrographer to the Royal Navy, was appointed IGY coordinator to establish these centres. Promptness in submitting material to the centres and in making it available to scientists of all countries were recognized as essential. Observations thus became assimilated into the general corpus of geophysical study and have appeared in publications over an extended period so that it is difficult to disentangle the Antarctic contribution.

The IGY itself was a period of exploration and accumulation of data in ionospherics rather than of theoretical advance and one should remind oneself that the concept of geospace had not emerged at this stage. The IGY took place at a maximum in the solar cycle and together with the International Year of the Quiet Sun in 1964 provided a mass of valuable information much of which remains important today. From 1961 onwards the IGY data provided for some remarkable developments. First amongst these from our point of view was that the Antarctic was confirmed as a region having unique advantages for research on geospace. These advantages do not, of course, stem from the extreme climatic conditions except in so far as these have discouraged human colonization and left the continent more or less open for the deployment of rockets, balloons, lengthy radio antennae and other hazards to life and limb. The strictly scientific advantages were precisely outlined by Piggott (1977b), the essence being that, whereas in the north the magnetic pole is linear and close to the geographic pole, in the south it is a point and much further away from the geographic pole. Dynamic phenomena of the ionosphere are mainly determined by the relation between the direction of the magnetic field in the ionosphere and the geographic co-ordinates. In the south the dip and invariant of corrected geomagnetic latitude systems have greater displacement from the geographic systems and also are almost equally displaced from each other. This allows critical determination of whether a factor is ordered in geographic, magnetic or invariant latitudes, so enabling theories to be tested and missing factors identified in a way not possible in the Arctic. Even the small number of stations in Antarctica and on the peri-Antarctic islands fall into chains with one of these latitudes constant but covering a wide range in the other two systems. The Japanese Syowa Station is a prime location for the study of auroral displays whereas Halley, Sanae and Siple lie at the foot of the geomagnetic field lines along the plasmapause. Faraday and Dumont d'Urville are at similar geographic latitudes but on diametrically opposed sides of the South Pole and the dip angles of their geomagnetic fields differ by more than 30°. From Faraday the field lines run up into the plasmasphere whereas at Dumont d'Urville, where the field lines are vertical, they run out into the magnetospheric tail. Magnetic field lines can be traced through the magnetosphere from one hemisphere to the other, linking 'conjugate' observing stations. It is a further advantage of the Antarctic that it provides locations on land, with

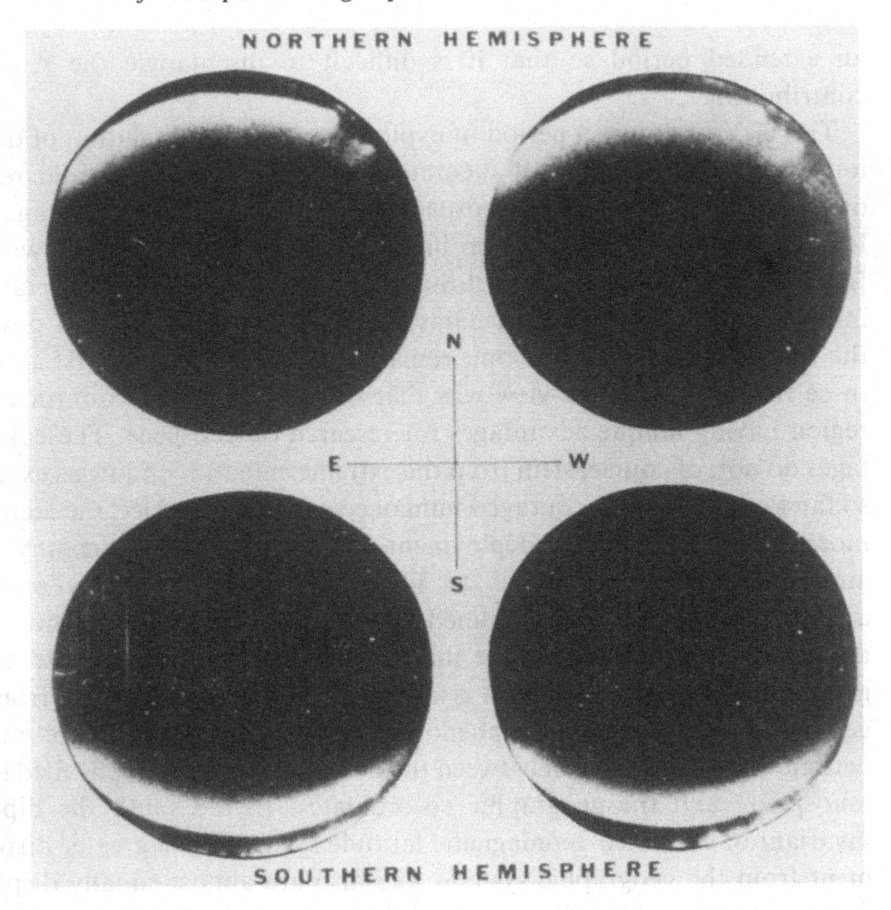

Fig. 9.15 Two pairs of all-sky photographs taken 14 March 1967, 10.47:40 UT at northern and southern conjugate points. In each pair the centres of the upper and lower photographs are connected in a single geomagnetic field line. From Akasofu, 1978, Fig. 4.1. (Copyright by the American Geophysical Union.)

low ambient noise levels of normal electromagnetic disturbances, which are conjugate with stations in North America sited where thunderstorms generate electromagnetic signals. A *tour de force* achieved in 1967 was to obtain paired all-sky photographs synchronously from aircraft flying over Alaska and over the conjugate point in the Weddell Sea, demonstrating the conjugacy of the Arctic and Antarctic aurorae (Fig. 9.15).

Dynamic processes in the ionosphere (diffusion, winds and electric currents) depend on dip whereas magnetospheric processes (precipitation of particles and polar winds) are linked within variant latitude. The solar input depends upon geographical latitude. This different ordering in space implies separation in time and three different time bases are used to characterize behaviour; universal time (UT) which is equivalent to Greenwich Mean Time, local solar time (LST) defined with respect to the local longitude, and local magnetic time (LMT) defined with respect to the local magnetic meridian. The discrepancies between LST and LMT are negligible for most of the world but become important near the magnetic poles and are greater in the Antarctic than in the Arctic. In the Weddell Sea area the spacing between lines of magnetic longitude is almost twice that of the lines of geographic longitude. This spreads out the sequence of events so that those in magnetic time are more easily investigated (Dudeney & Piggott, 1978).

The first hint of the unique situation in the Antarctic came in 1951 when ionosonde observations made by Argentina on Deception Island in the Peninsula sector showed that the maximum electron concentration in the ionosphere was at its lowest in the middle of the day in summer whereas all expectations were that it should have been at its largest (Walton, 1987, p. 240). During winter more intelligible variations were found, with concentration low at night and high around noon. Data obtained during IGY confirmed this pattern and showed that it was characteristic of the whole of the Peninsula and Weddell Sea sector (Rastogi, 1960). Study of ionosonde data from all over the Antarctic showed that the regular variations of electron concentration were fuctions of universal time rather than of local solar time, with the maximum occurring at 06.00 UT. This was called the Ross Sea anomaly. The Weddell Sea anomaly can be regarded as a special case of the Ross Sea anomaly, with UT control manifested only during the summer. Various theories, based on internationally co-ordinated work in Antarctica and on satellite observations carried out over 20 years, have been put forward to explain these anomalies (Dudeney & Piggott, 1978). For that in the Weddell Sea area it appears that neutral thermospheric wind, ordered in universal time, is the main agency modulating the observed F-region ion concentrations. The essential point is that the Antarctic anomalies result from combinations of geospace processes produced by the marked offset between geographic and magnetic poles.

9.12 Geospace research since the IGY

The IGY results suggested a whole new field of problems and a variety of new methods and instruments were brought to bear on their solution.

The ionosonde has continued to be a principal piece of equipment. The advanced ionospheric sounder, developed from it by the Space Environment Laboratory of the US National Oceanic and Atmospheric Administration, incorporated a small computer and associated microprocessors to exploit fully all the data available in the signals received and to carry out complex analysis in real time. It used radio waves in the frequency range 0.1–30 MHz for remote sounding of the ionosphere and was able to give a complete vector description of the ionospheric echoes obtained for each transmitted pulse, being thus able to track the movement of large scale ionospheric features. Of the first six of these built, one went in 1981 to BAS for use at its Halley Station where it was used to study the trough, a night-time minimum value of electron density at any given height in the F-region when plotted against latitude (Dudeney *et al.*, 1982). This is a major feature associated with the boundary between convecting and co-rotating plasma around 60° invariant latitude, unexpectedly ordered in UT and of practical importance for the complications which it produces in high frequency radio communication. A second advanced ionospheric sounder was deployed at Siple which, like Halley, was situated at the highest geographical latitude of any subauroral stations and thus had the longest period of winter darkness for the trough to develop.[6] The existence of variations in the earth's magnetic field, micropulsations lasting for a minute or two, had been known for a century but it was not until the IGY that they had been investigated intensively. These oscillations are generated by hydromagnetic waves – electromagnetic waves that can propagate in a highly conducting medium permeated by a magnetic field – in the magnetosphere and are associated with auroral phenomena such as optical and X-ray emissions and changes in riometer (see below) absorption measurements (Sato, 1965; Lanzerotti, 1978). The installation of very sensitive rubidium vapour magnetometers at Halley and Siple enabled these short period pulsations to be studied at both ends of the magnetic field lines, the other ends being at conjugate stations in the north. The relative phase differences show that the oscillating field lines are in first harmonic mode, like guitar strings. Under certain conditions hydromagnetic waves can mediate considerable energy transfer to the ionosphere (Lanzerotti, 1986). The riometer (relative ionospheric opacity meter) adapted a radio-astronomical technique for measurement of cosmic radio noise to measurement of radio wave absorption in the earth's atmosphere. Transient increases in absorption may be caused by additional ionization and differences in timing can provide information about movements. The riometer was first used in Antarctica in 1961 when a 30 MHz model was operated at Mirny. It was used later in studies of the precipitation of particles associated with auroral phenomena (Reid, 1971; Rosenberg & Barcus, 1978).

Rockets and satellites had been used during IGY but with the development of space technology they began to play a greater part in research in the Antarctic in the 1960s. It became possible to explore the top layers of the ionosphere using satellite-borne instruments and to measure total ionosphere density using satellite radio beacons. Spacecraft particle detectors were used, for example, by Gringauz and his colleagues in the USSR and revealed a plasma 'void' in the outer magnetosphere. Such techniques supplement, but cannot replace, ground-based observation because, among other things, in the passage of a spacecraft through a boundary regional spatial and temporal changes cannot be completely distinguished. Simultaneous observations at key locations throughout geospace by both satellite and ground-based techniques together with a soundly based programme of theoretical research offers the best prospect for advance. Such a project, involving perhaps as many as nine spacecraft in orbits designed to sample different regions of geospace and ground-based observations involving the Arctic and Antarctic, is planned for the early 1990s under the name Global Geospace Study.

There have also been advances in balloon technology, which offers a relatively inexpensive means of acquiring continuous data at altitudes up to 40 km. Large super pressure balloons can be sent on circumpolar flights lasting more than a year, the data from them being transmitted via systems such as the random access measuring system on *Nimbus* satellites. The advantages for balloons in the Antarctic are the absence of political boundaries such as restrict scientific flights in the Arctic and the minimal air traffic to which they might present a hazard.

The Antarctic ice-sheet is also ideal for setting out long radio antennae suitable for monitoring passive signals and broadcasting signals into the ionosphere and magnetosphere. It was the study of whistlers by such means that led Carpenter (1963) to locate the important boundary now known as the plasmapause. After preliminary trials at Eights Station up to 1965, then at Byrd Station between 1966 and 1971, a long dipole antenna, (initially 21 km then extended to 42 km) able to transmit coherent VLF (2–5 kHz) was established in 1973–75 at Siple Station (Fig. 9.16). This enabled controlled VLF waves to be used in wave-particle interaction experiments in the magnetosphere (Helliwell & Katsufrakis, 1978). Wave-particle interactions can cause trapped particles to be lost from the magnetosphere into the ionosphere where they produce X-ray and optical emissions and heating as well as perturbing magnetic and electrical fields at and below the site of precipitation into the upper atmosphere. The 'Trimpi effect', discovered in Antarctica data from the mid-1960s is a similar phenomenon in which whistler waves in the magnetosphere, generated by lightning, can trigger the precipitation of electrons which cause perturbations of man-

Fig. 9.16 The 21.2 km east leg of the dipole antenna, capable of transmitting coherent VLF (2 to 5 kHz) set up near Siple Station, Ellsworth Land, in 1973–75. (D. Carpenter, Stanford University.)

made VLF waves, used in navigational signals, propagating below the ionosphere (Helliwell *et al.*, 1973). Radiation from power networks in industrialized areas has also been found by satellite observations to be capable of causing similar precipitation of high energy particles – a new type of manmade pollution. Direct observation by satellite of electrons precipitated by manmade waves was reported in 1983 (Imhof *et al.*, 1983). This was not an Antarctic study but joint international experiments with VLF transmissions from Siple Station in conjunction with measurements from Canadian, Japanese, French and Soviet satellites have been carried out in continuation of this line of investigation (see Lanzerotti, 1986).

Studies in the magnetosphere, ionosphere and thermosphere cannot be separated from each other and there has been a shift towards co-ordinated multidisciplinary programmes. To this end it is important that workers should know what data are available and have easy access to them. This was met by annual reports from collaborating countries to SCAR and reports from SCAR working groups. Planning of synoptic observations in the Antarctic has been incorporated since 1976 within world-wide planning organized by the Special Committee on Solar Terrestrial Physics (SCOSTEP). With decreasing importance of individual research and budgetary restraints there has been increasing reliance on the automatic station, the unmanned geophysical observatory (UGO). After all, the upper atmosphere scientist, unlike his colleagues, the meteorologist, the glaciologist, geologist or biologist, has no need for contact with the immediate Antarctic environment and if he can do his work more comfortably and cheaply back at home, so much the better in some ways. By mid-1970s several prototype unmanned stations, ranging from expensive installations at McMurdo and Byrd Stations with real-time links by satellite to the US, to relatively cheap low-powered ones as favoured by Australia, were operating. The two major requirements are a reliable low cost power source which can be set up easily in the field and sensors which will give the required information accurately, be economical in the use of power and be unaffected by harsh conditions. An Australian unmanned station, near Casey, had an average power consumption of 0.75 W provided by secondary cells charged by a wind generator and solar panels. This allowed the operation of an all-sky camera, magnetometer, riometer and meteorological sensors recording on a digital logger with a crystal controlled chronometer (Piggott, 1977a). An unmanned VLF station near Halley ran into difficulties because it depended on a wind generator which, in spite of being set up to feather at high wind speeds and having a power dissipation unit, nevertheless caused serious overheating of the electronic systems.[7]

With these advances in observational techniques and the flow of data have come attempts at numerical modelling of aspects of geospace. In the thermosphere, neutral winds, which can be measured by a device known as the Faby Perot interferometer either from the ground or from a satellite, are generated by two processes: (1) global variations in neutral thermospheric pressure caused by solar heating, and (2) momentum transfer from ionospheric plasma, itself in motion because of the dawn–dusk electric field generated across the magnetic tail and extending down the magnetic field into the polar ionosphere. The first operates over the whole globe and dominates at latitudes lower than the auroral oval. The second acts within the oval and polar caps and can result in extreme winds. A model for the thermosphere and one for the ionosphere have been coupled to provide a valuable means of examining thermosphere–ionosphere interactions in the polar regions (Rees & Fuller-Powell, 1989). However, studies on interactions between the ionosphere and the lower atmosphere, which might link changes in solar radiation during the solar cycle with weather – the interest of which was realized by SCOSTEP in the mid-1970s – do not seem to have progressed far.[8] The two threads that have run through this chapter are not yet firmly tied together.

9.13 Cosmic ray studies and astronomy in the Antarctic

It seems tidy, if not strictly logical, to conclude this chapter with a brief remark on astronomy in the Antarctic. The study of cosmic radiation is, as we have seen, of long standing in the Antarctic. On Byrd's second expedition measurements of the intensity of cosmic radiation by means of an ionization chamber were begun on the voyage to Antarctica and continued at Little America for almost a year in 1934–35 by E. H. Bramhall and A. A. Zuhn. A. L. Kennedy with BANZARE, in 1930–31, had used the recently invented Geiger-Müller counter to measure cosmic radiation (Price, 1963, p. 161) and had found no variation of count rate with geomagnetic latitude. Bramhall and Zuhn, however, showed that although intensities are more or less constant at high latitudes there is a 'knee' at mid-latitudes. Their measurements during a flight from Little America up to 3400 m gave evidence of the charged nature of at least a proportion of the cosmic ray flux. These results were not published separately but as part of a major, world-wide, co-operative programme and were included in papers by A. H. Compton of the Carnegie Institution of Washington in the US. Compton (1934) concluded that there was a latitudinal effect in cosmic ray intensity – a point of conflict with the eminent physicist R. A. Millikan which was eventually resolved in Compton's favour (de Maria & Russo, 1989). His comparison of measurements from the Arctic and Antarctic showed that the intensities near the poles were the same within experimental error, evidence that in remote space cosmic rays must be isotropic and therefore have an origin outside the galaxy. Cosmic ray studies were continued on the US Antarctic Service Expedition in 1939 by E. T. Clarke and D. K. Bailey, who obtained the first continuous recordings of cosmic rays in Antarctica (Korff *et al.*, 1945).

Low energy cosmic radiation, which at the poles comes in along lines of magnetic force instead of being deflected into space as in the equatorial plane, was studied at ground level during IGY by neutron monitors of standard pattern. At Mawson large meson telescopes, which could be orientated at angles from the vertical and rotated to measure incident radiation at various azimuths, were used to detect asymmetries between radiation intensities from north and south and east and west (Law, 1959; Sullivan, 1961; Pomerantz, 1978). Further work on variations in cosmic radiation using special antennae at several Antarctic stations was carried out in the International Year of the Quiet Sun in 1964–65.

US studies on both solar and galactic cosmic rays have been concentrated at McMurdo and the South Pole stations. The most sensitive detection system in the world was installed at the South Pole in 1977

(Pomerantz, 1978). That at McMurdo, where the asymptotic cone of acceptance is nearly perpendicular to the ecliptic plane, together with those at Thule in Greenland, are particularly important for determining north–south anisotropy of cosmic ray fluxes. Observations of variations in intensity coupled with theoretical studies of the processes producing cosmic radiation provide insights into the large scale structure of the solar system as well as fundamental aspects of solar physics (Lanzerotti, 1986).

Apart from this the plateau has some distinct merits for the optical astronomer; the low temperature and high altitude give the atmosphere a unique transparency and the Pole itself stays in one place relative to the celestial object under study. In summer the sun is continuously above the horizon and clear weather enables it to be kept under observation for long periods. An astronomy programme was initiated at the Pole in 1979 to study southern hemisphere variable stars and dynamic solar phenomena.[9] A joint American and French team carried out investigations in the relatively recently developed field of solar seismology. Mechanical oscillations in the sun's surface produce Döppler shifts which can be detected by optical observation at particular wavelengths. Displacements of the solar diameter as small as 5 m in amplitude have been found and characteristic periods of oscillation identified (Walton, 1987). A conference at the University of Delaware in June 1989 considered what can be done to further astronomy in the Antarctic (Lindley, 1989).

Endnotes

1 Third CSAGI Antarctic Conference; 3 Resolutions, II Meteorology. *Annals of the IGY*, **2B**, p. 450.
2 Department of the Environment, Pollution Paper No. 15, London 1979.
3 *British Antarctic Survey Report* 1976–77, p. 18
4 *Ibid.*, 1979–80, p. 27.
5 The term 'geospace' originated during the International Magnetopheric Study (IMS) which was launched in 1978.
6 *British Antarctic Survey Report* 1977–78, p. 12 and 1979–80, p. 19–20.
7 *Ibid.*, 1978–79, p. 21.
8 *SCAR Bulletin* No. 61, 1979.
9 US Antarctic Program, Program Manual, Division of Polar Programs, National Science Foundation 1977.

10

Land-based biology

10.1 The natural history of the Antarctic

At the end of the nineteenth century, collection and classification were still the principal occupations for biologists and the major concerns of the leaders of biological thought were the cell theory, comparative anatomy and evolution. For the professional, therefore, unless he was a taxonomist anxious to add more twigs to the phylogenetic tree, there was no call to go to Antarctica and there was no mention of terrestrial biology in the otherwise comprehensive discussion on Antarctic science held by the Royal Society (Murray, 1898). Interesting work on penguin embryology was done on the *Scotia* expedition material (Waterstone & Campbell Geddes in Bruce, 1915) and it was Haeckel's doctrine that the development of the individual recapitulates the history of its race, which impelled Wilson and his two companions on their heroic winter journey in search of embryos of that supposedly primitive bird, the emperor penguin, but this yielded little of value concerning the evolution of birds. For the naturalist, however, the birds and seals of the far south, unafraid of man, were a joy and articles on these were included in the *Antarctic Manual* for the *Discovery* expedition (Murray, 1901). Botanists were not quite so obsessed with evolution as were zoologists, and botanical ideas that were subsequently to be important in Antarctic research were beginning to appear. Plant geography, which had started with men with Antarctic interests such as Humboldt and Hooker, had developed and its sister science, plant ecology, had made its appearance. E. Warming's *Oecology of Plants* was published in Danish in 1895 and A. F. W. Schimper's *Plant Geography upon a Physiological Basis* in German in 1898. These, however, had no immediate impact on work in the Antarctic and George Murray's article on botany for the *Antarctic Manual* (1901) was about collection and preservation techniques for taxonomic purposes. Limnology, which is also a topic to be dealt with in this chapter, was well established before the end of the century with freshwater laboratories at Plön-im-Holstein in Germany in 1892 and at Madison, Wisconsin, USA in 1893, but had yet to extend its compass beyond the temperate regions. Microbiology at this time was largely medically orientated and dominated by the plating techniques devised by Robert Koch for obtaining pure cultures. These methods were applied by M. W. Beijerinck in the 1890s to the study of micro-organisms, (algae as well as bacteria) in soil and water. At the same time S. Winogradski, who was less restricted by the ideal of pure cultures, was discovering the roles of bacteria in the cycling of elements such as nitrogen and sulphur. The basic importance of microbial ecology was thus becoming evident but it had scarcely developed far enough for there to be anything but the most preliminary work on it in so remote a place as Antarctica.

There was a reasonably good knowledge of the floras of the sub-Antarctic islands at the beginning of the new century (Kirk, 1891). Hooker (1847) had laid a solid foundation which had been built on by many subsequent visitors to Kerguelen, Macquarie and other islands. The flora of South Georgia, which Hooker had not visited, was studied by H. Will, botanist to the German International Polar Year Expedition based at Royal Bay. Will went beyond listing species, providing autecological notes on those he found. The vascular plant collection which he made with the assistance of the engineer E. Mosthaff was worked up by the eminent systematists A. Engler and K. Prantl. Their findings, together with those on mosses, liverworts and lichens, were published in the expedition report (Neumayer, 1890–91). It fell to the Bull expedition of 1894–95 to find lichen growing on the continent itself, at Cape Adare (Bull, 1896; see p. 108). E. Racovitza made the first record of one of the only two flowering plants indigenous to the continent when he found *Aira* sp. (*Deschampsia antarctica*, already described from the South Shetlands, p. 48), at 64° 35′ S, on the Danco Coast of the Peninsula in 1898 (Wildeman, 1905). All four expeditions which went south in 1901–2 had botany in their remit. On the *Discovery* the medical doctor, R. Koettlitz, was made responsible for this and had the benefit of the tutelage of George Murray, Keeper of Botany at the British Museum (Natural History), who went on the voyage as far as Cape Town as scientific adviser. Koettlitz demonstrated the fertility of Antarctic soil by growing a better crop of mustard and cress on it than on artificial substrata (Scott, 1929, p. 401). On the Swedish *Antarctic* expedition the botanist was Carl Skottsberg but he devoted much of his effort to the floras of Tierra del Fuego and the Falklands, as well as to South Georgia and the Trinity Peninsula. R. N. R. Brown on the *Scotia* studied the lichens, mosses and freshwater algae of Laurie Island, South Orkneys. His review (1906) gives us the best picture of Antarctic botany at this time. The chief interest lay in biogeography and he speculated on the reasons for the poverty of the phanerogamic flora – two species of flowering plants as compared with 400 in the Arctic – concluding that although seed dispersal by wind or birds seemed possible, the short summer, low temperatures and prolonged snow cover limited colonization. He sent seeds of 22 species of Arctic plants to be sown in a location which he thought favourable for their establishment at the Argentine Station on Laurie Island, but this was evidently not acted upon. He does not seem to have followed up Hooker's observations on microhabitats which attain temperatures far above ambient although he emphasized the importance of studying morphological and physiological features of Antarctic plants on the spot. Mosses, of which about 50 species had been recorded (Fig. 10.1), seemed to reproduce mainly by vegetative means. A rich lichen flora had been found, in which almost three-quarters of the species seemed also to be found in the Arctic. The terrestrial algal flora Brown considered to be poor in the number of species although not in quantity, and he mentioned the occurrence of red and yellow snow algae. Fritsch (1912) examined samples of snow algae, as well as other terrestrial algae collected by Brown and described several new species. Gain (1910, 1912) found he was able to grow forms in culture from green snow; this seems to have been the earliest use in the Antarctic of culture techniques in the study of algae. Finally, Brown (1906) considered that the northern South Sandwich Islands might have floras resembling that of South Georgia whereas the southern ones might be truly Antarctic so that their study might be of particular biogeographic significance – an idea that subsequently proved to be wrong. During the next 50 years taxonomic work on mosses, lichens and algae continued sporadically but otherwise Antarctic plant science remained much where it was. For example, the botany carried out on the second Byrd expedition contributed general descriptions of the habitats and distributions of lichens and mosses in Marie Byrd and King Edward VII Land and drew attention to the importance of nutrient enrichment from bird colonies on nunataks for the local vegetation (Siple, 1938) but detailed taxonomic treatment of the samples collected made up the bulk of the results (Bartram, 1938; Dodge & Baker, 1938).

Emile Racovitza of the Belgian Antarctic Expedition was the first to report the presence of free-living terrestrial invertebrate life in the Antarctic (Racovitza in Cook, 1900). Among mosses and lichens he found a small fly, *Belgica antarctica*, with rudimentary wings (Fig. 10.2), large numbers of springtails (*Collembola*) and several species of mite (*Acarina*). These were

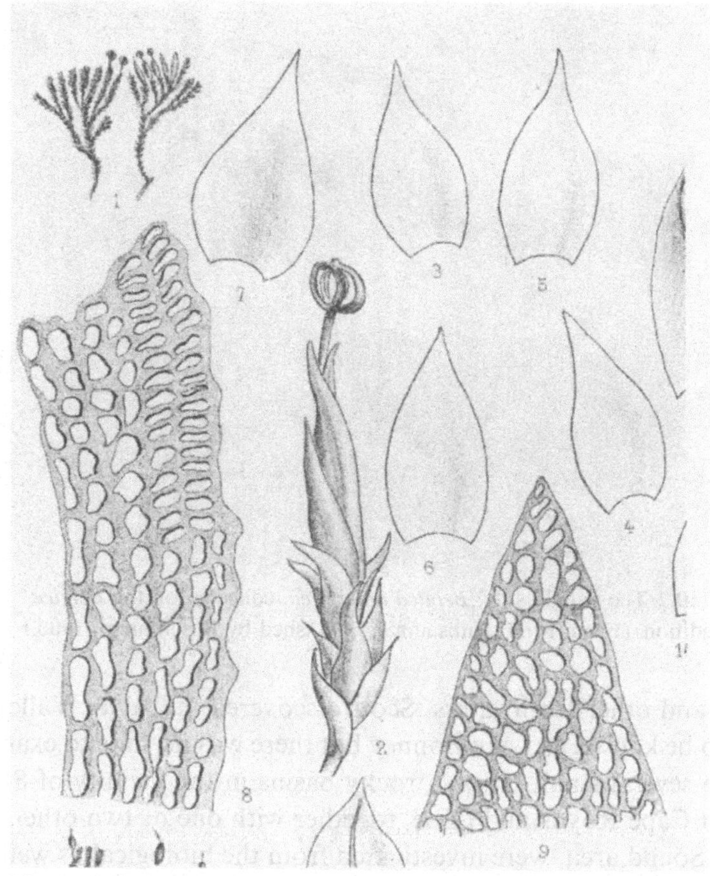

Fig. 10.1 A new species of Antarctic moss, *Andreaea gainii*, collected on the Antarctic Peninsula during Charcot's expedition in 1908–10. From Cardot, 1913, Plate III, Figs. 1–9. (Cardot, 1913.)

also found by other expeditions over the next few years. Thus the *Collembola* and *Acari* of the South Orkneys collected on the *Scotia* expedition were reported on by Carpenter (in Bruce, 1909) and Trouessart (in Bruce, 1912) respectively, and Carpenter (1908, 1921) also described the *Collembola* collected on Scott's first and second expeditions. About a third of the terrestrial arthropods now known were discovered at that time but then, for almost half a century, their study was neglected (Block, 1984a). There were also intestinal parasites of seals and sea-birds, some of which were studied from the *Scotia* collections, for example, nematodes by Linstow (in Bruce, 1909) and cestodes by Rennie and Reid (in Bruce, 1912).

The library of the *Discovery*[1] contained some half-dozen books, including the seminal works of Forel (1892, 1895) and Delebecque (1898), dealing

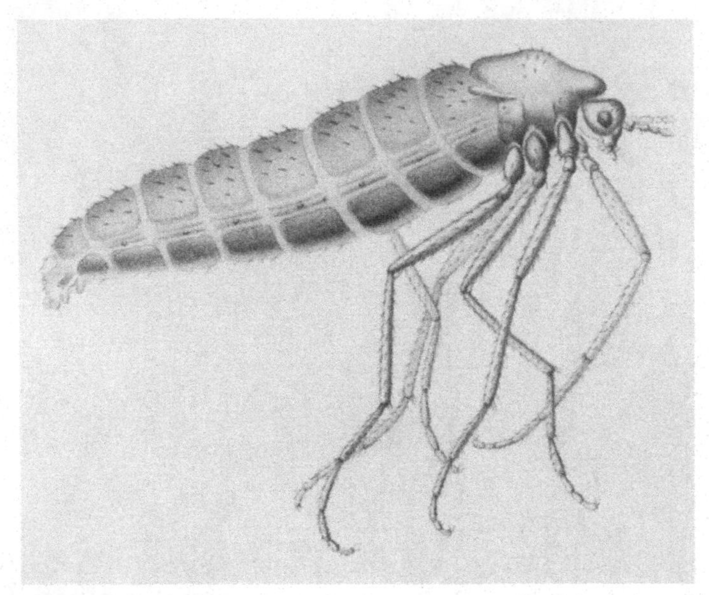

Fig. 10.2 The wingless fly, *Belgica antarctica*, collected on the *Belgica* expedition. (From E. H. Rubsaamen, published by Buschmann, 1906.)

with lakes and other freshwaters. Scott discovered in Taylor Valley what was later to be known as Lake Bonney but there was no time to examine it. There were several small lakes in rocky basins in the vicinity of Shackleton's hut at Cape Royds and these, together with one or two others in the McMurdo Sound area, were investigated from the biological as well as the physical point of view, providing the first real descriptions of Antarctic freshwaters (Hobbie, 1984). They were completely frozen for at least nine months in the year and one, Green Lake, was highly saline (Murray, 1910), but when they did thaw it became evident that they supported a surprising amount and variety of life. The algal flora differed from lake to lake but the most characteristic growth was a benthic felt of filamentous blue-green algae, mostly *Phormidium* spp. (West & West, 1911). Rotifers, tardigrades and other invertebrates, such as had already been reported from freshwater habitats on the Peninsula by Racovitza (in Cook, 1900), were found to be abundant in this growth and elsewhere, sometimes forming blood red patches more than an inch in diameter on stones. They could be recovered alive from frozen material and simple experiments showed that they remained viable after exposure to temperatures as low as -40 °C, as well as after repeated freezing and thawing, after drying, and after immersion in brine. On the other hand, they died if kept in water at moderate temperature. Two species, *Adineta* sp. and *Philodina gregaria*, were viviparous, the

Fig. 10.3 Young king penguins kept in captivity for purposes of study during the German expedition to South Georgia in 1882–83. The leather corsets were designed to tether the birds. From von den Steinen in Neumayer (1890).

first rotifers known to be other than oviparous. The rotifers and other animals were found generally distributed in all the lakes visited and it was inferred that they could be distributed on the feet of skuas or in wind-blown fragments of algal felt (Murray in Shackleton, 1909). Scott's second expedition also examined these lakes but got no specimens of the blood-red *P. gregaria*, although otherwise their observations tallied with those reported by Murray. Several more lakes, some of them on glaciers, were discovered but given no more than cursory examination (Huxley, 1913).

Because sea-birds and some seals come ashore to breed, most studies on them have been land-based. Since the time of Cook's voyage, when the Forsters described many new species, nearly all expeditions took note of the birds that they encountered and by the end of the nineteenth century the list was fairly comprehensive. K. Von den Steinen, medical officer and zoologist with the German First International Polar Year Expedition to South Georgia in 1882–83 began the broadening of Antarctic ornithology by a pioneer study of penguins in which he made observations of behaviour and estimations of rookery size (Von den Steinen in Neumayer, 1890; Barr, 1984; Fig. 10.3). The *Scotia* expedition contributed studies on anatomy, embryology, life and habits, as well as taxonomy, of the birds of the South

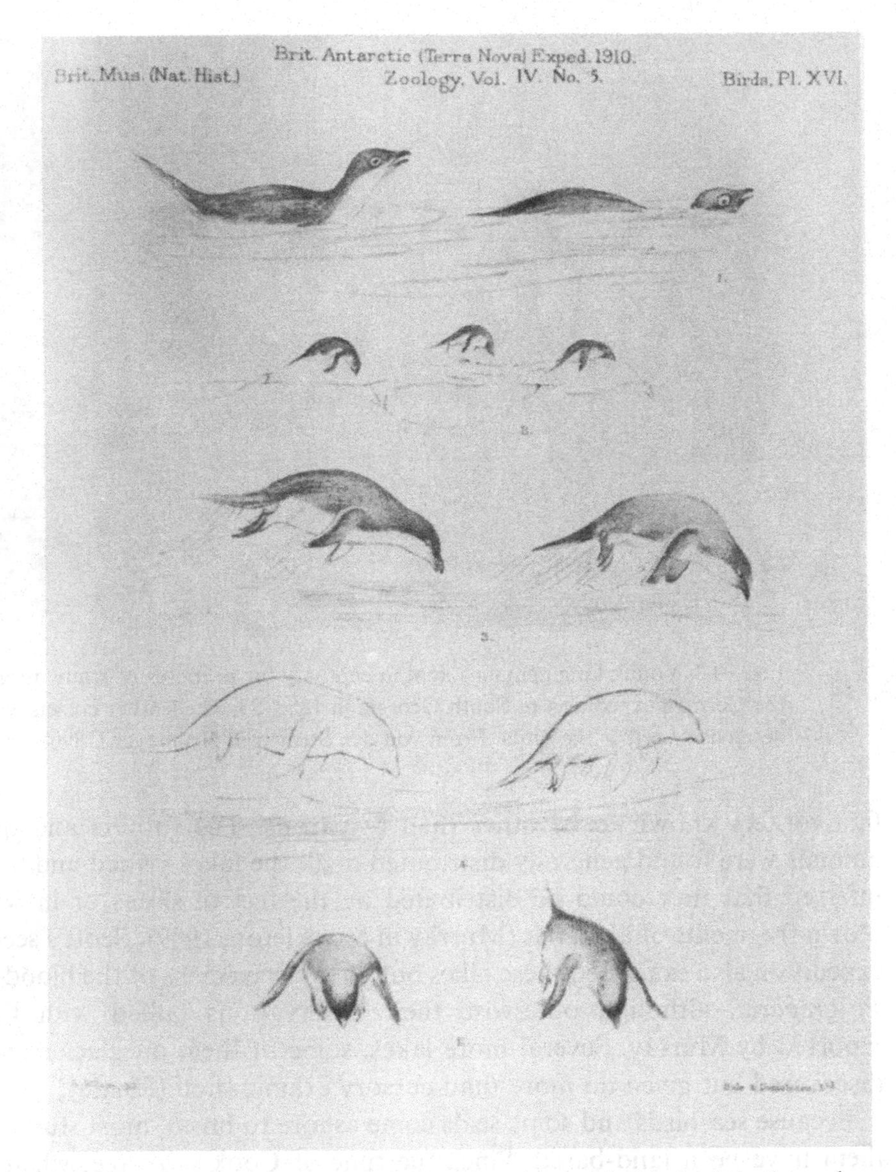

Brit. Antarctic (Terra Nova) Exped. 1910.
Brit. Mus. (Nat. Hist.)
Zoology. Vol. IV. No. 5.
Birds. Pl. XVI.

Fig. 10.4 Adélie penguin, *Pygoscelis adeliae*, 'porpoising' in leads in the pack. Drawing by E. A. Wilson from Lowe & Kinnear, 1930, Plate XVI.

Orkneys (Clarke *et al.*, Waterstone & Campbell Geddes, both in Bruce, 1915). Next came E. A. Wilson, who on both of Scott's expeditions carried out observations on birds, illustrating them with superb drawings (Roberts, 1967; Fig. 10.4). From his work with the *Discovery* expedition he produced a masterly account of the characteristics, life-history and behaviour of a number of Antarctic birds (Wilson, 1907). After the tragedy of the return from the Pole his notes and specimens went to the British Museum (Natural History) but the working up of these was delayed by the death of W. R. Ogilvie-Grant, who had been entrusted with this task, followed by the dispersal of the material, and an account did not appear until 1930 (Lowe & Kinnear, 1930). This put forward important conclusions about some supposed subspecies and their geographical distributions. Also on the *Terra Nova* expedition was Murray Levick, whose engaging account of the social habits of penguins (Levick, 1914) together with Herbert Ponting's beguiling photographs contributed to the clouding with anthropocentric interpretation of observations on the behaviour of these birds. In 1912 R. Cushman Murphy, a raw graduate, was dispatched to South Georgia in the whaling brig *Daisy* to make a bird collection for the American Museum of Natural History. His *Logbook for Grace* (1948), written for his wife, does not pretend to be a scientific account but gives a vivid picture of the conditions under which a naturalist had to work at that time. Later he reviewed the problems of Antarctic zoogeography, including other animals as well as birds, without, however, reaching any conclusions as to the existence of land bridges or of Gondwanaland (Murphy, 1928). His *Oceanic Birds of South America* (Murphy, 1936), to which the South Georgian sojourn contributed, is one of the monuments of ornithological scholarship. In the same tradition is B. B. Robert's (1940) monograph on Wilson's petrel, *Oceanites oceanicus* (Fig. 10.5), probably the most numerous of oceanic birds, in which he mapped its dispersal in the Atlantic month by month from observations made off the Peninsula, off South Georgia and on an Antarctic cruise of *Discovery II*. Roberts also carried out a study on the breeding behaviour of penguins and compiled a bibliography of Antarctic ornithology up to 1940 (Roberts, 1941).

Work on Antarctic seals followed something of the same pattern as that on birds. The species were all identified well before the end of the nineteenth century. From material obtained on the *Scotia* expedition, Hepburn (in Bruce, 1915) produced a series of anatomical studies of the Weddell seal, Rudmose Brown wrote about the habits and distribution of the different seal species of the Weddell Sea and W. S. Bruce (1915) reported on measurements and weights. Examples of Antarctic seals were collected in the Peninsula area by the *Pourquoi-Pas?* expedition and described by Trouessart (1906). Wilson (1907) also produced a general account of Antarctic seals. A more complete bibliography of work on seals dating from this period is given by Gilmore (1961). Little of the effort of the *Discovery* Investigations was devoted to seals but general accounts were given of the elephant seal (Matthews, 1929) and the leopard seal (Hamil-

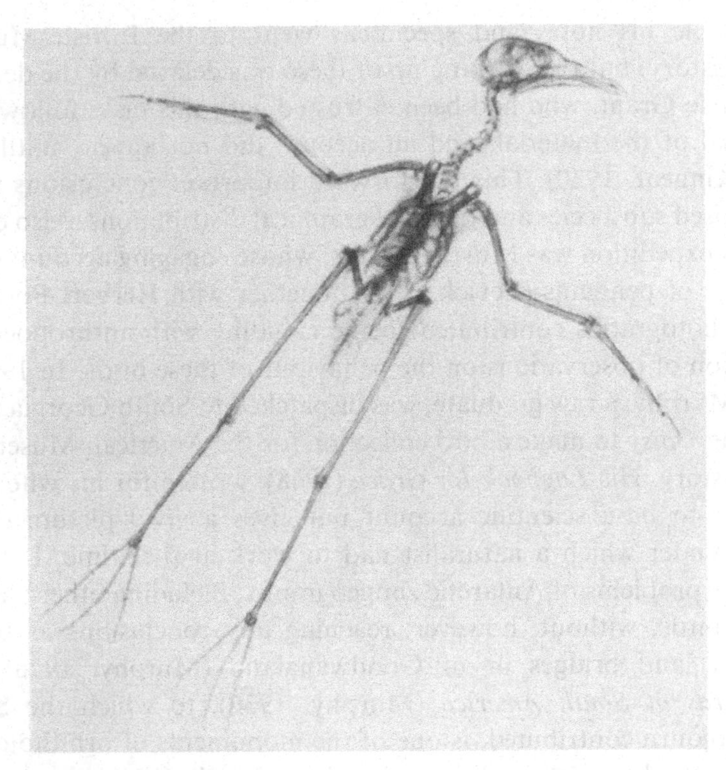

Fig. 10.5 X-ray photograph of a Wilson's petrel, *Oceanites oceanicus*, collected on the British Graham Land Expedition, 1934–37. From B. B. Roberts, 1940, Plate VII.

ton, 1934). On the British Graham Land Expedition, (1934–37), it was necessary to kill a large number of seals for dog food and advantage was taken of this by C. Bertram (1940) to obtain measurements, specimens and observations which resulted in a study of the two contrasting species, the crabeater, a gregarious animal of the pack ice feeding on krill, and the Weddell, an inshore animal, feeding on fish and squid. The former rarely had parasites whereas the latter was usually infected with a variety of forms.

Several expeditions at the beginning of the twentieth century included microbiology in their programmes. There was a preoccupation with plating out animal faeces and it was noted (e.g. by McLean, 1919) that bacteria were often only sparsely present, a fact which has been more recently ascribed to antibiotics derived from phytoplankton (Sieburth, 1961). Some important findings for microbial ecology were made. On Nordenskjöld's Antarctic expedition E. Ekelöf demonstrated the presence of a bacterial flora in surface soil, something which had not so far been done in the

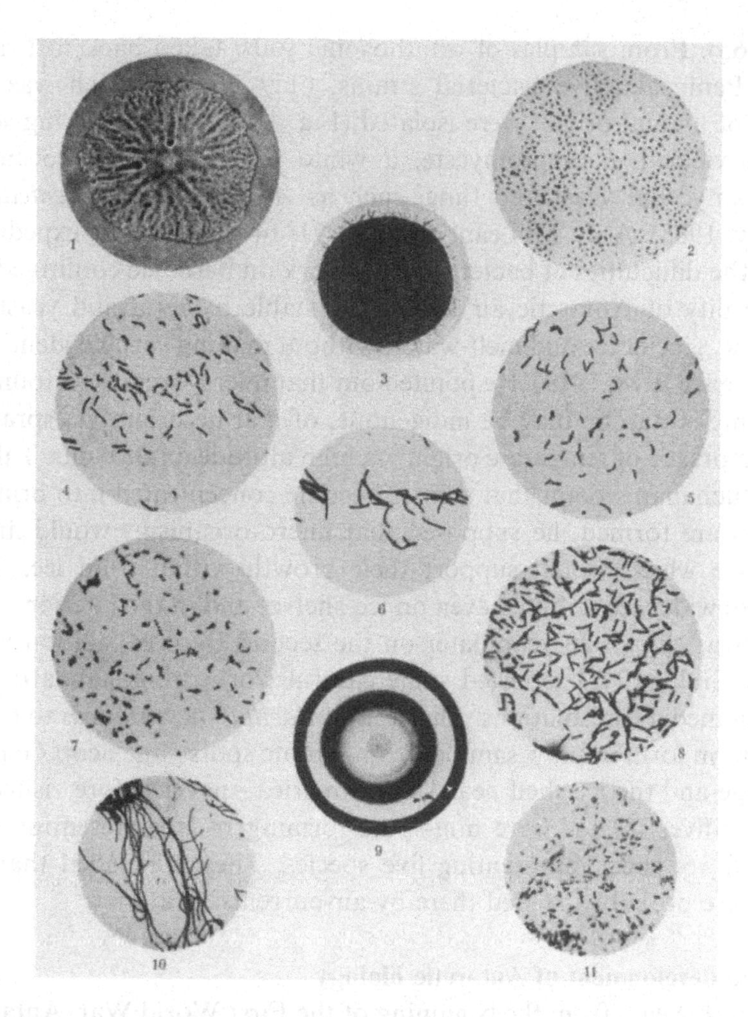

Fig. 10.6 Bacterial and fungal isolates from soil collected on the Danco Coast during Charcot's first expedition. From Tsiklinsky, 1908.

Arctic. He studied changes over a whole year on Snow Hill Island, Antarctic Peninsula, and isolated different kinds of bacteria, actinomycetes and a pseudomycelial yeast but commented that a Petri-dish of medium had to be exposed for two hours for one bacterium to settle on it (Ekelöf, 1908 a,b). J. H. Harvey Pirie (in Bruce, 1912) on the *Scotia* expedition exposed nutrient agar plates to the air in the vicinity of winter quarters on Laurie Island, South Orkneys, but obtained no bacterial growth. Both he and Gazert, on the *Gauss* expedition, had investigated microbial activity in seawater (p. 201). J.-B. Charcot on his first expedition carried out bacteriological work and also failed to detect bacteria in Antarctic air (Charcot,

1906, p. 465). From samples of ornithogenic soils, taken back to France from the Peninsula, five bacterial strains, only one of which was not obviously of animal origin, were isolated (Fig. 10.6). A soil bearing some vegetation yielded a streptomycete, a white psychrotolerant coccus, a red pseudomycelial yeast and fungi such as *Aspergillus* and *Penicillium* (Tsiklinsky, 1908). A. L. McLean, on the 1911–14 Australasian expedition described the difficulties of bacteriological work on base. He confirmed the virtual sterility of Antarctic air but found viable bacteria and yeasts in falling snow, glacier ice and melt-water, without making specific identifications (McLean, 1918, 1919). He pointed out that micro-organisms found in the maritime Antarctic may be indigenous, of marine origin via spray or animal vectors, or of temperate origin via high altitude air-streams. Taking up J. Y. Buchanan's point that solutes become concentrated into brine as ice crystals are formed, he supposed that micro-organisms would find a liquid phase which would support their growth within solid ice. That bacteria are widely distributed even on ice-shelves and in the interior of the continent was shown 20 years later on the second Byrd expedition when Darling & Siple (1941) reported some careful work using standard bacteriological media for isolations. They found *Bacillus mesentericus* to be the most common form in snow samples from remote spots – the Scott Glacier, Ford Range and the ice shelf near Little America – never before visited by man. Forty-five isolates were non-spore-forming rods representing nine species and 16, cocci representing five species. They concluded that the bacteria were probably carried there by air-currents.

10.2 The development of Antarctic biology

For 40 years from the beginning of the First World War, Antarctic terrestrial biology remained in the doldrums with little significant broadening of its scope. A review volume published as late as 1965 (Van Mieghem & Van Oye, 1965) was still largely concerned with floristics, faunistics and biogeography. The inventory of the larger plants and animals, at least, was reasonably complete but although there seemed to be a higher degree of endemism, Rudolph (1966) warned that this might not be true for all the flora and that further taxonomic work was necessary. The actively developing disciplines of experimental biology, biochemistry and genetics found ample material at home without seeking it at trouble and expense in remote parts. Ecology, which might have been a growing point for Antarctic biology, was still in its sociological stage when the potentialities of physiological ecology were unrealized and the scant communities of the southern fellfields and lakes seemed to present little of interest. Another factor was that the bigger expeditions of this period were concerned more with probing the interior of the continent, where the biological interest is minimal, rather than establishing the permanent stations, which favour the biologist, on its periphery. As already indicated, these expeditions did in fact take some biologists and they made their contribution but it was swamped by the more glamorous achievements of geographical exploration, glaciology, geology and geophysics.

Permanent stations on peri-Antarctic islands and the Peninsula gave biology an opportunity to develop. Whaling on South Georgia provided one means of access and enabled a private expedition led by N. Rankin to make a survey of the natural history of the island in 1946 (Rankin, 1951). Botanical and zoological work was carried out at the bases established from 1943 onwards in *Operation Tabarin* and later by its successor, the Falkland Islands Dependencies Survey. At this stage there was little organization or plan but good foundations were laid and some distinguished work emerged. I. Mackenzie Lamb, who had worked in the British Museum (Natural History) on the lichens collected by heroic age expeditions, was appointed botanist on *Operation Tabarin* in 1943. He collected extensively in the Peninsula area and produced important monographs on the taxonomy of Antarctic lichens (e.g. Lamb, 1948). Later he became curator of the Herbarium of Cryptogamic Botany at Harvard University, in the US and visited Antarctica again under US auspices (Taylor, 1990). R. M. Laws, who was appointed base leader on Signy Island in 1947 and later worked on South Georgia, produced a study of the elephant seal which was one of the first in mammal population dynamics (Laws, 1953a,b, 1956a,b). His demonstration that the age and time of first breeding of this animal could be determined from examination of incremental layers in teeth was an important advance in technique which has since been applied to other mammalian groups. The French base, Port-aux-Française, established on Kerguelen in 1951, provided for studies which initially were mainly of a floristic and faunistic nature (see the bibliography of Delépine, 1964).

The stimulus given to Antarctic research by the IGY extended to biology (Webber, 1986). Although biological work was not planned explicitly, except for some larger vertebrates, there were some good preliminary studies by amateurs, among which Falla (in Hatherton, 1958) singled out for special mention the lichen collection in Marie Byrd Land by the Byrd traverse party. In 1959 at the Third Meeting of SCAR, programmes in terrestrial biology and medical research were drafted and this was followed by the setting up at the Fourth Meeting of SCAR in 1960 of a permanent working group on biology. A report to the committee on Polar Research of the National Academy of Sciences acknowledged that the US expeditions had not so far distinguished themselves in biology and after reviewing the state of knowledge in various fields set out objectives which might be of most interest in the foreseeable future (Committee on Polar Research, 1961). These objectives were largely taxonomic as far as terrestrial biology was concerned. At this stage the Soviets did not appear to have much interest in biology in continental Antarctica (Somov, 1958) and their station, Bellingshausen on King George Island, which became a centre for biology, was not established until 1968. Meanwhile Sir Vivian Fuchs on his appointment as Director of the Falkland Islands Dependencies Survey (FIDS) in 1958 had considered how Antarctic science might be developed and concluded that a regular long-term biological programme, dealing with inshore, terrestrial and freshwater communities would be appropriate to complement the offshore work that had been done by *Discovery* Investigations. There were two obvious sites for this, Signy Island in the South Orkneys, and South Georgia, but it seemed best to concentrate on the former in the first instance as being more typically Antarctic, leaving the latter, with a climate having both temperate and polar characteristics, for later attention. An inspection in 1960 by a distinguished ecologist, Professor J. B. Cragg, confirmed that Signy was a suitable place in which to build up a substantial biological facility. Following a pattern of collaboration with universities which had been successful in geology and geophysics, a zoological section was set up in 1961 under Dr M. W. Holdgate in association with the Zoology Department of Queen Mary College, London, and a botanical section under Dr S. W. Greene, associated with the Botany Department of Birmingham University (Fuchs, 1982). A key step forward was the publication of R. I. Lewis Smith's (1972) account of the terrestrial vegetation of Signy Island, based on floristic and edaphic data. Actually, there was no long wait before biological studies on an almost equal scale could be started on South Georgia because with the cessation of whaling the UK government found it expedient to encourage BAS to establish a significant presence there and this was done in 1969. The British were thus the first to carry out a major continuous programme in terrestrial biology in the Antarctic. However, much work in terrestrial biology has also come from the South African station established on Marion Island in 1948, an important series of investigations on the role of birds and, later, seals, in transferring nutrients from the sea to terrestrial ecosystems being initiated there by W. R. Siegfried in 1973 (Fugler, 1985; Panagis, 1985). Another major step forward was the account of the vegetation of the sub-Antarctic islands, Marion and Prince Edward, by Gremmen (1982). The French base on Kerguelen also produced a considerable amount of work on sub-Antarctic ecology (Duchêne, 1989). The ANARE base on Macquarie Island, which has operated continuously since 1948, has provided for extensive studies on flora, fauna, microbiology and limnology (Selkirk *et al*., 1990). Palmer, the US station on the Peninsula built in 1967, is devoted to biology but the major emphasis has been marine rather than terrestrial. In the McMurdo area both the US and New Zealand have carried out much research in terrestrial biology but the bulk of this has been in the specialized environment of the dry valleys. Altogether, in terms of numbers of papers published, biology outstripped other disciplines in the Antarctic after the IGY (Budd, 1986).

10.3 The physiological ecology of plants

The International Biological Programme (IBP) devised in emulation of the IGY touched the Antarctic only marginally except in the Tundra Biome and Bipolar Botanical projects. The latter had field sites at Disko Island, West Greenland (69° 15′ N 53° 30′ W) and at Grytviken, South Georgia (54° 17′ S 36° 31′ W) and ran for some five years with the most intensive field work in 1969–71. It included detailed examination of the different environments, comparison of growth conditions in habitats of different exposure using temperate crop plants as phytometers, estimations of the primary productivity of ecologically important plants and communities, and investigations of adaptations in indigenous plants to their environments. Distinct differences were found between Arctic and Antarctic in that the short, but favourable, Greenland summers restricted development and reproduction whereas the long, cool and wet, South Georgian growing seasons limited vegetative growth. Thus, Arctic conditions have selected species capable of short periods of fast growth and opportunistic reproduction or rapid reproductive development. On the other hand the South Georgian climate has favoured species with slow but consistent rates of growth and reproduction and their higher incidence of sexual reproduction has allowed considerable genotypic differentiation, although some species with wide ecological tolerances show instead a high degree of plasticity. That South Georgia has an impoverished flora seems to be due to geographical situation rather than environmental constraints (Callaghan *et al*., 1976). In an IBP-sponsored discussion of tundra ecosystems, accounts of Signy Island, South Orkneys, and Macquarie Island were included as well as South Georgia (Rosswall & Heal, 1975). Signy, although more truly Antarctic and with a more limited flora and fauna than South Georgia, shows the same ecological characteristics. Even now it has been only slightly modified by human activity. Nearby specially protected areas (SPAs) were designated, including Moe Island, Lynch Island, Powell Island and the north coast of Coronation Island. A legacy of the Biopolar Botanical Project was that studies of mosses and lichens in the Arctic and Antarctic remained together. A recent synthesis of knowledge of the distribution, ecology, physiology, reproduction and evolution of these plants in polar regions uses information from both north and south (Longton, 1988). The bryophyte and lichen floras of the Antarctic are much less diverse than those of the Arctic, a difference that is even more obvious in the flowering plants. The vegetation-types are also different and direct comparisons between those of the two polar regions are difficult.

The general trends in ecology towards more sophisticated instrumentation, a more quantitative approach, and greater attention to processes, were quickly followed in Antarctica without, however, abandoning the still necessary work in taxonomy and community structure. Differences in approach arising from the kind of organization have been particularly evident in terrestrial ecology. Whereas the British, relying on young scientists serving down south for periods of two years, initiated long-term systematic surveys, the US, New Zealand and most other countries, carrying out research through more senior scientists making short summer visits, have produced more opportunistic short-term studies as individual interests or chance suggested. The British system is exemplified in the two Signy Island reference sites established in 1969 and which remained the focus of study until 1981 when effort was transferred to a new long-term research programme on fellfield – the open, unstable, community on exposed rock and glacial till subject to frost action and dominated by lichens and short-cushion mosses.[1] The reference sites were selected as typifying a well-drained moss turf community heavily colonized by lichens and a moss carpet on wet ground. It was decided that each must have a minimum area of 500 m^2 to allow for intensive sampling pressure, be as homogeneous in floristic composition as possible, and be as near to the station as feasible without running the risk of disturbance or contamination. Similar reference sites were established on South Georgia. Long-term monitoring of environmental factors were initiated by recording temperatures from arrays of thermistors and direct total radiation measurements (Tilbrook, 1973). Subsequent developments in measuring biological microclimates in polar regions have been summarized by Walton (1982a); it is noteworthy that among the increasingly sophisticated and robust equipment used for measuring temperature, solar radiation, humidity and wind speed the rate of disintegration of cotton flags, 'tatter flags', remained a good method of comparing the suitability of sites for plant growth under sub-Antarctic conditions. The operation of automatic instruments under conditions of icing still presented problems. An account of the operation of the equipment used on the Signy Island reference sites and examples of the records obtained have been given by Walton (1982b). A pioneer development was a synthesis of the results from the Signy Island reference sites, incorporating data on invertebrate and microbial activity mentioned below, by R. C. Davis. An important outcome of the IBP had been the recognition that the simple ecosystems of the Antarctic are more amenable to mathematical modelling than those of temperate, tropical, or even Arctic, ecosystems and Davis (1980) developed predictive equations for decomposer activity and accumulation of dead organic matter in moss-turf and moss-carpet communities. His estimates of consumption, egestion, assimilation and production together with efficiencies of organic matter transfer between trophic levels were used to construct flow diagrams representing the dynamics of the two communities (Davis, 1981). Application of similar process modelling to mineral nutrient cycling has not yet progressed as far. In the IBP most work on cycling under polar conditions had been in the Arctic but studies in the Antarctic, following the IPB guidelines, continued and the position was summarized by Smith (1985). Major investigations have been done on the sub-Antarctic Marion Island by the South Africans and on the Antarctic Signy Island by BAS but little of a comparable nature has been done on the continent except with the endolithic communities of the dry valleys (p. 361). From such studies it is clear that the presence or absence of large above-ground herbivores is of profound importance for the structure and function of terrestrial ecosystems. In the absence of these herbivores food chains are short, virtually ending at the invertebrate herbivore level, and decomposers have the dominant role in the biological cycling of nutrients. The seasonal activity of critical groups of micro-organisms has therefore been the subject of study both in the field and in the laboratory. The considerable input of nitrogen and phosphorus from sea-birds and seals is largely responsible for the sometimes high productivity of maritime communities.

10.4 Invertebrate ecology and physiology

The soils of the maritime Antarctic harbour large numbers of a relatively few species of invertebrates – protozoans, rotifers, nematodes, tardigrades, springtails and mites – the biomass density of this community decreasing polewards. After several decades of neglect, studies of these invertebrates revived at the time of the IGY, interest then being directed to questions of distribution, dispersal and general ecology, particularly by workers from the Bishop Museum in Hawaii (Gressitt, 1965). It was soon realized that these communities, being comparatively simple in terms of species numbers and therefore interactions, are particularly suitable for study of basic ecological processes. There is little evidence of grazing on mosses although some mites feed on crustose lichens and the macro-alga *Prasiola crispa*. Mainly the animals subsist on micro-algae, other micro-flora and detritus. Functional analysis of arthropod communities in the Signy reference sites in relation to flux of energy and predation played an integral part in the ecosystem analysis by Davis, already mentioned, and has been considered in particular by Block (1985a). Invertebrates other than protozoa accounted for less than one per cent of the total heterotroph respiration. Predators, of which there were few, took little, less than five per cent, of the primary consumer production. However, on the sub-Antarctic islands invertebrates are an important food source for certain overwintering birds. The population dynamics of these various groups have been studied and investigations of adaptations to the physiologically testing environment, combined with information of life cycles, give insight into overall survival strategies (Block, 1984a).

Much attention has been directed since 1977 to cold-hardiness, a crucial feature of the physiology of these animals. Earlier work in this field was mainly on arthropods from the Arctic, sub-Arctic and Canada. Of the two possible strategies, avoiding damaging ice crystallization in body fluids by supercooling, rather than that of being freezing tolerant, and surviving extracellular ice formation, has been found to be adopted by most Antarctic terrestrial invertebrates (Block, 1984b). An example of freezing tolerance with accumulation of a complex of cryoprotectants was found, however, in the larva of the Antarctic midge, *Belgica antarctica*, in studies at Palmer Station (Baust & Edwards, 1979). The adults, which do not feed and are short-lived, on the other hand, show freezing avoidance. Such work, carried out mainly on peri-Antarctic islands, was extended to invertebrates from further south – Ross Island and South Victoria Land. Again, freezing avoidance by supercooling was found, going as low as −24 to −30 °C (Block, 1985b) or even to −50 °C (Pryor, 1962). It was of interest that a chironomid midge and enchytraeid worm, both accidentally introduced on to Signy Island on plant material, established themselves and survived for 17 years. Both are capable of supercooling and seem to have been pre-adapted for survival under much harsher conditions than they experienced in their original habitat, by extension of existing physiological mechanisms (Block *et al.*, 1984). This suggests that the main obstacle to colonization of suitable Antarctic land areas by soil-dwelling invertebrates is geographical. There do not appear to be any truly unique adaptations to cold in Antarctic animals (Webber, 1986). After a phase concerned mainly with determination of lethal temperatures, attention was directed to mechanisms. Studies of the patterns of cold-hardiness arising from the interaction of biochemical, physiological and behavioural features have merged with those on organisms in Arctic and alpine environments (Block, 1990). The special contribution of Antarctic investigations is that they allow testing of hypotheses developed with organisms in less extreme environments or with a basis of purely physical considerations.

10.5 Microbiology

The importance of decomposers as against herbivores in the cycling of material in Antarctic soils focused interest on the microbiology. Although questions of identification, distribution and origins of the microflora remained on the agenda, the nature and extent of microbial activities came to the fore. During the first half of the twentieth century microbiology had been generally restricted in outlook, being mainly a laboratory subject, concentrating on medical bacteriology and having contact with the natural environment chiefly through agricultural soils. After the Second World War the subject broadened, the activities of micro-organisms other than bacteria were included to a greater extent in the scope of microbiology and under the leadership of people such as T. D. Brock in the US, interest developed in the microbiology of extreme environments. The first international symposium in the field of microbial ecology was held in New Zealand in 1977. In Antarctica the revival began with surveys in a classical mould by Argentinian microbiologists. Corte & Daglio (1964) exposed nutrient plates to the air at three sites on the Peninsula and obtained colonies of various filamentous fungi, among which *Penicillium* spp. were most abundant, and concluded that the species and genera found were cosmopolitan and unlikely to develop mycelia under Antarctic conditions. In the same region Margni & Castrelos (1964) found the predominant bacteria in air, soil and snow to be gram-negative bacilli of the genera *Pseudomonas*, *Achromobacter* and *Alcaligenes*. Boyd *et al.* (1966) concluded that substrate is more important than climate in determining the nature of the micro-flora. BAS began a new trend with a survey of soil microbiology on Signy Island with quantitative enumerations of bacteria and fungi at different seasons (Baker, 1970; Bailey & Wynn-Williams, 1982). Direct bacterial counts correlated with fungal counts and soil factors such as moisture, nitrogen and phosphorus contents but viable bacteria did not, except with total phosphorus. All fungal counts correlated to some extent with soil variables. The microbial floras of Signy were found to be distinctive but comparable with other maritime Antarctic areas and Arctic sites studied during the IBP. The fungal isolates in general varied from those obtained from other Antarctic sites but some had been reported from the Arctic and thus appeared to be bipolar in distribution. Others, about half of the total, were non-sporing and therefore could not be identified. Various algae – green, yellow-green and diatoms, together with the cyanobacterium *Nostoc* – make up an important part of the microbial community in moss turfs and carpets (Broady, 1979). The study of processes had begun with *in situ* measurements of rates of biological nitrogen fixation using the isotope ^{15}N as a tracer. Fixation by *Nostoc*, either free-living or in symbiosis in lichens, seemed to make an appreciable contribution to the biological productivity of certain sites (Fogg & Stewart, 1968; Horne, 1972). The oxygen uptake of the community is mainly due to the mosses but the microbial population, particularly yeasts, have a significant oxygen demand in spring when freeze-thaw cycles bring about the liberation of soluble carbohydrates from both living and dead plant material (Wynn-Williams, 1985). Peat is found on peri-Antarctic islands, often to depths of over 2 m, which radiocarbon dating has shown to have been the accumulation of about 2000 years. Cellulose decomposition, measured by loss of weight in litter bags and loss in strength of cotton strips, does occur in these soils. On South Georgia it is brought about by fungi, in contrast to the usual situation in the Arctic, where bacteria are the active agents (Walton, 1985). Under the anaerobic conditions of the wet moss-carpet on Signy, microbial metabolism is sufficient to result in appreciable methane production even at the prevailing low temperature (Yarrington & Wynn-Williams, 1985).

A microbial community which sometimes may be extensive and of importance in determining changes in extent of ice-cover, since it decreases the albedo, is that of the snow algae. Red, yellow or green colourations occur where snow undergoes freeze-thaw cycles and has long attracted attention in the northern hemisphere. A cartoon[3] by George Cruikshank, depicting John Ross's return from his North-west Passage expedition of 1818, figures two sailors carrying a barrel labelled 'Red snow 4 BM' [i.e. for British Museum]. Bauer (1820) made observations on this material but misconceived the situation by supposing that the red cells which he saw were fungal. Many taxonomic studies have been made, particularly by the Hungarian, Erzsébet Kol, who published *inter alia* on the snow algae from

Signy Island (Kol, 1972). No investigations of the growth and physiology of these organisms seem to have been made until Fogg (1967) measured rates of photosynthesis in snow on Signy Island and showed that the sudden appearance of patches of coloured snow was caused by concentration of algal cells during ablation rather than by rapid growth. This was followed by some more detailed physiological work on North American snow-fields (Thomas, 1972). Long-distance dispersal of snow algae must be by wind and Benninghoff & Benninghoff (1985) showed that positively charged particles such as algal cells tend to be collected by negatively charged surfaces such as melting snow banks. Along with algae, 'dirt', including rock dust accumulates, Parker & Zeller (1979) reported relatively high concentrations on nitrate and ammonium in snow samples from various sites across the continent, so that nutrient supply can be sufficient. Apart from the community of snow algae and associated heterotrophs, ice offers a variety of habitats for living organisms (see Vincent, 1988).

Although other habitats in continental Antarctica have not been entirely neglected, the main interest here has centred on the dry valleys and oases which show an entirely different kind of microbial ecology. The physiographic features and soils of these have been described in chapter 8. In 1966 to 1970, not long after the interest of geologists in these cold deserts had revived, ecological studies began in the Antarctic Peninsula, Southern Victoria Land and in the Trans-Antarctic Mountains to establish the habitats, number, distribution, kinds and activities of micro-organisms in relation to physical, chemical and topographic factors (Cameron 1971). A subsidiary motive for this research was to gain experience for a search for evidence of life on Mars – although the Antarctic environment differs in many respects from that on Mars it provided a useful approximation for the testing of ideas and equipment, as first recognized by N. Horowitz. R. E. Cameron of the Jet Propulsion Laboratory, California Institute of Technology, in the US, was a principal investigator at this stage and his previous experience with the microbiology of hot deserts provided useful pointers for approaches and techniques. One approach to relating microbial abundance to the physico-chemical conditions was to attempt to correlate it to the natural thermoluminescence of the soils (Cameron, 1971) but this does not seem to have been of much use. The presence of a variety of bacteria – mostly aerobic or micro-aerophilic heterotrophs – fungi, algae and protozoa, was established, with the algae contributing most of the biomass. Heterotrophic bacteria, in low numbers, were the only organisms to persist under the harshest conditions, with actinomycetes, algae, fungi and protozoa, lichens and finally mosses coming in successively as conditions became more favourable. Cameron *et al.* (1970) summarized favourable conditions as being a north–south valley orientation with slope, drainage and exposure giving maximum duration, frequency and quantity of insolation, available moisture and protection from wind. Horowitz *et al.* (1972) concluded that dry valley soils are essentially abiotic and that micro-organisms isolated from them must be chance contaminants. The microbial population of a given soil according to this view depends on influx brought by wind from more favourable environments, balanced by mortality on the ground. This was disputed by W. V. Vishniac (Vishniac & Mainzer, 1972) – a microbiologist who met his death in the Wright Valley in 1973 – and H. S. Vishniac & Hempfling (1979) later reported a previously undescribed psychrophilic yeast from dry valley soils. The idea that certain genetically distinct micro-organisms may be indigenous in Antarctica has been reinforced by the discovery, albeit in an ice-core in central Antarctica rather than in a dry valley situation, of a new actinomycete *Nocardiopsis antarcticus* (Abyzov *et al.*, 1983). In view of the apparent presence, reported by Cameron & Morelli (1974), of viable organisms in drill cores from dry valleys in positions where they must have been for thousands of years, this controversy turns on rather a fine point. In any case, as a study of the impact of the Dry Valley Drilling Project on the microbiology of the area showed, a large proportion of the species found seem to be introduced (Cameron *et al.*, 1977). However, one argument for there being no truly indigenous micro-flora in extreme dry valley soils, namely the apparent absence of primary producers, was destroyed by the discovery of endolithic autotrophs – eukaryotic algae and cyanobacteria – within rocks in the dry valleys of southern Victoria Land (Friedmann & Ocampo, 1976). Friedmann had previously demonstrated the presence of an appreciable micro-flora of algae and associated bacteria in the microscopic air–space system in certain porous rocks from hot deserts and looked for something similar in cold deserts. In the dry valleys of southern Victoria Land he found an abundant microbial flora colonizing minute cracks in weathering rocks or, more conspicuously, growing in porous rocks as a thin green or dark brown layer underneath and parallel to the surface. Water is provided by the melting of the snow that occasionally falls (Friedman, 1978). Both in morphology and survival strategies, notably the ability to switch metabolic activities off and on in response to changes in the environment, the endolithic micro-organisms of hot and cold deserts show common features. The biomasses in the two extreme habitats are also comparable. However, whereas the microbial flora from the hot desert is entirely prokaryotic, in keeping with the high temperatures experienced, eukaryotic micro-organisms predominate in the cold desert (Friedmann, 1980). After the recognition of the remarkable nature of these lithic communities Friedmann organized a programme of investigation involving a wide variety of specialists. This included isolation and study in culture of the various micro-organisms, study of their ability to survive environmental stresses, modelling of production from physiological data and microclimate records, and comparison with the lithic communities of hot deserts. This is perhaps the best and most innovative long-term integrated study on the terrestrial biology of continental Antarctic.

10.6 Limnology

The modern phase of Antarctic limnology began after the IGY, with surveys of which the first was that of Armitage & House (1962) from the University of Kansas, USA, in the McMurdo Sound area. This included, *inter alia*, measurements of pH and chlorinity, some measurements of primary production, and sampling of fauna from melt-water ponds. Some lakes in the dry valleys of southern Victoria Land were also investigated; an expedition from the Victoria University of Wellington, New Zealand, had visited the same area in 1959–60 and made water analyses and collected plankton samples. The University of Kansas party the following year found Lake Vida to be solid ice but Lakes Bonney and Vanda, although permanently covered with about 4 m of ice, had stratified waters with highly saline non-mixing bottom waters, that is, they were meromictic. The high temperature of the water below 60 m in Lake Vanda has been discussed in chapter 8. The mass of information now available on the plankton and microbiology of these lakes has been summarized by Heywood (1984) and Vincent (1987). Whereas the dry valley lakes usually have relatively high concentrations of inorganic nitrogen they are poor in phosphate and Lake Vanda is one of the most oligotrophic freshwaters known. The primary producers are mainly flagellates and, in the benthic algal felts, cyanobacteria and pennate diatoms. These benthic felts, which have been looked at in detail by divers from the Virginia Polytechnic Institute, USA, vary between smooth or flocculose prostrate mats to 'pinnacle' mats with columnar structures several centimetres high – these must be the 'water plants' glimpsed through the ice of Lake Bonney by Scott's western party in 1911 (Huxley, 1913, vol. II, p. 193). Accumulation of gas as a result of photosynthesis causes pieces of mat to float up and become embedded in the ice. As new ice accumulates below and old ice above is lost by ablation these pieces work their way up and after five to 10 years reach the surface and are blown away, a phenomenon first noted by members of the Shackleton expedition (Murray, 1910). This may be an important means of dispersal of organisms and also of appreciable loss of nutrients from these lakes (Parker *et al.*, 1982). Apart from the algal felts there is also an aquatic moss, resembling *Bryum algens*, in Lake Vanda but, curiously, not in other dry valley lakes in this area (Vincent, 1987). The largest animals in these lakes are rotifers and tardigrades; aquatic insects and crustacea being totally absent (Vincent, 1987). The nearly self-contained nature of the ecosystem, the unusual pattern of mineral cycling arising from the low diversity and biomass of consumer animals, as well as the different biochemical regimes maintained in discrete layers by the stabilization of the water column, combine to make these lakes as unique biologically as they are in other respects. This was recognized by the establishment of a permanent station on the shore of Lake Vanda in 1967 by New Zealand. There is now a substantial body of information about these lakes (Vincent, 1987, 1988) but there is a gap as far as seasonality is concerned, the logistics restricting access to a brief November–January field season which may be too late for the algal production maximum and too early for the maximum in decomposition activity. The biology of flowing waters in continental Antarctica have, not surprisingly since they are infrequent and intermittent, received little attention but the river Onyx flowing into Lake Vanda has recently been investigated (Vincent & Howard-Williams, 1986). Again, mats of cyanobacteria are the dominant feature. Their tolerance to periodic freezing and extreme dehydration together with the absence of animal grazers seem to be important for their success but they show no particular adaption to the highly intermittent flow and sometimes intensive scouring by glacial sediment.

The Polish worker Opaliński (1972a) studied the flora and fauna of 10 lakes in the Thala Hills oasis at 67° 40′ S 45° 50′ E. Some of the several hundred lakes scattered among the Vestfold Hills were later investigated by Australian limnologists. These lakes, of marine origin, support similar biota of planktonic and benthic coccoid and flagellated green algae, pennate and centric diatoms, cyanobacteria and bacteria to those of the southern Victoria Land lakes. Deep Lake, with a salinity of approximately 280 g l^{-1} has never been observed to freeze and is apparently devoid of invertebrate predators, so that it has biota even more impoverished than that of the Dead Sea (Kerry *et al.*, 1977).

Another remarkable type of lake is that formed where an ice shelf or glacier tongue dams the mouth of a valley and melt-water accumulating at the seaward end becomes connected with the sea and shows tidal movement. Such lakes were first discovered by Simonov in 1960–61 in the

Schirmachervatna (70° 45′ S 11° 20′ to 11° 55′ E). Following a prediction by C. W. M. Swithinbank, based on inspection of aerial photographs, another was found in 1973–74 in the Ablation Point area of Alexander Island. Here the freshwater lies over seawater and the surface, which is permanently frozen, moves up and down in response to tidal movement. An impoverished freshwater flora and fauna exists in the freshwater but the saline layer contains marine animals (Heywood, 1977).

Inside the Antarctic circle freshwater phytoplankton is liable to be subject to continuous high intensity illumination. Armitage & House (1962) noticed indications of photoinhibition of photosynthesis and this was followed up by Goldman *et al.* (1963) who were the first to use the radiocarbon method for determining primary productivity in Antarctic freshwaters. In two small lakes near Cape Evans, Ross Island, they found the diel variation in photosynthesis to be completely out of phase with that of light intensity, the effect being more marked in the surface waters than at depth. Inhibition of photosynthesis by high light intensities was occurring but experiments showed that recovery could occur if the plankton was transferred to low intensity.

Outside the Antarctic circle a systematic and long-term investigation of water bodies on Signy Island has been carried out by BAS. These are typical of the vast majority of lakes in Antarctica rather than remarkable for their peculiarities. Heywood (1967, 1968) carried out the basic survey of catchment areas, drainage systems, morphometry, physics and chemistry of a series of small lakes on Signy Island, and Light (1977) was the first to make an extended series of observations, over two and a half years, on the periodicity and production of Antarctic freshwater phytoplankton. The importance of year-round measurements was evident in the finding that maximum development of the phytoplankton occurred in early spring in low light intensities under the ice. An unexpected finding by SCUBA diving was that many of the Signy lakes surveyed had deep-water growths of mosses, *Calliergon sarmentosum* and *Drepanocladus* sp., although phytoplankton was minimal. Together with their considerable load of epiphytes these mosses, in dense stands and up to 40 cm in height, were found to have a considerable biomass but as their photosynthesis is light-limited this probably represents 20 to 50 years growth (Light & Heywood, 1973). Full seasonal studies on the bacteria of an oligotrophic, a mesotrophic and a eutrophic lake, carried out by J. C. Ellis-Evans, showed increasing activity and proportions of anaerobic forms through this series and a seasonal cycle paralleling that of algal activity. The fauna is extremely limited as far as the larger forms go, with only three species of planktonic crustacea and no fish.

The data from these and other studies on two of these lakes have been brought together to give annual balance sheets for the carbon cycles (Heywood, 1984). Lake Amos is eutrophic because it is in a favourite hauling out area for seals and the water column becomes anaerobic in early winter with a hydrogen sulphide concentration going up to 5 mg l^{-1}. It has proved impossible to set up experiments in this lake because of the curiosity of the fur seals. The transfer of nutrients from the sea to freshwater in the maritime Antarctic by the excretion and moulting of animals is considerable and, as anyone who has been near a penguin colony can testify, it occurs in the gaseous as well as in the aqueous phase (Wodehouse & Parker, 1981). The Polish worker, Opaliński (1972b), quoted the estimate of Gollerbach and Syroyechkovsky that birds alone brought 100 tonnes per year of organic matter on to Haswell Island (66° 32′ S 93° 00′ E). Roughly, this would provide 13 g nitrogen and 3 g phosphorus per square metre per year, and Opaliński noted that the algae were corresponding 'nitrophile'.

Sub-Antarctic lakes on Macquarie and Kerguelen were reconnoitred by Tyler (1972). More recently information on their limnology has been summarized by Selkirk *et al.* (1990) and Duchêne (1989) respectively. Grobbelaar (1978) studied freshwaters on Marion Islands but the South Georgia lakes remain almost uninvestigated. As might be expected the macrophyte flora and zooplankton fauna are richer than those of truly Antarctic lakes but there are still no indigenous fish. Various salmonid fish were introduced into Kerguelen rivers and lakes from 1954 onwards. Brown trout, *Salmo trutta*, and brook trout, *Salvelinius fontinalis*, have established themselves successfully and research directed to development of a salmon fishery has been instituted (Duchêne, 1989).

In the 1980s Antarctic limnologists felt that the volume of data available permitted some generalizations. Priddle & Heywood (1980) attempted to fit together the wide variety of freshwater ecosystems of Antarctica into an evolutionary pattern. They concluded that the Antarctic freshwater environment is relatively favourable although temperature and light-limited growing seasons slow down the rate of ecosystem functioning. Most of the selection pressure is encountered in colonization from across the Southern Ocean and the ice sheets of Antarctica itself. Thus even the freshwater environments with the most diversity have relatively few specialist organisms and the over-all assemblage is a hotch-potch of opportunistic organisms in an ill-defined trophic structure. This contrasts with the Arctic situation, in which there are no sea barriers to dispersal, large areas are ice-free and have been for longer than anywhere in the Antarctic, and the freshwater biota are richer in species and with a more highly evolved trophic structure (Heywood, 1984; Hobbie, 1984). Priddle *et al.* (1986) brought together information, from both marine and freshwater studies, on the physiological ecology of phytoplankton. The two environments have many features in common, including strong vertical mixing which has important effects on phytoplankton performance. Loss of phytoplankton biomass during the ice-free period by physical removal via outflow was identified as a dominant factor in Antarctic lakes.

10.7 Ornithology

Ecosystems have indefinite boundaries and those of the sea and land intergrade both physically and via organisms which commute between them. The considerable biomass of sea-birds and seals derives from marine sources but the crucial reproductive stages, of all birds and most of the seals, are conducted on land or on fast-ice and the impact of their presence on terrestrial ecosystems can be overwhelming. This provides some justification for including them in this chapter but the effective reason for doing so is that these animals are so much more easily studied when they are on a solid substratum, so that historically the bulk of research on them has been carried out from the same bases as the other work described in this chapter. It must not, however, be forgotten that, as emphasized in chapter 7, their roles in the marine food web are substantial.

An appraisal written by W. J. L. Sladen – who had the harrowing experience when with FIDS of having his two companions burnt to death in the base hut while he was out observing penguins – and H. Friedmann for the Committee on Polar Research (1961) gives a starting point for discussion of post-IGY research on Antarctic birds. They began by making the point that because many of these birds are large, long-lived, mostly unafraid of man, and breed in large numbers in accessible colonies, they are ideal subjects for many kinds of research. Providing interest in the sometimes monotonous Antarctic existence they have been the subject of much spare-time observation which, however, had not always been as objective and well-organized as it might have been. Identification was straightforward for competent amateurs but knowledge of distribution was still fragmentary at this time. Thus it was possible to be fairly certain that new records of the chinstrap penguin indicated a spread of this species (Fig. 10.7) but for the emperor penguin, the rookeries of which tend to be in inaccessible places, new records evidently filled in large gaps in a very sketchy picture of a long established breeding range. A little work had been done on fossil history, embryology, anatomy and physiology, mainly of penguins, but the most interesting aspects of the physiology of these birds

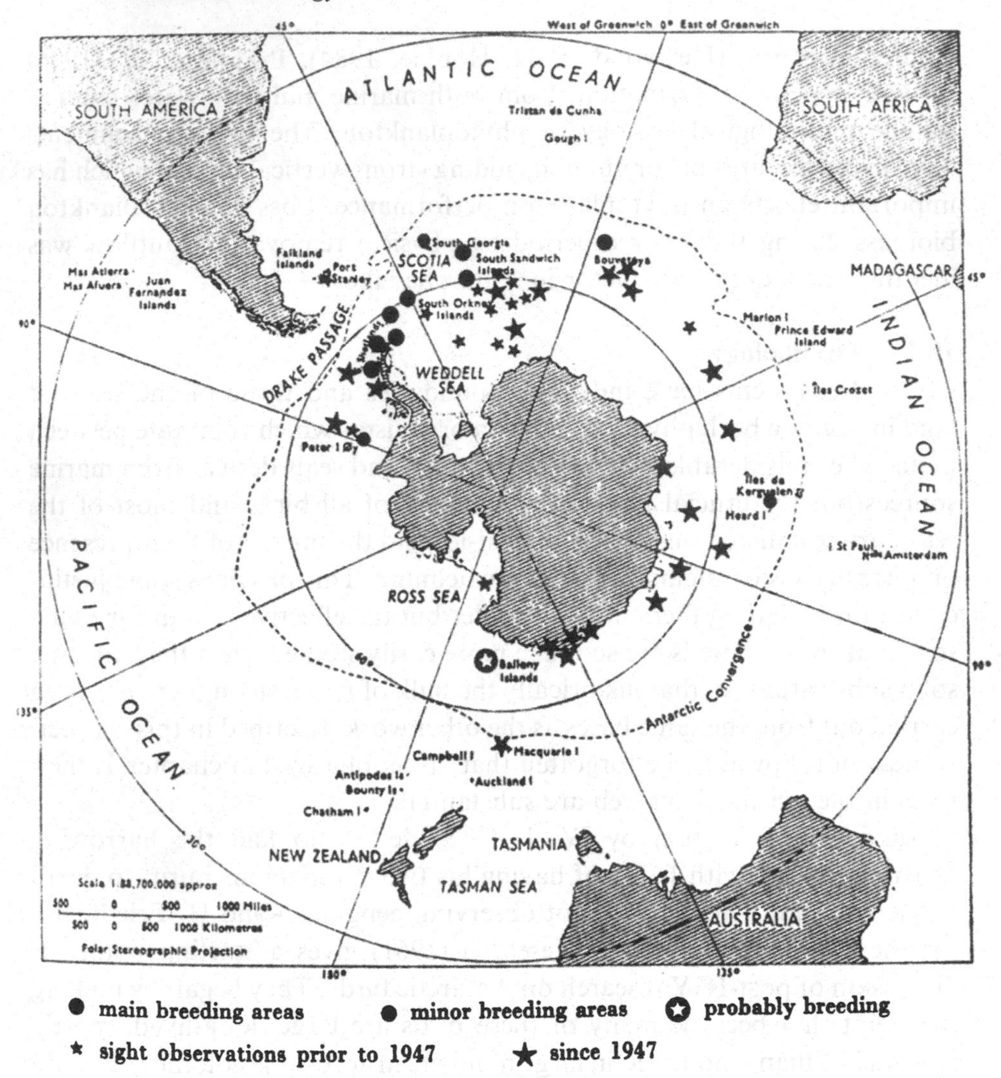

Fig. 10.7 The distribution of chinstrap penguins, *Pygoscelia antarctica*, from Sladen in Carrick *et al.*, 1964, Fig. 1.

are the ability to resist cold and to dive to considerable depths, which had scarcely been touched. Adaptation to cold in Arctic mammals and birds had been investigated by Scholander *et al.* (1950). Eklund (1945) had done some similar work with the south polar skua and French workers with penguins had been assisted by the discovery of a colony of emperors in Adélie Land near a recently established station. Prévost and Bourlière (1957) showed that these birds when in a huddle maintain a lower cloacal temperature and suffer less weight loss than when isolated. The absence of territorialism makes possible social thermoregulation. Studies on ecology and behaviour were much improved by marking and a formal scheme for banding was introduced in 1947. Gain (Charcot, 1911) had started this in the Antarctic using celluloid rings and Eklund (1945) branded skuas but durable metal bands, often in special designs such as flipper bands, carrying numbers specific for the individual and an address to which recapture should be reported were now used (Sladen, 1958). Among the future research objectives envisaged by Sladen and Friedmann (1961) were comparative physiology and detailed behaviour studies. The importance of birds as the main link other than man by which dispersal of invertebrates and micro-organisms may take place between Antarctica and other continents was emphasized and the need for conservation measures pointed out.

During the decade after the IGY most bird studies continued to rely mainly on visual observation with the main interests lying in problems of distribution and behaviour. The work summarized by Stonehouse (1965), the papers in Carrick *et al.* (1964) and in Austin (1968) are mostly of this type. Amateur observers, guided by experienced ornithologists, continued to accumulate useful information on distribution, for example that gathered on BAS supply ships during the periods 1954–64 as summarized by Tickell and Woods (1972) and Thurston (1982). Simple but effective methods based on banding were introduced to study the interaction of two species, McCormick's skua and the Adélie penguin (Young, 1970). A major step forward was taken in 1978 when the SCAR subcommittee on Bird Biology agreed on international co-ordination of Antarctic and sub-Antarctic banding programmes and on the use of standard recording cards for observations at sea.[4] South Africa offered the services of its bird-ringing unit to run a central data bank in connexion with this.[5] Studies of a more physiological nature began; Tickell (1964) contributed a quantitative investigation of the feeding preferences of albatrosses, taking advantage of the habit of the nestlings of regurgitating their stomach contents when alarmed. Feeney and his co-workers at McMurdo Station worked on egg and blood serum proteins of the Adélie penguin and at Cape Hallet Douglas studied the water and salt metabolism of the same species (in Austin, 1968). More quantitative approaches to population dynamics (Carrick & Ingham, 1970) and sophisticated ecological and physiological techniques were being brought into use (Sladen & Leresche, 1970). This included improved methods of capture, marking and identification of sex (a matter of peculiar difficulty with penguins) air photography for population analysis, and biotelemetry. This last technique was first used in the Antarctic by Eklund & Charlton (1959) to obtain measurements of incubation temperatures of Adélie penguins and south polar skuas. The transmitter inserted in the egg was one designed and constructed by the Office of Naval Research for investigating human physiology and its signal was picked up by a loop antenna mounted on a pole above the nest. Penney (in Austin, 1968) used transmitters (see Fig. 10.8) attached to harnesses on the backs of Adélie penguins for orientation experiments, and Sladen *et al.* (1966) used a small transmitter, weighing less than 20 g, implanted into the abdominal cavities of Adélie and emperor penguins, to study the body temperatures. The use of telemetry allied with satellite tracking as a means of following the migrations of wide-ranging polar sea-birds, perhaps even those of penguins during the winter, was envisaged at this stage (Sladen & Leresche, 1970) and its importance was reiterated in 1978 by the SCAR subcommittee.[6] Work was on a small scale and confined to the US programme. The first successful tracking of any bird by satellite telemetry was achieved with wandering albatrosses nesting on the Crozet Islands, which were fitted with 180 g transmitters. This showed that these birds cover between 3600 and 15,000 km at speeds up to 80 km h^{-1} in a single foraging trip and that wind appeared to have a major influence on foraging strategy (Jouventin & Weimerskirch, 1990).

Bird Island, off the northern tip of South Georgia, was justifiably so named by James Cook in 1775 but it was not until 1958 that it was used as a base for ornithological studies when B. Stonehouse and L. Tickell, on a private expedition, built a small hut there. Another hut was put up by Tickell and H. Dollman in 1962 for work under American auspices on albatrosses. The banding programme started by Tickell (1968) on Bird Island albatrosses has provided the best set of population data for long-lived sea-birds anywhere in the world. The data-base which has developed from it is now used to look at such things as differential mortality between cohorts, breeding success with age, and population losses associated with long-line squid fishery. Such monitoring of sea-bird populations is of great value for assessing the state of the Southern Ocean ecosystem.[7] Following the handing over of King Edward Point on South Georgia to the BAS in 1969, the Bird Island hut was used from 1971 onwards by groups investigating fur seals and birds (Fuchs, 1982, p. 255) and soon it became the base for a continuing research programme. After the fracas of 1982 it remained as the only permanently occupied BAS station in the South Georgia group and the focus for long-term work by J. P. Croxall and P. A. Prince in collaboration with American workers (see Parmelee, 1980; Kooyman *et al.*, 1982) on population dynamics coupled with quantitative studies on food and feeding ecology. Artificial nests incorporating auto-

Fig. 10.8 Chinstrap penguin, *Pygoscelis antarctica*, with radio transmitter, 1985. (Courtesy of S. Pickering, the British Antarctic Survey.)

matic weighing equipment (Fig. 10.9), at-sea activity recorders, depth recorders and radioisotope techniques for assessment of energy consumption were some of the devices used to give precision to this work. The several different species using krill as their staple food were shown to be able to co-exist because of differences in feeding techniques and foraging ranges. Other active sea-bird programmes, all showing some tendency towards physiology, more precise relation of breeding cycle to food availability and economy of resources, and more quantitative investigation of populations, have been carried out on a broad range of species. The work on the significant effects of sea-birds on terrestrial ecosystems by the South Africans on Marion Island has already been referred to. The French have made ecological studies of fulmars, petrels and albatrosses on Iles Crozet, and on bioenergetics, reproduction, adaption and behaviour of various species on Kerguelen, as well as massive contributions to all aspects

Fig. 10.9 Albatross chick on an artificial nest incorporating automatic weighing equipment, 1981. (P. A. Prince, 1981, Courtesy of the British Antarctic Survey.)

of knowledge of emperor penguins from Terre Adélie (Prévost, 1961; Duchêne 1989; see also Croxall, 1984). Australians on Macquarie Island have investigated several species of penguin, albatross, petrels and other birds (Selkirk *et al.*, 1990). New Zealanders and Americans in the Ross Sea area have produced important work on penguins, especially on their diving, and Americans in the Peninsula sector on feeding and distributions of skuas and other birds (see Croxall, 1984). The importance of this work in filling a substantial gap in our knowledge of the Antarctic marine ecosystem has been mentioned already (p. 239). Although studies on the feeding ecology and ecological segregation of sea-birds began in north temperate and tropical regions it may now be said that the Antarctic work, for example that of Croxall & Prince (1980), leads the field (Brown, 1987). The cycle linking the diet, behaviour, food distribution and breeding success is now complete for some Antarctic sea-birds.

10.8 Seal studies

For logistic reasons research on Antarctic seals have usually gone on side by side with those on birds, with which they share the useful feature of approachability, and again, the greater part of our knowledge of them is derived from land-based studies. Distribution of pelagic seals must, of course, be studied at sea and has already been touched on (chapter 7). Gilmore (1961) and Carrick (1964) summarized the information available about seals at the beginning of the post-IGY era and listed areas for future research. Even at this stage the physical characters of some species had not been adequately described. Scarcely anything was known of the physiology of swimming and the associated problems of respiration, maintenance of body temperature, senses and orientation. The value of a land base for studies was shown in the particular paucity of information about the feeding and reproductive behaviour, growth and longevity, of the three almost entirely pelagic species – the crabeater, the Ross and the leopard seals. A major item of interest around the time of the IGY was the recovery of fur seal populations following their near extermination in the nineteenth century. Between 1910 and 1915 regular inspections of the shores of South Georgia had failed to reveal any of these seals but after this occasional groups were reported from Bird Island and in 1956 a systematic search found well established breeding colonies on Bird Island and on the adjacent Willis Islands. Since then populations have increased dramatically, and repopulation of the South Orkneys and South Shetlands from South Georgia has taken place (Bonner, 1964). The population dynamics, growth and social organization of fur seals on Bird Island have been continuing subjects for study (Bonner, 1968). The impact of fur seals on terrestrial ecology has already been indicated (p. 355). Studies on the elephant seal have been carried out on South Georgia (McCann, 1980) and on Macquarie Island, where the population had been stable for some time after sealing ceased in 1919 and known-age branded seals were kept under observation (Carrick & Ingham, 1960). There were differences in age structure and social organization of the breeding herds from the different regions, some of which could be attributed to the effects of sealing on the South Georgian populations. Laws in 1960 had constructed life tables for the elephant seal, probably the first for a pinniped seal, and McCann later produced revised tables, giving values for survival, mortality and mortality rate (see Laws, 1984). Laws's prediction of maximum sustainable yield of male elephant seals on South Georgia was almost exactly realized in practice. Variations in activity patterns of elephant seals during their prolonged fast in the breeding season have been related by McCann (1983) to the partitioning of limited resources of energy available from their blubber reserves. Data, of varying degrees of reliability on population dynamics, abundance, biomass, food consumption and ecological efficiency have been obtained for all six species of Antarctic seal (Laws,

1984). Investigations of the sounds emitted by submerged Weddell seals showed that they have some directionality (Schevill & Watkins in Burt, 1971). A biochemical approach, including the then new electrophoretic technique for characterizing proteins and enzymes first used in the Antarctic by R. E. Feeney and his collaborators, was employed by Seal *et al.* (1971a,b in Burt, 1971) to investigate phylogenetic relationships amongst the Antarctic seal species. The Weddell seal seemed to have at least two isolated breeding populations whereas the crabeater population examined appeared homogeneous. Biochemical, as well as histological, studies were also carried out on a supposedly 1400-year-old mummified Weddell seal from southern Victoria Land. Many components, particularly neutral lipids, were well preserved but of the several enzymes assayed only two – triose phosphate isomerase and catalase – survived with detectable activity (Hsu & Coe in Burt, 1971). However, because of irregularity in upwelling of ancient seawater in this area, radiocarbon dating, on which the estimate of the age of the mummy was based, is suspect and it may be that it was not more than a few decades old (Dort, 1981). The elegant studies on the Weddell seal carried out by G. L. Kooyman (1981) have made the diving behaviour of this seal better known than that of any other marine mammal. The advantages for such work of this seal are that it has no fear of man so that time-depth recorders can be attached without difficulty and that it uses fixed breathing and haul-out holes in thick stable ice (Fig. 10.10). Dives normally last for less than 25 minutes and usually go down to between 200 and 400 m, but sometimes to 600 m or more, in search of fish. Similar studies have been done on the fur seal (Fig. 10.11). The information from these and other investigations give an idea of the ecological separation, in terms of habitat and food preferences, feeding and diving behaviour, between the various seal species and between them and the sea-birds (Laws, 1984). Besides this there have been numerous studies of a more classical type on anatomy, growth, development, reproductive cycles and, to a lesser extent, bioenergetics of Antarctic seals contributed by most of the nations engaged in Antarctic research (Burt, 1971; Laws, 1984).

10.9 Conclusions

Investigations of the terrestrial biology of the Antarctic have been mostly of a broadly ecological nature. The natural history – the term is not used in any derogatory sense – characteristic of the first half of the twentieth century provided a sound basis from which in the second half modern quantitative, physiological and biochemical approaches could be applied to the understanding of the functioning, adaptations and be-

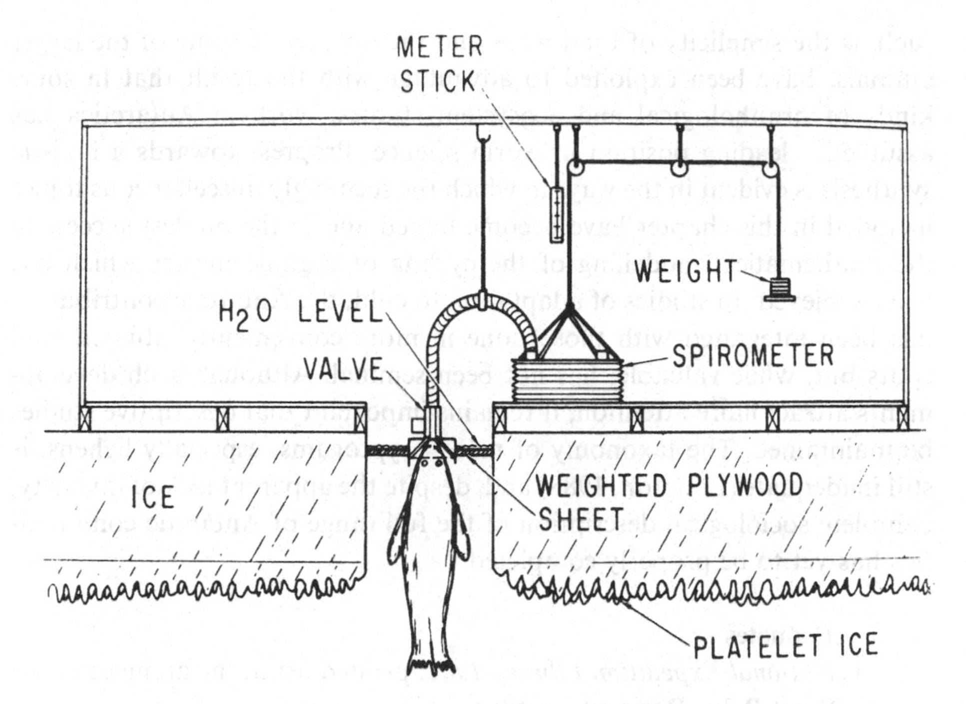

Fig. 10.10 Determination of the metabolic rate of a seal under nearly natural conditions. The seal's breathing hole in the ice is completely covered with a weighted sheet of plywood with a small hole at its centre. This hole is covered with a lucite (perspex) dome, upon which is mounted a valve with inspiratory and expiratory ports. The seal's expirations are collected in the spirometer and the volume and composition of the gas are used for calculation of the metabolic rate. The whole arrangement is housed in a hut. From Kooyman, 1981, Fig. 7.2. (Courtesy of Cambridge University Press.)

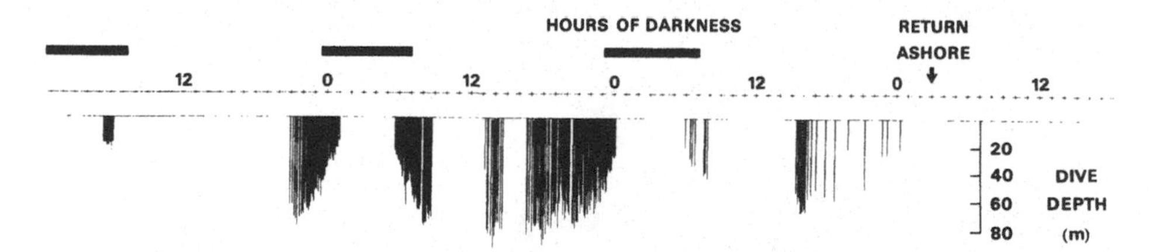

Fig. 10.11 Pattern of seal activity at sea as monitored by an instrument package, measuring the time, duration and depth of each dive, attached to the seal. (Courtesy of Dr G. L. Kooyman.)

haviour of the micro-organisms, plants and animals of habitats more extreme than any elsewhere on earth. It has become evident that communities really are different from those in the Arctic and that it is not always easy to make direct comparisons. Some special features of the Antarctic, such as the simplicity of food webs and the docility of some of the larger animals, have been exploited to advantage with the result that in some kinds of ornithological and ecosystem studies, work in Antarctica has assumed a leading position in world science. Progress towards a holistic synthesis is evident in the ways in which the seemingly miscellaneous topics included in this chapter have become linked and in the modest success in the mathematical modelling of the cycling of organic matter which has been achieved. In studies of adaptation to cold, the Antarctic contribution has been integrated with those done in more conveniently situated cold spots but, while valuable, has not been seminal. Although such developments attract more attention, it remains important that descriptive studies be maintained. The taxonomy of many cryptogams, especially lichens, is still inadequate (Longton, 1988) and, despite the apparent lack of diversity, complete sociological description of the full range of Antarctic communities has yet to be properly completed.

Endnotes

1 *National Expedition Library 1901*: printed list in the archives of the Scott Polar Research Institute.
2 British Antarctic Survey Annual Report 1980–81, pp. 48–9; 1981–82, pp. 60–1.
3 In the Scott Polar Research Institute, Cambridge.
4 SCAR Bulletin, No. 60 (September 1978), pp. 304–7.
5 SCAR Bulletin, No. 68 (May 1981), p. 77.
6 SCAR Bulletin, No. 60 (September 1978), p. 43.
7 British Antarctic Survey Annual Report 1988–89, p. 71.

11

Man and the Antarctic environment

The reactions of the human body and mind to the peculiarities of the Antarctic environment and the impact of man's activities on the natural phenomena of the region are matters of scientific interest. Medical care was, of course, necessary from the start but was largely empirical and systematic research was slow to develop. Although, as was seen in chapter 2, there was almost from the beginning some realization that man might have profound effects on the fauna of the Antarctic, here too it was a long time before such impact was assessed and studied scientifically.

11.1 Heroic age medicine

The expeditions of around the beginning of the twentieth century did not include animal or human physiology in their programmes. Physiology was developing rapidly at this time and as long ago as 1821 a Royal Society committee[1] had considered that experiments on effects of freezing on muscle activity, survival of animals in the cold, respiration and pulse rates in cold air, and other physiological investigations, were of interest and should be carried out in the Arctic. Nevertheless, there was no mention of the need for physiological or medical research in the Antarctic at the 1898 Royal Society meeting (Murray, 1898) nor in the 1901 *Antarctic Manual* (Murray, 1901).

These expeditions took medical men, of course, and symptoms that appeared to be related to Antarctic conditions were noted, for example, by Cook (1900; Fig. 11.1) and medical problems were summarized by Ekelöf (1920). Systematic recording of such things as men's weights, measurements, pulse rates and blood counts was done on some expeditions (see, for example, Drygalski, 1904, p. 141; Wilson, 1966) but there was little that could be dignified by the name of medical research. Scurvy appeared on the *Discovery* expedition but neither of the doctors, Wilson or Koettlitz, was much interested in clinical research and, anyway, the circumstances were

Fig. 11.1 Three members of the *Belgica* expedition, F. A. Cook, R. Amundsen and E. Racovitza, before and after overwintering in 1898–99. From Cook, 1900, facing p. 360.

scarcely propitious for any useful sort of experimentation. Koettlitz firmly believed that scurvy was caused by ptomaine poisoning and spent much time checking tinned provisions. Wilson was presumably not writing as a scientist when he attributed red swollen gums to nothing more than 'clay pipes, strong tobacco, coarse feeding, neglect of the toothbrush and the constant use of foul language' (Wilson, 1966, p. 278). The disease cleared up when allowances of fresh meat and bottled fruits were increased. In 1912 Holst and Fröhlich demonstrated that scurvy in guinea-pigs was definitely a deficiency disease but it was not until 1932 that King and Waugh recognized vitamin C – ascorbic acid – as the specific curative agent. An empirical approach was also taken to avoid snow blindness, Charcot being the first to use yellow glass. McLean, the doctor on Mawson's 1911–14 expedition, made a systematic investigation of human adaptation to the Antarctic environment. He found that the haemoglobin content in the blood of the members of the expedition increased dramatically on arrival in Adélie Land, blood pressure became slightly more marked, weight increased but resistance to infection decreased. After a few months *Staphylococcus aureus*, a common germ of civilization, could not be isolated from throat, nose or skin of six individuals subjected to regular bacteriological sampling. On the other hand, the number of colonies of bacteria, yeasts and moulds, appearing on plates exposed within the living-hut continuously increased (McLean, 1915). There were no psychological investigations in the expeditions we have discussed so far. A paper written in 1914 by Priestley, a geologist, entitled *The Psychology of Exploration*[2] is no more than a common sense and general discussion of the effects of over-strain and starvation during sledging, sources of irritation, and dreams. A section headed 'Psychology' in a report by McLean (1919) is concerned with his own religious feelings about Antarctica.

11.2 Medical research before and during the IGY

Planned medical studies with wintering parties began with the US Antarctic Service Expedition 1939–41 when R. G. Frazier and E. E. Lockhard (1945) carried out exploratory studies of blood changes, blood pressure, pulse rate, metabolism, upper respiratory tract, cold tolerance, onset of skin freezing and occurrence of cold injury. It was also on this expedition that a biologist, P. A. Siple, devised an index for expressing the impact of the Antarctic environment, the wind-chill factor, which related loss of heat and consequent effects on the body to wind velocity and temperature (Siple & Passel, 1945). This has been much used, particularly by meteorologists (e.g. Dalrymple, 1966) but it fails to take radiation into account and later work has not always confirmed its physiological value; Wilson (1973) carried out an experimental investigation of the correlation of freezing of finger skin with wind-chill index and found that freezing did not occur at any well-defined value of the index, varying degrees of supercooling, cold-induced vasodilation and skin moisture complicating the situation. A. R. C. Butson (1949) studied basal metabolism and fasting blood sugar in 11 men with FIDS on the Peninsula. Observations on clinical problems, cold injury, physiological and psychological adjustments and bioclimatology were made by J. Sapin-Jaloustre in Adélie Land in 1949–52 and O. Wilson examined changes in blood and basal metabolism during the Norwegian–British–Swedish Expedition to Queen Maud Land, 1949–52, finding a decline in basal metabolism which he related to cold adaption and the sedentary life at base (see Rodahl, 1961). A general comment on the work of this period is that it was not well structured, often being devised to fill in the time which doctors had on their hands rather than to tackle basic problems. Sample sizes were small and psychological work took little account of the fact that assessments were made by someone who was a member of the group, or of the need for adequate controls.

Thus far medical men worked much on their own without the backing of an established group concerned with medical or physiological effects of environmental stress. However, around this time such a nucleus had its beginnings when the Medical Research Council persuaded the organizers of the British North Greenland Expedition, 1952–54, to include both a medical officer and a physiologist among the 25 men. After their return these two were centred in the Division of Human Physiology of the National Institute of Medical Research where they were in a good position to advise and to collect information from other polar medical men (Lewis & Masterton, 1963).

During the IGY there was gain in general experience of the Antarctic situation by both medical men and their patients. At the Soviet station, Myrny, consultations fell from 609 per 100 men in 1957–58 to 128 per 100 men in 1966–67 and sickness went down to a fifth of its initial level (Tikhomirov, 1973). Apart from routine medical work the US, Chile, the UK and Japan carried out medical and physiological investigations although it was no part of the official IGY programme (Rodhal, 1961; Wilson, 1965). These were concerned with much the same things as 10 years before: cold adaptation, diurnal rhythms, work loads and food consumption, metabolic and hormonal changes, cardiovascular functions, and relationships between weather and microclimate and skin temperature. Usually the instrumentation was crude and measurements were often made under difficult conditions. Soviet work was particularly interesting since men at Vostok, the station near the Pole of Inaccessibility, faced the combination of the extremely severe cold and hypoxia due to the altitude. They did not appear to be fully acclimatized even after a year (Tikhomirov, 1973). The value of reinforcing physiological adaptation by deliberate training was emphasized and the state of ionization of the air was thought to be important for non-specific human resistance, an idea also investigated by the French (Rivolier, 1973). On the Commonwealth Trans-Antarctic Expedition, Goldsmith had followed up a suggestion made by Butson (1949) that the clothing worn was more correlated with temperature than with wind-chill and from statistical studies on the massive amount of data obtained, Rogers (1973) concluded that there was no evidence of acclimatization and that polar man creates and controls his own microclimate. At the Scott Base end of this expedition an integrating motor pneumotachograph was used to obtain data on energy expenditure in man-hauling, and a sledge used by Wilson and his companions on their winter journey was found to require relatively twice the energy expenditure to move than did a modern sledge (Lewis & Masterton, 1963). Work on the psychology of wintering parties at this time was very tentative (Perrier, 1964; see also Delépine, 1964).

11.3 Medical and psychological research after the IGY

The IBP included a Human Adaptability section and stimulated the meeting of interested medical workers as a sub-committee of the SCAR Working Group on Biology. They recommended a variety of standard techniques and measurements for use in the Antarctic. However, for various reasons effective co-ordination of research at the different national bases was not achieved. Some papers on human biology were included in a meeting organized by the SCAR Working Group on Biology held in Paris in 1962 (Carrick *et al.*, 1964). In the same year the first meeting to deal specifically with human aspects of polar research, the Conference on Medicine and Public Health in the Arctic and Antarctic, was arranged by the World Health Organization. Collaboration between workers in this field was further improved when, following a SCAR/IUPS/IUBS symposium on *Human Biology and Medicine in the Antarctic* held in Cambridge in 1972 (Edholm & Gunderson, 1973), the SCAR sub-committee became a full Working Group in 1974.

Systematic physiological research in the years immediately following the IGY was done by BAS, which in 1956 had entered into an agreement with the Medical Research Council to have suitable medical officers trained in the Council's Division of Human Physiology, the director of which since 1949 had been O. G. Edholm. Edholm was extensively involved in research on human performance in extreme environments, polar, tropical and alpine, and fought for recognition by SCAR of the importance of Antarctic human physiology and medicine. He realized that this could serve a two-fold object, to learn about adaptation or acclimatization of man to Antarctic conditions and to exploit a situation where regular observations could be made on an isolated group of healthy young men over a period of a year or more. The reaction of the men to being used as subjects in this way is indicated in an account given by Herbert (1968). In 1956, Dr Hugh Simpson carried out a study of changes in eosinphilic count in blood in relation to stress. After establishing normal levels for each man during a period at the base at Hope Bay he continued his observations during a sledging trip pioneering a new route at the time of the equinoctial storms. Herbert relates:

> Both tents were flattened by a stupendous gust of wind, in which two tent poles of each tent were snapped, food boxes and personal gear blown over the horizon, and two fully-grown huskies lifted twenty feet through the air. Roger was outside at the time and fell to the ground just before the gust hit or he might have gone the way of the dogs. He and the others spent the rest of the night exposed to vicious winds and were very weak by the morning. They made one good tent out of the two broken ones and huddled together for days in cramped quarters and in constant fear of being blown away.
>
> Hugh's physiological research programme had a gap in the records at that time, which even his tent mates in retrospect admit was a pity. They had all been under severe stress, and the fall in their graph of eosinophils up to the time their tents were crushed had in some cases been most dramatic. On arriving back at base a few weeks later, their eosinophil count soared way above their normal base level as if suddenly released from tremendous tension.
>
> (Herbert, 1968, p. 76, by courtesy of the author.)

At Halley, studies on food intake, energy expenditure, and weight loss and gain, were made at base and on sledging journeys undertaken specifically for physiological observation, the energy requirements for maintenance of body weight being found to be around 3600 and 5000 kcal day^{-1}, respectively. Measurements of energy expenditure in various representative Antarctic activities were made (Brotherhood, 1973; Fig. 11.2). Weight was lost on sledging journeys but regain was found not to be simply a matter of rehydration and not necessarily accompanied by increase in skinfold thickness (Lewis & Masterton, 1963). The energy intakes during

Fig. 11.2 Measurements of energy expenditure of men under Antarctic conditions, Signy Island, 1966. Photograph by G. E. Fogg.

sledging have subsequently been found to be overestimates when an individual weighed–diet survey was done instead of inferring intakes from the known energy contents of food boxes (Campbell, 1981). The invention by H. Wolff in 1958 of a vest knitted with a continuous wire covered with plastic that could be used as a resistance thermometer enabled the measurement of integrated sub-clothing temperatures. No difference was found between men even though they had different histories of exposure to cold. The wind-chill factor and solar radiation were found to be more important in determining whether the Antarctic environment is tolerable than is its actual temperature. Lewis & Masterton (1963) concluded in their review that the effects of cold are small, usually identifiable, and not of great importance for a successful polar expedition. Nevertheless, observations on peripheral blood flow later indicated decreased flow during winter months, giving evidence for cold acclimatization in fingers and hands (Edholm in Gunderson, 1974).

Lewis & Masterton (1963) commented that the time had come to replace the relatively crude methods used hitherto with more advanced techniques. A step of this sort was taken by a team of medical research scientists from the University of Oklahoma, USA, in a psychophysiological study of the sleeping and dreaming behaviour of men at the South Pole Station in the consecutive austral winters of 1967 and 1968. The polar regions are obvious places to investigate the effects of unusual schedules or time zone shifts on a group of persons, without undue confinement to a controlled environment room, and studies of this sort in the Arctic go back as far as 1910. The Soviets and the Japanese studied seasonal and circadian rhythms in a restricted range of functions of subjects at their Antarctic stations (Edholm & Gunderson, 1973). The central equipment used by the Oklahoma group was a Beckman biomedical system which displayed and recorded electrophysiological data including electroencephalograms (Brooks *et al.*, 1973). Sleep and activity patterns, haematological and cardiopulmonary data, erythropoietin level measurements in blood and urine, and the acute effects of hypoxia on sleep psychophysiology were examined. A major and surprising finding was the virtually complete loss of stage 4 (slow-wave) sleep at the end of the austral winter and its failure to return six months after the subject was back in the US (Shurley in Gunderson, 1974). This, however, conflicted with the findings of Dr R. A. H. Paterson working at Halley Bay, a difference that might be due to latitude or, possibly, to differences in morale at the two stations.[3] Perhaps the most important contribution, however, was to demonstrate that sophisticated medical research can be conducted under the difficult conditions of an isolated Antarctic base. In recent years France and Australia have conducted notably well organized and long-continued research in fields such as immunology.

Psychological studies were initiated at US stations in the Antarctic during the IGY and evaluation by psychologists and psychiatrists, by means of biographical information and attitude and personality tests, have been used in the selection of US Antarctic personnel since 1963 (Gunderson, 1973, 1974). Other countries – Australia, France and New Zealand – followed suit in the next 10 years (Edholm & Gunderson, 1973; Taylor, 1987). This has achieved reasonable success in avoiding misfits; a comparison between predicted and actual performance in the Antarctic showed a statistically significant correlation, with only one prediction in the group of 12 badly out (Rivolier *et al.*, 1988). BAS, however, has not used psychological testing in selecting personnel and has relied on interview alone (Paterson, 1978).

Shurley (1973) saw Antarctica as providing unique opportunities for the investigation of human behaviour. He ascribed the fact that no major or systematic studies in this field had been carried out there not only to the comparatively late development of this branch of science but also to logistic difficulties, and a deep-seated and irrational suspicion on the part of other scientists. Psychological investigations in Antarctica have been carried out particularly at US stations. Gunderson (1974) summarized the results of a series of studies using supervisor ratings and peer nominations as criteria of performance by saying that effective individual performance includes three essential behavioural components: emotional stability, task motivation and social compatibility. Occupational role was important during long-term isolation, navy men showing significant deterioration in morale or satisfaction during winter whereas civilians showed little or no change. One opportunity which the Antarctic offers for psychological research, pointed out by Edholm (1973) but evidently not yet taken, is for examination of different national groups, isolated from each other but all exposed to isolation and a harsh environment. Motivation would be an important factor to take into account here because it can be different according to nationality; it has been said that a builder recruited into BAS will have applied because he wanted to go south, whereas in the US the high rate of pay would be the attraction and an Argentine would be obeying orders. Macpherson (1977) a psychologist who was a base leader with the BAS concluded that adaptation to the isolation of Antarctic working conditions involves the evolution of a social system which has a more prominent informal component than in other, more normal, group situations. Management in the home country, he considered, cannot be privy to the hidden, irrational arena of this social system in which men's attitudes are formed and he tended to put difficulties down to effects of the Antarctic winter on the isolated individual. Edholm, in a survey using a standard 'likes and interests' questionnaire, found large differences between individuals in BAS, the outstanding men being highly tolerant of others whereas those who were failures were intolerant (de Monchaux *et al.*, 1979). A. J. W. Taylor (1987), who had extensive experience in the field with New Zealand parties gave a general review of methods of working and results of Antarctic psychology. Difficulties of many sorts have been encountered. The exigencies of base life may disrupt research programmes; researches may have adverse effects on morale; subjects are insufficient in numbers and cannot always be allocated to experimental and control groups. Thus it may be necessary for the psychologist to combine observational, clinical, psychometric, quasi-experimental and experimental methods. Taylor made a virtue of this necessity and considered that Antarctic psychology has broken fresh ground by using a loosely interlocking network of methods which is more appropriate to real-life situations than laboratory techniques.

11.4 The International Biomedical Expedition

In 1977 the SCAR Working Group on Human Biology and Medicine put forward the idea of an international multi-disciplinary project of field work on various aspects of human biology. It was envisaged that the expedition should be in a region sufficiently cold and remote, and the logistic support sufficiently spartan to provide an environment in which specific experiments could be conducted. Isolation, life in a small international group, physical difficulties of life in the field and pressure of work were aspects that called for study. There were recurring difficulties in obtaining the necessary logistic support but these were eventually resolved when Expeditions Polaires Françaises agreed to provide transport for the expedition, together with a small glaciological party, to undertake a 800 km journey inland from the French base of Dumont d'Urville. The first International Biomedical Expedition to the Antarctic (IBEA) was thus able to spend 10 weeks travelling by motorized toboggans and living in tents on the plateau in the austral summer of 1980–81 (Rivolier *et al.*, 1988). The 12 participants, representing France, Australia, the UK, New Zealand and Argentina, all acted both as subjects and experimenters, an arrangement which, while economical, had the methodological disadvantage that no 'single-blind' or 'double-blind' procedures such as are commonly used in medical research could be employed. The programme included projects in physiology, biochemistry, microbiology, immunology, psychology and behavioural adaptation, sleep and epidemiology, but the main objective was to look on man as a whole in conditions in which he was subjected to climatic, social and metabolic stresses. Undoubtedly the level of these stresses was greater than expected. Acclimatization in the field was found to increase resistance to cold. A preliminary regime of cold baths such as has been suggested as the reason for 'Birdie' Bowers' extraordinary resistance to cold (Stroud, 1988), was of little use quite apart from it having an adverse effect on morale. Another interesting point was that immune response was not weakened. However, it was generally difficult to determine whether a stimulus for an observed change was climate, attitude, activity, food or social environment. The results of studies of the whole man in relation to a stressful environment have a value beyond that for Antarctic science. Rivolier *et al.* (1988) point out that both the USSR and the US have in the past drawn on Antarctic biomedical science in designing their space programmes and that the findings of the International Biomedical Expedition to the Antarctic (IBEA) should be equally useful for these programmes and the European Space Agency projects in the future. Apart from space-specific problems such as that of micro-gravity, the stresses, particularly the psychological ones, imposed by Antarctic and space situations are analogous. To consider the possibility of carrying forward research in this field the SCAR Working Group on Human Biology and Medicine set up an *ad hoc* group on space-related human factors research in 1989.[4]

11.5 Sledge dog physiology

When dogs provided an important means of transport in the Antarctic it was worth carrying out research on their nutrition and physiology. The community of dogs maintained by FIDS, and later by BAS, on the Peninsula offered ideal material for breeding and research. The original dogs came from Labrador, Canada, in 1945–46 and apart from additions from Canada and Greenland the stock remained isolated until it was phased out in 1974–75. Taylor spent the years 1954–55 at Hope Bay on the Peninsula where as dog physiologist he studied problems of breeding, work output and nutrition (Taylor, 1957). He made the comment that the maximum amount of 'useful' work output from his dog team was similar, weight for weight, to that of the rowing eight that won its event in the 1924 Olympic Games. Information on sledge dog physiology was brought together by Orr (1964) with the conclusion that the most striking feature of the husky dog is its capacity for sustained physical work under trying conditions and the very high nutritional requirements that this exertion demands. Disease was scarcely mentioned in this review. An epidemic in which afflicted dogs had ulcerated pads, loss of hair on feet and legs, and a pink inflammation of the skin occurred in 1971–73. A veterinary scientist recruited to investigate the problem identified the causal organism as a *Trichophyton*, related to the ringworm fungus but not infecting man (Fuchs, 1982).

11.6 Introduced organisms

Man has introduced animals, plants and micro-organisms into the Antarctic both deliberately and inadvertently. Rudolph & Benninghoff (1977) pointed out that at the microbiological level the continent is the last region on earth where man's role in assisting invasions can be studied essentially from the beginning and they made a plea that this should be monitored in an organized way, but little seems to have been done in response. The alien flowering plant flora of South Georgia, amounting to 25 species as against 17 native species, was listed and the species distributions plotted by Greene (1964). Rudolph (1966) showed that even under the more severe continental conditions at Cape Hallett the alien grass *Poa pratensis* was capable of germination and growth. Drygalski (1904) had commented on the virtual elimination of the Kerguelen cabbage by the rabbits deliberately introduced onto the island by the *Challenger* expedition in 1874. A list of animal species introduced on to the sub-Antarctic island groups, with dates of introduction, has been given by Leader-Williams (1985). More details of the histories of introductions by sealers and whalers of rats, rabbits, cats, pigs and cattle on the French sub-Antarctic islands and on Macquarie Island were given, respectively, by Dorst & Milon (1964) and Cumpston (1968). Only Heard and Macdonald Islands amongst the sub-Antarctic groups are free of introduced mammals. Control measures for such pests – for so we must call them if we deplore the profound changes they have caused in the indigenous floras and faunas – as rats, rabbits and feral cats have not been successful and, if they had been, it seems doubtful whether the native floras and faunas would have reverted to their original states. One of the most intensively studied of introductions has been that of the reindeer, which was taken to South Georgia by the whalers in 1911 and again in 1926. Being confined to limited areas by sea and glaciers, the three herds of these animals have not affected the ecology of the island as a whole. They constitute the largest reindeer population in the southern hemisphere and differ from northern reindeer in having no parasites and no predators. They are not managed or culled in any systematic way. The change in diet and reversal of breeding season occasioned by their transfer from Norway and their narrow genetic base are further unique characteristics. Leader-Williams and his associates have published on their population ecology, sexual physiology, forage selection and growth (Leader-Williams & Ricketts, 1981; Leader-Williams, 1988). From this sound basis of knowledge further investigations on South Georgian reindeer should lead to valuable contributions on such problems as interactions between populations and carrying capacity and evolutionary trends in isolated populations. Genetical changes related to age have been described in animal species introduced into an environment in which they are subject to physiological stress by cold. Electrophoretic studies of gene loci products in a population of house mice on South Georgia, for example, showed strong evidence of stabilizing selection for characters which promote resistance (Berry *et al.*, 1979).

11.7 Conservation

We have just seen that man's impact on the Antarctic can give rise to situations with intrinsic scientific interest. However, a non-scientific sense of values impelling us to conserve the natural scene makes the study of this impact, which, of course includes many things besides the introduction of alien organisms, of great practical importance. In the early days of exploration a few individuals were uneasy about the effects of sealing in the Antarctic and soon after the beginning of whaling in the Southern Ocean the UK Government took steps to conserve stocks (Bonner, 1987b). Apart from this there was little concern about pollution or conservation amongst the generality of explorers or scientists – in the immensity of the Antarctic a handful of men would scarcely be expected to have any lasting effect. Consciousness of the need to control pollution and conserve the natural environment developed gradually in the more prosperous countries after the Second World War. In Britain, for example, the Nature Conservancy was established in 1949 and the Royal Commission on Environmental Pollution in 1970. Until the popular 'green' movement burgeoned in the 1980s the Antarctic community followed rather than led in conservation, applying lessons learnt elsewhere to comparatively virgin soil. The politics of conservation in Antarctica have already been discussed in chapter 6 and will be touched on again; here the more immediate practical aspects must be considered. The brief reference to measures for the 'preservation and conservation of living resources in Antarctica' in Article IX of the Antarctic Treaty applied specifically to land and ice shelves. The necessity to be more comprehensive and explicit was immediately realized by SCAR, which at its third meeting, in March 1959, began discussions, the history of which has been summarized by Llano (1972). The following year it was recommended that a leaflet on the preservation of wildlife should be circulated to all persons landing in Antarctica, that information be collected to enable sanctuaries to be designated, and that the problems presented by the sub-Antarctic islands be considered. These were incorporated in the 'Agreed measures for the conservation of Antarctic fauna and flora' formulated by the Antarctic Treaty organization in 1964. These enjoined participating governments, among other things, to prohibit the killing of animals save for scientific purposes or in cases of necessity, to minimize disturbance of bird and seal colonies, to prohibit the collection of plants except for scientific purposes, and to alleviate pollution. It also listed a number of areas of outstanding scientific interest to be designated 'Specially Protected Areas'.[5] Around the same time, the Convention for the Conservation of Antarctic Seals, which has already been discussed (p. 190), was agreed. This too was an exercise in forethought and has, on the face of it, proved entirely effective although it is probable that logistic difficulties and consumer resistance to seal products have been the deterrent to commercial exploitation rather than the Convention (Heap & Holdgate, 1986). The sixth Antarctic Treaty consultative meeting in Tokyo in 1970 made further recommendations about such things as the use of radio-isotopes in the Antarctic, the effects of tourists and non-government expeditions and the collection of data on the conservation of fauna and flora. The Convention on the Conservation of Antarctic Marine Living Resources, concluded in 1980, was an ambitious exercise but has been notably unsuccessful in respect to finfish. This failure can be ascribed to the exceptional circumstance of the redeployment of distant-water fishing fleets following the imposition of 200-mile fishing zones elsewhere (Heap & Holdgate, 1986).

In 1980 a *World Conservation Strategy* was published under the auspices of the International Union for the Conservation of Nature and Natural Resources. It was recognized in SCAR that the general needs expressed in this applied to Antarctica as much as to other parts of the world.[6] The Twelfth Antarctic Treaty consultative meeting in Canberra, Australia, in 1983 called for an assessment of the impact of mankind on the Antarctic environment, in response to which SCAR produced a report (Benningoff & Bonner, 1985). The general points made in this were that assessment should go beyond biological effects and on to the changes they cause in the physical and chemical environment and that it must be a continuing and developing process. The Antarctic comprises two kinds of system. On the one hand there are the extensive and broadly uniform marine and ice-cap areas capable of absorbing substantial human activity with little or no change. On the other are small, but numerous, terrestrial areas, including inland waters, where even modest human activity may have considerable impact (Heap & Holdgate, 1986). Such areas, free of snow or ice, with vegetation and attractive to birds and seals, are precisely the places on which it is most convenient for expeditions to establish their bases as well as for tourists to visit. Research stations thus have a disproportionate impact and for this reason the setting aside of areas for protection is essential. In 1982 the Subcommittee on Conservation invited the committees of the member nations of SCAR to propose extensions in area and number of conservation sites. In 1985, SCAR published an atlas of *Conservation Areas in the Antarctic*, setting out Specially Protected Areas (SPA) and Sites of Special Scientific Interest (SSSI). The designation of an SPA should virtually isolate it from active scientific research whereas an SSSI receives a measure of protection while research is in progress (Fifield, 1987). These areas and the means taken to protect them are under frequent review and are being added to.[5]

One of the latest of the instruments devised by the Treaty powers for conservation of the continent, CRAMRA,[7] (see 'A note for the reader') ran into trouble (see p. 412) and has been replaced by the Madrid Protocol of 1991 which prohibits mining and drilling for oil for at least the next 50 years. In framing CRAMRA claimants and non-claimants to Antarctic territory argued not so much about mineral rights as environmental protection (Heap & Holdgate, 1986) and the objections to CRAMRA have become linked with the idea of Antarctica as a 'world park'. This was advocated within the Treaty organization by France and Australia. The UK no longer opposes the suggestion[8] but misgivings remain. Antarctica as a world park or wilderness reserve may well be an unenforceable ideal and it seems likely that the restrictions it would entail would bear most heavily on science – which would certainly be counterproductive for conservation on the global scale.

Meanwhile, in 1988, the Fourteenth Antarctic Treaty Consultative Meeting[9] made recommendations for guide-lines for environmental impact assessment to improve the value of this as a means of limiting damage. Steps were also taken to improve the selection of representative areas to ensure conservation of all types of situation. New categories of SSSI were introduced to include, for example, marine systems and areas of particular value for the study of natural electromagnetic fields which need to be protected from man-made electromagnetic interference. Detailed plans for active management were now required for all such sites. Detailed protocols for waste disposal were recommended and governments enjoined to ensure compliance with them.[10] Recognition of the wide importance of conservation in Antarctica was manifest in the transmogrification in 1988 of the Subcommittee on Conservation of the SCAR Working Group on Biology into a new Group of Specialists on Antarctic Environmental Affairs and Conservation.[11]

There has thus been much unexceptionable precept but implementation has had to be left to individual governments, with varying results. The US National Science Foundation in 1971 sponsored a major colloquium on problems of conservation in the Antarctic (Parker, 1972). This discussed topics such as litter, waste disposal, pollution, impacts on ecosystems, the psychology of pollution, and the interference with science itself that arises from pollution and contamination. Following this the US passed its Antarctic Conservation Act in 1978 with the object of ensuring the protection of flora and fauna from any interference, except from scientists under a restrictive permit system (Auburn, 1982; Schatz, 1988). The record of US conservation measures in practice is mixed. On the one hand those who have visited the McMurdo area agree that there is mess and pollution (Auburn, 1982, p. 269, Sun, 1988) although except in scale it is no worse

Fig. 11.3 All toxic waste is removed from UK stations in special drums for disposal outside the Treaty area. (Courtesy of C. Gilbert, British Antarctic Survey.)

than on many Peninsula bases belonging to other nations, such as Argentina. Greenpeace in particular has voiced much criticism of the effects of US activities on the Antarctic environment (Bogart, 1988). On the other hand, measures have been taken by the US to economize on fuel and so reduce atmospheric emissions, to minimize the impact of sewage outfall to the sea, to clean up old stations, and to employ meticulous safeguards to avoid contamination of especially sensitive areas such as the dry valleys (Schatz, 1988). The US NSF has sought funds for a major clean-up programme (Austin, 1989). Australia introduced legislation similar to that of the US in 1980 but it too has come under severe criticism by conservationists (Bogart, 1988). New Zealand and the UK do not have such legal enforcement but distribute *A Visitor's Introduction to the Antarctic and its Environment*, prepared by the Subcommittee on Conservation of the SCAR Working Group on Biology, to their staffs and other visitors. The UK has been criticized by Greenpeace for the inadequacy of its waste disposal arrangements at bases (Fig. 11.3) and for the construction of an airstrip at Rothera Station.

With growing world concern about the extent of environmental damage,

Antarctica has been viewed as the last unspoilt continent and particularly to need protection. Three particular cases of environmental damage which have received publicity may be discussed briefly to illustrate what has happened. The construction begun in the early 1980s by the French of an airstrip at Pointe Geologie in Adélie Land attracted considerable adverse comment, both internationally and from within the French scientific community, since it involved destruction of an area with fine sea-bird colonies (Boart, 1988). Greenpeace, which sent a team to observe – and hamper – the construction, produced a report which was discussed by the SCAR Group of Specialists on Environmental Affairs and Conservation.[12] While concluding that this report made a useful contribution and accepting that it was right in drawing attention to shortcomings in the environmental impact assessment procedure, the Group pointed out that guide-lines for such an assessment had not been recommended by the Treaty organization at the time when the project was initiated. There has also been great alarm over the oil-spill resulting from the sinking of the Argentine supply ship *Bahia Paraiso* about a mile (1.6 km) away from the US Palmer Station on 28 January, 1989. More than 150,000 gallons (681,894 l), primarily diesel fuel arctic, were released. Within four days some 100 km^2 of sea surface were covered in oil and mortality among littoral invertebrates was obvious. The greatest impact was on sea-birds but since most of the affected population moved out to sea shortly after the accident it was not possible to estimate the numbers killed. The US NSF responded within 36 hours by dispatching boats and equipment to control the damage and followed this up with a programme of damage and ecological impact assessment (Kennicutt *et al.*, 1990). This is no doubt a foretaste of many similar incidents to come. It is on King George Island that conservation measures have broken down to a disastrous extent. The history of the beginnings of this has been summarized by Auburn (1982). Fildes Peninsula on this island had been designated as a SPA but it was there that the Soviets established their Bellingshausen Station in 1968. There is nothing in the Agreed Measures which says that stations may not be sited in SPAs, although it is obvious that conservation and the activities of a research station are incompatible, but it is not clear that the Soviet authorities were aware that the Agreed Measures were to apply to this area. In the same year Chile announced its intention of setting up a station near the Soviet one. Instead of pointing out to these two countries the inconsistency of siting research stations in this manner the Consultative Meeting in 1968 regularized the situation by reducing the size of the SPA. Even the more circumscribed site was completely changed by the presence of the stations –

one of the biologically important lakes was used as a rubbish dump (Bogart, 1988) – and its status was terminated in 1975.[13] Since then King George Island has been the popular place for governments, anxious to adhere to the Antarctic Treaty with the minimum of expense and logistic difficulty, to set up their token stations. Eight nations now have stations there and Chile has built an airstrip and hotel complex. Some of these establishments have complete disregard for the environment (Bogard, 1988). The right of inspection given under the Treaty might be used to bring pressure to bear on these miscreants but this has not been exercised. The SCAR Group of Specialists on Environmental Affairs and Conservation considering the situation in 1990 concluded that such problems were often highly site-specific and might be best addressed by consultation between station representation in the field.[8] Auburn's (1982) summing up of the situation when he wrote 'Sovereignty, logistic convenience and the facilitation of scientific research prevail over environmental considerations' still, alas, seems true.

Conservation is largely a matter of education and politics; for the scientific management of the problems it presents the basic requirement is for broad but sound understanding of the systems involved. Refinement of monitoring techniques and their judicious application in field programmes is necessary and research directed to limited problems can be useful, but successful management of conservation and pollution problems is at present at least, more a matter of judgment than rigorous science. Research publications on conservation in the Antarctic have in fact been few. Symposia on Antarctic biology (Carrick *et al.*, 1964; Holdgate, 1970; Llano, 1977; Siegfried *et al.*, 1985) have each included two or three papers on conservation but these have been concerned with the impact of human activities rather than with methods of reducing damage. Thus Thomson (1977) assembled information which showed the effect of disruption by humans on the Cape Royds Adélie penguin rookery and its recovery after remedial measures had been taken (Fig. 11.4) and Cameron *et al.* (1977) showed that man's influence extends to the microbiological as well as the macroscopic level.

The record of Antarctic science in conservation and avoidance of pollution has been commendable in producing protocols in advance of serious damage being done but somewhat uneven in practice and not at all to the satisfaction of the more aggressive environmentalists. However, looked at in a broader context its contribution to environmental protection has been immense. The striking evidence of the ozone 'hole' and the sobering thought of the melting of Antarctic ice has brought home to

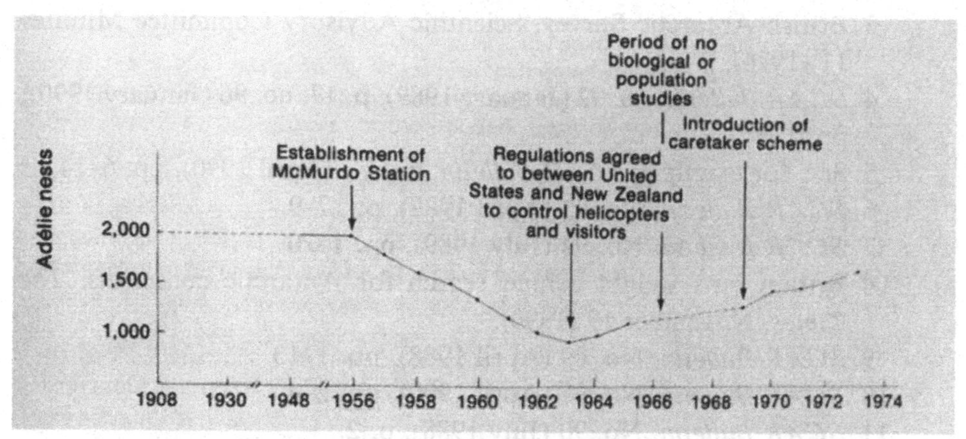

Fig. 11.4 The effect of disruption by humans on the Cape Royds Adélie penguin rookery and its recovery after remedial measures were taken. From Thomson 1977, Fig. 1. (Courtesy of R. B. Thomson.)

everyone the potentially catastrophic disruption that unthinking expansion of industry and domestic consumption may bring to the processes which regulate our physical environment (see Fig. 12.2). Knowledge of the circulation patterns of atmosphere and ocean and accompanying transfers of energy in the south polar region is essential in constructing the mathematical models with which we endeavour to predict what may happen. The samples of air and its impurities stored in chronological order in the ice-cap provide one of the best means at our disposal for finding out about changes in the past and for monitoring what goes on at present. The demonstration by ionosphericists of the effect that power transmission systems have on geospace has broadened our concept of pollution. The detailed knowledge of the workings of the marine ecosystem which is emerging from BIOMASS provides the only sure basis for avoiding ecological disaster by overfishing of krill and other marine resources and the data on the dynamics of bird populations gathered by ornithologists gives a measure of the health of the marine ecosystem. Long-term studies in Antarctica and the holistic approach in science have come into their own in conservation.

Endnotes

1 Royal Society, Minutes of Committees (1817–37), CMBI, p. 25.
2 Priestley, R. E., *The Psychology of Exploration*; photocopy of typescript annotated 'written in 1914 and published in 1921', 11 pp., in library of Antarctic Division, Christchurch, New Zealand.

3 British Antarctic Survey, Scientific Advisory Committee Minutes 11 (1976).
4 *SCAR Bulletin*, no. 92 (January 1989), p. 17; no. 96 (January 1990), pp. 1–3.
5 See, for example, *SCAR Bulletin*, no. 97 (April 1990), pp. 6–14.
6 *SCAR Bulletin*, No. 93 (April 1989), pp. 7–9.
7 *SCAR Bulletin*, No. 94 (July 1989), pp. 1–20.
8 Britain puts weight behind search for Antarctic consensus, *The Times*, November 19 (1990).
9 *SCAR Bulletin*, No. 89 (April 1988), pp. 1–13.
10 *SCAR Bulletin*, No. 97 (April 1990), pp. 2–4.
11 *SCAR Bulletin*, No. 90 (July 1988), p. 2.
12 *SCAR Bulletin*, No. 96 (January 1990), p. 5.
13 *SCAR Bulletin*, No. 49 (January 1975); *Polar Record 17*, p. 443.

12

Some concluding comments

12.1 The persistent features of Antarctic science

A feature of Antarctic science that emerges from this survey is that several of its main characteristics have been present from the beginning. At a time when nearly all scientific effort was individual and small scale it started as 'big science', dependent on substantial funding from the state and therefore a matter of politics. This has continued to be so; governments' attitudes have often been vacillating but usually crucial. Interest first in sealing and then in whaling led to notable support from commerce for exploration but only the most meagre voluntary provision from industry itself for strictly scientific work in the Antarctic. For a time in the heroic age public interest was aroused to the extent that a large part of the necessary finance for expeditions could be raised from private sources by appealing to patriotism and glamourizing geographical discovery. This phase did not last long and today almost all significant scientific work in Antarctica is government supported in the same way as are space research and oceanography – with which it has to compete for public funds. One difference between Antarctic science and these other major fields of scientific endeavour is that Antarctic 'big science' has always carried 'little science' along with it. Thus, expeditions concerned primarily with magnetic survey had naturalists with them and the IGY provided stimulus and support for individual biological projects in the Antarctic (Webber, 1986).

It is equally obvious that the reach, compass and nature of Antarctic science has always been dependent on the kind of technological support available. It was the perfection of the sailing ship and the control of scurvy that enabled Halley and Cook to initiate investigation of Antarctica and, successively, liquid fuel and the Nansen cooker, the dog sledge, the aeroplane, mechanized surface transport, radio communications and satellite surveillance have been promptly employed to extend its scope. Different constraints have operated at different stages and when two have been removed at the same time, as were those of logistics and instrumentation in the IGY, there has been a tremendous leap forward.

It is more difficult to compare personalities and decide whether Antarctic science has always been led by a particular kind of person. There is no difficulty in imagining Halley taking part in a SCAR meeting or enjoying the conviviality in the bar of a BAS station. Some of the leaders, like Halley, have been distinguished as scientists quite apart from any fame they acquired as Antarctic explorers – here one thinks of Bruce, Drygalski, Nordenskjöld, Mawson, and, in recent years, Siple, Gould, Laws and Hempel. These represented, one may note, some widely differing disciplines. Others have not been scientists by training but have had instinctive sympathy for research and have been highly successful as entrepreneurs in promoting it – for example, Cook, Ross, Scott, Shackleton and Byrd. There are those, such as Humboldt and Neumayer, who, although they never visited Antarctica nevertheless exerted considerable influence on the kind of research conducted there. Some highly successful explorers of the Antarctic, on the other hand, have left no appreciable legacy of science, amongst these are Amundsen and Ronne. There are no obvious examples of scientists who have been disastrous as leaders of Antarctic expeditions. In the past, although less so at present, a scientist had need to be highly motivated towards work in the field to think of going to Antarctica and perhaps the obstacles to be overcome to get an expedition launched have eliminated those without the ability to inspire the necessary enthusiasm or having tenacity of purpose under difficult conditions. As to the rank and file, psychological studies at Antarctic stations have shown them to be inclined to be introverted, reserved and trusting (Paterson, 1978). However, a comparison with Arctic personnel has shown the average Antarctic man to be the more extrovert, with more breadth of interest and organization (Mocellin & Suedfeld, 1990).

Dependence on common logistics has demanded co-operation and joint planning between scientists of different disciplines but beyond this there has always been more of a holistic approach than can be found in most other areas of science. Such an approach to understanding the world was advocated by Humboldt but, although he had considerable influence on the early development of Antarctic science, he was not followed explicitly in this by any of his contemporaries who worked in the south. With the triumphs of the reductionist philosophy in science in the later part of the nineteenth century Humboldtian ideas became unfashionable and Maury, who urged the necessity of a view of oceanography and meteorology extending into the Antarctic, met with a similar lack of response during his life. This disregard of the inescapable interrelations of phenomena in the natural world was an underlying reason for the period of 'averted interest', lasting some 50 years, during which only two scientific expeditions visited the Antarctic. Nevertheless, when interest revived at the end of the century it was assumed without question that discussions of Antarctic science should include representatives of all the disciplines which seemed relevant. At the meeting organized by the Royal Society in 1898 (Murray, 1898) Neumayer remarked:

> Understanding of the importance of Antarctic research requires an unusual amount of knowledge, and not in one branch of science only, but in the whole complex of natural philosophy and natural science.

Reviewing the expedition which he led, Drygalski (1904) wrote:

> It contributed to the sum total of knowledge and of human endeavour, and should be judged on those grounds. It is precisely there that the greatest excitement of such an undertaking lies, that it links together the various disciplines of knowledge, and engages theory with practice.

While, of course, individual scientists and their projects are usually as narrowly specialist as in any other area and impenetrable jargon and unexplained acronyms impede interdisciplinary communication, this wide view of Antarctic science has nevertheless persisted. The comment was already made in chapter 1 that different disciplines are still treated side by side in books on Antarctic science; Antarctic bibliographies likewise cover the whole range, and the complicated way in which the sciences interrelate is all too evident to the writer on Antarctic science who endeavours to arrange material into clear-cut sections. The value of the holistic approach in conservation has been mentioned in the preceding chapter. Benninghoff & Bonner (1985) summed up the present day outlook in truly Humboldtian terms;

> The inter-relatedness of the global environment's components is real, not just theoretical or philosophical, although the couplings of the myriad components vary from pathways of major forcing functions to associations between environmental facets of immeasurably small significance. Perturbations in one part of an environmental system usually create disturbances in other parts of the system and can influence other environmental processes.

There has been a roughly even and continuing balance between the various scientific disciplines practised in Antarctica in recent years. The percentages which they contribute to the total of publications change over the years in no statistically significant manner except for 'expeditions', that

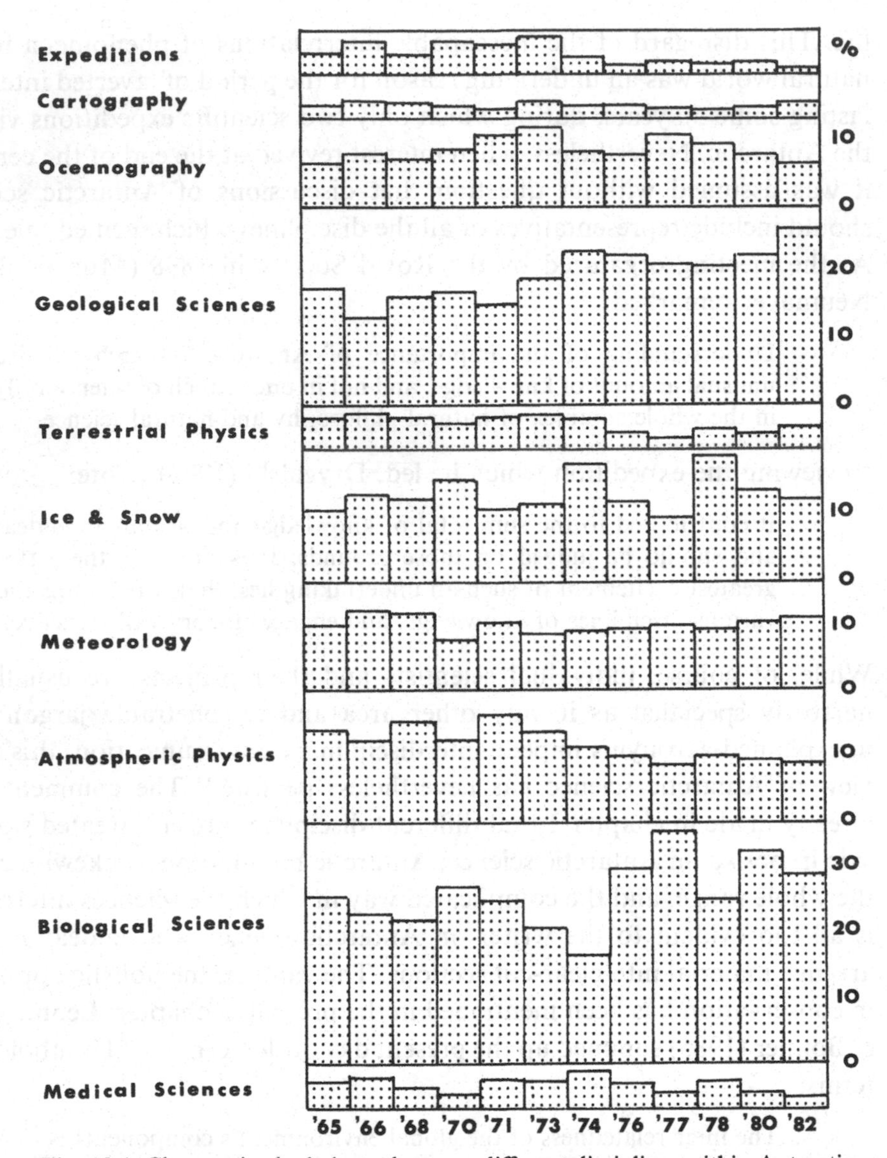

Fig. 12.1 Changes in the balance between different disciplines within Antarctic science as shown in percentages of total number of publications per volume listed in the first 12 volumes of *Antarctic Bibliography*. Note that this publication was not consistently annual and that over the period covered the number of scientific papers listed per volume rose from 1,586 to 2,167.

is general accounts, which have diminished (Fig. 12.1; see also Budd, 1986). Geophysics, which from the time of Halley onwards, has been the spearhead of Antarctic science, now makes up a relatively small proportion of the scientific output and may even be decreasing somewhat. Links between the various disciplines are clear. It is unnecessary to labour the point that atmospheric physics has close connections on the one hand with meteorology and on the other, through magnetism, with terrestrial physics. Air–sea interactions link meteorology and oceanography, the latter in turn taking into account glaciological phenomena and linking with biological oceanography. The geology of the continent merges on the one hand with glaciology and on the other with palaeontology. The latter links with biology, of which the medical sciences are part. The borderlines between these subjects are all under active investigation in the Antarctic. It is satisfying, in some respects, at least, that the circle should be closed by the involvement of atmospheric physics and chemistry and meteorology with biology through investigations of the occurrence and significance of ozone depletion and the 'greenhouse effect' arising from an increase in atmospheric carbon dioxide concentrations. Apart from these interdisciplinary links there is a powerful non-scientific factor making for the unity of Antarctic science – the strong bond of comradeship which develops between those who have worked in that remote and hostile environment.

12.2 The contribution to science in general

We should now turn to the contribution of the Antarctic to the general advancement of science. In the nineteenth century it was undoubtedly terrestrial magnetism, physical geography and, to a lesser extent, meteorology which were the beneficiaries of expeditions south. A crudely objective method of assessing contributions in more recent times is by bibliometric analysis (Price, 1986; Budd, 1986). An indication of the wide spread of Antarctic science, and its contribution to science in general, may be seen in the statistic that 34 per cent of the 17,154 items in this field published between 1970 and 1981, appeared in publications not specializing in Antarctic science, of which there were 1386. Of the sources examined, each publishing more than four articles per year on Antarctic science, 44 might be considered to give some preference in this area. The *Antarctic Bibliography* and the *Science Citation Index* are two machine-readable data-bases, the overlap between which provides a starting point for more detailed citation analysis (Cozzens, 1981). Between 1961 and 1978 a total of 2942 publications included in the *Antarctic Bibliography* were cited at least once and for these the average citation rate per paper was about 10 (Budd, 1986). The qualitative impact of these publications may be evaluated by means of a statistical technique called co-citation clustering, a cluster being a group of frequently cited papers which tend to occur together in the bibliographies of later publications. Such sets of papers have been found to be associated with the appearance and development of new lines of research. This technique picks out Antarctic research on atmospheric processes and plate tectonics as having played a crucial role in the rapid growth of these two areas of science in the 1970s (Cozzens, 1981). This supports the conclusion one reaches subjectively by reading reviews and listening to informed opinion, that it is Antarctic scientists who have provided some of the most original recent work in these fields. Indeed, our knowledge of geospace would still be rudimentary without the contributions made from the Antarctic. According to Guthridge (1983), 84 Antarctic papers published betweeen 1961 and 1978 achieved the status of 'citation classics', each being cited more than 50 times during that period. Table 12.1, listing the first half of the 26 topics covered by these papers, demonstrates the wide spread over which Antarctic studies have made significant contributions. Such statistical techniques may however fail to identify seminal contributions in fields with comparatively low publication rates. A marine biologist might think that the new insights into marine ecosystems which have arisen from studies of krill harvesting and the role of sea-birds, were of as much significance as the developments in physical science.

To a great extent these advances have been possible because of special conditions prevailing in the Antarctic. Investigation of the upper atmosphere and geospace has been favoured by the configuration of the geomagnetic field in relation to the geographic co-ordinates and, as the central fragment of Gondwanaland, Antarctica is inevitably a focus for study of plate tectonics. The effectiveness of the movements of ice-fields in providing a collecting mechanism is a happy chance that has made Antarctica the most profitable part of the world for finding uncontaminated meteorites. For biologists it is the comparative simplicity of food chains, a result of geographical isolation, which makes the Antarctic a good place for ecosystem studies. However, returning to the holistic theme, it is perhaps as a key area for the regulation of global environmental processes that Antarctica is beginning to assume greatest scientific significance. The realizations by Humboldt that measurements from the Antarctic were necessary for the understanding of geomagnetism and by Maury that data from the South Polar regions were needed for the forecasting of weather elsewhere have been amply borne out by subsequent work. The Antarctic as the earth's major heat sink is a principal driving force for atmospheric and oceanic circulation. Modern meteorology, oceanography and geospace studies would not make sense without the contribution of Antarctic science. With growing awareness of the extent of the effects of

Table 12.1. *The first 26 of 52 specialities in which citation analysis demonstrated that Antarctic research had a significant role between 1961 and 1978.*

Speciality	Speciality
Atmospheric pollution	Stratigraphy of deep-sea carbonates
Magnetosphere physics	Models of climatic change
Aurora and electric fields	Genetic variation in marine invertebrates
Biological antifreeze mechanisms	Geomagnetic exclusions and reversals
Reproductive strategies in arctic and Antarctic birds	Carbon dioxide and climate
Thermospheric dynamics	Radionuclides and suspended matter in seawater
Stable isotope geochemistry	Life on Mars: Viking results
Asymmetric sea-floor spreading	Biochemical adaptations to temperature
Geothermal dynamics of ocean ridges	Palaeoclimatology of the Antarctic region
Glaciation models; arctic data	Magnetospheric wave phenomena
Very-low-frequency waves in the magnetosphere	Hydromagnetic waves and geomagnetic pulsations
Trace-elements and isotope geochemistry	Tectonics of the Scotia Sea
Mid-ocean ridges	Stratospheric warmings

Source: (After Budd, 1986, from Guthridge, 1983. Courtesy of Dr G. G. Guthridge.)

man's activities on the global environment (Fig. 12.2), Antarctica has become even more of a key area.

12.3 Arctic and Antarctic

On one facet, Antarctic science has not been as closely in contact with other sciences as might be expected; the connections between Arctic and Antarctic studies are uneven and often slight. The greatest of early Arctic scientists, William Scoresby, had little influence on Antarctic investigations during his lifetime but there was direct transfer of polar experience from north to south through several later nineteenth and early twentieth century scientists. James Clark Ross had 15 years of Arctic work behind him when he sailed for the Antarctic. Nansen, another Arctic scientist of Scoresby's calibre, was generous with advice to the Antarctic expeditions of the heroic age but the influence on Antarctic investigations of the new Scandinavian school of scientific exploration of the Arctic, of

Fig. 12.2 Cartoon from the San Francisco Chronicle, 1988. (© San Francisco Chronicle. Reprinted by permission.)

which Nansen and Nordenskiöld were the founders, did not extend to providing personnel. However, Drygalski had four years experience in Greenland before going south. Scott had with him on the *Discovery* as surgeon and botanist, Reginald Koettlitz who had been on the Jackson-Harmsworth expedition in the Arctic. Geomagnetism, of course, had always integrated observations from the two poles, as was particularly evident in the First Polar Year, 1882–83, and the IGY, 1957. Comparisons of the climates of the two poles should obviously be made, as was done, for example, by Maury (1860) and more recently by Sugden (1982). In classical geology there is little scientific reason for linking studies in diametrically opposed parts of the world. Glaciology began in the European Alps before extending to the Arctic and Antarctic. Sea-ice studies advanced first in the Arctic (Weeks, 1968). Several distinguished glaciologists, e.g. C. Swithinbank, G. de Q. Robin and D. J. Drewry, have worked in both these regions. It is in biology that cross-fertilization between investigations around the two poles is not as extensive as might be expected. Ross (1847, vol. 1, p. 202) overdid the equation of Arctic and Antarctic zoology but in work in more recent times on the cold adaptation of animals profitable ideas have been transferred from north to south. The Bipolar Botanical Project of the IBP was a well organized comparative investigation of ecology, adaptation and production in Greenland and South Georgia (Callaghan *et al.*, 1976). An idea of the present position is given by a simple bibliometric study of *Polar Biology*, a journal which embraces both Arctic and Antarctic studies. Volume 8 (1988) included 56 articles of which 44 related to the Antarctic, 11 to the Arctic, and only one dealt with both. Of the 1220 references quoted by the Antarctic authors only 4.5 per cent were of Arctic work and of the 254 references in the Arctic papers, 4.7 per cent referred to the Antarctic. Perhaps as a result of co-operation during the IBP it is the plant ecologists (see Rosswall & Heal, 1975), freshwater biologists (Hobbie, 1984),[1] bryologists and lichenologists (Longton, 1988), who tend most to take the bipolar view although a zoologist, Stonehouse, has recently (1989) produced a synthesis of knowledge of polar ecology from both regions. More generally, of the 44 journals listed by Thuronyi (1982) which average more than four articles per year on Antarctic matters, only five are specifically devoted to studies from both Arctic and Antarctic. Lack of contact between scientists working in the two polar regions may be partly a matter of nationality. They naturally tend to associate with and quote from their own countrymen rather more than they do with colleagues from abroad. Canada, whilst a leader in Arctic research has little interest in the Antarctic, whereas Argentina, Chile, South Africa, Australia and New Zealand are not much concerned with the Arctic. The Scott Polar Research Institute, the Norsk Polarinstitutt, the Soviet Arctic and Antarctic Research Institute and the West German Alfred-Wegener-Institut für Polar und Meeresforschung all deal with both areas but the US does not have a common focus for studies from the two polar regions. The Soviet system effectively integrates Arctic and Antarctic science and logistics. Work in the Antarctic has benefited from the vast Soviet experience in the north but no doubt there will be a reverse flow of information which will be of use under the extreme climates and difficult living conditions of northern USSR (Brigham, 1988).

12.4 Internationalism

Antarctic science in the eighteenth and early nineteenth centuries partook of the free exchange of ideas between men of learning of different nationalities which prevailed in those times but this did not extend to closer co-operation between states. Humboldt and the German Magnetic Association saw the necessity of international collaboration if progress was to be made in geomagnetism and Maury made the same point for meteorology. The ideas of these men slowly percolated into polar science and received their first concrete expression in the First International Polar Year of 1882, in which the Antarctic was only marginally involved, set up mainly through the efforts of Weyprecht, an Arctic scientist. The revival of Antarctic exploration, and with it Antarctic science, may be dated from the International Geographical Congress of 1895 when the feelings of geographers from many nations came to a focus. There was some exchange of information, some consultation about the apportionment of areas of interest and some inclusion of foreigners in the national expeditions which followed but, on the whole, rivalry, particularly for the attainment of the Pole, meant that international co-operation was not extensive. At this time international collaboration in oceanographic research was being established but this did not extend to joint cruises. The Norwegian–British–Swedish Expedition of 1949–52 to the Norwegian sector of Antarctica was the first large-scale expedition to any part of the world to be planned and staffed on a truly international basis. It was most successful. After the IGY of 1957–58 international co-operation in the Antarctic became the mode, including overall discussion by the international committee, SCAR, exchange of personnel, some joint logistic facilities, pooling of data, and international research projects. This co-operation has been, however, within a select coterie of nations. Price (1986) has ranked the nations in order of numbers of scientific authors in the period 1967–73 and finds that there is a highly significant correlation of this number with the size and development of these nations as measured by the kilowatt-hours of electricity used. The top three in his list, the US, the UK, and the USSR, are also the leaders in Antarctic research and the 17 signatory nations to the Antarctic Treaty up to 1973 are found within the first 35 of the list. If one excludes nations which for geographical reasons have no interest, for example Canada, or, on the other hand, have strong reasons for participating in Antarctic activities, for example Argentina, Chile, New Zealand and South Africa, and adds West Germany, which acceded to the Treaty in 1974, then the correlation between Price's list and the list of the Treaty powers becomes excellent. The Antarctic Treaty has undeniably been a rich nations' club but, on the other hand, its members have showed notable willingness to assist those who wish to join. The language distribution of Antarctic literature shows another aspect of internationalism. Analysis by Thuronyi (1977) of data for the period 1971–76 showed that of the total number of publications (7248), 74.02 per cent were in English, 14.58 per cent in Russian, 3.64 per cent in Japanese, 3.46 per cent in French, 2.95 per cent in Spanish, 0.57 per cent in German, 0.19 per cent in Norwegian, and a few in Polish, Italian, Bulgarian, Afrikaans, Czech, Dutch, Flemish, Portugese,

Rumanian and Swedish. These figures are not truly representative in that a significant proportion of the publications in English represent translations from Russian. If an adjustment is made to allow for this, the percentages become 69.68 and 19.64 for English and Russian, respectively. The high proportion of papers in English, of course, reflects not only the activity of the English-speaking nations but the international status of the language. It may be noted here that resource-sharing between different libraries is particularly effective in this field and that the *Antarctic Bibliography* is unique in giving complete coverage of one area of the earth.

12.5 Antarctic science and politics

The relationship between Antarctic science and politics has changed over the years. For the early expeditions general instructions and advice were usually drawn up by government in consultation with scientific bodies but once ships were in Antarctic waters they were beyond the control of officialdom and, provided the main object was not neglected, their commanders could do, and did, whatever they saw fit, so that the approach to science was flexible and opportunistic. An object was usually to take formal possession of any unoccupied territory but this was a task which took little time and there was opportunity for some excellent science. During the Second World War and immediately after it, it became a major purpose of scientific expeditions to establish a political presence and in consequence some of the science was trivial and of indifferent quality. Just when it appeared that Antarctica would inevitably become yet another subject of international friction the suggestion made by scientists for purely scientific reasons for an international geophysical study provided an acceptable expedient for the politicians to get round the problems which were accumulating. To perpetuate a regime something like that of the IGY of 1957–58 seemed necessary and out of this the Antarctic Treaty was born. There were, fortunately, individuals who were at home on both sides of the fence between politics and science and able to act as interpreters in the way which Dahrendorf (1988) sees as essential if two such disparate partners are to work together. The 'contrived ambiguity' in the Treaty on the sovereignty issue proved to be an inspired and effective means of lessening tensions (Triggs, 1987) and peace was established south of the 60° S parallel at the expense of supporting scientific research there. The prohibition of military activities (Article I.1) and of disposal of radioactive waste (Article V) and the agreement to exchange scientific information (Article III.1) and allow inspection of bases and facilities (Article VII) have resulted in there being little pressure from governments for the science to be utilitarian in its objectives. Scientists have, in fact, been allowed remarkable freedom in deciding what they should do in Antarctica. In his discussion of the relations of science and politics, Dahrendorf (1988) has made the point that for the two to have a sensible relationship it is necessary that, on the one hand, scientists should have created order within their own domain and that, on the other, state authorities must fund science without creating any political commitments or some kind of bias. These requirements were met under the Antarctic Treaty.

The Science that has developed under the aegis of the Treaty has been good, sometimes in the forefront, and, as frequently happens when economic application is not in mind, of great significance for mankind. The surprising thing is that beyond the science something of immense humanitarian potential has emerged. Politicans have inadvertently allowed science to become a political force and in Antarctic matters political discourse was brought to a worthy standard. The Antarctic Treaty has enabled nations of opposed ideologies to work peaceably and co-operatively side by side for nearly 30 years. It is something new in the way of treaties, not only in approach but in being short, simple and adaptable. It has probably been actually perused by more people than any other treaty document, it having been almost obligatory for recent books on Antarctica to include it as an appendix, so that it is readily available in at least a dozen publications.[2] It has also been the subject of many discussions and assessments.[3] It is unique at the international level in having led to the formulation of measures for the conservation of natural resources before those resources have been seriously depleted. It has served as a precedent for the Outer Space and Seabed Treaties. In the Antarctic Treaty science seems to have given a hope of a movement towards more rationality and more peaceful solutions to international problems. Bernal's vision of science as a major social force (Ziman, 1983) has been given a new twist in Antarctica and it may be noted that this has happened not through the knowledge gained by science but through the attitude of scientists themselves.

Outside the Treaty itself there has been another beneficial influence of Antarctica on the discourse between nations. The international conference on the ozone layer called together by the UK Prime Minister, Margaret Thatcher, in March 1989 was in accord beyond the most sanguine hopes in its determination to take steps to reduce the threat. If such unanimity and rational discussion of world problems can be sustained it will be a major step forward for humanity.

It would be naive to suppose that the Antarctic Treaty system is anything but fragile and that there are not serious threats to its continuing success.

The continuing increase in the number of adhering states is undoubtedly bringing about a dilution of the scientific commitment within the Treaty organization. As Child's (1988) analysis of South American involvement with Antarctica shows, the overriding factor there is geopolitical – geopolitics being a peculiarly Latin-American obsession with the relation between power politics and geography, with immense complications arising from interstate rivalries and suspicions of the rest of the world – with science playing no more than a token role. Apart from the geopolitical motive there is the conviction among some of the more recently joined states that Antarctica has immense resources of oil and minerals. Although thought has been given to what should happen if valuable mineral deposits are found in Antarctica the system would be put under an immense strain if this possibility became a reality. One crisis point was safely passed at the Ninth Antarctic Treaty Consultative Meeting in 1977 when possible exploitation of resources was discussed at length with the outcome that the signatories did not choose to pursue their individual national interests in isolation but to co-operate under the Treaty system.[4] However, the situation remains precarious. The Convention for the Conservation of Antarctic Marine Living Resources (1980) has been characterized as being theoretically faultless but operationally ineffective (Bonner, 1987a) and the corresponding convention on mineral expoitation (Beck, 1989a) has been replaced by a more idealistic but less practical arrangement (p. 388). The existence of South Africa as a Consultative Party has given offence outside, but not to any serious degree inside, the Treaty system. The consensus rule in the Treaty allowed no legal way which South Africa could be suspended or excluded without its consent (Parsons, 1987, p. 114). Non-Treaty states are critical of the rights of the Consultative Parties to legislate for Antarctica (Zain-Azraai, 1987) and, annually in recent years, the United Nations have debated 'the question of Antarctica' (Beck, 1988a, 1989b). The concept of a 'common heritage of mankind', which is what Antarctica is to the non-Treaty states, seems in this case to be verbiage masking an intention to reap without having sown. Legally, the concept is not applicable to Antarctica (Triggs, 1987) and given that the two superpowers are Consultative Parties it does not seem that the United Nations could affect any drastic revision of the Treaty. The Antarctic Treaty Consultative Parties now represent some four-fifths of the world's population and should undoubtedly prevail unless they differ among themselves – as perhaps they are beginning to over the exploitation of mineral resources. For the conservation movement Antarctica is a heritage to be left inviolate as a wilderness or world park. This is an admirable idea except that its proponents mistrust the Treaty system and have allied themselves with the non-Treaty states in agitating for United Nations jurisdiction. The ineffectiveness of the International Whaling Commission stands as a warning as to how this might work. To introduce a large majority of states, with no experience of Antarctic conditions but mainly preoccupied with irresponsible exploitation of imagined resources – Dr Mahathir, Malaysian Prime Minister and a leading opponent of the Treaty is reported to have said 'I have heard that the South Pole is made of gold and I want my piece of it'[5] – into the decision making process would be a recipe for disaster. Bonner (1987a) considered that, however well intentioned, conservation organizations such as Greenpeace presented the greatest threat to Antarctica. However, a recent movement in Greenpeace towards putting itself on a firmer scientific base may reduce this threat. There is no absolute deadline for a decision about Antarctic affairs – the common assumption that the Antarctic Treaty came to an end in 1991 is not correct. At that date any Consultative Party might call for its revision, any modification having to be approved at a specially called conference by a majority of the Parties, but otherwise it continues as before. Meanwhile there is a proliferation of studies analysing current and future possibilities by organizations outside the Treaty (Beck, 1988b).

12.6 The effects of bureaucracy on Antarctic science

If, indeed, the strength of the Treaty lies in it having been devised for science and being operated under advice from scientists, we have to consider whether it will continue to work so well when the nature of science itself is changing. As Ziman (1983) has argued, science is being transformed from an individualistic activity into a homogeneous collective enterprise. Antarctic science has always been collective but was well able in the past to provide for individual lines of research within the larger enterprise. Systems such as those of the US, New Zealand and Germany, which rely mostly on universities to make up their programmes, continue to provide scope for individualism. The BAS now supports university research and, although it is very much a unitary undertaking it functions on a block grant system such as Ziman (1983) considers to offer the best prospect under present circumstances of allowing individual scientists to make their fullest contribution. In the Antarctic the imposition of deadlines – an increasing tendency in modern science – which might inhibit the free development of an investigation, happens not because results are needed especially urgently but because weather and logistics dictate the timing and extent of field investigations and these are factors which have always been imperative.

The most serious threat to science comes from the growing bureaucracy which it accumulates. Ziman (1983) asked:

> In a highly collectivized and rationalized system, how long could a wild conjecture – for example Wegener's absurd theory of drifting continents – be allowed to linger on, against all the authorative opinion of its times. Would a modern Wegener, or one of his misguided disciples, get funded for a research project associated with such a ridiculous notion?

The bureaucracy which has developed in the US Antarctic programme has already been mentioned and other countries' organizations seem to be going the same way. Straws in the wind are the assumption of responsibility for the UK's achievements in Antarctic research by the Natural Environment Research Council[6] and a tendency to have 'managers' rather than 'directors' at the head of Antarctic organizations. There is, of course, a natural inclination for scientists who are better at administration than original research to head Antarctic organizations. There are also temptations to go for projects which are politically impressive and to prefer work to be done by biddable technicians rather than scientists of independent mind. There remains the reluctance of funding agencies, which was evident already in the case of the *Challenger* expedition, to provide for the evaluation and writing up of work which has been done. The ingenuity of the explorer and scientist in circumventing the bureaucrat, which Cook showed to a high degree, may reduce the negative effects of red tape but only at an expense of energy which might be used to better purpose. The proportion of funds allocated to research which is diverted to administration is not always easy to ascertain from balance sheets but it undoubtedly has increased with the years. However, one should not be dogmatic; the Council of Managers of National Programmes (MNAP, see p. 189) is another bureaucratic layer added to Antarctic affairs but its contribution is potentially a positive and constructive one which is to be welcomed.

12.7 Science and the humanist view of Antarctica

Antarctic science had its beginnings at a time when the confident anthropocentrism of the Judaeo-Christian tradition was being eroded by the thought that all parts of creation have a right to exist and nature itself has an intrinsic spiritual value. Scientists in general have been slow to adopt the new outlook, maintaining the tradition 'that, Nature being known, it may be master'd, managed, and used in the services of human life' and that their task is morally innocent (Thomas, 1983). The macho image cultivated by some explorers of Antarctica is a manifestation of something of the same attitude. Nevertheless, a confused sense of responsibility towards the natural environment is now widespread in the developed countries and is being given point by the realization that material advance has to be paid for at the price of environmental imbalance manifest in such things as the 'greenhouse effect' and the thinning of the ozone layer. Since the IGY, Antarctic science has played a prominent role in the detection and monitoring of such changes and is in the forefront in providing knowledge of the processes involved in these and other threats to the environment. It was the evidence of depletion in stratospheric ozone over Antarctica in the spring which led to the Montreal Agreement of 1987 to limit the production of CFCs, the first attempt at limiting pollution on a global scale. Perhaps even more significantly, it led to the co-operation of competing industrial concerns in research to find harmless substitutes for these substances (Anderson, 1988).

There has long been a contradiction in approach to the history of science between externalists, who look to the general intellectual and sociological milieau, and internalists, who focus on the filiation of ideas (Pyenson, 1989). Surely both must be involved in the development of science. For Antarctic science the intellectual climate, politics, commerce and availability of the necessary technology have played their essential parts but so also has the imagination and motivation of the scientists themselves.

For the individual, Antarctica has had something beyond its strictly scientific interest to heighten awareness and to give it a powerful hold on the imagination. Most scientists in Antarctica find deep satisfaction in the combination of intellectual stimulus, modest physical challenge and the select company of like-minded people; Cherry-Garrard (1965) wrote 'Exploration is the physical expression of the Intellectual Passion'. For many there has been a more deep-seated feeling which has given meaning and purpose to their scientific work. The impact of this lonely, fearsome and yet compellingly beautiful expanse is tremendous and the vast Antarctic prospect, fashioned from little more than a single pure substance and entirely aloof from man, seems for these a window to realms beyond science. For early explorers such as Ross and Hooker, the Antarctic was an awesome manifestation of the majesty of a god in whom they had implicit faith. Scott, and even more so, Wilson, were devout men and it is clear from their diaries that for both of them science was an expression of religion and that the Antarctic intensified this feeling. Another of the same persuasion was Sir Alister Hardy, who spent his formative years as a biologist working on the Southern Ocean in the *Discovery* Investigations and saw his work as a continuum extending from the purely practical object of benefiting the fishing industry, to the establishment of an ecological outlook in the conduct of life which would include the recognition of man's spiritual as well as physical nature (Hardy, 1966). It remains to be seen whether the Religious Experience Research Unit which he set up in Oxford in his later years will lead to any significant extension in the compass of science. If it does, the Antarctic may have made some contribution. For others, without attachment to any particular religious faith, Antarctica nevertheless gives powerful feelings of the numinous. Richard Byrd was one of these (Byrd, 1938, especially p. 183). More recently Pyne (1986) has ordered the geophysical facts into a humanistic picture of Antarctica as a place remote and alien to the ordinary world of human experience, a symbol simultaneously of the strength of the natural world and of our modern sense of isolation.

The human spirit expands in wilderness and we may hope that Antarctica's remote and hostile nature will ensure that most of it will remain unsullied and a place for contemplation. Over-organization and blind faith in technology may nevertheless cut Antarctic scientists off from the source of their inspiration. In the past Antarctic science has drawn much of its strength and unique character from the attitude of mind which the wilderness engenders but perhaps scientists may be becoming overweening in their confidence to extract information from this, the most unyielding and unfriendly environment on earth. Technological arrogance in the setting of some of the world's most sublime scenery seems a portent of a loss of humility and a separation from the natural environment which could lead to the decay of Antarctic science.

Endnotes

1 See *Verh. Internat. Verein. Limnol.* (1988) **23**, 629–36.

2 See, for example, Hatherton (1965); King (1969); Lewis & Smith (1973); Auburn (1982); Charney (1982); Quigg (1983); Mickleburgh (1987); Walton (1987).

3 See Auburn (1982); Charney (1982); Vicuña (1983); Beck (1986); Polar Research Board (1986); Parsons (1987); Triggs (1987); Kimball (1988); Beck (1989a,b).

4 BAS Scientific Advisory Committee minutes 14, 1977.

5 *The Guardian*, 26th November, 1988.

6 *Antarctic 2000: NERC Strategy for Antarctic Research* Natural Environment Research Council 1989.

13

Postscript

While this book has been in the final stages of preparation the thirtieth anniversary of the signing of the Antarctic Treaty – at which point a drastic revision might have been proposed – has passed without major upset. Attack on the Treaty has, however, intensified. A well financed group of non-governmental organizations, the Antarctic and Southern Ocean Coalition (ASOC), has been successful in handling the media and lobbying governments on environmental matters with the aim of achieving 'World Wilderness Park' status for Antarctica. This activity has been double edged in its effects; it has led to increasing public awareness of the global importance of Antarctic research but the initiatives and efforts of SCAR over a thirty-year period have been ignored, minor shortcomings being emphasized and achievements belittled.[1] Perversely, science has become branded as opposed to conservation although the Environmental Protocol[2] promises, given some extra effort by all operators, to maintain the continent in an essentially pristine cleanliness in the foreseeable future. On another front, critics within the United Nations, still continuing their campaign for the exclusion of South Africa from the Treaty but dropping their demands for a share of Antarctic mineral resources to shift to an anti-mining stance, have proposed an international scientific research station, to be run under United Nations auspices, which would prevent proliferation of stations and eliminate duplication of research.[3] Within the Treaty organization differences over the regulation of mining have been resolved by abandonment of the scientifically based convention, CRAMRA, in favour of a blanket prohibition as demanded by the conservation lobby. Thus have immediate threats to the consensus been averted but there are more insidious forces at work eroding the position of science in Antarctic affairs. The majority of delegates at Treaty meetings over the last 15 to 20 years have not been interested in science and this is a tendency which seems set to increase. As the supple leanness of the original Treaty document becomes encrusted with accessory verbiage intended to give it more precision, so pessimists fear that soon the only growth area in Antarctic affairs will be for lawyers. Concomitantly, financial contributions to the work of SCAR – negligible in comparison with the costs of logistics – have become increasingly inadequate. This indifference of policy makers was very evident at an international conference on Antarctic Science – Global Concerns held in Bremen, Germany, in September 1991 under the auspices of SCAR. This meeting was intended to foster awareness in the general public of the intimate connexions of Antarctica with the global systems of earth, oceans and atmosphere and its controlling influence on the environment in which we live. The science presented was impressive[4] and so was the spirit of common purpose and friendship among the numerous nationalities represented but policy makers, the media and interested laymen were conspicuous by their absence. Once again scientists had been lacking in expertise in public relations. The chief point to be made, that the understanding derived from scientific research in Antarctica is crucial to the proper management of the global environment, was reiterated by Dr David Drewry, Director of BAS, in his closing address to the conference.[4] He voiced two concerns; firstly that the need to protect and manage the Antarctic environment has compromised the freedom of scientific enquiry so that innovative research may be replaced by routine monitoring and, secondly, that increased diplomatic activity within the Antarctic Treaty regime is being conducted by representatives with little or no direct experience of work in Antarctica and with scant regard for advice from SCAR. The role of SCAR in the Treaty system needs to be clarified. It will presumably continue to co-ordinate the activities of scientists but it cannot continue to make any important contribution to the operation of the Treaty on its present inadequate funding. Science in Antarctica has given us, not only insights, which could not have been obtained by any other means, into the processes which determine the environment in which we live, alerting us to previously unrecognized dangers, but has been the driving force in establishing peace within an entire continent. Mankind cannot afford an estrangement of science and politics in Antarctica.

Endnotes

1 R. Laws (1991). Guest editorial: Antarctic politics and science are coming into conflict. *Antarctic Science* **3**, 231.

2 *SCAR Bulletin*, no. 101 (April 1991), pp. 5–7.

3 P. J. Beck (1991). Antarctica, Viña del Mar and the 1990 UN debate. *Polar Record* **27**, 211–16.
4 *Antarctic Science – Global Concerns*: Programme, Lecture abstracts, SCAR poster abstracts. Alfred Wegener Institute for Polar and Marine Research, Bremerhaven; Bremen 1991.

References

Abyzov, S. S., Filippova, S. N. & Kuznetsov, V. D. (1983). *Nocardiopsis antarcticus* – isolated from a glacial thickness in central Antarctica. *Bulletin of the Academy of Sciences of the USSR*, **4**, 559–68.

Academy of Sciences, Paris (1837). *Comptes Rendus Académie des Sciences, Paris*, **5**, 133–55.

Adie, R. J. (1964). Geological history. In *Antarctic Research*, ed. R. Priestley, R. J. Adie & G. deQ. Robin, pp. 118–62. Butterworth, London.

Agassiz, L. (1840). *Études sur les glaciers*, 2 vols. H. Nicolet, Neuchâtel.

Ahlgren, J. A. & DeVries, A. L. (1984). Comparison of antifreeze glycopeptides from several Antarctic fishes. *Polar Biology*, **3**, 93–7.

Aiken, J. (1981). The undulating oceanographic recorder mark 2. *Journal of Plankton Research*, **3**, 551–60.

Akasofu, S.-I. (1978). Recent progress in Antarctic auroral studies. In *Upper Atmosphere Research in Antarctica*, ed. L. J. Lanzerotti & C. G. Park, *Antarctic Research Series*, no. *29*, pp. 157–99. American Geophysical Union, Washington, D.C.

Alley, R. B., Blankenship, D. D., Bentley, C. R. & Rooney, S. T. (1986). Deformation of till beneath ice stream B, West Antarctica. *Nature*, **322**, 57–9.

Allison, I. (Ed.) (1983). *Antarctic Climate Research*. SCAR ICSU.

Amundsen, R. (1912). *The South Pole*. Translated from the Norwegian by A. G. Chater. In 2 vols. John Murray, London. Reprinted 1976 by C. Hurst & Co. (Publishers), London.

Anderson, A. (1988). Depletion of ozone layer drives competitors to cooperate. *Nature*, **331**, 201.

Anderson, P. J. (1974). United States aircraft losses in Antarctica. *Antarctic Journal of the United States*, **9**, 1–21.

Anon. (1875). Deutsche Entdeckungen am Südpol. *Petermanns Geographische Mitteilungen*, **21**, 312.

Anon. (1880). On Halley's Mount. *Nature*, **21**, 303–4.

Anon. (1901a). The Antarctic Expedition. *Nature*, **64**, 233–4.

Anon. (1901b). Notes. *Nature*, **64**, 656–7.

Anon. (1974). International Antarctic Glaciological Project. *Antarctic Journal of the United States*, **9**, 187–189.

Anon. (1978). Mawson veteran meets last of 'magnetic ladies'. *Antarctic*, **8**, 171.

Antarctic Meteorology (1960). *Proceedings of the Symposium held in Melbourne, February 1959*, Pergamon Press, Oxford.
Appleby, J. H. (1990). Daniel Dumaresq, D. D., F.R.S. (1712–1805) as a promoter of Anglo-Russian science and culture. *Notes and Records of the Royal Society of London*, **44**, 25–50.
Appleman, D. E. (1985). James Dwight Dana and Pacific Geology. In *Magnificent Voyagers*, ed. H. J. Viola & C. Margolis, pp. 89–117. Smithsonian Institution Press, Washington, D.C.
Appleton, E. V. (1963). Radio and the ionosphere. In *Clerk Maxwell and Modern Science*, ed. C. Domb, pp. 70–88. Athlone Press, University of London.
Arçtowski, H. (1901a). Exploration of Antarctic lands. *Geographical Journal*, **17**, 150–80.
Arçtowski, H. (1901b). The Antarctic voyage of the 'Belgica' during the years 1897, 1898 and 1899. *Geographical Journal*, **18**, 353–94.
Armitage, A. B. (1905). *Two Years in the Antarctic: being a Narrative of the British National Antarctic Expedition*. Edward Arnold, London.
Armitage, A. (1966). *Edmond Halley*. Thomas Nelson & Sons, London.
Armitage, K. B. & House, H. B. (1962). A limnological reconnaisance in the area of McMurdo Sound, Antarctica. *Limnology and Oceanography*, **7**, 36–41.
Armstrong, T. (1971). Bellingshausen and the discovery of Antarctica. *Polar Record*, **15**, 887–91.
Asahina, K. (1973). Japanese Antarctic expedition of 1911–12. In *Polar Human Biology*, ed. O. G. Edholm & E. K. E. Gunderson, pp. 8–14. William Heinemann Medical Books, London.
Atkinson, A. & Peck, J. M. (1988). A summer-winter comparison of zooplankton in the oceanic area around South Georgia. *Polar Biology*, **8**, 463–73.
Auburn, F. M. (1982). *Antarctic Law and Politics*. C. Hurst, London.
Austin, O. L., Jr. (Ed.) (1968). *Antarctic Bird Studies. Antarctic Research Series*, vol. 12. American Geophysical Union, Washington, D.C.
Austin, P. (1989). Antarctic environment: keeping the South Pole clean. *Nature*, **341**, 93.
Azam, F., Beers, J. R., Campbell, L., Carlucci, A. F., Holm-Hansen, O. & Reid, F. M. H. (1979). Occurrence and metabolic activity of organisms under the Ross Ice Shelf, Antarctica, at Station J9. *Science*, **203**, 451–3.
Bagshawe, T. W. (1939). *Two Men in the Antarctic: an Expedition to Graham Land 1920–1922*. Cambridge University Press, Cambridge.
Bailey, A. D. & Wynn-Williams, D. D. (1982). Soil microbiological studies at Signy Island, South Orkney Islands. *British Antarctic Survey Bulletin*, no. 51, 167–91.
Bakaev, V. G. (Ed.) (1966). *Atlas of Antarctica*. USSR Academy of Sciences, Moscow. (In Russian.)
Baker, A. de C. (1954). The circumpolar continuity of Antarctic plankton species. *Discovery Reports*, **27**, 203–17.
Baker, F. W. G. (1982). The first international polar year, 1882–83. *Polar Record*, **21**, 275–85.
Baker, J. H. (1970). Quantitative study of yeasts and bacteria in a Signy Island peat. *British Antarctic Survey Bulletin*, no. 23, 51–5.

Balleny, J. (1839). Discoveries in the Antarctic Ocean, in February 1839. *Journal of the Royal Geographical Society*, **9**, 517–28.

Bardin, V. I. & Suyetova, I. A. (1967). Basic morphometric characteristics for Antarctica and budget of the Antarctic ice cover. In *Proceedings of the Symposium on Pacific–Antarctic Science*, ed. T. Nagata, pp. 92–100. National Science Museum, Tokyo.

Barker, P. F. & Griffiths, D. H. (1972). The evolution of the Scotia Ridge and Scotia Sea. *Philosophical Transactions of the Royal Society of London* A, **271**, 151–83.

Barker, P. F. & Griffiths, D. H. (1977). Towards a more certain reconstruction of Gondwanaland. *Philosophical Transactions of the Royal Society of London* B, **279**, 143–59.

Barnola, J. M., Raynaud, D., Korotkevich, Y. S. & Lorius, C. (1987). Vostok ice core provides 160,000-year record of atmospheric CO_2. *Nature*, **329**, 408–14.

Barr, W. (translator) (1984). Zoological observations, Royal Bay, South Georgia 1882–1883. Part I. Karl von den Steinen. *Polar Record*, **22**, 57–71. Part II: Penguins. Karl von den Steinen. *Polar Record*, **22**, 145–58.

Barr, W. (1985). The expeditions of the First International Polar Year, 1882–3. *The Arctic Institute of North America, Technical Paper*, no. 29, Calgary.

Barraclough, D. R. (1985). Halley's Atlantic magnetic surveys. In *Historical Events and People in Geosciences*, ed. W. Schröder, pp. 163–83. Peter Lang, Frankfurt am Main.

Barrett, P. J. (1968). How Labyrinthodont was found. *Antarctic*, **5**, 107–8.

Barrow, C. J. (1983). Palynological studies in South Georgia: III. Three profiles from near King Edward Cove, Cumberland East Bay. *British Antarctic Survey Bulletin*, no. 58, 43–60.

Bartlett, H. H. (1940). The reports of the Wilkes Expedition, and the work of the specialists in science. *Proceedings of the American Philosophical Society*, **82**, 601–705.

Bartram, E. B. (1938). Botany of Second Byrd Antarctic Expedition. III Mosses. *Annals of the Missouri Botanical Garden*, **25**, 719–24.

Bauer, F. (1820). Some experiments on the fungi which constitute the colouring matter of the red snow discovered in Baffin's Bay. *Philosophical Transactions of the Royal Society of London*, **110**, 165–73.

Baust, J. G. & Edwards, J. (1979). Mechanisms of freezing tolerance in an Antarctic midge, *Belgica antarctica. Physiological Entomology*, **4**, 1–5.

Beaglehole, J. C. (Ed.) (1968). *The Journals of Captain James Cook. I. The Voyage of the Endeavour, 1768–1771.* Cambridge University Press for the Hakluyt Society.

Beaglehole, J. C. (Ed.) (1969). *The Journals of Captain James Cook. II. The Voyage of the Resolution and Adventure, 1772–1775.* Cambridge University Press for the Hakluyt Society.

Beaglehole, J. C. (1974). *The Life of Captain James Cook.* The Hakluyt Society, London.

Beagley, J. W. & King, G. A. M. (1965). The polar ionosphere. In *Antarctica*, ed. T. Hatherton, pp. 429–62. Methuen, London.

Beck, P. J. (1984). Britain's role in the Antarctic: some recent changes in organization. *Polar Record*, **22**, 85–7.

Beck, P. J. (1985). Preparatory meetings for the Antarctic Treaty 1958–59. *Polar Record*, **22**, 653–64.

Beck, P. J. (1986). *The International Politics of Antarctica*. Croom Helm, London.

Beck, P. J. (1988a). Another sterile annual ritual? The United Nations and Antarctica 1987. *Polar Record*, **24**, 207–12.

Beck, P. J. (1988b). A continent surrounded by advice: recent reports on Antarctica. *Polar Record*, **24**, 285–91.

Beck, P. J. (1989a). Convention on the regulation of Antarctic mineral resource activities: a major addition to the Antarctic Treaty system. *Polar Record*, **25**, 19–32.

Beck, P. J. (1989b). Antarctica at the UN 1988: seeking a bridge of understanding. *Polar Record*, **25**, 329–34.

Beloussov, V. V. (1964). The upper mantle project. In *Research in Geophysics*, vol. 2, *Solid Earth and Interface Phenomena*, ed. H. Odishaw, pp. 555–63. M.I.T. Press, Cambridge, Massachusetts.

Belov, M. I. (1962–63). Report map of the first Russian Antarctic expedition. *Soviet Antarctic Expedition Information Bulletin*, **4** (31), 1–6.

Belov, M. (1966). Englishman Ross must share with Bellingshausen the credit for the first precise computation of the location of the South Magnetic Pole. *Nauka i zhizn*, August 1966, no. 8, 21–3. (In Russian.)

Benninghoff, W. S. & Benninghoff, A. S. (1985). Wind transport of electrostatically charged particles and minute organisms in Antarctica. In *Antarctic Nutrient Cycles and Food Webs*, ed. W. R. Siegfried, P. R. Condy & R. M. Laws, pp. 592–6. Springer-Verlag, Berlin.

Benninghoff, W. S. & Bonner, W. N. (1985). *Man's Impact on the Antarctic Environment: a Procedure for Evaluating Impacts from Scientific and Logistic Activities*. SCAR, Scott Polar Research Institute, Cambridge.

Bentley, C. R. (1964). The structure of Antarctica and its ice cover. In *Research on Geophysics*, vol. 2, *Solid Earth and Interface Phenomena*, ed. H. Odishaw, pp. 335–89. M.I.T. Press, Cambridge, Massachusetts.

Bentley, C. R. (1965). The land beneath the ice. In *Antarctica*, ed. T. Hatherton, pp. 259–77. Methuen, London.

Bentley, C. R. (1983). Crustal structure of Antarctica from geophysical evidence – a review. In *Antarctic Earth Science*, ed. R. L. Oliver, P. R. James & J. B. Jago, pp. 491–7. Cambridge University Press, Cambridge.

Bentley, C. R. (1984). The Ross Ice Shelf geophysical and glaciological survey (RIGGS): Introduction and summary of measurements performed. *Antarctic Research Series*, no. 42, 1–20. American Geophysical Union, Washington, D.C.

Berg, T. E. & Black, R. F. (1966). Preliminary measurements of growth of nonsorted polygons. In *Antarctic Soils and Soil Forming Processes*, ed. J. C. F. Tedrow, *Antarctic Research Series*, no. 8, pp. 61–108. American Geophysical Union, Washington, D.C.

Berkowitz, E. A. (1986). *Frozen in Place: American Policy and Practice in Antarctica*. Thesis for Master of Philosophy, University of Cambridge.

Bernacchi, L. C. (1901). Meteorological observations on the 'Southern Cross' expedition to the Antarctic, 1899–1900. In *The Antarctic Manual*, ed. G. Murray, pp. 50–6. Royal Geographical Society, London.

Bernacchi, L. C. (1938). *Saga of the 'Discovery'*. Blackie, London.

Bernal, J. D. (1965). *Science in History*, Watts, London.
Bernstein, R. E. (1985). The Scottish National Antarctic Expedition 1902–4. *Polar Record*, **22**, 379–92.
Berry, R. J., Bonner, W. N. & Peters, J. (1979). Natural selection in house mice (*Mus musculus*) from South Georgia (South Atlantic Ocean). *Journal of Zoology, London*, **189**, 385–98.
Bertram, C. (1987). *Antarctica, Cambridge, Conservation and Population: a Biologist's Story*. Published privately, Petworth, Sussex.
Bertram, G. C. L. (1940). The biology of the Weddell and crabeater seals. With a study of the comparative behaviour of the Pinnepedia. *Scientific Reports of the British Graham Land Expedition, 1934–37*, **1**, 1–139. Trustees of the British Museum, London.
Bertrand, K. J. (1971). *Americans in Antarctica 1775–1948*. American Geographical Society, Special Publication no. 39, New York.
Biermann, L. (1985). On the history of the solar wind concept. In *Historical Events and People in Geosciences*, ed. W. Schröder, pp. 39–47. Peter Lang, Frankfurt-am-Main.
Bishop, A. C. (1981). The development of the concept of continental drift. In *The Evolving Earth: Chance, Change and Challenge*, ed. L. R. M. Cocks, pp. 155–64. British Museum (Natural History) & Cambridge University Press.
Bishop, J. F. & Walton, J. W. L. (1977). Problems encountered when monitoring tidal movement in extremely cold conditions. *Polar Record*, **18**, 502–5.
Black, R. F. (1973). Cryomorphic processes and micro-relief features, Victoria Land, Antarctica. In *Research on Polar and Alpine Geomorphology; Proceedings of the third Guelph Symposium in Geomorphology, Norwich*, ed. B. D. Fahey & R. D. Thompson, pp. 11–24.
Blay, S. K. N. & Tsamenyi, B. M. (1990). Australia and the Convention for the Regulation of Antarctic Mineral Resource Activities (CRAMRA). *Polar Record*, **26**, 195–202.
Block, W. (1984a). Terrestrial microbiology, invertebrates and ecosystems. In *Antarctic Ecology*, ed. R. M. Laws, vol. 1, pp. 163–236. Academic Press, London.
Block, W. (1984b). A comparative study of invertebrate supercooling at Signy Island, maritime Antarctic. *British Antarctic Survey Bulletin*, no. 64, 67–76.
Block, W. (1985a). Arthropod interactions in an Antarctic community. In *Antarctic Nutrient Cycles and Food Webs*, ed. W. R. Siegfried, P. R. Condy & R. M. Laws, pp. 614–19. Springer-Verlag, Berlin.
Block, W. (1985b). Ecological and physiological studies of terrestrial arthropods in the Ross Dependency 1984–85. *British Antarctic Survey Bulletin*, no. 68, 115–22.
Block, W. (1990). Cold tolerance of insects and other arthropods. *Philosophical Transactions of the Royal Society of London*, B, **326**, 613–33.
Block, W., Burn, A. J. & Richard, K. J. (1984). An insect introduction to the maritime Antarctic. *Biological Journal of the Linnean Society*, **23**, 33–9.
Bloxham, J., Gubbins, D. & Jackson, A. (1989). Geomagnetic secular variation. *Philosophical Transactions of the Royal Society of London*, A, **329**, 415–502.
Bogart, P. S. (1988). Environmental threats in Antarctica. *Oceanus*, **31**, 104–7.
Bone, D. G. (1986). An LHPR system for adult Antarctic krill (*Euphausia superba*). *British Antarctic Survey Bulletin*, no. 73, 31–46.

Bonner, W. N. (1964). Population increase in the fur seal *Arctocephalus tropicalis gazella* at South Georgia. In *Biologie Antarctique*, ed. R. Carrick, M. Holdgate & J. Prévost, pp. 433–43. Hermann, Paris.

Bonner, W. N. (1968). The fur seal of South Georgia. *British Antarctic Survey Scientific Reports*, no. 56, 1–82.

Bonner, W. N. (1981). The krill problem in Antarctica. *Oryx*, **16**, 31–7.

Bonner, W. N. (1987a). Recent developments in Antarctic conservation. In *The Antarctic Treaty Regime: Law, Environment and Resources*, ed. G. D. Triggs, pp. 143–9, Cambridge University Press, Cambridge.

Bonner, W. N. (1987b). Antarctic science and conservation – the historical background. *Environment International*, **13**, 19–25.

Bonner, W. N. (1988). Review of 'International Research in the Antarctic' by R. Fifield. *Polar Record*, **24**, 254–5.

Borchgrevink, C. E. (1901). *First on the Antarctic Continent*. George Newnes, London.

Boutron, C. F. & Patterson, C. C. (1983). The occurrence of lead in Antarctic recent snow, firn deposited over the last two centuries and prehistoric ice. *Geochimica et Cosmochimica Acta*, **47**, 1355–68.

Bowen, M. J. (1970). Mind and nature: the physical geography of Alexander von Humboldt. *Scottish Geographical Magazine*, **86**, 222–33.

Bower, F. O. (1913). Sir Joseph Dalton Hooker, 1817–1911. In *Makers of British Botany*, ed. F. W. Oliver, pp. 302–23. Cambridge University Press, Cambridge.

Bowman, K. P. (1988). Global trends in total ozone. *Science*, **239**, 48–50.

Boyd, W. L., Staley, J. T. & Boyd, J. W. (1966). Ecology of soil microorganisms of Antarctica. In *Antarctic Soils and Soil Forming Processes*, ed. J. C. F. Tedrow, pp. 125–59, *Antarctic Research Series* no. 8, American Geophysical Union, Washington, D.C.

Brennecke, W. (1921). The oceanographic work of the German Antarctic Expedition 1911 to 1912. Aus dem *Archiv der Seewarte*, **39**, no. 1. Hamburg.

Bretterbauer, K. (1985). J. Payer, C. Weyprecht, H. Wilczek: the promoters of international polar research. In *Historical Events and People in Geosciences*, ed. W. Schröder, pp. 49–57. Peter Lang, Frankfurt-am-Main.

Brigham, L. W. (1988). The Soviet Antarctic program. *Oceanus*, **31**, 87–92.

British Museum (1914–35). *British Antarctic ('Terra Nova') Expedition, 1910. Natural History Reports*, 8 vols. Trustees of the British Museum, London.

British Museum (1964). *British Antarctic ('Terra Nova') Expedition, 1910. Natural History Reports, Geology*. British Museum (Natural History), London.

Broady, P. A. (1979). The Signy Island terrestrial reference sites: IX. The ecology of the algae of site 2, a moss carpet. *British Antarctic Survey Bulletin*, no. 47, 13–29.

Brooks, R. E., Natani, K., Shurley, J. T., Pierce, C. M. & Joern, A. T. (1973). An Antarctic sleep and dream laboratory. In *Polar Human Biology*, ed. O. G. Edholm & E. K. E. Gunderson, pp. 322–41. William Heinemann Medical Books, London.

Brotherhood, J. R. (1973). Studies on energy expenditure in the Antarctic. In *Polar Human Biology*, ed. O. G. Edholm & E. K. E. Gunderson, pp. 182–92. William Heinemann Medical Books, London.

Brown, R. G. B. (1987). Oceangoing animals. *Science*, **238**, 222–3.

Brown, R. N. R. (1906). Antarctic botany: its present state and future problems. *Scottish Geographical Magazine*, **22**, 473–84.

Brown, R. N. R. (1923). *A Naturalist at the Poles: the Life, Work and Voyages of Dr W. S. Bruce the Polar Explorer*. Seeley, Service & Co., London.

Brown, R. N. R., Pirie, J. H. H. & Mossman, R. C. (1906). *The Voyage of the 'Scotia'*. William Blackwood, Edinburgh. Reprint 1978, C. Hurst (Publishers), London.

Bruce, W. S. (1896). Cruise of the 'Balaena' and the 'Active' in the Antarctic seas, 1892–3. *Geographical Journal*, **7**, 502–21.

Bruce, W. S. (Ed.) (1907–1915). *Report on the Scientific Results of the Voyage of S. Y. 'Scotia' during the Years 1902, 1903, and 1904*. The Scottish Oceanographic Laboratory, Edinburgh.

Bruce, W. S., King, A. & Wilton, D. W. (1916–17). The temperatures, specific gravities and salinities of the Weddell Sea and of the North and South Atlantic Ocean. *Transactions of the Royal Society of Edinburgh*, **51**, 71–169.

Buck, K. R. & Garrison, D. L. (1983). Protists from the ice-edge region of the Weddell Sea. *Deep-Sea Research*, **30**, 1261–77.

Budd, W. F. (1986). The Antarctic Treaty as a scientific mechanism (post-IGY) – contributions of Antarctic scientific research. In *Antarctic Treaty System: An Assessment*, pp. 103–51. Polar Research Board, National Research Council, National Academy Press, Washington, D.C.

Budd, W. F., Dingle, W. R. J. & Radok, U. (1966). The Byrd snowdrift project; outline and basic results. In *Studies in Antarctic Meteorology*, ed. M. Rubin, pp. 71–134, *Antarctic Research Series*, no. 9. American Geophysical Union, Washington, D.C.

Bugayev, V. A. (1963). Climatic zones of Antarctica. *Meteorological Researches, II Section of IGY Program* no. 5, 5–38, (Moscow, 1963).

Bull, H. J. (1896). *The Cruise of the 'Antarctic' to the South Polar Regions*. Edward Arnold, London. (Reprint 1984 by Bluntisham Books and The Paradigm Press, Bungay, Suffolk.)

Bunt, J. S. (1963). Diatoms of antarctic sea-ice as agents of primary production. *Nature*, **199**, 1255–7.

Bunt, J. S. (1967). Some characteristics of microalgae isolated from Antarctic sea ice. In *Biology of the Antarctic Seas III*, ed. G. A. Llano & W. L. Schmitt, pp. 1–14, *Antarctic Research Series*, no. 11. American Geophysical Union, Washington, D.C.

Bunt, J. S., Owens, O. van H. & Hoch, G. (1966). Exploratory studies on the physiology and ecology of a psychrophilic marine diatom. *Journal of Phycology*, **2**, 96–100.

Burkholder, P. R. & Sieburth, J. M. (1961). Phytoplankton and chlorophyll in the Gerlache and Bransfield Straits of Antarctica. *Limnology and Oceanography*, **6**, 45–52.

Burstyn, H. L. (1968). Science and government in the nineteenth century: the Challenger Expedition and its report. *Bulletin de l'Institute Océanographique, Monaco*, special number 2, 603–13.

Burt, W. H. (Ed.) (1071). *Antarctic Pinnipedia. Antarctic Research Series* no. 18. American Geophysical Union, Washington, D.C.

Bushnell, V. C. (Ed.) (1975). *History of Antarctic Exploration and Scientific Investigation. Antarctic Map Folio Series*, no. 19. American Geographical Society, New York.

Butler, R. (1988). *Breaking the Ice*. Albatross Books, Sutherland, New South Wales.
Butson, A. R. C. (1949). Acclimatization to cold in the Antarctic. *Nature*, **163**, 132–3.
Byrd, R. E. (1930). *Little America: Aerial Exploration in the Antarctic: the Flight to the South Pole*. G. P. Putnam's Sons, New York.
Byrd, R. E. (1935). *Discovery: the Story of the Second Byrd Antarctic Expedition*. G. P. Putnam's Sons, New York.
Byrd, R. E. (1938). *Alone*. G. P. Putnam's Sons, New York.
Calkin, P. E. (1964). Geomorphology and glacial geology of the Victoria Valley system, southern Victoria Land, Antarctica. *Ohio State University Institute of Polar Studies*, report no. 10.
Callaghan, T. V., Smith, R. I. L. & Walton, D. W. H. (1976). The I.B.P. Bipolar Botanical Project. *Philosophical Transactions of the Royal Society of London* B, **274**, 315–19.
Calman, W. T. (1937). James Eights, a pioneer antarctic naturalist. *Proceedings of the Linnean Society*, Session 149, 171–84.
Cameron, R. E. (1963). The role of soil science in space exploration. *Space Science Review*, **2**, 297–312.
Cameron, R. E. (1971). Antarctic soil microbial and ecological investigations. In *Research in the Antarctic*, ed. L. O. Quam, pp. 137–89. American Association for the Advancement of Science, Washington, D.C.
Cameron, R. E., Honour, R. C. & Morelli, F. A. (1977). Environmental impact studies of Antarctic sites. In *Adaptations within Antarctic Ecosystems*, ed. G. A. Llano, pp. 1157–76. Smithsonian Institution, Washington, D.C.
Cameron, R. E., King, J. & David, C. N. (1970). Microbiology, ecology and microclimatology of soil sites in dry valleys of southern Victoria Land, Antarctica. In *Antarctic Ecology*, ed. M. W. Holdgate, pp. 702–16. Academic Press, London.
Cameron, R. E. & Morelli, F. A. (1974). Viable microorganisms from ancient Ross Island and Taylor Valley drill core. *Antarctic Journal of the United States*, **9**, 113–16.
Campbell, I. B. & Claridge, G. G. C. (1987). *Antarctica: Soils, Weathering Processes and Environment*, Elsevier, Amsterdam.
Campbell, I. T. (1981). Energy intakes on sledging expeditions. *British Journal of Nutrition*, **45**, 89–94.
Campbell, P. (1987). The Antarctic cornucopia. *Nature*, **329**, 387.
Cardot, J. (1913). *Deuxième Expédition antarctique française (1908–1910) commandée par le Dr. Jean Charcot. Mousses*. Masson, Paris.
Carlucci, A. F. & Cuhel, R. L. (1977). Vitamins in the south polar seas: distribution and significance of dissolved and particulate vitamin B_{12}, thiamine, and biotin in the southern Indian Ocean. In *Adaptations within Antarctic Ecosystems*, ed. G. A. Llano, pp. 115–28. Smithsonian Institution, Washington, D.C.
Carpenter, D. L. (1963). Whistler evidence of a 'knee' in the magnetospheric ionization density profile. *Journal of Geophysical Research*, **68**, 1675–82.
Carpenter, G. H. (1908). Aptera. In *National Antarctic Expedition 1901–1904 Natural History*, vol. IV, *Zoology (Various Invertebrata)*, pp. 1–5. British Museum (Natural History), London.
Carpenter, G. H. (1921). Insecta. Part I – Collembola. In *British Antarctic*

('Terra Nova') Expedition, 1910. Natural History Reports, vol. III, *Zoology*, pp. 259–66. British Museum (Natural History), London.

Carpenter, K. J. (1986). *The History of Scurvy and Vitamin C.* Cambridge University Press, Cambridge.

Carpenter, W. B. (1868). Preliminary report of dredging operations in the seas to the north of the British Islands, carried on in H.M.S. *Lightning*. *Proceedings of the Royal Society of London*, **17**, 168–200.

Carr, D. J. (1983). The books that sailed with the Endeavour. *Endeavour*, N.S. **7**, 194–201.

Carrick, R. (1964). Southern seals as subjects for ecological research. In *Biologie Antarctique*, ed. R. Carrick, M. Holdgate & J. Prévost, pp. 421–32. Hermann, Paris.

Carrick, R., Holdgate, M. & Prévost, J. (Eds.) (1964). *Biologie Antarctique.* Hermann, Paris.

Carrick, R. & Ingham, S. E. (1960). Ecological studies of the southern elephant seal, *Mirounga leonina* (L.), at Macquarie Island and Heard Island. *Mammalia*, **24**, 325–42.

Carrick, R. & Ingham, S. E. (1970). Ecology and population dynamics of Antarctic sea birds. In *Antarctic Ecology*, ed. M. W. Holdgate, vol. 1, pp. 505–25. Academic Press, London.

Carroll, J. J. (1982). Long-term means and short-time variability of the surface energy balance components at the South Pole. *Journal of Geophysical Research*, **87** (C6), 4277–86.

Carter, P. A. (1979). *Little America: Town at the End of the World.* Columbia University Press, New York.

Cartwright, D. E. (1983). Detection of large-scale ocean circulation and tides. *Philosophical Transactions of the Royal Society of London* A, **309**, 361–70.

Cawood, J. (1977). Terrestrial magnetism and the development of international collaboration in the early nineteenth century. *Annals of Science*, **34**, 551–87.

Cawood, J. (1979). The magnetic crusade: science and politics in early Victorian Britain. *Isis*, **70**, 493–518.

Chambers, M. J. G. (1970). Investigations of patterned ground at Signy Island, South Orkney Islands: IV. Long-term experiments. *British Antarctic Survey Bulletin* no. 23, 93–100.

Charcot, J.-B. (1906). *Le 'Française' au Pôle Sud: Journal de l'Expédition Antarctique Française 1903–1905.* Ernest Flammarion, Paris.

Charcot, J. (1911). *The Voyage of the 'Why Not?' in the Antarctic.* Hodder & Stoughton, London.

Charney, J. I. (Ed.) (1982). *The New Nationalism and the Use of Common Spaces.* Allenheld, Osmun; Totowa, New Jersey.

Charnock, H. (1985). George Edward Deacon, 21 March 1906 to 16 November 1984. *Biographical Memoirs of Fellows of the Royal Society*, **31**, 111–42.

Cherry-Garrard, A. (1965). *The Worst Journey in the World.* Chatto & Windus, London. (New edition.)

Child, J. (1988). *Antarctica and South American Geopolitics: Frozen Lebensraum.* Praeger, New York.

Chipman, E. (1986). *Women on the Ice: a History of Women in the Far South.* Melbourne University Press.

Chree, C. (Ed.) (1909). *National Antarctic Expedition 1901–1904. Magnetic Observations.* Royal Society of London.

Chree, C. (1923). Magnetic phenomena in the region of the South Magnetic Pole. *Proceedings of the Royal Society of London* A, **104**, 165–91.

Christensen, L. (1935). *Such is the Antarctic*. Hodder & Stoughton, London.

Christensen, L. (1939). Recent reconnaissance flights in the Antarctic. *Geographical Journal*, **94**, 192–203.

Christie, E. W. H. (1951). *The Antarctic Problem: An Historical and Political Study*. George Allen & Unwin, London.

Chun, C. (Ed.) (1902–1940). *Wissenschaftliche Ergebnisse der Deutschen Tiefsee-Expedition auf dem Dampfer 'Valdivia' 1898–1899*, 24 vols. Gustav Fischer, Jena. (Edited after Chun's death by A. Brauer, E. Vanhöffen & C. Apstein.)

Clapham, A. R. (1976). Introductory remarks to a review of the United Kingdom contribution to the International Biological Programme. *Philosphical Transactions of the Royal Society of London* B, **274**, 277–81.

Clarke, A. (1980). A reappraisal of the concept of metabolic cold adaptation in polar marine invertebrates. *Biological Journal of the Linnean Society*, **14**, 77–92.

Clarke, G. K. C. (1987a). A short history of scientific investigations on glaciers. *Journal of Glaciology, Special Issue 1987*, 4–24.

Clarke, G. K. C. (1987b). Fast glacier flow: ice streams, surging, and tidewater glaciers. *Journal of Geophysical Research*, **92** (B9), 8835–41.

Clarke, J. M. (1916). The reincarnation of James Eights, antarctic explorer. *Scientific Monthly*, **2**, 189–202.

Clarke, M. R. (1969). A new midwater trawl for sampling discrete depth horizons. *Journal of the Marine Biological Association of the United Kingdom*, **49**, 945–60.

Clarke, M. R. (1985). Marine habitats – Antarctic cephalopods. In *Key Environments: Antarctica*, ed. W. N. Bonner & D. W. H. Walton, pp. 193–200. Pergamon Press, Oxford.

Clarkson, P. D. (1983). The reconstruction of Lesser Antarctica within Gondwana. In *Antarctic Earth Science*, ed. R. L. Oliver, P. R. James & J. B. Jago, p. 599, Cambridge University Press, Cambridge.

Clerk, H. (1846). Meteorological observations made on Board Her Majesty's (hired) Bark Pagoda, from January 10 to June 20, 1845, between −20° and −68° Latitude, and 0° and 120° East Longitude. *Philosophical Transactions of the Royal Society of London*, **136**, 433–40.

Clowes, A. J. (1933). Influence of the Pacific on the circulation in the southwest Atlantic Ocean. *Nature*, **131**, 189–91.

Colbert, E. H. (1971). Antarctic fossil vertebrates and Gondwanaland. In *Research in the Antarctic*, ed. L. O. Quam, pp. 685–701, American Association for the Advancement of Science, Washington, D.C.

Colebrook, J. M. (1979). Continuous plankton records: monitoring the plankton of the North Atlantic and the North Sea. In *Monitoring the Marine Environment*, ed. D. Nichols, pp. 87–102, Institute of Biology, London.

Coleman-Cooke, J. (1963). *Discovery II in the Antarctic: the Story of British Research in the Southern Seas*. Odhams Press, London.

Collins, N. J., Baker, J. H. & Tilbrook, P. J. (1975). Signy Island, maritime Antarctic. In *Structure and Function of Tundra Ecosystems*, ed. T. Rosswall & O. W. Heal, *Ecological Bulletin, Stockholm*, **20**, 345–74.

Committee on Polar Research (1961). *Science in Antarctica. Part I. The Life*

Sciences in Antarctica. National Academy of Sciences – National Research Council, Publication no. 839.

Compton, A. H. (1934). Studies of cosmic rays. *Carnegie Institution of Washington Yearbook*, **33**, 316–21.

Conklin, E. G. (1940). Connection of the American Philosphical Society with our first national exploring expedition. *Proceedings of the American Philosophical Society*, **82**, 519–41.

Cook, F. A. (1900). *Through the First Antarctic Night 1898–1899*. William Heinemann, London.

Cook, J. (1776). The method taken for preserving the health of the crew of His Majesty's Ship the *Resolution* during her late voyage round the World. *Philosophical Transactions of the Royal Society of London*, **66**, 402–6.

Cook, J. (1777). *A Voyage towards the South Pole, and Round the World*. 2 vols. London.

Corte, A. & Daglio, C. A. N. (1964). A mycological study of the Antarctic air. In *Biologie Antarctique*, ed. R. Carrick, M. Holdgate & J. Prévost, pp. 115–20. Hermann, Paris.

Cotter, C. H. (1981). Captain Edmond Halley R.N., F.R.S. *Notes & Records of the Royal Society of London*, **36**, 61–77.

Court, A. (1942). Tropopause disappearance during the Antarctic winter. *Bulletin of the American Meteorological Society*, **23**, 220–238.

Court, A. (1951). Antarctic atmospheric circulation. In *Compendium of Meteorology*, ed. T. F. Malone, pp. 917–41. American Meteorological Society, Boston, Massachusetts.

Cozzens, S. E. (1981). Citation analysis of antarctic research. *Antarctic Journal of the United States*, **16**, 233–5.

Crary, A. P. (1982). International Geophysical Year: Its evolution and U.S. participation. *Antarctic Journal of the United States*, **17**, 1–6.

Croll, J. (1897). *Climate and Time*. 4th. edition, London.

Croxall, J. P. (1984). Seabirds. In *Antarctic Ecology*, ed. R. M. Laws, pp. 533–619. Academic Press, London.

Croxall, J. P. & Prince, P. A. (1980). Food, feeding ecology and ecological segregation of seabirds at South Georgia. In *Ecology in the Antarctic*, ed. W. N. Bonner & R. J. Berry, pp. 103–31. Academic Press, London.

Cumpston, J. S. (1968). *Macquarie Island*. Antarctic Division, Department of External Affairs, Australia.

Cushing, D. H. (1959). The seasonal variation in oceanic production as a problem in population dynamics. *Journal du Conseil Permanent International pour l'Exploration de la Mer*, **24**, 455–64.

Dahrendorf, R. (1988). Science and politics: expectations, errors and clarifications. *Interdisciplinary Science Reviews*, **13**, 12–17.

Dalrymple, P. C. (1966). A physical climatology of the Antarctic plateau. In *Studies in Antarctic Meteorology*, ed. M. J. Rubin, pp. 195–231, *Antarctic Research Series* no. 9. American Geophysical Union, Washington, D.C.

Dalrymple, P. C., Lettau, H. H. & Willaston, S. H. (1966). South Pole micrometeorology program: data analysis. In *Studies in Antarctic Meteorology*, ed. M. J. Rubin, pp. 13–57, *Antarctic Research Series* no. 9, American Geophysical Union, Washington, D.C.

Dalton, J. (1828). On the height of the aurora borealis above the surface of the earth; particularly one seen on the 29th of March, 1826. *Philosphical Transactions of the Royal Society of London*, **118**, 291–302.

Daniels, P. C. (1973). The Antarctic Treaty. In *Frozen Future: a Prophetic Report from Antarctica*, ed. R. S. Lewis & P. M. Smith, pp. 31–45. Quadrangle Books, New York.

Darling, C. A. & Siple, P. A. (1941). Bacteria of Antarctica. *Journal of Bacteriology*, **42**, 83–98.

Darlington, J. (1957). *My Antarctic Honeymoon: a Year at the Bottom of the World.* Frederick Muller, London.

Darwin, C. (1839). Note on a rock seen on an iceberg in 61° south latitude. *Journal of the Royal Geographical Society*, **9**, 528–9.

Darwin, G. (1910). The tidal observations of the British Antarctic Expedition, 1907. *Proceedings of the Royal Society of London* A, **84**, 403–22.

Davenport, J. & Fogg, G. E. (1989). The invertebrate collections of the Erebus and Terror Antarctic expedition: a missed opportunity. *Polar Record*, **25**, 323–7.

David, P. M. (1966). Pelagic organisms in the superficial layers. In *Symposium on Antarctic Oceanography*, ed. R. I. Currie, pp. 24–9. SCAR, Scott Polar Research Institute, Cambridge.

David, T. W. E. & Priestley R. E. (1914). *British Antarctic Expedition 1907–9. Reports on the Scientific Investigations. Geology*, vol. I. *Glaciology, Physiography, Stratigraphy, and Tectonic Geology of South Victoria Land.* (With short notes on palaeontology by T. Griffith Taylor & E. J. Goddard.) Heinemann, London.

David, T. W. E., Smeeth, W. F. & Schofield, J. A. (1895). Notes on Antarctic rocks collected by Mr C. E. Borchgrevink. *Journal and Proceedings of the Royal Society of New South Wales*, **29**, 461–92.

David, M. (1937). *Professor David.* Edward Arnold, London.

Davis, R. C. (1980). Peat respiration and decomposition in Antarctic terrestrial moss communities. *Biological Journal of the Linnean Society*, **14**, 39–49.

Davis, R. C. (1981). Structure and function of two Antarctic terrestrial moss communities. *Ecological Monographs*, **51**, 125–43.

Day, A. (1967). The Admiralty Hydrographic Service 1795–1919. H.M.S.O., London.

Deacon, G. E. R. (1933). A general account of the hydrology of the South Atlantic Ocean. *Discovery Reports*, **7**, 171–238.

Deacon, G. E. R. (1934). Die Nordgrenzen antarktischen und subantarktischen Wassers im Weltmeer. *Annalen der Hydrographie und Maritimen Meteorologie*, **4**, 129–36. (Translated in *Oceanography: Concepts and History*, ed. M. B. Deacon, pp. 89–97, Dowden, Hutchinson & Ross, Stroudsburg, Pennsylvania, 1978.)

Deacon, G. E. R. (1937). The hydrology of the Southern Ocean. *Discovery Reports*, **15**, 1–124.

Deacon, G. E. R. (1958). Antarctic meteorology. *Nature*, **182**, 900–1.

Deacon, G. E. R. (1968). Early scientific studies of the Antarctic Ocean. *Bulletin de l'Insitut Océanographique, Monaco*, No. spécial 2, 269–79.

Deacon, G. E. R. (1975). The oceanographical observations of Scott's last expedition. *Polar Record*, **17**, 391–419.

Deacon, G. (1984). *The Antarctic Circumpolar Ocean.* Cambridge University Press, Cambridge.

Deacon, M. (1971). *Scientists and the Sea 1650–1900: a Study of Marine Science*. Academic Press, London.

Debenham, F. (ed.) (1945). *The Voyage of Captain Bellingshausen to the Antarctic Seas 1819–1821*. Translated from the Russian. Hakluyt Society, London.

Debenham, F. (1948). The problem of the Great Ross Barrier. *Geographical Journal*, **112**, 196–218.

Delépine, R. (1964). *La biologie depuis 1945 dans les Iles Australes Français*. CNFRA–Biologie **1** (11), 173–203.

Dell, R. K. (1966). Benthic faunas of the Antarctic. In *Symposium on Antarctic Oceanography*, ed. R. I. Currie, pp. 110–18. SCAR, Scott Polar Research Institute, Cambridge.

De Maria, M. & Russo, A. (1989). Cosmic ray romancing: the discovery of the latitude effect and the Compton-Millikan controversy. *Historical Studies in the Physical and Biological Sciences*, **19**, 211–66.

De Monchaux, C., Davis, A. & Edholm, O. G. (1979). Psychological studies in the Antarctic. *British Antarctic Survey Bulletin* no. 48, 93–7.

De Pradel de Lamase, M. (1950). Rappel historique de la mission de Dumont d'Urville. *La Revue Maritime, Paris*, **49**, 657–64.

DeVries, A. L. & Somero, G. N. (1971). The physiology and biochemistry of low temperature adaptation in Antarctic marine organisms. In *Symposium on Antarctic Ice and Water Masses*, ed. G. E. R. Deacon, pp. 101–13. SCAR, Scott Polar Research Institute, Cambridge.

Dibbern, J. S. (1976). The first attempts at motor transport in Antarctica, 1907–1911. *Polar Record*, **18**, 259–67.

Dickman, S. (1989). Women accept polar challenge. *Nature*, **341**, 273.

Dieckmann, G., Hemleben, C. & Spindler, M. (1987). Biogenic and mineral inclusions in a green iceberg from the Weddell Sea, Antarctica. *Polar Biology*, **7**, 31–3.

Dietz, R. S. (1948). Deep scattering layer in the Pacific and Antarctic Oceans. *Journal of Marine Research*, **7**, 430–42.

Dietz, R. S. (1961). Ocean-basin evolution by sea-floor spreading. *Nature*, **190**, 854–7.

Doake, C. S. M. (1976). Land beneath the Antarctic ice. *Geographical Magazine*, **48**, 670–4.

Dodge, C. W. & Baker, G. E. (1938). Botany of Second Byrd Antarctic Expedition. II. Lichens and lichen parasites. *Annals of the Missouri Botanical Gardens*, **25**, 515–718.

Dorst, J. & Milon, Ph. (1964). Acclimatation et conservation de la nature dans les iles subantarctiques Française. In *Biologie Antarctique*, ed. R. Carrick, M. Holdgate & J. Prévost, pp. 579–88, Hermann, Paris.

Dort, W., Jr. (1981). The mummified seals of southern Victoria Land, Antarctica. In *Terrestrial Biology*, III, ed. B. C. Parker, pp. 123–54, *Antarctic Research Series* no. 30. American Geophysical Union, Washington, D.C.

Drewry, D. J. (1975). Radio echo sounding map of Antarctica (~ 90° E to 180°). *Polar Record*, **17**, 359–74.

Drewry, D. (1990). R. R. S. James Clark Ross: A new vessel for the British Antarctic Survey. *Ocean Challenge*, **1**, 26–30.

Drewry, D. J., Jordan, S. R. & Jankowski, E. (1982). Measured properties of the Antarctic ice sheet: surface configuration, ice thickness, volume and bedrock characteristics. *Annals of Glaciology*, **3**, 83–91.

Drewry, D. J., McIntyre, N. F. & Cooper, P. (1985). The Antarctic ice sheet: a surface model for satellite altimeter studies. In *Models in Geomorphology*, ed. M. J. Woldenberg, pp. 1–23. Allen & Unwin, Boston.

Drewry, D. J., McIntyre, N. F., Cooper, A. P. R. & Novotny, E. (1984). Modelling the surface of the Antarctic ice sheet for satellite radar altimeter studies (abstract). *Annals of Glaciology*, **5**, 202.

Drygalski, E. von (1904). *Zum Kontinent des eisigen Südens*. Georg Reimer, Berlin.

Drygalski, E. von (1911–1931). *Deutsche Südpolar-Expedition 1901–1903*, vols I–XV, Georg Reimer, Berlin; vols XVI–XX, Walter de Gruyter, Berlin.

Drygalski, E. von (1928). The oceanographical problems of the Antarctic. In *Problems of Polar Research*, ed. W. L. G. Joerg, pp. 269–83. American Geographical Society Special Publication no. 7, New York.

Duchêne, J. C. (1989). *Kerguelen: Recherches au Bout du Monde*. Territoire des Terres Australes et Antarctiques Françaises, Mission de Recherche, Paris.

Dudeney, J. R., Jarvis, M. J., Kressman, R. I., Pinnock, M., Rodger, A. S. & Wright, K. H. (1982). Ionospheric troughs in Antarctica. *Nature*, **295**, 307–8.

Dudeney, J. R. & Piggott, W. R. (1978). Antarctic ionosphere research. In *Upper Atmosphere Research in the Antarctic*, ed. L. H. Lanzerotti & C. G. Park, pp. 200–35, *Antarctic Research Series* no. 29. American Geophysical Union, Washington, D.C.

Dumoulin, V. & Coupvent-Desbois, A. E. (1842). *Voyage au Pole Sud et dans l'Océanie* . . . etc. *Physique, Tome Premier*. Gide, Paris.

Dunbar, M. J. (Ed.) (1977). *Polar oceans*. Proceedings of the SCOR/SCAR Conference, Montreal, May 1974. Arctic Institute of North America, Calgary.

D'Urville, M. J. Dumont (1847). *Voyage au Pole Sud et dans l'Océanie sur les Corvettes l'Astrolabe et la Zélée*, 10 vols., ed. M. Jacquinot. Gide, Paris.

Du Toit, A. L. (1937). *Our Wandering Continents*. Oliver & Boyd, Edinburgh.

Earland, A. (1935). Foraminifera. Part III. The Falklands sector of the Antarctic (excluding South Georgia). *Discovery Reports*, **10**, 3–208.

Eddie, G. (1977). *The Harvesting of Krill*. Southern Ocean Fisheries Survey Programme: GLO/SO/77/2. FAO, Rome.

Edholm, O. G. (1973). Introduction. In *Polar Human Biology*, ed. O. G. Edholm & E. K. E. Gunderson, pp. 1–7. William Heinemann Medical Books, London.

Edholm, O. G. & Gunderson, E. K. E. (Eds.) (1973). *Polar Human Biology*. William Heinemann Medical Books, London.

Edwards, W. N. (1928). The occurrence of *Glossopteris* in the Beacon Sandstone of Ferrar Glacier, South Victoria Land. *Geological Magazine*, **65**, 323–7.

Ehrenberg, C. G. (1844). Resultate über das Verhalten des kleinsten Lebens in den Oceanen und den grössten bisher zuganglichen Tiefen des Weltmeers. *Verhandlungen der Königliche Preussische Akademie der Wissenschaften zu Berlin*, 1844, 182–207.

Eights, J. (1833a). Description of a new animal belonging to the crustacea. *Transactions of the Albany Institute*, **2**, 331–4.

Eights, J. (1833b). Description of a new crustaceous animal found on the shores of the South Shetland Islands. *Transactions of the Albany Institute*, **2**, 53–69.

Eights, J. (1835). Description of a new animal belonging to the Arachnides of Latreille: discovered in the sea along the shores of the New South Shetland Islands. *Boston Journal of Natural History*, **1**, 203–6.

Eights, J. (1846). On the ice-bergs of the Ant-Arctic Sea. *American Quarterly Journal of Agricultural Science*, **4**, 20–4.

Ekelöf, E. (1908a). Bakteriologische Studien während der Schwedischen Südpolar-Expedition 1901–1903. *Wissenschaftliche Ergebnisse der Schwedischen Südpolar-Expedition 1901–1903*, **4**, 1–120.

Ekelöf, E. (1908b). Studien über den Bakteriengehalt der Luft und des Erdbodens der antarktischen Gegenden, augeführt während der Schwedischen Südpolar Expedition 1901–1903. *Zeitschrift für Hygiene und Infektion, Leipzig*, **56**, 344–70.

Ekelöf, E. (1920). Die Gesundheits- und Krankenpflege während der schwedischen Südpolar-Expedition 1901–1904. *Wissenschaftliche Ergebnisse der Schwedischen Südpolar-Expedition 1901–1903*, Band I, Lief. 3, 1–30.

Eklund, C. R. (1945). Condensed ornithology report, East Base, Palmer Land. *Proceedings of the American Philosophical Society, Philadelphia*, **89**, 299–304.

Eklund, C. R. (1964). Population studies of Antarctic seals and birds. In *Biologie Antarctique*, ed. R. Carrick, M. Holdgate & J. Prévost, pp. 415–19, Hermann, Paris.

Eklund, C. R. & Charlton, F. E. (1959). Measuring the temperatures of incubating penguin eggs. *American Scientist*, **47**, 80–6.

El-Sayed, S. Z. (1966). Prospects of primary productivity studies in Antarctic waters. In *Symposium on Antarctic Oceanography*, ed. R. I. Currie, pp. 227–39. SCAR, Scott Polar Research Institute, Cambridge.

El-Sayed, S. Z. (Ed.) (1981). *Biological Investigations of Marine Antarctic Systems and Stocks*, vol. II, *Selected Contributions to the Woods Hole Conference on Living Resources in the Southern Ocean 1976*. SCAR & SCOR, Scott Polar Research Institute, Cambridge.

El-Sayed, S. Z. & Green, K. A. (1974). Use of remote sensing in the study of Antarctic marine resources. In *Approaches to Earth Survey Problems through Use of Space Techniques*, ed. P. Bock, pp. 47–63. Akademie-Verlag, Berlin.

Emery, W. J. (1980). The *Meteor* expedition, an ocean survey. In *Oceanography: the Past*, ed. M. Sears & D. Merriman, pp. 690–702. Springer-Verlag, New York.

Ennis, C. C. (1934). Magnetic results of the United States Exploring Expedition, 1838–1842, Lieutenant Charles Wilkes, Commander. *Terrestrial Magnetism & Atmospheric Electricity*, **39**, 91–101.

Eugster, O. (1989). History of meteorites from the moon collected in Antarctica. *Science*, **245**, 1197–202.

Evans, H. B. & Jones, A. G. E. (1975). A forgotten explorer: Carsten Egeberg Borchgrevink. *Polar Record*, **17**, 221–35.

Evans, S. (1969). Glacier soundings in the polar regions: I. The VHF radio echo technique. *Geographical Journal*, **135**, 547–8.

Everson, I. (1977). *The Living Resources of the Southern Ocean*. Southern Ocean Fisheries Survey Programme: GLO/SO/77/1. FAO, Rome.

Everson, I. & Bone, D. G. (1986). Effectiveness of the RMT-8 system for sampling krill (*Euphausia superba*) swarms. *Polar Biology*, **6**, 83–90.

Everson, I., Watkins, J. L., Bone, D. G. & Foote, K. G. (1990). Implications of a new acoustic target strength for abundance estimates of Antarctic krill. *Nature*, **345**, 338–40.

Ewing, M., Ewing, J. I., Houtz, R. E. & Leyden, R. (1966). Sediment distribution in the Bellingshausen Basin. In *Symposium on Antarctic Oceanography*, ed. R. I. Currie, pp. 89–100. SCAR, Scott Polar Research Institute, Cambridge.

Eyde, R. H. (1985). Expedition botany: the making of a new profession. In *Magnificent Voyagers*, ed. H. J. Viola & C. M. Margolis, pp. 25–41. Smithsonian Institution Press, Washington, D.C.

Falla, R. A. (1964). Ships of the Southern Ocean. *Antarctic*, **3**, 517–20.

Fanning, E. (1924). *Voyages and Discoveries in the South Seas 1792–1832*. Marine Research Society, Salem, Massachusetts.

Farman, J. C. (1977). Ozone measurements at British Antarctic Survey stations. *Philosophical Transactions of the Royal Society of London* B, **279**, 261–71.

Farman, J. C., Gardiner, B. G. & Shanklin, J. D. (1985). Large losses of total ozone in Antarctica reveal seasonal $C1O_x/NO_x$ interaction. *Nature*, **315**, 207–10.

Farquharson, J. (1830). Experiments on the influence of the aurora borealis on the magnetic needle. *Proceedings of the Royal Society of London*, **2**, 391–2.

Feeney, R. (1974). *Professor on the Ice*. Pacific Portals, Davis, California.

Ferrar, H. T. (1907). *Report on the field geology of the region explored during the 'Discovery' Antarctic Expedition, 1901–4. National Antarctic Expedition 1901–4, Natural History*, I. *Geology*, British Museum, London.

Fiennes, R. (1983). *To the Ends of the Earth*. Hodder & Stoughton, London.

Fifield, R. (1987). *International Research in the Antarctic*. Oxford University Press.

Filchner, W. (1922). *Zum Sechsten Erdteil: die Zweite Deutsche Südpolar-Expedition*. Ullstein, Berlin.

Fogg, G. E. (1967). Observations on the snow algae of the South Orkney Islands. *Philosophical Transactions of the Royal Society of London* B, **252**, 279–87.

Fogg, G. E. (1975). Primary productivity. In *Chemical Oceanography*, ed. J. P. Riley & G. Skirrow, vol. 2, pp. 385–453. Academic Press, London.

Fogg, G. E. (1990). Our perceptions of phytoplankton: an historical sketch. *British Phycological Journal*, **25**, 103–15.

Fogg, G. E. & Stewart, W. D. P. (1968). *In situ* determinations of biological nitrogen fixation in Antarctica. *British Antarctic Survey Bulletin* no. **15**, 39–46.

Forbes, J. D. (1846). Illustrations of the viscous theory of glacier motion – Part III. *Philosophical Transactions of the Royal Society of London*, **136**, 177–210.

Forchhammer, G. (1865). On the composition of sea-water in the different parts of the ocean. *Philosophical Transactions of the Royal Society of London*, **155**, 203–62.

Foreman, S. J. (1989). Experiences with a coupled global model. *Philosophical Transactions of the Royal Society of London* A, **329**, 275–88.

Forster, J. R. (1778). *Observations made during a Voyage Round the World, on Physical Geography, Natural History, and Ethnic Philosophy*. London.

Fowler, H. W. (1940). The fishes obtained by the Wilkes Expedition 1838–1842. *Proceedings of the American Philosophical Society*, **82**, 733–800.

Frakes, L. A. (1983). Problems in Antarctic marine geology: a review. In *Antarctic Earth Science*, ed. R. L. Oliver, P. R. James & J. B. Jago, pp. 375–8. Cambridge University Press, Cambridge.

Frakes, L. A. & Crowell, J. C. (1971). The position of Antarctica in Gondwanaland. In *Research in the Antarctic*, ed. L. O. Quam, pp. 731–45. American Association for the Advancement of Science, Washington, D.C.

Frängsmyr, T. (Ed.) (1989). *Science in Sweden: The Royal Swedish Academy of Sciences 1739–1989*. Canton MA, Science History Publications, USA.

Frankel, H. (1976). Alfred Wegener and the specialists. *Centaurus*, **20**, 305–24.

Frazier, R. G. & Lockhard, E. E. (1945). Acclimatization and effects of cold on the human body as observed at Little America III. U.S. Antarctic Service Expedition, 1939–41. *Proceedings of the American Philosophical Society*, **89**, 249–55.

Freitag, D. R. & Dibbern, J. S. (1986). Dr Poulter's Antarctic snow cruiser. *Polar Record*, **23**, 129–41.

Friedmann, E. I. (1978). Melting snow in the dry valleys is a source of water for endolithic microorganisms. *Antarctic Journal of the United States*, **13**, 162–3.

Friedmann, E. I. (1980). Endolithic microbial life in hot and cold deserts. *Origins of Life*, **10**, 223–35.

Friedmann, E. I. & Ocampo, R. (1976). Endolithic blue-green algae in the dry valleys: primary producers in the antarctic desert system. *Science*, **193**, 1247–9.

Fritsch, F. E. (1912). *Freshwater algae. National Antarctic Expedition, Natural History*, **6**, British Museum, London.

Fu, L.-L. & Chelton, D. B. (1984). Temporal variability of the Antarctic Circumpolar Current observed from satellite altimetry. *Science*, **226**, 343–6.

Fuchs, V. E. (1973). Evolution of a venture in Antarctic science: Operation Tabarin and the British Antarctic Survey. In *Frozen Future*, ed. R. S. Lewis & P. M. Smith, pp. 233–48. Quadrangle Books, New York.

Fuchs, V. E. (1982). *Of Ice and Men: the Story of the British Antarctic Survey 1943–73*. Anthony Nelson, Oswestry.

Fuchs, V. E. & Laws, R. M. (Eds.) (1977). Scientific research in Antarctica. *Philosophical Transactions of the Royal Society of London* B, **279**, 1–288.

Fugler, S. R. (1985). Chemical composition of guano of burrowing petrel chicks (Procellariidae). In *Antarctic Nutrient Cycles and Food Webs*, ed. W. R. Siegfried, P. R. Condy & R. M. Laws, pp. 169–72. Springer-Verlag, Berlin.

Gain, L. (1910). Rapport sur les travaux de zoologie et de botanique. In *Rapports préliminaires sur les travaux exécutés dans l'Antarctique*, pp. 73–100. Gauthier-Villars, Paris, for Académie des Sciences.

Gain, L. (1912). *La flore algologique des régions antarctiques et subantarctiques. Deuxième Expedition Antarctique Française (1908–1910)*. Masson, Paris.

Gazert, H. (1901). The bacteriological work of the German South Polar Expedition. *Scottish Geographical Magazine*, **17**, 470–3.

Gazert, H. (1927). Untersuchungen über Meeresbakterien und ihren Einfluss auf den Stoffwechsel im Meere. In *Deutsche Südpolar-Expedition 1901–1903*, ed. E. von Drygalski, *VII Bakteriologie, Ozeanographie*, pp. 235–96. Walter de Gruyter, Berlin.

Genthon, C., Barnola, J. M., Raynaud, D., Lorius, C., Jouzel, J., Barkov, N. I.,

Korotkevich, Y. S. & Kotlyakov, V. M. (1987). Vostok ice core: climatic response to CO_2 and orbital forcing changes over the last climatic cycle. *Nature*, **329**, 414–18.

Gerlache, A. (1943). *Quinze Mois dans l'Antarctique*. Librairie Générale, Brussels.

Giaever, J. (1954). *The White Desert: the Official Account of the Norwegian–British–Swedish Antarctic Expedition*. Chatto & Windus, London.

Gilbert, J. R. & Erickson, A. W. (1977). Distribution and abundance of seals in the pack ice of the Pacific Sector of the Southern Ocean. In *Adaptations within Antarctic Ecosystems*, ed. G. A. Llano, pp. 703–40. Smithsonian Institution, Washington, D.C.

Gill, A. E. (1973). Circulation and bottom water production in the Weddell Sea. *Deep-Sea Research*, **20**, 111–40.

Gillmor, C. S. (1978). The early history of upper atmosphere physics research in Antarctica. In *Upper Atmosphere Research in Antarctica*, ed. L. J. Lanzerotti & C. G. Park, pp. 236–62, *Antarctic Research Series* no. 29. American Geophysical Union, Washington, D.C.

Gilmore, R. M. (1961). Research on Antarctic seals. In *Science in Antarctica: The Life Sciences in Antarctica. Part I of a Report by the Committee on Polar Research*, pp. 54–61. National Academy of Sciences, National Research Council, Washington, D.C.

Glen, J. W. (1955). The creep of polycrystalline ice. *Proceedings of the Royal Society of London* A, **228**, 519–38.

Gold, E. (1965). George Clarke Simpson. *Biographical Memoirs of Fellows of the Royal Society*, **11**, 157–75.

Goldman, C. R., Mason, D. T. & Wood, B. J. B. (1963). Light injury and inhibition in Antarctic freshwater phytoplankton. *Limnology and Oceanography*, **8**, 313–22.

Gordon, A. L. (1971a). Oceanography of Antarctic waters. In *Antarctic Oceanology* I, ed. J. L. Reid, pp. 169–203, *Antarctic Research Series* no. 15. American Geophysical Union, Washington, D.C.

Gordon, A. L. (1971b). Antarctic polar front zone. In *Antarctic Oceanology* I, ed. J. L. Reid, pp. 205–21, *Antarctic Research Series* no. 15. American Geophysical Union, Washington, D.C.

Gordon, A. L. (1972). Introduction: Physical oceanography of the Southeast Indian Ocean. In *Antarctic Oceanology* II, *The Australian–New Zealand Sector*, ed. D. E. Hayes, pp. 3–9, *Antarctic Research Series* no. 19. American Geophysical Union, Washington, D.C.

Gordon, A. L. (1986). Physical oceanography. *Antarctic Journal of the United States*, **21**, 12–13.

Gordon, A. L. (1988). The Southern Ocean and global climate. *Oceanus*, **31**, 39–46.

Gordon, A. L., Molinelli, E. J. & Baker, T. N. (1982). *Southern Ocean Atlas*. Columbia University Press, New York.

Gough, B. M. (1986). British–Russian rivalry and the search for the Northwest Passage in the early 19th Century. *Polar Record*, **23**, 301–17.

Gould, L. M. (1931a). Some geographical results of the Byrd Antarctic Expedition. *The Geographical Review*, **21**, 177–200.

Gould, L. M. (1931b). *Cold: the Record of an Antarctic Sledge Journey*. Brewer, Warren & Putnam, New York.

Gould, L. M. (1940). The glaciers of Antarctica. *Proceedings of the American Philosophical Society*, **82**, 835–76.
Gould, L. M. (1973). Emergence of Antarctica, the mythical land. In *Frozen Future: a Prophetic Report from Antarctica*, ed. R. S. Lewis & P. M. Smith, pp. 11–45. Quadrangle Books, New York.
Gould. R. T. (1924). The Ross Deep. *Geographical Journal*, **63**, 237–41.
Gow, A. J. (1965). The ice sheet. In *Antarctica*, ed. T. Hatherton, pp. 221–58. Methuen, London.
Gran, H. H. (1931). On the conditions for the production of plankton in the sea. *Rapports et Procès-Verbaux des Réunions, Conseil Permanent International pour l'Exploration de la Mer*, **75**, 37–46.
Grange, J. (1848). *Voyage au Pole Sud et dans l'Océanie* etc. *Géologie, Minéralogie et Geographie Physique du Voyage d'après les matériaux recueillis par MM. les chirurgiens naturalistes de l'expedition.* Gide, Paris.
Grange, J. (1854). *Voyage au Pole Sud et dans l'Océanie etc. Géologie, Minéralogie et Géographie Physique du Voyage d'après les matériaux recueillis par MM. les chirurgiens naturalistes de l'expedition.* Gide et J. Baudry, Paris.
Grantham, G. J. (1977). *The Utilization of Krill.* Southern Ocean Fisheries Survey Programme: GLO/SO/77/3. FAO, Rome.
Green, J. R. (1965). A different way of life. *Antarctic*, **4**, 106–8.
Greene, S. W. (1964). The vascular flora of South Georgia. *British Antarctic Survey, Scientific Reports* no. 45.
Greene, M. T. (1982). *Geology in the Nineteenth Century: Changing Views of a Changing World.* Cornell University Press, Ithaca.
Gremmen, N. J. M. (1982). *The Vegetation of the Subantarctic Islands Marion and Prince Edward.* Junk, The Hague.
Gressitt, J. L. (1965). Terrestrial animals. In *Antarctica*, ed. T. Hatherton, pp. 351–71, Methuen, London.
Grimminger, G. (1941). Meteorological results of the Byrd Antarctic expeditions 1928–30, 1933–35: Summaries of data. *Monthly Weather Review, Washington*, supplement no. 42.
Grimminger, G. & Haines, W. C. (1939). Meteorological results of the Byrd Antarctic expeditions 1928–30, 1933–35: Tables. *Monthly Weather Review, Washington*, supplement no. 41.
Griscom, Professor (1823). Foreign literature and science. *American Journal of Science and Arts Conducted by Benjamin Silliman*, **6**, 197–200.
Grobbelaar, J. U. (1978). Mechanisms controlling the composition of fresh waters on the sub-antarctic island Marion. *Archiv für Hydrobiologie*, **83**, 145–57.
Grossi, S. M., Kottmeier, S. T. & Sullivan, C. W. (1984). Sea-ice microbial communities. III. Seasonal abundance of microalgae and associated bacteria, McMurdo Sound, Antarctica. *Microbial Ecology*, **10**, 231–42.
Gulland, J. A. (1987). The Antarctic Treaty system as a resource management mechanism. In *The Antarctic Treaty Regime: Law, Environment and Resources*, ed. G. D. Triggs, pp. 116–27. Cambridge University Press, Cambridge.
Gunderson, E. K. E. (1973). *Psychological studies in Antarctica: a Review*. In *Polar Human Biology*, ed. O. G. Edholm & E. K. E. Gunderson, pp. 352–61. William Heinemann Medical Books, London.

Gunderson, E. K. E. (Ed.) (1974). *Human Adaptability to Antarctic Conditions, Antarctic Research Series* no. 22. American Geophysical Union, Washington, D.C.

Guthridge, G. G. (1983). Citation of Research Literature. *Antarctic Journal of the United States*, **18**, 12–13.

Guymer, L. B. & Le Marshall, J. F. (1980). Impact of FGGE buoy data on southern hemisphere analysis. *Australian Meteorological Magazine*, **28**, 19–42.

Hackmann, W. (1984). *Seek and Strike: Sonar, anti-submarine warfare and the Royal Navy 1914–54*. HMSO, London.

Hall, K. (1988). Daily monitoring of a rock tablet at a maritime Antarctic site: moisture and weathering results. *British Antarctic Survey Bulletin* no. 79, 17–25.

Hall, D. H. (1976). *History of the Earth Sciences during the Scientific and Industrial Revolutions with Special Emphasis on the Physical Geosciences*. Elsevier, Amsterdam.

Hall, M. B. (1984). *All Scientists Now: The Royal Society in the Nineteenth Century*. Cambridge University Press, Cambridge.

Hallam, A. (1973). *A Revolution in the Earth Sciences: from Continental Drift to Plate Tectonics*. Clarendon Press, Oxford.

Halpern, M. (1971). Evidence for Gondwanaland from a review of West Antarctic radiometric ages. In *Research in the Antarctic*, ed. L. O. Quam, pp. 717–30. American Association for the Advancement of Science, Washington, D.C.

Hamilton, J. E. (1934). The southern sea lion, *Otaria byronia* (De Blainville). *Discovery Reports*, **8**, 269–318.

Hamre, I. (1933). The Japanese South Polar Expedition of 1911–1912: a little-known episode in Antarctic exploration. *Geographical Journal*, **82**, 411–23.

Hanson, R. B., Lowery, H. K., Shafer, D., Sorocco, R. & Pope, D. H. (1983). Microbes in Antarctic waters of the Drake Passage: vertical patterns of substrate uptake, productivity and biomass in January 1980. *Polar Biology*, **2**, 179–88.

Hardy, A. C. (1926). A new method of plankton research. *Nature*, **118**, 630–2.

Hardy, A. C. (1936a). The continuous plankton recorder. *Discovery Reports*, **11**, 457–510.

Hardy, A. C. (1936b). Observations on the uneven distribution of oceanic plankton. *Discovery Reports*, **11**, 511–38.

Hardy, A. (1966). *The Divine Flame: Natural History and Religion*. Collins, London.

Hardy, A. (1967). *Great Waters*. Harper & Row, New York.

Hardy, A. C. & Gunther, E. R. (1935). The plankton of the South Georgia whaling grounds and adjacent waters, 1926–27. *Discovery Reports*, **11**, 1–146.

Harmer, S. F. (1928). History of Whaling, *Proceedings of the Linnean Society of London*, 140th session, 51–95.

Harmer, S. F. (1931). Southern Whaling. *Proceedings of the Linnean Society of London*, session 142, 85–163.

Harrison, E. (1987). Whigs, prigs and historians of science. *Nature*, **329**, 213–14.

Harrowfield, D. L. (1981). *Sledging into History*. Macmillan Company of New Zealand, Aukland.

Hart, T. J. (1934). On the phytoplankton of the south-west Atlantic and the Bellingshausen Sea, 1929–31. *Discovery Reports*, **8**, 1–268.
Haskell, D. C. (1942). *The United States Exploring Expedition 1838–1842 (Wilkes) and its Publications 1844–1874. A Bibliography*. New York Public Library, New York.
Hatherton, T. (1958). Antarctic research. *Nature*, **182**, 285–9.
Hatherton, T. (Ed.) (1965). *Antarctica*. Methuen, London.
Hatherton, T. (1967). The birth of VUWAE. *Tuatara: Journal of the Biological Society, Victoria University of Wellington, New Zealand*, **15**, 100–1.
Hatherton, T. (1986). Antarctica prior to the Antarctic Treaty – a historical perspective. In *Antarctic Treaty System: an Assessment*, pp. 15–32. National Academy Press, Washington, D.C.
Hayes, J. G. (1928). *Antarctica: a Treatise on the Southern Continent*. The Richards Press, London.
Hayes, P. K., Whitaker, T. M. & Fogg, G. E. (1984). The distribution and nutrient status of phytoplankton in the Southern Ocean between 20° and 70° W. *Polar Biology*, **3**, 153–65.
Hayter, A. (1968). *The Year of the Quiet Sun: One Year at Scott Base, Antarctica: a Personal Impression*. Hodder & Stoughton, London.
Headland, R. (1984). *The Island of South Georgia*. Cambridge University Press, Cambridge.
Headland, R. K. (1989). *Chronological List of Antarctic Expeditions and Related Historical Events*. Cambridge University Press, Cambridge.
Heap, J. A. & Holdgate, M. W. (1986). The Antarctic Treaty System as an environmental mechanism – an approach to environmental issues. In *Antarctic Treaty System: an Assessment*, pp. 195–210. National Academy Press, Washington, D.C.
Hedgpeth, J. W. (1971). James Eights of the Antarctic (1798–1882). In *Research in the Antarctic*, ed. L. O. Quam, pp. 3–4. American Association for the Advancement of Science, Washington, D.C.
Heezen, B. C., Tharp, M. & Hollister, C. D. (1966). Illustrations of the marine geology of the Southern Ocean. In *Symposium on Antarctic Oceanography*, ed. R. I. Currie, pp. 101–9. SCAR, Scott Polar Research Institute, Cambridge.
Heirtzler, J. R. (1971). The evolution of the Southern Oceans. In *Research in the Antarctic*, ed. L. O. Quam, pp. 667–84. American Association for the Advancement of Science, Washington, D.C.
Helliwell, R. A. & Katsufrakis, J. P. (1978). Controlled wave-particle interaction experiments. In *Upper Atmosphere Research in Antarctica*, ed. L. J. Lanzerotti & C. G. Park, pp. 100–29, *Antarctic Research Series* no. 29. American Geophysical Union, Washington, D.C.
Helliwell, R. A., Katsufrakis, J. P. & Trimpi, M. L. (1973). Whistler-induced amplitude perturbation in VLF propagation. *Journal of Geophysical Research*, **78**, 4679–88.
Helmholtz, H. von (1888). Über atmosphaerische Bewegungen. *Sitzungberichte der Königliche Preussische Akademie der Wissenschaften zu Berlin*, Jahrgang 1888, Demi-Bd. I, 647–63.
Hempel, G. (1988). Antarctic marine research in winter: the winter Weddell Sea Project 1986. *Polar Record*, **24**, 43–8.
Hendy, C. H., Healy, T. R., Rayner, E. M., Shaw, J. & Wilson, A. T. (1979).

Late Pleistocene glacial chronology of the Taylor Valley, Antarctica, and the global climate. *Quaternary Research*, **11**, 172–84.

Hepworth, M. W. C. (1908). Climatology of South Victoria Land and the neighbouring seas. *National Antarctic Expedition 1901–1904. Meteorology*, Part I, pp. 417–51. Royal Society of London.

Herbert, W. (1968). *A World of Men: Exploration in Antarctica.* Eyre & Spottiswoode, London.

Herdman, H. R. P. (1959). Some notes on sea-ice observed by Captain James Cook, R.N., during his circumnavigation of Antarctica. *Journal of Glaciology*, **3**, 534–41.

Herrmann, E. (1941). *Deutsche Forscher im Südpolarmeer*. Safari-Verlag, Berlin.

Heyburn, H. R. (1980). Profile: William Lamond Allardyce, 1861–1930: Pioneer Antarctic conservationist. *Polar Record*, **20**, 39–42.

Heywood, R. B. (1967). Ecology of the fresh-water lakes of Signy Island, South Orkney Islands: I. Catchment areas, drainage systems and lake morphology. *British Antarctic Survey Bulletin* no. 14, 25–43.

Heywood, R. B. (1968). Ecology of the fresh-water lakes of Signy Island, South Orkney Islands: II. Physical and chemical properties of the lakes. *British Antarctic Survey Bulletin* no. 18, 11–44.

Heywood, R. B. (1977). A limnological survey of the Ablation Point area, Alexander Island, Antarctica. *Philosophical Transactions of the Royal Society of London* B, **279**, 39–54.

Heywood, R. B. (1984). Antarctic inland waters. In *Antarctic Ecology*, vol. 1, ed. R. M. Laws, pp. 279–344. Academic Press, London.

Heywood, R. B. & Light, J. J. (1975). First direct evidence of life under Antarctic shelf ice. *Nature*, **254**, 591–2.

Heywood, R. B. & Priddle, J. (1987). Retention of phytoplankton by an eddy. *Continental Shelf Research*, **7**, 937–55.

Hirst, J. F. (1989). Measuring meltwater patterns around icebergs. *Polar Record*, **25**, 252–4.

Hoare, M. E. (Ed.) (1982). *The* Resolution *Journal of Johann Reinhold Forster* 1772–1775. In four volumes, Hakluyt Society, London.

Hobbie, J. E. (1984). Polar limnology. In *Lakes and Reservoirs*, ed. F. S. Taub, pp. 63–105. Elsevier, Amsterdam.

Hobbs, W. H. (1926). *The Glacial Anticyclones: The Poles of the Atmospheric Circulation*. University of Michigan Studies: Scientific Series, vol. 4. Macmillan Company, New York.

Hobbs, W. H. (1940). The discovery of Wilkes Land, Antarctica. *Proceedings of the American Philosophical Society*, **82**, 561–82.

Hodson, R. E., Azam, F., Carlucci, A. F., Fuhrman, J. A., Karl, D. M. & Holm-Hansen, O. (1981). Microbial uptake of dissolved organic matter in McMurdo Sound, Antarctica. *Marine Biology*, **61**, 89–94.

Hoelzel, A. R. & Amos, W. (1988). DNA fingerprinting in 'scientific whaling'. *Nature*, **333**, 305.

Hoinkes, H. C. (1964). Glacial meteorology. In *Research in Geophysics*, vol. 2, *Solid Earth and Interface Phenomena*, ed. H. Odishaw, pp. 391–424. Massachusetts Institute of Technology, Cambridge, Massachusetts.

Holdgate, M. W. (1964a). National organizations for Antarctic biological research. In *Biologie Antarctique*, ed. R. Carrick, M. Holdgate & J. Prévost, pp. 59–68. Hermann, Paris.

Holdgate, M. W. (1964b). Terrestrial ecology in the maritime Antarctic. In *Biologie Antarctique*, ed. R. Carrick, M. Holdgate & J. Prévost, pp. 181–94. Hermann, Paris.

Holdgate, M. (Ed.) (1970). *Antarctic Ecology*, 2 vols. Academic Press, London.

Holdgate, M. W., Allen, S. E. & Chambers, M. J. G. (1967). A preliminary investigation of the soils of Signy Island, South Orkney Islands. *British Antarctic Survey Bulletin* no. 12, 53–71.

Holmes, A. (1944). *Principles of Physical Geology*. Thomas Nelson & Sons, London.

Holm-Hansen, O., El-Sayed, S. Z., Franceschini, G. A. & Cuhel, R. L. (1977). Primary production and the factors controlling phytoplankton growth in the Antarctic seas. In *Adaptations within Antarctic Ecosystems*, ed. G. A. Llano, pp. 11–50. Smithsonian Institution, Washington, D.C.

Home, E. (1822). On the difference in the appearance of the teeth and the shape of the skull in different species of Seals. *Philosophical Transactions of the Royal Society of London*, **112**, 239–40.

Hooker, J. D. (1847). *The Botany of the Antarctic Voyage of H.M. Discovery Ships* Erebus *and* Terror, *in the Years 1839–1843 under the Command of Captain Sir James Clark Ross, Kt., R.N., F.R.S., etc.* Reeve Brothers, London. Reprint 1963, J. Cramer, Weinheim.

Horne, A. J. (1972). The ecology of nitrogen fixation on Signy Island, South Orkney Islands. *British Antarctic Survey Bulletin* no. 27, 1–18.

Horowitz, N. H., Cameron, R. E. & Hubbard, J. S. (1972). Microbiology of the dry valleys of Antarctica. *Science*, **176**, 242–5.

Horsburgh, J. (1830). Remarks on several icebergs which have been met with in unusually low latitudes in the southern hemisphere. *Philosophical Transactions of the Royal Society of London*, **120**, 117–20.

Humboldt, A. von (1824). Account of the discoveries of the Russians in the Southern Polar Seas, by M. Simonoff. *Literary Gazette and Journal of the Belles Lettres*, January 1824, 26–7.

Huntford, R. (1979). *Scott and Amundsen*. Hodder & Stoughton, London.

Hurley, P. M. (1968). The confirmation of continental drift. *Scientific American*, **218**, 52–64.

Huxley, E. (1977). *Scott of the Antarctic*. Weidenfeld & Nicolson, London.

Huxley, L. (Ed.) (1913). *Scott's Last Expedition*. 2 vols. Smith, Elder & C., London.

Imhof, W. L., Reagan, J. B., Voss, H. D., Gaines, E. E., Datlowe, D. W., Mobilia, J., Helliwell, R. A., Inan, U. S. & Katsufrakis, J. (1983). The modulated precipitation of radiation belt electrons by controlled signals from VLF transmitters. *Geophysical Research Letters*, **10**, 615–18.

Interdepartmental Committee (1920). *Report on Research and Development in the Dependencies of the Falkland Islands*. HMSO, London.

IOC/WMO (1978). *Oceanographic Aspects of the First GARP Global Experiment (FGGE)*. UNESCO, Paris.

Irish, J. D. & Snodgrass, F. E. (1972). Australian-Antarctic tides. In *Antarctic Oceanology II, The Australian-New Zealand Sector*, ed. D. E. Hayes, pp. 101–16, *Antarctic Research Series* no. 19. American Geophysical Union, Washington, D.C.

Jacka, F. & Jacka, E. (1988). *Mawson's Antarctic Diaries*. Unwin Hyman, London.

Jacques, G. (1983). Some ecophysiological aspects of the Antarctic phytoplankton. *Polar Biology*, **2**, 27–33.

Jensen, H. I. (1916). Report on Antarctic soils. *British Antarctic Expedition 1907–1909. Reports on Scientific Investigations. Geology* 2, Part VI, pp. 89–92. William Heinemann, London.

Joerg, W. L. G. (Ed.) (1928). *Problems of Polar Research: a Series of Papers by Thirty-one Authors.* American Geographical Society, Special Publication no. 7, New York.

John, D. D. (1934). The second Antarctic commission of the R.R.S. *Discovery II. Geographical Journal*, **83**, 381–98.

Johnson, R. S. & Mollendorf, J. C. (1984). Transport from a vertical ice surface melting in saline water. *International Journal of Heat Mass Transfer*, **27**, 1928–32.

Johnston, K. (1987). First steps in ozone protection agreed. *Nature*, **329**, 189.

Jones, A. G. E. (1969). New light on John Balleny. *Geographical Journal*, **135**, 55–61.

Jones, A. G. E. (1974). Dr W. H. B. Webster, 1793–1875: Antarctic scientist. *Polar Record*, **17**, 143–5.

Jones, A. G. E. (1982). *Antarctica Observed: Who discovered the Antarctic Continent?* Caedmon of Whitby.

Jouventin, P. & Weimerskirch, H. (1990). Satellite tracking of wandering albatrosses. *Nature*, **343**, 746–8.

Jouzel, J., Lorius, C., Petit, J. R., Genthon, C., Barkov, N. I., Kotlyakov, V. M. & Petrov, V. M. (1987). Vostok ice core: a continuous isotope temperature record over the last climatic cycle (160,000 years). *Nature*, **329**, 403–8.

Joyner, C. C. (1988). Comparison of Soviet arctic and antarctic policies. *Woods Hole Oceanographic Institution Technical Report*, January 1988, WHOI-88-5, pp. 22–3. (Extended abstract.)

Karsten, G. (1905). Das Phytoplankton des Antarktischen Meeres nach dem Material der deutschen Tiefsee-Expedition, 1898–1899. *Wissenschaftiche Ergebnisse der Deutschen Tiefsee-Expedition*, vol. 2, part 2, pp. 1–136. Gustav Fischer, Jena.

Kennicutt, M. C. II, *et al.* (26 names) (1990). Oil spillage in Antarctica. *Environmental Science and Technology*, **24**, 620–4.

Kerr, R. A. (1987). Winds, pollutants drive ozone hole. *Science*, **238**, 156–8.

Kerry, K. R., Grace, D. R., Williams, R. & Burton, H. R. (1977). Studies on some saline lakes of the Vestfold Hills, Antarctica. In *Adaptations within Antarctic Ecosystems*, ed. G. A. Llano, pp. 839–58. Smithsonian Institution, Washington, D.C.

Keynes, R. D. (ed.) (1979). *The Beagle Record.* Cambridge University Press, Cambridge.

Keys, J. R. (1990). Ice. In *Antarctic Sector of the Pacific*, ed. G. P. Glasby, pp. 95–123. Elsevier, Amsterdam.

Keys, J. R. & Williams, K. (1981). Origin of crystalline, cold desert salts in the McMurdo Region, Antarctica. *Geochimica Cosmochimica Acta*, **45**, 2299–309.

Keys, J. R. & Williams, K. L. (1984). Rates and mechanisms of iceberg ablation in the D'Urville Sea, Southern Ocean. *Journal of Glaciology*, **30**, 218–22.

Kidson, E. (1946). Discussions of observations at Adélie Land, Queen Mary

Land and Macquarie Island. *Australasian Antarctic Expedition 1911–1914, Scientific Reports*, Series B, vol. 6.
Kidson, E. (1947). Daily weather charts extending from Australia and New Zealand to the Antarctic continent. *Australasian Antarctic Expedition 1911–1914, Scientific Reports*, Series B, vol. 7.
Kimball, L. A. (1988). The Antarctic Treaty system. *Oceanus*, **31**, 14–19.
King, H. G. R. (1964–65). An early proposal for conserving the southern seal fishery. *Polar Record*, **12**, 313–16.
King, H. G. R. (1969). *The Antarctic*. Blandford Press, London.
King, H. G. R. (1980). Polar studies in Cambridge. *Cambridge* no. 7, 38–45.
Kirk, T. (1891). On the botany of the Antarctic islands. *Report of the Third Meeting of the Australasian Association for the Advancement of Science, Christchurch, New Zealand, 1891*, pp. 213–31.
Kirwan, L. P. (1962). *A History of Polar Exploration*. Penguin Books, Harmondsworth, Middlesex.
Klyashtorin, L. G. (1961). Primary production in the Atlantic and Southern Oceans according to the data obtained during the fifth Antarctic voyage of the diesel-ekectric Ob'. *Doklady Akademii Nauk SSSR*, **141**, 1204–7. (In Russian.)
Knox, G. A. (1966). Tides and intertidal zones. In *Symposium on Antarctic Oceanography*, ed. R. I. Currie, pp. 131–46. SCAR, Scott Polar Research Institute, Cambridge.
Kol, E. (1972). Snow algae from Signy Island (South Orkney Islands, Antarctica). *Annales Historico-Naturales Musei Nationalis Hungarici*, **64**, 63–70.
Koopmann, G. (1953). Entstehung und Verbreitung von Divergenzen in der oberflächennahen Wasserbewegung der antarktischen Gewässer. *Deutsche Hydrographische Zeitschrift*, **2**, 1–38.
Kooyman, G. L. (1981). *Weddell Seal, Consumate Diver*. Cambridge University Press, Cambridge.
Kooyman, G. L., Davis, R. W., Croxall, J. P. & Costa, D. P. (1982). Diving depths and energy requirements of king penguins. *Science*, **217**, 726–7.
Korff, S. A., Bailey, D. K. & Clarke, E. T. (1945). Report on cosmic-ray observations made on the United States Antarctic Service Expedition, 1939–1941. *Proceedings of the American Philosophical Society*, **89**, 316–23.
Kort, V. G. (1964). Antarctic oceanography. In *Research in Geophysics*, vol. 2, *Solid Earth and Interface Phenomena*, ed. H. Odishaw, pp. 309–33. Massachusetts Institute of Technology Press, Cambridge, Massachusetts.
Kort, V. G. (1968). Frontal zones of the Southern Ocean. In *Symposium on Antarctic Oceanography*, ed. R. I. Currie, pp. 3–7. SCAR, Scott Polar Research Institute, Cambridge.
Kragh, H. (1987). *An Introduction to the Historiography of Science*. Cambridge University Press, Cambridge.
Kriss, A. E., Mishustina, I. E., Mitskevich, I. N. & Zemtsova, E. V. (1967). *Microbial Population of Oceans and Seas*. Edward Arnold, London. (Translation from Russian edition 1964.)
Kucherov, I. P. (1962). Navigation charts of the Antarctic compiled by the Bellingshausen-Lazarev Expedition in 1819–1821. *Antarctica. Commission Reports* 1962, pp. 159–71. Israel Programme for Scientific Translations, Jerusalem 1969.

Kuhn, M., Lettau, H. H. & Riordan, A. J. (1977). Stability-related wind spiraling in the lowest 32 meters. In *Meteorological Studies at Plateau Station*, ed. A. J. Businger, pp. 93–111, *Antarctic Research Series* no. 25. American Geophysical Union, Washington, D.C.

Lamb, H. H. (1961). Melbourne symposium on Antarctic meteorology. *Nature*, **191**, 210–11.

Lamb, I. M. (1948). Antarctic pyrenocarp lichens. *Discovery Reports*, **25**, 1–30.

Land, B. (1981). *The New Explorers: Women in Antarctica*. Dodd, Mead & Company, New York.

Landy, M. P. & Peel, D. A. (1981). Short-term fluctuations in heavy metal concentrations in Antarctic snow. *Nature*, **291**, 144–6.

Langway, C. C., Jr., Gow, A. J. & Lyle Hansen, B. (1971). Deep drilling into polar ice sheets for continuous cores. In *Research in the Antarctic*, ed. L. O. Quam, pp. 351–65. American Association for the Advancement of Science, Washington, D.C.

Lanzerotti, L. J. (1978). Studies of geomagnetic pulsations. In *Upper Atmosphere Research in Antarctica*, ed. L. J. Lanzerotti & C. G. Park, pp. 130–56, *Antarctic Research Series* no. 29. American Geophysical Union, Washington, D.C.

Lanzerotti, L. J. (1986). Advances in antarctic geophysical sciences from the IGY to the present. *Antarctic Journal of the United States*, **21**, 1–24.

Lanzerotti, L. J. & Park, C. G. (Eds.) (1978). *Upper Atmosphere Research in Antarctica. Antarctic Research Series* no. 29. American Geophysical Union, Washington, D.C.

Larsen, C. A. (1894). The voyage of the 'Jason' to the Antarctic regions. *Geographical Journal*, **4**, 333–44.

Law, P. (1959). The IGY in Antarctica. *Australian Journal of Science*, **21**, 285–94.

Law, P. (1979). Obituary: Brian Birley Roberts, CMG. *Polar Record*, **19**, 399–404.

Law, P. (1983). *Antarctic Odyssey*. Heinemann, Melbourne.

Law, P. (1985). A future policy for the Antarctic. *Interdisciplinary Science Review*, **10**, 336–48.

Laws, R. M. (1953a). A new method of age determination for mammals with special reference to the elephant seal, *Mirounga leonina* Linn. *Falkland Islands Dependencies Survey Scientific Report* no. 2. HMSO, London.

Laws, R. M. (1953b, 1956a,b). The elephant seal (*Mirounga leonina* Linn.) *Falkland Islands Dependencies Survey Scientific Reports*: I. Growth and age, no. 8; II. General, social and reproductive behaviour, no. 13; III. The physiology of reproduction, no. 15. HMSO, London.

Laws, R. M. (1984). Seals. In *Antarctic Ecology*, ed. R. M. Laws, vol. 2, pp. 621–715. Academic Press, London.

Laws, R. M. (1985). International stewardship of the Antarctic: Problems, successes and future options. *Marine Pollution Bulletin*, **16**, 49–55.

Laws, R. (1989). Dangerous crack in the ice pact. *The Times*, 13 September 1989.

Leader-Williams, N. (1985). The sub-Antarctic Islands – introduced species. In *Key Environments: Antarctica*, ed. W. N. Bonner & D. W. H. Walton, pp. 318–28. Pergamon Press, Oxford.

Leader-Williams. N. (1988). *Reindeer on South Georgia.* Cambridge University Press, Cambridge.

Leader-Williams, N. & Ricketts, C. (1981). Seasonal and sexual patterns of growth and condition of reindeer introduced into South Georgia. *Oikos*, **38**, 27–39.

Lebedev, V. (1959). *Antarctica.* Foreign Languages Publishing House, Moscow. (Translated from the Russian by G. P. Ivanov-Mumjiev.)

Lebedev, V. L. (1961a). Geographical observations in the Antarctic made by the expeditions of Cook 1772–1775 and Bellingshausen-Lazarev 1819–1821. *Antarctica. Commission Reports 1960*, pp. 1–19. Israel Programme for Scientific Translations, Jerusalem, 1966.

Lebedev, V. L. (1961b). Who discovered Antarctica? *Antarctica. Commission Reports 1961*, pp. 161–8. Israel Programme for Scientific Translations, Jerusalem, 1966.

Lebedev, V. L.(1963). Different interpretations of certain excerpts from documents of the Bellingshausen-Lazarev Expedition. *Antarctica. Commission Reports 1963*, pp. 182–7, Israel Programme for Scientific Translations, Jerusalem, 1966.

LeGrand, H. E. (1988). *Drifting Continents and Shifting Theories.* Cambridge University Press, Cambridge.

Le Jehan, S. & Treguer, P. (1983). Uptake and regeneration ΔSi/ΔN/ΔP ratios in the Indian sector of the Southern Ocean. Originality of the biological cycle of silicon. *Polar Biology*, **3**, 127–36.

Lemke, J. L., Nitecki, M. H. & Pullman, H. (1980). Studies of the acceptance of plate tectonics. In *Oceanography: the Past*, ed. M. Sears & D. Merriman, pp. 614–21. Springer-Verlag, New York.

Lenz, W. (1980). The Forster's offences against convention during and after Capt. Cook's second voyage around the world and the governmental reprisals. In *Oceanography: the Past*, ed. M. Sears & D. Merriman, pp. 682–9. Springer-Verlag, New York.

Leslie, J. (1820). *Elements of Geometry and Plane Trigonometry with an Appendix, and very copious notes and illustrations.* 4th edition. W. & C. Tait, Edinburgh.

Lester, M. C. (1923). An expedition to Graham Land, 1920–1922. *Geographical Journal*, **62**, 174–94.

Lettau, H. (1971). Antarctic atmosphere as a test tube for meteorological theories. In *Research in the Antarctic*, ed. L. O. Quam, pp. 443–75. American Association for the Advancement of Science, Washington, D.C.

Levick, G. M. (1914). *Antarctic Penguins: a Study of their Social Habits.* William Heinemann, London.

Lewis, E. L. & Weeks, W. F. (1971). Sea ice: some polar contrasts. In *Symposium on Antarctic Ice and Water Masses*, ed. G. E. R. Deacon, pp. 23–34. SCAR, Scott Polar Research Institute, Cambridge.

Lewis, H. E. & Masterson, J. P. (1963). Polar physiology: its development in Britain. *Lancet*, **1** (*7289*), 1009–14.

Lewis, R. S. & Smith, P. M. (Eds.) (1973). *Frozen Future: A Prophetic Report from Antarctica.* Quadrangle Books, New York.

Light, J. J. (1977). Production and periodicity of Antarctic freshwater phytoplankton. In *Adaptations within Antarctic Ecosystems*, ed. G. A. Llano, pp. 829–37. Smithsonian Institution, Washington, D.C.

Light, J. J. & Heywood, R. B. (1973). Deep-water mosses in Antarctic lakes. *Nature*, **242**, 535–6.

Liljequist, G. H. (1956–57). Energy exchange of an Antarctic snow-field. *Norway–Britain–Sweden Antarctic Expedition 1949–1952, Scientific Results*, vol. 2, part 1, 1–289. Norsk Polarinstitutt, Oslo.

Lindley, D. (1989). Antarctic astronomy in the clear. *Nature*, **340**, 192.

Lipschutz, M. E. (1985). Meteorite studies: terrestrial and extraterrestrial applications, 1985. *Antarctic Journal of the United States, 1985 Review*, **20**, 55–6.

Lister, H. (1960). Glaciology I. Solid precipitation and drift snow. *Trans-Antarctic Expedition 1955–1958. Scientific Reports*, **5**. Trans-Antarctic Expedition Committee, London.

Llano, G. (1972). Antarctic conservation: prospects and retrospects. In *Conservation Problems in Antarctica*, ed. B. C. Parker, pp. 1–11. Virginia Polytechnic Institute and State University, Blacksburg, Virginia.

Llano, G. A. (Ed.) (1977). *Adaptations within Antarctic Ecosystems*. Smithsonian Institution, Washington, D.C.

Llano, G. A. (1988). Book review: Antarctic Science. *BIOMASS Newsletter*, **10**, 14–15.

Lloyd, P. (1989). An anniversary for atmospheric ozone. *Nature*, **338**, 113.

Loewe, F. (1956). Études de glaciologie en Terre Adélie, 1951–1952. *Expeditions Polaires Françaises*, no. 62. Hermann, Paris.

Logan, H. F. M. (1979). *Cold Commitment: the Development of New Zealand's Territorial Role in Antarctica 1920–1960*. M.A. Thesis in History. University of Canterbury, Christchurch, New Zealand.

Longton, R. E. (1988). *Biology of Polar Bryophytes and Lichens*. Cambridge University Press, Cambridge.

Lovelock, J. E., Maggs, R. J. & Wade, R. J. (1973). Halogenated hydrocarbons in and over the Atlantic. *Nature*, **241**, 194–6.

Lowe, P. R. & Kinnear, N. B. (1930). Birds. In *British Antarctic ('Terra Nova') Expedition, 1910. Natural History Reports, Zoology*, vol. 4 (no. 5), pp. 103–93. British Museum (Natural History), London.

Lugg, D. (1984). Obituary: Major Eric N. Webb. *Polar Record*, **22**, 205–6.

MacDonald, E. A. (1973). The role of icebreakers in the Antarctic. In *Frozen Future: a Prophetic Report from Antarctica*, ed. R. S. Lewis & P. M. Smith, pp. 385–97. Quadrangle Books, New York.

Mackintosh, N. A. (1946). The Antarctic convergence and the distribution of surface temperature in Antarctic waters. *Discovery Report*, **23**, 177–212.

Mackintosh, N. A. (1966). The swarming of krill and problems of estimating the standing stock. In *Symposium on Antarctic Oceanography*, ed. R. I. Currie, pp. 259–60. SCAR, Scott Polar Research Institute, Cambridge.

Mackintosh, N. A. & Wheeler, J. F. G. (1929). Southern blue and fin whales. *Discovery Reports*, **1**, 257–540.

Macpherson, N. (1977). The adaptation of groups to Antarctic isolation. *Polar Record*, **18**, 581–5.

MacPike, E. F. (Ed.) (1932). *Correspondence and papers of Edmond Halley: Preceded by an Unpublished Memoir of his Life by One of his Contemporaries and the 'Éloge' by D'Ortous de Mairan*. Clarendon Press, Oxford.

Maksimov, I. V. (1958). Oceanographic research of Soviet Antarctic

expeditions. *Soviet Antarctic Expedition Information Bulletin*, **1**, 4–8. (Elsevier, Amsterdam, 1964.)
Malaurie, J. (1989). J.-B. Charcot: father of French polar research. *Polar Record*, **25**, 191–6.
Mangin, L. (1915). Phytoplancton de l'Antarctique. *Deuxième Expédition Antarctique Française (1908–1910) commandé par le Dr Jean Charcot*, pp. 1–95, Paris.
Mangin, L. (1922). Phytoplancton Antarctique. Expedition Antarctique de la 'Scotia', 1902–4. *Mémoires de l'Académie des Sciences, Paris*, **57**, 1–134.
Margni, R. A. & Castrelos, O. D. (1964). Quelques aspects de la bactériologie Antarctique. In *Biologie Antarctique*, ed. R. Carrick, M. Holdgate & J. Prévost, pp. 121–39. Hermann, Paris.
Markham, C. R. (1986). *Antarctic Obsession: A Personal Narrative of the Origins of the British National Antarctic Expedition 1901–1904*. Edited and introduced by C. Holland. Bluntisham Books, Erskine Press.
Marr, J. W. S. (1962). The natural history and geography of the Antarctic krill (*Euphausia superba* Dana). *Discovery Reports*, **32**, 33–464.
Marschall, H.-P. (1988). The overwintering strategy of Antarctic krill under the pack-ice of the Weddell Sea. *Polar Biology*, **9**, 129–35.
Marshall, N. B. (1964). Fish. In *Antarctic Research*, ed. R. Priestley, R. J. Adie & G. deQ. Robin, pp. 206–18. Butterworths, London.
Martin, L. (1940). Early explorers of southern America from the United States. *Nature*, **146**, 238–9.
Maslanik, J. A. & Barry, R. G. (1990). Remote sensing in Antarctica and the Southern Ocean: applications and developments. *Antarctic Science*, **2**, 105–21.
Matthews, L. H. (1929). The natural history of the elephant seal. *Discovery Reports*, **1**, 233–56.
Matthews, L. H. (1931). *South Georgia: the British Empire's Subantarctic Outpost*. John Wright & Sons, Bristol.
Maury, M. F. (1855). *The Physical Geography of the Sea*. 2nd edn, Sampson Low, Son & Co., London.
Maury, M. F. (1860). On the physical geography of the sea in connection with the Antarctic regions. *Proceedings of the Royal Geographical Society*, **5**, 22–6.
Maury, M. F. (1883). *The Physical Geography of the Sea*. 6th edn. Thomas Nelson and Sons, London.
Mawson, D. (1915). *The Home of the Blizzard: being the Story of the Australasian Antarctic Expedition 1911–1914*. In 2 vols. William Heinemann, London.
Mawson, D. (1916). A contribution to the study of ice-structures. *British Antarctic Expedition 1907–9. Reports on the Scientific Investigations. Geology*, vol. II, pp. 1–24. William Heinemann, London.
Mawson, D. (1928). Unsolved problems of Antarctic exploration and research. In *Problems of Polar Research*, ed. W. L. G. Joerg, pp. 253–66. American Geographical Society, Special Publication no. 7, New York.
Mawson, P. (1964). *Mawson of the Antarctic*. Longmans, London.
May, R. M., Beddington, J. R., Clark, C. W., Holt, S. J. & Laws, R. M. (1979). Management of multispecies fisheries. *Science*, **205**, 267–77.
McCann, T. S. (1980). Population structure and social organization of southern

elephant seals, *Mirounga leonina* (L.). *Biological Journal of the Linnean Society*, **14**, 133–50.

McCann, T. S. (1983). Activity budgets of southern elephant seals, *Mirounga leonina*, during the breeding season. *Zeitschrift für Tierpsychologie*, **61**, 111–26.

McConnell, A. (1978). Historical methods of temperature measurement in Arctic and Antarctic waters. *Polar Record*, **19**, 217–31.

McConnell, A. (1980). Six's thermometer: a century of use in oceanography. In *Oceanography: the Past*, ed. M. Sears & D. Merriman, pp. 252–65. Springer-Verlag, New York.

McConnell, A. (1985). Nineteenth-century geomagnetic instruments and their makers. In *Nineteenth-Century Scientific Instruments and their Makers*, ed. P. R. de Clercq, pp. 29–52. Rodopi, Holland.

McCormick, R. (1841). Geological remarks on Kerguelen's Land. *Proceedings of the Royal Society of London*, **4**, 299.

McCormick, R. (1884). *Voyages of Discovery in the Arctic and Antarctic Seas.* In 2 vols., London.

McGinnis, L. D. (Ed.) (1981). *Dry valley drilling project. Antarctic Research Series* no. 33. American Geophysical Union, Washington, D.C.

McIntyre, N. (1988). *90° South: South Pole Expedition 1986/87. Final Report.* English edition. Antarctic Foundation, Blindern, Oslo.

McLean, A. L. (1915). Medical reports: Main Base (Adelie Land). In *The Home of the Blizzard* by D. Mawson, vol. II, pp. 308–10. William Heinemann, London.

McLean, A. L. (1918). Bacteria of ice and snow in Antarctica. *Nature*, **102**, 35–9.

McLean, A. L. (1919). Bacteriological and other researches. *Australasian Antarctic Expedition 1911–1914. Scientific Report* **7**, 1–128.

McWhinnie, M. A. (1964). Temperature responses and tissue respiration in Antarctic crustacea with particular reference to the krill *Euphasia superba*. In *Biology of the Antarctic Seas*, ed. M. O. Lee, pp. 63–72, *Antarctic Research Series* no. 1. American Geophysical Union, Washington, D.C.

Mear, R. & Swan, R. (1987). *In the Footsteps of Scott*. Jonathan Cape, London.

Medawar, P. (1984). *Pluto's Republic*. Oxford University Press, Oxford.

Meinardus, W. (1909). Meteorologische Ergebnisse der Deutschen Südpolar Expedition 1901–1903. *Deutsche Südpolar Expedition*. III. *Meteorologie* I (1). Georg Reimer, Berlin.

Meinardus, W. (1923). Meteorologische Ergebnisse der Deutschen Südpolar-Expedition 1901–1903. In *Deutsche Südpolar-Expedition 1901–1903*, ed. E. von Drygalski, III Band, Meteorologie. Walter de Gruyter, Berlin.

Meinardus, W. (1938). Klimakunde der Antarktis. In *Handbuch der Klimatologie*, ed. W. Köppen & R. Geiger, Band 4, Teil U. Borntraeger, Berlin.

Mellor, M. (1959). Ice flow in Antarctica. *Journal of Glaciology*, **3**, 377–85.

Menzies, R. J. (1964). Improved techniques for benthic trawling at depths greater than 2000 meters. In *Biology of the Antarctic Seas*, ed. M. O. Lee, pp. 93–109, *Antarctic Research* Series no. 1. American Geophysical Union, Washington, D.C.

Mercer, J. H. (1978). West Antarctic ice sheet and CO_2 greenhouse effect: a threat of disaster. *Nature*, **271**, 321–5.

Michel, R. L., Linick, T. W. & Williams, P. M. (1979). Tritium and carbon-14 distributions in seawater from under the Ross Ice Shelf Project ice hole. *Science*, **203**, 445–6.

Mickleburgh, E. (1987). *Beyond the Frozen Sea: Visions of Antarctica.* The Bodley Head, London.

Miers, J. (1820). Account of the discovery of New South Shetland, with observations on its importance in a geographical, commercial, and political point of view. *Edinburgh Philosophical Journal*, **3**, 367–80.

Mill, H. R. (1900). The Pettersson-Nansen insulating water-bottle. *Geographical Journal*, **16**, 469–71.

Mill, H. R. (1905). *The Siege of the South Pole. The Story of Antarctic Exploration.* Alston Rivers, London.

Mill, H. R. (1926). Preface. In *The Glacial Anticyclones: The Poles of the Atmospheric Circulation*, by W. H. Hobbs, pp. ix–xiv, University of Michigan Studies, Scientific Series vol 4. Macmillan, New York.

Miller, C. & Goldsmith, N. (1980). James Eights, Albany naturalist: New evidence. *New York History*, **61**, 23–42.

Miller, D. G. M., & Hampton, I. (1989). *Biology and ecology of the Antarctic krill (Euphausia superba Dana): A review.* BIOMASS Scientific Series no. 9. SCAR & SCOR, Scott Polar Research Institute, Cambridge.

Miller, S. & Schwerdtfeger, W. (1972). Ice crystal formation and growth in the warm layer above the Antarctic temperature inversion. *Antarctic Journal of the United States*, **7**, 170–1.

Mills, E. L. (1989). *Biological Oceanography: An Early History, 1870–1960.* Cornell University Press.

Mills, W. (1983). Darwin and the iceberg theory. *Notes & Records of the Royal Society of London*, **38**, 109–27.

Mitterling, P. I. (1959). *America in the Antarctic to 1840.* University of Illinois Press, Urbana.

Mocellin, J. & Suedfeld, P. (1990). Polar personality. Paper presented at the *8th International Union on Circumpolar Health*, Whitehorse, Canada, May 20–25, 1990.

Molina, M. J. & Rowland, F. S. (1974). Stratospheric sink for chlorofluoromethanes: chlorine atom-catalysed destruction of ozone. *Nature*, **249**, 810–12.

Morgan, E. D. (1891). Antarctic Exploration. Proceedings of the Geographical Section of the British Association, Monday August 24th. *Proceedings of the Royal Geographical Society, New Monthly Series*, **13**, 632.

Morgan, W. J., Tyler, D. B., Leonhart, J. L. & Loughlin, M. F. (Eds.) (1978). *Autobiography of Rear Admiral Charles Wilkes, U.S. Navy, 1798–1877.* Naval History Division, Department of the Navy, Washington, D.C.

Morita, R. Y., Griffiths, R. P. & Hayasaka, S. S. (1977). Heterotrophic activity of microorganisms in Antarctic waters. In *Adaptations within Antarctic Ecosystems*, ed. G. A. Llano, pp. 99–113. Smithsonian Institution, Washington, D.C.

Morley, J. P. (1964). Polar ships and navigation. In *Antarctic Research*, ed. R. Priestley, R. J. Adie & G. deQ. Robin, pp. 28–38. Butterworths, London.

Mosby, H. (1934). The waters of the Atlantic Antarctic Ocean. *Norwegian Antarctic Expedition 1927–1928 et SQQ., Scientific Results*, vol. 1, no. 11. Norske Videnskapelig Akademi, Oslo.

Mosby, H. (1956). The Norwegian Antarctic expedition in the 'Brategg' 1947–1948. Publication no. 17, from Komm. Chr. Christensens Hvalfangstmuseum. *Scientific Research 'Brategg' Expedition, 1947–48*, no. 1, 3–10.

Mott, P. (1986). *Wings over Ice: the Falkland Islands and Dependencies Aerial Survey Expedition.* Peter Mott, Long Sutton, Somerset.

Mullan, A. B. & Hickman, J. S. (1990). Meteorology. In *Antarctic Sector of the Pacific*, ed. G. P. Glasby, pp. 21–54. Elsevier, Amsterdam.

Multhauf, R. P. & Good, G. (1987). A brief history of geomagnetism and a catalog of the collections of the National Museum of American History. *Smithsonian Studies in History and Technology*, no. 48. Smithsonian Institution Press, Washington, D.C.

Murphy, R. C. (1922). South Georgia, an outpost of the Antarctic. *National Geographic Magazine*, **41**, 409–44.

Murphy, R. C. (1928). Antarctic zoögeography and some of its problems. In *Problems of Polar Research*, ed. W. L. G. Joerg, pp. 355–79. American Geographical Society, New York.

Murphy, R. C. (1936). *Oceanic Birds of South America.* 2 vols. Macmillan & American Museum of Natural History, New York.

Murphy, R. C. (1948). *Logbook for Grace.* Robert Hale, London.

Murray, G. (Ed.) (1901). *The Antarctic Manual for the Use of the Expedition of 1901.* Royal Geographic Society, London.

Murray, G. & Barton, E. S. (1895). A comparison of the Arctic and Antarctic marine floras. *Phycological Memoirs of the British Museum*, part iii, 88–98. Dulau & Co., London.

Murray, J. (1886). The exploration of the Antarctic regions. *Scottish Geographical Magazine*, **2**, 527–47.

Murray, J. (1894a). The renewal of Antarctic exploration. *Geographical Journal*, **3**, 1–42.

Murray, J. (1894b). Notes on an important geographical discovery in the Antarctic regions. *Scottish Geographical Magazine*, **10**, 195–9.

Murray, J. (1898). The scientific advantages of an Antarctic expedition. *Proceedings of the Royal Society of London*, **62**, 424–51.

Murray, J. (Ed.) (1910). *British Antarctic Expedition, 1907–9. Reports on the Scientific Investigations.* Heinemann, London.

Murray, J. (1911). The observation of tides and seiches in frozen seas. *Internationale Revue der gesamte Hydrobiologie*, **4**, 129–35.

Nairne, E. (1776). Experiments on water obtained from the melted ice of sea-water, to ascertain whether it be fresh or not; and to determine its specific gravity with respect to other water. Also experiments to find the degree of cold in which sea-water begins to freeze. *Philosophical Transactions of the Royal Society of London*, **66**, 249–56.

Naish, G. P. B. (1957). Ships and ship-building. In *A History of Technology, vol. III. From the Renaissance to the Industrial Revolution c. 1500 to c. 1750*, ed. C. Singer, E. J. Holmyard, A. R. Hall & T. I. Williams, pp. 471–500. Oxford University Press.

Nansen, F. (1897). *Farthest North.* 2 vols. Archibald Constable, London.

Nathansohn, A. (1906). Über die Bedeutung vertikaler Wasserbewegungen für die Produktion des Planktons im Meere. *Konigliche sächsische Gesellschaft der Wissenschaften, Leipzig, Abhandlungen der Mathematische-Physicalische Klasse*, Band *29*, no. 5, 357–441.

Nathorst, A. G. (1904). Sur la flore fossil des régions antarctiques. *Comptes rendues hebdomadaires des Séances de l'Académie de Science, Paris*, **138**, 1447–50.

Neal, V. T. & Nowlin, W. D. (1979). International Southern Ocean studies of circumpolar dynamics. *Polar Record*, **19**, 461–70.

Neider, C. (1974). *Edge of the World: Ross Island, Antarctica*. Doubleday, New York.

Nemoto, T. (1966). Feeding of baleen whales and krill, and the value of krill as a marine resource in the Antarctic. In *Symposium on Antarctic Oceanography*, ed. R. I. Currie, pp. 240–53. SCAR, Scott Polar Research Institute, Cambridge.

Neshyba, S. (1977). Upwelling by icebergs. *Nature*, **267**, 507–8.

Neumayer, G. (Ed.) (1890–91). *Die Internationale Polarforschung 1882–83. Die deutschen Expeditionen und ihre Ergebnisse*. 2 vols., Asher, Berlin.

Neumayer, G. (1895). Ueber Südpolarforschung. *Report of the 6th International Geographical Congress*, pp. 109–62.

Neumayer, G. (1901). *Auf zum Südpol – 45 Jahre Wirkens zur Förderung der Erforschung der Südpolar-Region 1855–1900*. Vita Deutsches Verlaghaus, Berlin.

Neushul, M. (1961). Diving in Antarctic waters. *Polar Record*, **10**, 353–8.

Neushul, M. (1964). Diving observations of sub-tidal Antarctic marine vegetation. *Botanica Marina*, **8**, 234–43.

Newell, H. E., Jr. & Townsend, J. W., Jr. (1959). IGY Conference in Moscow. *Science*, **129**, 79–84.

Nicol, S. (1989). Who's counting on krill? *New Scientist* no. 1690, 11 November 1989, 38–41.

Nicolson, M. (1987). Alexander von Humboldt, Humboldtian science and the origins of the study of vegetation. *History of Science*, **25**, 167–94.

Nobile, U. (1928). The dirigible and polar exploration. In *Problems of Polar Research*, ed. W. L. G. Joerg, pp. 419–25, American Geographical Society Special Publication no. 7, New York.

Nordenskjöld, O. (Ed.) (1908–1920). *Wissenschaftliche Ergebnisse der Schwedischen Südpolar-Expedition, 1901–1903*, 6 vols. Stockholm.

Nordenskjöld, N. O. G. & Andersson, J. G. (1905). *Antarctica*. Hurst & Blackett, London.

Nye, J. F. (1959). The motion of ice sheets and glaciers. *Journal of Glaciology*, **3**, 493–507.

Odell, N. E. (1952). Antarctic glaciers and glaciology. In *The Antarctic Today*, ed. F. A. Simpson, pp. 25–55. Reed & New Zealand Antarctic Society, Wellington.

O'Hara, J. G. (1983). Gauss and the Royal Society: the reception of his ideas on magnetism in Britain (1832–1842). *Notes & Records of the Royal Society of London*, **38**, 17–78.

Oliphant, M. (1983). Mawson the man. In *Antarctic Earth Science*, ed. R. L. Oliver, P. R. James & J. B. Jago, pp. xix–xx. Cambridge University Press, Cambridge.

Oliver, R. L., James, P. R. & Jago, J. B. (Eds.) (1983). *Antarctic Earth Science*. Cambridge University Press, Cambridge.

Ommanney, F. D. (1938). *South Latitude*. Longmans, Green & Co., London.

Opaliński, K. W. (1972a). Flora and fauna in freshwater bodies of the Thala

Hills Oasis (Enderby Land, Eastern Antarctica). *Polskie Archiwum Hydrobiologii*, **19**, 383–98.

Opaliński, K. W. (1972b). Freshwater fauna and flora in Haswell Island (Queen Mary Land, Eastern Antarctica). *Polski Archwum Hydrobiologii*, **19**, 377–81.

Opdyke, N. D. (1968). The palaeomagnetism of oceanic cores. In *The History of the Earth's Crust: A Symposium*, ed. R. A. Phinney, pp. 61–72. Princeton University Press, New Jersey.

Orr, N. W. M. (1964). Physiology of sledge dogs. In *Antarctic Research*, ed. R. Priestley, R. J. Adie & G. deQ. Robin, pp. 61–70, Butterworths, London.

Palm, L. C. (1989). Leeuwenhoek and other Dutch correspondents of the Royal Society. *Notes and Records of the Royal Society of London*, **43**, 191–207.

Panagis, K. (1985). The influence of elephant seals on the terrestrial ecosystem of Marion Island. In *Antarctic Nutrient Cycles and Food Webs*, ed. W. R. Siegfried, P. R. Condy & R. M. Laws, pp. 173–90. Springer-Verlag, Berlin.

Panzarini, R. N. M. (1968). Official organizations for scientific co-operation in the Antarctic before the IGY, 1957–58. *Polar Record*, **14**, 438–40.

Parfit, M. (1988). *South Light: a Journey to Antarctica*. Bloomsbury, London.

Parker, B. C. (Ed.) (1972). *Proceedings of the Colloquium on Conservation Problems in Antarctica*. Virginia Polytechnic Institute & State University.

Parker, B. C., Simmons, G. M., Jr., Wharton, R. A., Jr., Seaburg, K. G. & Love, F. G. (1982). Removal of organic and inorganic matter from Antarctic lakes by aerial escape of bluegreen algal mats. *Journal of Phycology*, **18**, 72–8.

Parker, B. C. & Zeller, E. J. (1979). Nitrogenous chemical composition of antarctic ice and snow. *Antarctic Journal of the United States*, **14**, 80–2.

Parmelee, D. F. (1980). *Bird Island in Antarctic Waters*. University of Minnesota Press, Minneapolis.

Parsons, A. (chairman) (1987). *Antarctica: the Next Decade: Report of a Study Group*. Cambridge University Press, Cambridge.

Parsons, C. W. (1934). Penguin embryos. *British Antarctic ('Terra Nova') Expedition, 1910. Natural History Reports. Zoology*, vol. IV, no. 7, pp. 253–62. Trustees of the British Museum, London.

Paterson, R. A. H. (1978). Personality profiles in a group of Antarctic men. *International Review of Applied Psychology*, **27**, 33–7.

Paton, J. & Evans, S. (1964). Aurorae. In *Antarctic Research*, ed. R. Priestley, R. J. Adie & G. deQ. Robin, pp. 318–32. Butterworths, London.

Pearsall, A. W. H. (1973). Bomb vessels. *Polar Record*, **16**, 781–8.

Pearse, J. S. (1965). Reproductive periodicities in several contrasting populations of *Odontaster validus* Koehler, a common Antarctic asteroid. In *Biology of the Antarctic Seas* II, ed. G. A. Llano, pp. 39–85, *Antarctic Research Series* no. 5. American Geophysical Union, Washington, D.C.

Peel, D. A. (1975). The study of global atmospheric pollution in Antarctica. *Polar Record*, **17**, 639–43.

Perkins, R. (1986). *Operation Paraquat: the Battle for South Georgia*. Picton Publishing (Chippenham), Beckington, Bath.

Perrier, F. (1964). Psychopathologie en expédition polaire. In *Biologie Antarctique*, ed. R. Carrick, M. Holdgate & J. Prévost, pp. 565–6, Hermann, Paris.

Pettersson, O. (1904). On the influence of ice-melting upon oceanic circulation. *Geographical Journal*, **24**, 285–333.

Philippi, E. (1910). Die Grundproben der Deutschen Südpolar-Expedition 1901–1903. *Deutsche Südpolar-Expedition 1901–1903*, ed. E. von Drygalski, vol. 2, part VI. Berlin.

Piggott, W. R. (1977a). The importance of the Antarctic in atmospheric sciences. *Philosophical Transactions of the Royal Society of London* **B, 279**, 275–85.

Piggott, W. R. (1977b). The advantages of the Antarctic for ionospheric research. In *Dynamical and Chemical Coupling*, ed. B. Grandal & J. A. Holtet, pp. 367–72. D. Reidel Publishing Company, Dordrecht, Holland.

Pitman, W. C., III & Heirtzler, J. R. (1966). Magnetic anomalies over the Pacific-Antarctic ridge. *Science*, **154**, 1164–71.

Pledge, H. T. (1939). *Science since 1500: A Short History of Mathematics, Physics, Chemistry, Biology*. HMSO, London.

Poesch, J. (1961). Titian Ramsay Peale, 1799–1885, and his journals of the *Wilkes Expedition*. American Philosophical Society, Philadelphia.

Polar Research Board (1986). *Antarctic Treaty System: An Assessment*. National Academy Press, Washington, D.C.

Pomerantz, M. A. (1978). Cosmic Rays. In *Upper Atmosphere Research in Antarctica*, ed. L. J. Lanzerotti & C. G. Park, pp. 12–41, *Antarctic Research Series* no. 29. American Geophysical Union, Washington, D.C.

Potocsky, G. J. & Kniskern, F. E. (1970). *Satellite Sea Ice Reconnaissance. Antarctica* November 1968 to March 1969. Informal Report no. 70–46. Naval Oceanographic Office, Washington, D.C.

Poulter, T. C. (1950). *Geophysical Studies in the Antarctic*. Stanford Research Institute, Palo Alto, California.

Poulton, E. B. (1901). The National Antarctic Expedition. *Nature*, **64**, 83–6.

Pound, R. (1966). *Scott of the Antarctic*. Cassell, London.

Pounder, E. R. (1962). The physics of sea-ice. In *The Sea*, vol. 1, *Physical Oceanography*, ed. M. N. Hill, pp. 826–38. Interscience Publishers, New York.

Powell, G. (1822). *Notes on South Shetland ... printed to Accompany the Chart of these Newly Discovered Lands, which has been constructed from the explorations of the sloop Dove by her Commander George Powell*. Laurie, London.

Pratt, J. G. D. (1960a). A gravity traverse of Antarctica. *Trans-Antarctic Expedition 1955–1958. Scientific Reports* no. 2. Trans-Antarctic Expedition Committee, London.

Pratt, J. G. D. (1960b). Seismic soundings across Antarctica. *Trans-Antarctic Expedition 1955–1958. Scientific Reports* no. 3. Trans-Antarctic Expedition Committee, London.

Pratt, J. G. D. (1960c). Tides at Shackleton, Weddell Sea. *Trans-Antarctic Expedition 1955–1958. Scientific Reports* no. 4. Trans-Antarctic Expedition Committee, London.

Prévost, J. (1961). *Écologie du manchot empereur*. Hermann, Paris (Expéditions Polaires Françaises, Missions Paul-Émile Victor, Publication no. 222).

Prévost, J. & Bourlière, F. (1957). Vie sociale et thermorégulation chez le Manchot Empereur *Aptenodytes forsteri*. *Alauda*, **25**, 167–73.

Price, A. G. (1963). *The Winning of Australian Antarctica: Mawson's B.A.N.Z.A.R.E. Voyages 1929–1931.* Angus & Robertson, London.

Price, D. J. de Solla (1964). The Science of Science. In *The Science of Science*, ed. M. Goldsmith & A. MacKay. Penguin, London. Republished in *Science and Public Policy*, **16**, 152–8 (1989).

Price, D. J. de Solla (1986). *Little Science, Big Science . . . and Beyond.* Columbia University Press, New York.

Priddle, J., Hawes, I., Ellis-Evans, J. C. & Smith, T. J. (1986). Antarctic aquatic ecosystems as habitats for phytoplankton. *Biological Reviews*, **61**, 199–238.

Priddle, J. & Heywood, R. B. (1980). Evolution of Antarctic lake ecosystems. *Biological Journal of the Linnean Society*, **14**, 51–66.

Priestley, R. E. (1923). *Physiography (Robertson Bay and Terra Nova Bay regions). British ('Terra Nova') Antarctic Expedition 1910–1913.* Harrison, London.

Priestley, R. E. & Tilley, C. E. (1928). Geological problems of Antarctica. In *Problems of Polar Research*, ed. W. L. G. Joerg, pp. 315–28. American Geographical Society Special Publication no. 7, New York.

Priestley, R. E. & Wright, C. S. (1928). Some ice problems of Antarctica. In *Problems of Polar Research*, ed. W. L. G. Joerg, pp. 331–41. American Geographical Society Special Publication no. 7, New York.

Pringle, J. (1776). A discourse upon some late improvements of the means for preserving the health of mariners. *Philosophical Transactions of the Royal Society of London*, **66**, suppl. (1–44).

Prior, G. T. (1898). Petrographical notes on the rock specimens collected in the Antarctic regions during the voyage of H.M.S. *Erebus* and *Terror* under Sir James Clark Ross in 1839–43. *Mineralogical Magazine*, **12**, 69–91.

Pryor, M. E. (1962). Some environmental features of Hallett Station, Antarctica, with special reference to soil arthropods. *Pacific Insects*, **4**, 681–728.

Purnell, C. W. (1878). On Antarctic exploration. *Transactions and Proceedings of the New Zealand Institute*, **11**, 31–8.

Pyenson, L. (1989). What is the good of history of science? *History of Science*, **27**, 353–389.

Pyne, S. J. (1986). *The Ice: a Journey to Antarctica.* University of Iowa Press.

Quam, L. O. (Ed.) (1971). *Research in the Antarctic.* American Association for the Advancement of Science, Washington, D.C.

Quartermain, L. B. (1967). *South to the Pole: the Early History of the Ross Sea Sector, Antarctica.* Oxford University Press.

Quartermain, L. B. (1971). *New Zealand and the Antarctic.* A. R. Shearer, Government Printer, Wellington, New Zealand.

Quigg, P. W. (1983). *A Pole Apart: the Emerging Issue of Antarctica.* McGraw-Hill Book Company, New York.

Racovitza, E. G. (1903). Cétacés. *Résultats du voyage du S.Y. Belgica en 1897–99. Zoologie. Vertebrata* (3), Anvers.

Rainaud, A. (1893). *Le Continent Austral: Hypothèses et Découvertes.* Armand Colin, Paris.

Rankin, N. (1951). *Antarctic Isle: Wild Life in South Georgia.* Collins, London.

Rastogi, R. G. (1960). Abnormal features of the F_2 region of the ionosphere at some southern high-latitude stations. *Journal of Geophysical Research*, **65**, 585–92.

Ratcliffe, J. A. (1966). Edward Victor Appleton 1892–1965. *Biographical Memoirs of Fellows of the Royal Society*, **12**, 1–21.
Rayner, G. W. (1940). Whale marking, progress and results to December 1939. *Discovery Reports*, **19**, 245–84.
Rees, D. & Fuller-Rowell, T. J. (1989). The response of the thermosphere and ionosphere to magnetospheric forcing. *Philosophical Transactions of the Royal Society of London* A, **328**, 139–71.
Rehn, J. A. G. (1940). Connection of the Academy of Natural Sciences of Philadelphia with our first national exploring expedition. *Proceedings of the American Philosphical Society*, **82**, 543–9.
Reid, G. C. (1971). Particle precipitation in the Antarctic ionosphere. In *Research in the Antarctic*, ed. L. O. Quam, pp. 553–64. American Association for the Advancement of Science, Washington, D.C.
Rice, A. L. (1986). *British Oceanographical Vessels 1800–1950*. The Ray Society, London.
Richardson, J. & Gray, J. E. (Eds.) (1844–75). *The Zoology of the Voyage of H.M.S.* Erebus *and* Terror, *under the Command of Captain Sir James Clark Ross, R.N., F.R.S., during the Years 1839 to 1843*. Vol. 1, *Mammalia, Birds*. E. W. Janson, London.
Riiser-Larsen, H. J. (1930). *Mot Ukjent Land: Norvegia-Ekspedisjonen 1929–1930*. Gyldendal Norsk Forlag, Oslo.
Rivolier, J. (1973). Review of medical research performed in the French Antarctic territories. In *Polar Human Biology*, ed. O. G. Edholm & E. K. E. Gunderson, pp. 48–53. William Heinemann Medical Books, London.
Rivolier, J., Goldsmith, R., Lugg, D. J. & Taylor, A. J. W. (1988). *Man in the Antarctic: the Scientific Work of the International Biomedical Expedition to the Antarctic (IBEA)*. Taylor & Francis, London.
Roberts, B. B. (1940). The life cycle of Wilson's petrel *Oceanites oceanicus* (Kuhl). *British Graham Land Expedition 1934–37, Scientific Reports*, **1**, 141–94. British Museum (Natural History), London.
Roberts, B. B. (1941). A bibliography of Antarctic ornithology. *British Graham Land Expedition 1934–37, Scientific Reports*, **1**, 337–67. British Museum (Natural History), London.
Roberts, B. (Ed.) (1967). *Edward Wilson's Birds of the Antarctic*. Blandford Press, London.
Robin, G. de Q. (1953). Measurements of ice thickness in Dronning Maud Land, Antarctica. *Nature*, **171**, 55–8.
Robin, G. de Q. (1958). Seismic shooting and related investigations. *Norwegian–British–Swedish Antarctic Expedition 1949–52. Scientific Results*, **5**, *Glaciology* III. Norsk Polarinstitutt, Oslo.
Robin, G. de Q. (1972). Polar ice sheets: a review. *Polar Record*, **16**, 5–22.
Robin, G. de Q. (1989). Defence forces and the global environment. *Interdisciplinary Science Reviews*, **14**, 345–7.
Robin, G. de Q., Doake, C. S. M., Kohnen, H., Crabtree, R. D., Jordan, S. R. & Möller, D. (1983). Regime of the Filchner-Ronne ice shelves, Antarctica. *Nature*, **302**, 582–6.
Robin, G. de Q., Swithinbank, C. W. M. & Smith, B. M. E. (1968). Radio echo exploration of the Antarctic ice sheet. *ISAGE Symposium*, Hanover, USA, September 1968, 97–115.

Robinson, I. S. (1985). *Satellite Oceanography: an Introduction for Oceanographers and Remote-sensing Scientists*. Ellis Horwood, Chichester.

Rodahl, K. (1961). Man in Antarctica. In *Science in Antarctica. Part I. The Life Sciences in Antarctica. Report by the Committee on Polar Research*, pp. 156–62. National Academy of Sciences, National Research Council, Washington, D.C.

Rodger, A. S. (1982). Union Radio Mark II ionosondes in Antarctica. *British Antarctic Survey Bulletin* no. 57, 69–77.

Rogers, A. F. (1973). Antarctic climate, clothing and acclimatization. In *Polar Human Biology*, ed. O. G. Edholm & E. K. E. Gunderson, pp. 265–89. William Heinemann Medical Books, London.

Ronan, C. A. (1968). Edmond Halley and early geophysics. *Geophysical Journal*, **15**, 241–8.

Ronne, F. (1949). *Antarctic Conquest: the Story of the Ronne Expedition 1946–48*. G. P. Putnam's Sons, New York.

Rose, L. A. (1980). *Assault on Eternity: Richard E. Byrd and the Exploration of Antarctica 1946–47*. Naval Institute Press, Annapolis, Maryland.

Rosenberg, T. G. & Barcus, J. R. (1978). Energetic particle precipitation from the magnetosphere. In *Upper Atmosphere Research in Antarctica*, ed. L. J. Lanzerotti & C. G. Park, pp. 42–71, *Antarctic Research Series* 29. American Geophysical Union, Washington, D.C.

Rosenman, H. (translator & ed.) (1987). *Two Voyages to the South Seas by Jules S-C Dumont d'Urville*, in 2 vols. Melbourne University Press.

Ross, J. C. (1847). *A Voyage of Discovery and Research in the Southern and Antarctic Regions during the Years 1839–43*, in 2 vols. John Murray, London. (Reprint by David & Charles (Holdings), Newton Abbot, Devon, 1969.)

Ross, M. J. (1982). *Ross in the Antarctic: the Voyages of James Clark Ross in Her Majesty's Ships* Erebus & Terror *1839–1843*. Caedmon of Whitby.

Rosswall, T. & Heal, O. W. (Eds.) (1975). *Structure and Function of Tundra Ecosystems*. Ecological Bulletin no. 20. Swedish Natural Science Council, Stockholm.

Rowley, P. D., Vennum, W. R., Kellogg, K. S., Laudon, T. S., Carrara, P. E., Boyles, J. M. & Thomson, M. R. A. (1983). Geology and plate tectonic setting of the Orville Coast and eastern Ellsworth Land, Antarctica. In *Antarctic Earth Science*, ed. R. L. Oliver, P. R. James & J. B. Jago, pp. 245–50. Cambridge University Press, Cambridge.

Royal Society (1840). *Instructions for the use of the magnetic and meteorological observatories and for the magnetic surveys*. Revised edition in 1842, Royal Society of London.

Rubin, M. J. (1962). Atmospheric advection and the Antarctic mass and heat budget. In *Geophysical Monograph* no. 7: *Antarctic Research*, pp. 149–59. American Geophysical Union, Washington, D.C.

Rubin, M. J. (1964). Antarctic weather and climate. In *Research in Geophysics*, vol. 2, *Solid Earth and Interface Phenomena*, ed. H. Odishaw, pp. 461–78. Massachusetts Institute of Technology, Cambridge, Massachusetts.

Rubin, M. J. (Ed.) (1966). *Studies in Antarctic Meteorology, Antarctic Research Series* no. 9. American Geophysical Union, Washington, D.C.

Rubin, M. J. (1982a). James Cook's scientific programme in the Southern Ocean, 1772–75. *Polar Record*, **21**, 33–49.

Rubin, M. J. (1982b). Thaddeus Bellingshausen's scientific programme in the Southern Ocean, 1818–21. *Polar Record*, **21**, 215–29.

Rubin, M. J., Ostapoff, F. & Weyant, W. S. (1968). Surface winds and currents of the Southern Ocean. In *Symposium on Antarctic Oceanography*, ed. R. I. Currie, pp. 8–23. SCAR, Scott Polar Research Institute, Cambridge.

Rudolph, E. D. (1966). Terrestrial vegetation of Antarctica: past and present studies. In *Antarctic Soils and Soil Forming Processes*, ed. J. C. F. Tedrow, pp. 109–24, *Antarctic Research Series* no. 8. American Geophysical Union, Washington, D.C.

Rudolph, E. D. & Benninghoff, W. S. (1977). Competitive and adaptive responses of invading versus indigenous biotas in Antarctica – a plea for organized monitoring. In *Adaptations within Antarctic Ecosystems*, ed. G. A. Llano, pp. 1211–25. Smithsonian Institution, Washington, D.C.

Runcorn, S. K. (1956). Palaeomagnetic comparison between Europe and North America. *Proceedings of the Geologist's Association of Canada*, **8**, 77–85.

Rusin, N. P. (1961). Meteorological and radiational regime of Antarctica. *Gidrometeorologicheskoe Izadel'stvo, Leningrad.* (In Russian.) Translated by Israel Programme for Scientific Translations 1964.

Russell, H. C. (1895). Icebergs in the Southern Ocean. *Journal and Proceedings of the Royal Society of New South Wales*, **29**, 286–315.

Ruud, J. T. (1930). Nitrates and phosphates in the Southern Seas. *Journal du Conseil Permanent International pour l'Exploration de la Mer*, **5**, 347–60.

Rycroft, M. J. (1989). Solar-terrestrial physics: an overview. *Philosphical Transactions of the Royal Society of London* A, **328**, 39–42.

Rymill, J. (1939). *Southern Lights: The Official Account of the British Graham Land Expedition 1934–1937*. Travel Book Club, London.

Sabine, E. (1846). Contributions to terrestrial magnetism. *No. VIII. Philosophical Transactions of the Royal Society of London*, **136**, 337–432.

Sabine, E. (1851). On periodical laws discoverable in the mean effects of the larger magnetic disturbances. *Philosophical Transactions of the Royal Society of London*, **141**, 123–39.

Saijo, Y. & Kawashima, T. (1964). Primary productivity in the Antarctic Ocean. *Journal of the Oceanographical Society of Japan*, **19**, 190–6.

Sanderson, M. (1988). *Griffith Taylor: Antarctic Scientist and Pioneer Geographer*. Carleton University Press, Ottowa.

Sarton, G. (1919). War and civilization. *Isis*, **2**, 315–21.

Sato, T. (1965). Long-period geomagnetic oscillations in southern high latitudes. In *Geomagnetism and Aeronomy*, ed. A. H. Waynick, pp. 173–88, *Antarctic Research Series*, no. 4. American Geophysical Union, Washington, D.C.

Savin, S. M. (1977). The history of the earth's surface temperature during the past 100 million years. *Annual Review of Earth & Planetary Science*, **5**, 319–55.

Savours, A. (1983). John Biscoe, Master Mariner 1794–1843. *Polar Record*, **21**, 485–91.

Savours, A. (Mrs Shirley) & McConnell, A. (1982). The history of the Rossbank Observatory, Tasmania. *Annals of Science*, **39**, 527–64.

SCAR (1968). *Symposium on Antarctic Oceanography*. Scott Polar Research Institute, Cambridge.

Schatz, G. S. (1988). Protecting the Antarctic environment. *Oceanus*, **31**, 101–3.
Schlee, S. (1973). *A History of Oceanography*. Robert Hale, London.
Scholander, P. F., Hock, R., Walters, V. & Irving, L. (1950). Adaptation to cold in Arctic and tropical mammals and birds in relation to body temperature, insulation, and basal metabolic rate. *Biological Bulletin*, **99**, 259–71.
Schröder, W. (1984). *Das Phänomen des Polarlichts*. Wissenschaftliche Buchgesellschaft, Darmstadt.
Schwerdtfeger, W. (1970). Climate of the Antarctic. In *Climates of the Polar Regions*, ed. S. Orvig, pp. 253–355, *World Survey of Climatology*, **14**, Elsevier, Amsterdam.
Schwerdtfeger, W. (1984). *Weather and Climate of the Antarctic*. Elsevier, Amsterdam.
Scoresby, W. (1815). On the Greenland or Polar Ice. *Memoirs of the Wernerian Society*, **2**, 328–36. Caedmon Reprints, Whitby, 1980.
Scoresby, W. (1820). *An Account of the Arctic Regions with a History and Description of the Northern Whale-Fishery*, vol. 1. *The Arctic*. Archibald Constable, Edinburgh. (Reprinted 1969 by David & Charles (Holdings), Newton Abbot, Devon.)
Scoresby, W. (1980). *The Polar Ice and the North Pole*. Caedmon of Whitby Press.
Scott, R. F. (1905). *The Voyage of the 'Discovery'*. In 2 vols. Smith Elder, London.
Scott, R. F. (1929). *The Voyage of the 'Discovery'*. New edition in 1 vol., John Murray, London.
Seal, U. S., Erickson, A. W., Siniff, D. B. & Hofman, R. J. (1971a). Biochemical, population genetic, phylogenetic and cytological studies of Antarctic seal species. In *Symposium on Antarctic Ice and Water Masses*, ed. G. Deacon, pp. 77–95. SCAR, Scott Polar Research Institute, Cambridge.
Seal *et al.* (1971b). See Burt (1971).
Seligman, G. (1980). *Snow Structure and Ski Fields*. 3rd edn. International Glaciological Society, Cambridge.
Selkirk, P. M., Seppelt, R. D. & Selkirk, D. R. (1990). *Subantarctic Macquarie Island: Environment and Biology*. Cambridge University Press, Cambridge.
Seward, A. C. (1914). Antarctic fossil plants. *British Antarctic (Terra Nova) Expedition, 1910. Natural History Report, Geology*, **1** (1), 1–49. British Museum (Natural History), London.
Shackleton, E. H. (1909). *The Heart of the Antarctic*. In 2 vols. William Heinemann, London.
Shackleton, E. (1919). *South: the Story of Shackelton's 1914–1917 Expedition*. William Heinemann, London.
Shaw, N. (Ed.) (1908). *Meteorology, Part I. National Antarctic Expedition 1901–4*. Royal Society of London.
Shea, J. H. (Ed.) (1985). *Continental Drift*. Benchmark Papers in Geology Series. Van Nostrand Reinhold Company, New York.
Shirley, C. C. (1945). Photographic accomplishments and photographic technique at West Base, Antarctica. *Proceedings of the American Philosophical Society*, **89**, 382–5.
Shulenberger, E. (1983). Water-column studies near a melting Arctic iceberg. *Polar Biology*, **2**, 149–58.

Shurley, J. T. (1973). Antarctica is also a prime natural laboratory for the behavioural sciences. In *Polar Human Biology*, ed. O. G. Edholm & E. K. E. Gunderson, pp. 430–5. William Heinemann Medical Books, London.

Sieburth, J. McN. (1961). Antibiotic properties of acrylic acid, a factor in the gastro-intestinal antibiosis of polar marine animals. *Journal of Bacteriology*, **82**, 72–9.

Siegfried, W. R., Condy, P. R. & Laws, R. M. (Eds.) (1985). *Antarctic Nutrient Cycles and Food Webs*. Springer-Verlag, Berlin.

Simmons, D. A. (1987). Measurement of declination at South Georgia 1700–1984. *Polar Record*, **23**, 419–26.

Simpson, F. A. (Ed.) (1952). *The Antarctic Today: a Mid-Century Survey by the New Zealand Antarctic Society*. A. H. & A. W. Reed in conjunction with the New Zealand Antarctic Society, Wellington.

Simpson, G. C. (1919). *Meteorology, Part I. Discussion. British Antarctic Expedition 1910–1913*. Calcutta.

Siple, P. A. (1938). The Second Byrd Antarctic Expedition – Botany I. Ecology and geographical distributions. *Annals of the Missouri Botanical Garden*, **25**, 467–514.

Siple, P. A. (1945). General principles governing selection of clothing for cold climates. *Proceedings of the American Philosophical Society*, **89**, 200–34.

Siple, P. (1959). *90° South: the Story of the American South Pole Conquest*. G. P. Putnam's Sons, New York.

Siple, P. A. & Passel, C. F. (1945). Measurements of dry atmospheric cooling in subfreezing temperatures. *Proceedings of the American Philosophical Society*, **89**, 177–99.

Sisler, F. D. (1961). Geomicrobiology of Antarctica. *Science in Antarctica. Part I, the Life Sciences in Antarctica*, pp. 147–50. Committee on Polar Research, Publication no. 839. National Academy of Sciences – National Research Council, Washington, D.C.

Skottsberg, C. (1941). Communities of marine algae in sub-Antarctic and Antarctic waters. *Kungliga Svenska Vetenskapsakademiens Handlingar*, Series 3, vol. 19, no. 4, 1–92.

Sladen, W. J. L. (1958). *The Pygoscelid penguins*. I. *Methods of study*. II. The Adélie penguin. *Falkland Islands Dependencies Survey Scientific* Reports no. 17. HMSO, London.

Sladen, W. J. L., Boyd, J. C. & Pedersen, J. M. (1966). Biotelemetry studies on penguin body temperatures. *Antarctic Journal of the United States*, **1**, 142–3.

Sladen, W. J. L. & Friedmann, H. (1961). Antarctic ornithology. In *Science in Antarctica: The Life Sciences in Antarctica*, Committee on Polar Research, pp. 62–76. Publication 839. National Academy of Sciences – National Research Council, Washington D.C.

Sladen, W. J. L. & Leresche, R. E. (1970). New and developing techniques in Antarctic ornithology. In *Antarctic Ecology*, vol. 1, ed. M. W. Holdgate, pp. 585–96, Academic Press, London.

Small, G. L. (1971). *The Blue Whale*. Columbia University Press, New York.

Smith, B. M. E. (1972). Automatic plotting of airborne geophysical survey data. *Journal of Navigation*, **25**, 162–75.

Smith, P. M. (1981). The role of the dry valley drilling project in Antarctic and international science policy. In *Dry Valley Drilling Project*, ed. L. D.

McGinnis, pp. 1–5, *Antarctic Research Series* no. 33. American Geophysical Union, Washington, D.C.

Smith, R. I. L. (1972). Vegetation of the South Orkney Islands with Particular Reference to Signy Island. *British Antarctic Survey Scientific Report* No. 68, London.

Smith, R. I. L. (1981). The earliest report of a flowering plant in the Antarctic? *Polar Record*, **20**, 571–2.

Smith, R. I. L. (1985). Nutrient cycling in relation to biological productivity in Antarctic and sub-Antarctic terrestrial and freshwater ecosystems. In *Antarctic Nutrient Cycles and Food Webs*, ed. W. R. Siegfried, P. R. Condy & R. M. Laws, pp. 138–55. Springer-Verlag, Berlin.

Solopov, A. V. (1967). Oases in Antarctica. Meteorology no. 14. Izdatel'stvo 'Nauka' Moskva, 1967. (Translated from Russian by Israel Programme for Scientific Translations, Jerusalem, 1969).

Somov, M. M. (1958). Soviet research on the Antarctic continent. *Soviet Antarctic Expedition Information Bulletin*, **1**, 1–4. Elsevier, Amsterdam.

Stamp, T. & Stamp, C. (1976). *William Scoresby – Arctic Scientist*. Caedmon of Whitby Press.

Stamp, T. & Stamp, C. (1978). *James Cook – Maritime Scientist*. Caedmon of Whitby Press.

Stamp, T. & Stamp, C. (1983). *Greenland Voyager*. Caedmon of Whitby Press.

Stauffer, B., Fischer, G., Neftel, A. & Oeschger, H. (1985). Increase of atmospheric methane recorded in Antarctic core ice. *Science*, **299**, 1386–8.

Steemann Nielsen, E. & Hansen, V. K. (1959). Light adaptation in marine phytoplankton populations and its interrelations with temperature. *Physiologia Plantarum*, **12**, 353–70.

Stolarski, R. S. (1986). Nimbus 7 satellite measurements of the springtime Antarctic ozone decrease. *Nature*, **322**, 808–11.

Stolarski, R. S. (1988). The Antarctic ozone hole. *Scientific American*, **258**, 20–6.

Stommel, H. & Arons, A. B. (1960). On the abyssal circulation of the world ocean. II. An idealised model of the circulation pattern and amplitude in oceanic basins. *Deep-Sea Research*, **6**, 217–33.

Stone, I. R. (1984). Profile: Edward Sabine, polar scientist 1788–1883. *Polar Record*, **22**, 305–9.

Stone, I. R. (1988). The Arctic portraits of Stephen Pearce. *Polar Record*, **24**, 55–8.

Stonehouse, B. (1965). Birds and mammals. In *Antarctica*, ed. T. Hatherton, pp. 153–86. Methuen, London.

Stonehouse, B. (1989). *Polar Ecology*. Blackie, Glasgow.

Störmer, C. (1955). *The Polar Aurora*. Clarendon Press, Oxford.

Streten, N. A. & Troup, A. J. (1973). A synoptic climatology of satellite observed cloud vortices over the Southern Hemisphere. *Quarterly Journal of the Royal Meteorological Society*, **99**, 56–72.

Stroud, D. A. (1988). Whence came Bowers' great heat supply? *Polar Record*, **24**, 245.

Suess, E. (1904–9). *The Face of the Earth*. Translated by H. B. C. Sollas, in 4 vols., Clarendon Press, Oxford. First published in German as *Das Antlitz der Erde* (1883–1904).

Sugden, D. (1982). *Arctic and Antarctic: A Modern Geographical Synthesis*. Basil Blackwell, Oxford.

Sullivan, C. W. & Palmisano, A. C. (1984). Antarctic sea-ice microbial communities. II. Distribution, diversity and production of ice bacteria in McMurdo Sound. *Applied Environmental Microbiology*, **47**, 788–95.

Sullivan, W. (1957). *Quest for a Continent*. Secker & Warburg, London.

Sullivan, W. (1961). *Assault on the Unknown: the International Geophysical Year*. McGraw-Hill Book Company, New York.

Sun, M. (1988). NSF and Antarctic wastes. *Science*, **241**, 897.

Sverdrup, H. U. (1933). On vertical circulation in the ocean due to the action of the wind with application to conditions within the Antarctic Circumpolar Current. *Discovery Reports*, **7**, 139–70.

Sverdrup, H. U. (1934). Wie entsteht die Antarktische Konvergenz? *Annalen der Hydrographie und Maritimische Meteorologie*, **62**, 315–17.

Sverdrup, H. U. (1953). On conditions for the vernal blooming of phytoplankton. *Journal du Conseil Permanent International pour l'Exploration de la Mer*, **18**, 287–95.

Swithinbank, C. W. M. (1972). Radio-echo sounding by the British Antarctic Survey, 1972. *Polar Record*, **16**, 411.

Swithinbank, C. (1988a). Antarctica. In *Satellite Image Atlas of Glaciers of the World*, ed. R. S. Williams, Jr. & J. G. Ferrigno, pp. 1–278, US Geological Survey Professional Paper 1386-B. US Government Printing Office, Washington, D.C.

Swithinbank, C. (1988b). Antarctic Airways: Antarctica's first commercial airline. *Polar Record*, **24**, 313–16.

Tasch, P. (1971). Invertebrate fossil record and continental drift. In *Research in the Antarctic*, ed. L. O. Quam, pp. 703–16. American Association for the Advancement of Science, Washington, D.C.

Taylor, A. (1990). Obituary: Dr Ivan Mackenzie Lamb, *Polar Record*, **26**, 343.

Taylor, A. J. W. (1987). *Antarctic Psychology*. New Zealand Department of Scientific & Industrial Research, Bulletin no. 244, SIPC, Wellington.

Taylor, F. J. R. (1980). Phytoplankton ecology before 1900: supplementary notes to the 'Depths of the Ocean'. In *Oceanography: the Past*, ed. M. Sears & D. Merriman, pp. 509–21. Springer-Verlag, New York.

Taylor, G. (1922). *British Antarctic ('Terra Nova') Expedition 1910–13: The Physiography of McMurdo Sound and Granite Harbour Region*. Harrison, London.

Taylor, G. (1958). *Journeyman Taylor*. Robert Hale, London.

Taylor, R. J. F. (1957). The physiology of sledge dogs. *Polar Record*, **8**, 317–21.

Tedrow, J. C. F. (Ed.) (1966). *Antarctic Soils and Soil Forming Processes. Antarctic Research Series* no. 8. American Geophysical Union, Washington, D.C.

Thomas, K. (1983). *Man and the Natural World: Changing Attitudes in England 1500–1800*. Allen Lane, London.

Thomas, W. H. (1972). Observations on snow algae in California. *Journal of Phycology*, **8**, 1–9.

Thomson, C. W. & Murray, J. (1885). *Report on the Scientific Results of the Voyage of H.M.S.* Challenger *during the Years 1873–76. Narrative*, vol. I (in 2 parts). HMSO, London.

Thomson, R. B. (1977). Effects of human disturbance on an Adélie penguin rookery and measures of control. In *Adaptations within Antarctic Ecosystems*, ed. G. A. Llano, pp. 1177–80. Smithsonian Institution, Washington, D.C.

Thrower, N. J. W. (Ed.) (1981). *The Three Voyages of Edmond Halley in the Paramore 1698–1701*. Hakluyt Society, London.

Thuronyi, G. T. (1977). Language distribution of antarctic literature. *Antarctic Journal of the United States*, **12**, 202–3.

Thuronyi, G. T. (1982). Some statistics on antarctic serial literature. *Antarctic Journal of the United States*, **17**, 254–5.

Thurston, M. H. (1982). Ornithological observations in the south Atlantic Ocean and Weddell Sea, 1959–64. *British Antarctic Survey Bulletin* no. 55, 77–103.

Tickell, W. L. N. (1964). Feeding preferences of the albatrosses *Diomedea melanophris* and *D. chrysostoma* at South Georgia. In *Biologie Antarctique*, ed. R. Carrick, M. Holdgate & J. Prévost, pp. 383–7. Hermann, Paris.

Tickell, W. L. N. (1968). The biology of the great albatrosses, *Diomedea exulans* and *Diomedea epomorphora*. In *Antarctic Bird Studies*, ed. O. L. Austin, Jr., pp. 1–55, *Antarctic Research Series* no. 12. American Geophysical Union, Washington, D.C.

Tickell, W. L. N. & Woods, R. W. (1972). Ornithological observations at sea in the South Atlantic Ocean. *British Antarctic Survey Bulletin* no. 31, 63–84.

Tikhomirov, I. I. (1973). The main trends of Soviet medical investigations in Antarctica. In *Polar Human Biology*, ed. O. G. Edholm & E. K. E. Gunderson, pp. 41–53. William Heinemann Medical Books, London.

Tilbrook, P. J. (1973). The Signy Island terrestrial reference sites: I. An introduction. *British Antarctic Survey Bulletin* nos 33 and 34, 65–76.

Tingey, R. J. (1983). Heroic age geology in Victoria Land, Antarctica. *Polar Record*, **21**, 451–7.

Torii, T. & Yamagata, N. (1981). Limnological studies of saline lakes in the dry valleys. In *Dry Valley Drilling Project*, ed. L. D. McGinnis, pp. 141–59, *Antarctic Research Series* no. 33. American Geophysical Union, Washington, DC.

Torrey, J. (1823). Description of a new species of Usnea, from New South Shetland. *American Journal of Science*, **6**, 104–6.

Triggs, G. D. (1987). The Antarctic Treaty System: some jurisdictional problems. In *The Antarctic Treaty Regime: Law, Environment and Resources*, ed. G. D. Triggs, pp. 88–109. Cambridge University Press, Cambridge.

Trouessart, E. L. (1906). Mammifères pinnipèdes. In *Expédition Antarctique Française (1903–1905)*, ed. J. B. A. E. Charcot, no. 4, pp. 1–27. Masson et Cie, Paris.

Tsiklinsky, Mlle. (1908). La flore microbienne dans les régions du Pôle Sud. *Expédition Antarctique Française 1903–1905*, no. 3, pp. 1–33. Masson et Cie, Paris.

Tucker, D. G. (1978). Electrical communication. In *A History of Technology. Vol. VII. The Twentieth Century c. 1900 to c. 1950, Part II*, ed. T. I. Williams, pp. 1220–67. Oxford University Press, Oxford.

Turrill, W. B. (1963). *Joseph Dalton Hooker: Botanist, Explorer, and Administrator*. Thomas Nelson & Sons, London.

Tyler, D. B. (1968). *The Wilkes Expedition.* American Philosophical Society, Philadelphia.

Tyler, P. A. (1972). Reconnaissance limnology of sub-Antarctic islands. I. Chemistry of lake waters from Macquarie Island and the Iles Kerguelen. *International Revue der gesamte Hydrobiologie*, **57**, 759–78.

Tyndall, J. & Huxley, T. H. (1857). On the structure and motion of glaciers. *Philosophical Transactions of the Royal Society of London*, **147**, 327–46.

Van Heurck, H. (1909). Diatomées. *Résultats du Voyage du S.Y. 'Belgica' en 1897–9 sous le commandement de A. de Gerlache de Gomery. Rapports Scientifiques – Botaniques*, pp. 3–126. Anvers.

Van Mieghem, J. & Van Oye, P. (Eds.) (1965). *Biogeography and Ecology in Antarctica.* Junk, The Hague.

Vicuña, F. O. (Ed.) (1983). *Antarctic Resources Policy: Scientific, Legal and Political Issues.* Cambridge University Press, Cambridge.

Vincent, W. F. (1987). Antarctic limnology. In *Inland Waters of New Zealand*, ed. A. B. Viner, pp. 379–412. SIPC, New Zealand.

Vincent, W. F. (1988). *Microbial Ecosystems of Antarctica.* Cambridge University Press, Cambridge.

Vincent, W. F. & Howard-Williams, C. (1986). Microbial ecology of Antarctic streams. In *Perspectives in Microbial Ecology, Proceedings of the 4th International Symposium on Microbial Ecology*, ed. F. Megušar & M. Gantar, pp. 201–6. Slovene Society for Microbiology, Ljubljana.

Vine, F. J. & Matthews, D. H. (1963). Magnetic anomalies over oceanic ridges. *Nature*, **199**, 947–9.

Viola, H. J. (1985). The story of the US Exploring Expedition. In *Magnificent Voyagers*, ed. H. J. Viola & C. Margolis, pp. 9–23. Smithsonian Institution Press, Washington, D.C.

Viola, H. J. & Margolis, C. (Eds.) (1985). *Magnificent Voyagers: The US Exploring Expedition, 1838–1842.* Smithsonian Institution Press, Washington, D.C.

Vishniac, H. S. & Hempfling, W. P. (1979). Evidence of an indigenous microbiota (yeast) in the dry valleys of Antarctica. *Journal of General Microbiology*, **112**, 301–14.

Vishniac, W. V. & Mainzer, S. E. (1972). Soil microbiology studied *in situ* in the dry valleys of Antarctica. *Antarctic Journal of the United States*, **7**, 88–9.

Wade, F. A. (1945a). An introduction to the symposium on scientific results of the United States Antarctic Service Expedition, 1939–1941. *Proceedings of the American Philosophical Society*, **89**, 1–3.

Wade, F. A. (1945b). The physical aspects of the Ross Ice Shelf. *Proceedings of the American Philosophical Society*, **89**, 160–73.

Wadhams, P. & Kristensen, M. (1983). The response of Antarctic icebergs to ocean waves. *Journal of Geophysical Research*, **88**, 6053–65.

Wales, W. & Baly, W. (1777). *Astronomical Observations. The Original Astronomical Observations, Made in the Course of A Voyage Towards the South Pole, and Round the World.* Board of Longitude, London.

Walton, D. W. H. (1982a). Instruments for measuring biological microclimates for terrestrial habitats in polar and high alpine regions: a review. *Arctic & Alpine Research*, **14**, 275–86.

Walton, D. W. H. (1982b). The Signy Island terrestrial reference sites: XV. Micro-climate monitoring, 1972–74. *British Antarctic Survey Bulletin* no. 55, 111–26.

Walton, D. W. H. (1985). Cellulose decomposition and its relationship to nutrient cycling at South Georgia. In *Antarctic Nutrient Cycles and Food Webs*, ed. W. R. Siegfried, P. R. Condy & R. M. Laws, pp. 192–9. Springer-Verlag, Berlin.

Walton, D. W. H. (Ed.) (1987). *Antarctic Science*. Cambridge University Press, Cambridge.

Walton, E. W. K. (1955). *Two Years in the Antarctic*. Lutterworth Press, London.

Warren, G. (1965). Geology of Antarctica. In *Antarctica*, ed. T. Hatherton, pp. 279–320. Methuen, London.

Watson, R. (1782). *Chemical Essays*. 2nd edn, vol. 2. London.

Webb, E. N. (1975). The magnetograph hut: a historic scientific laboratory of the Australasian Antarctic Expedition 1911–14. *Polar Record*, **17**, 694–7.

Webb, E. N. (1977). Location of the South Magnetic Pole. *Polar Record*, **18**, 610–11.

Webb, P. N. (1983). Climatic, palaeo-oceanographic and tectonic interpretation of Palaeogene-Neogene biostratigraphy from MSSTS-1 drillhole, McMurdo Sound, Antarctica. In *Antarctic Earth Science*, ed. R. L. Oliver, P. R. James & J. B. Jago, p. 560. Cambridge University Press, Cambridge.

Webb, P. N. & McKelvey, B. C. (1959). Geological investigations in South Victoria Land, Antarctica. Part I – Geology of Victoria Dry Valley. *New Zealand Journal of Geology & Geophysics*, **2**, 120–36.

Webber, P. J. (1986). Terrestrial biology in Antarctica since the International Geophysical Year. *Antarctic Journal of the United States*, **21**, 1–6.

Webster, W. H. B. (1834). *Voyage to the Southern Atlantic Ocean*. 2 vols. London. (Reprint by Dawsons of Pall Mall, Folkestone & London, 1970.)

Weddell, J. (1825). *A Voyage towards the South Pole performed in the Years 1822–24*. London. (Reprint of 2nd edition, 1827, containing an examination of the Antarctic Sea, by David & Charles, Newton Abbot, Devon, 1970.)

Weeks, W. F. (1968). Understanding the variations of the physical properties of sea ice. In *Symposium on Antarctic Oceanography*, ed. R. I. Currie, pp. 173–90. SCAR, Scott Polar Research Institute, Cambridge.

Weertman, J. (1976). Glaciology's grand unsolved problem. *Nature*, **260**, 284–6.

Weertman, J. (1987). Impact of the International Glaciological Society on the development of glaciology and its future role. *Journal of Glaciology, Special Issue*, 86–90.

Weihaupt, J. G. (1984). Historic cartographic evidence for holocene changes in the Antarctic ice cover. *Eos*, **65**, 493–501.

Weller, G. (1974). Polar meteorology: a review of some recent research. *Polar Record*, **17**, 277–94.

Weller, G. (1986). Meteorology. In Advances in antarctic geophysical sciences from the IGY to the present. *Antarctic Journal of the United States*, **21**, 7–12.

Weller, S. (1903). The Stokes collection of Antarctic fossils. *Journal of Geology*, **11**, 413–19.

Wellman, H. W. & Wilson, A. T. (1962). Stored solar heat at Lake Vanda. *Antarctic*, **3**, 102–3.

West, W. & West, G. S. (1911). Freshwater algae. *British Antarctic Expedition 1907–9. Reports on the Scientific Investigations, vol. 1, Biology*, ed. J. Murray, Part VII, pp. 263–98. William Heinemann, London.

Wexler, H. (1961). Ice budgets for Antarctica and changes in sea level. *Journal of Glaciology*, **3**, 867–72.

Wexler, H., Rubin, M. J. & Caskey, J. E., Jr. (Eds.) (1962). *Antarctic Research*. The Matthew Fontaine Maury Memorial Symposium. American Geophysical Union, Geophysical Monograph no. 7.

Weyant, W. S. (1966). The antarctic atmosphere: climatology of the troposphere and lower stratosphere. *Antarctic Map Folio Series, folio 4*. American Geographical Society, New York.

Weyant, W. S. (1967). The antarctic atmosphere: climatology of the surface environment. *Antarctic Map Folio Series, folio 8*. American Geographical Society, New York.

Weyprecht, K. (1875). Scientific work of the second Astro-Hungarian Polar Expedition, 1872–4. *Royal Geographical Society Journal*, **45**, 19–33.

Whitaker, T. M. (1982). Primary production of phytoplankton off Signy Island, South Orkneys, the Antarctic. *Proceedings of the Royal Society of London* B, **214**, 169–89.

White, M. G. (1984). Marine benthos. In *Antarctic Ecology*, vol. 2, ed. R. M. Laws, pp. 421–61. Academic Press, London.

Whitworth, T., III (1988). The Antarctic Circumpolar Current, *Oceanus*, **31**, 53–8.

Whymper, E. (1870). The veined structure of glaciers. *Nature*, January 6, 1870.

Wiederkehr, K. H. (1985). The 'Göttinger magnetische Verein' (Magnetic Association or Magnetic Union) and the Antarctic expedition of James Clark Ross 1839–1843. In *Historical Events and People in Geosciences*, ed. W. Schröder, pp. 73–9. Peter Lang, Frankfurt-am-Main.

Wiener, M. A. (1945). Results of auroral observations at West Base, Antarctica, April to September 1940. *Proceedings of the American Philosophical Society*, **89**, 364–78.

Wild, F. (1923). *Shackleton's Last Voyage. The Story of the Quest*. Cassell, London.

Wildeman, E. de (1905). Les phanérogames des Terres magéllaniques. *Expédition Antarctique Belge, 1897–99. Résultats du Voyage de la S.Y.* Belgica . . . *Rapports Scientifiques*, (5) *Botanique*. Buschman, Anvers.

Wilkes, C. (1845). *Narrative of the United States Exploring Expedition during the Years 1838, 1839, 1840, 1841, 1842*. 5 vols and an Atlas. Philadelphia.

Willett, H. C. (1949). Significant correlation between meridional heat transport and the zonal character of the general circulation. *Journal of Meteorology*, **6**, 370.

Williams, P. J. le B. (1984). Bacterial production in the marine food chain: the emperor's new suit of clothes? In *Flows of Energy and Materials in Marine Ecosystems*, ed. M. J. R. Fasham, pp. 271–99. Plenum Press, New York.

Wilson, E. A. (1907). *National Antarctic Expedition 1901–1904. Natural History*, vol. II. *Zoology (Vertebrata: Mollusca: Crustacea)*. I. Mammalia (Whales and Seals), pp. 1–69; II. Aves, pp. 1–121. British Museum (Natural History), London.

Wilson, E. (1966). *Diary of the Discovery Expedition*. Edited by A. Savours, Blandford Press, London.

Wilson, O. (1965). Human adaptation to life in Antarctica. In *Biogeography and Ecology in Antarctica*, ed. J. van Mieghem & P. van Oye, pp. 690–752. Junk, The Hague.

Wilson, O. (1973). Experimental freezing of the finger: a review of studies. In *Polar Human Biology*, ed. O. G. Edholm & E. K. E. Gunderson, pp. 246–55. Heinemann Medical Books, London.

Wodehouse, E. B. & Parker, B. C. (1981). Atmospheric ammonia nitrogen: a potential source of nitrogen eutrophication of freshwater Antarctic ecosystems. In *Terrestrial Biology* III, ed. B. C. Parker, pp. 155–67, *Antarctic Research Series* no. 30. American Geophysical Union, Washington, D.C.

Wohlschlag, D. E. (1964). Respiratory metabolism and ecological characteristics of some fishes in McMurdo Sound, Antarctica. In *Biology of the Antarctic Seas*, ed. M. O. Lee, pp. 33–62, *Antarctic Research Series*, no. 1. American Geophysical Union, Washington, D.C.

Woollard, G. P. (1959). Crustal structure from gravity and seismic measurements. *Journal of Geophysical Research*, **64**, 1521–44.

Woollard, G. P. (1962). Crustal structure in Antarctica. *Geophysical Monographs*, **7**, 53–73.

Woolley, R. (1969). Captain Cook and the transit of Venus. *Notes and Records of the Royal Society of London*, **24**, 19–32.

Wordie, J. M. (1918). The drift of the 'Endurance'. *Geographical Journal*, **51**, 216–37.

Wordie, J. M. (1921a). Shackleton Antarctic Expedition, 1914–1917: Geological observations in the Weddell Sea area. *Transactions of the Royal Society of Edinburgh*, **53**, 17–27.

Wordie, J. M. (1921b). Shackleton Antarctic Expedition, 1914–1917: The natural history of pack-ice as observed in the Weddell Sea. *Transactions of the Royal Society of Edinburgh*, **52**, 795–829.

Wordie, J. M. & Kemp, S. (1933). Observations on certain Antarctic icebergs. *Geographical Journal*, **81**, 428–34.

Wright, C. S. (1940). The transmission of wireless signals in relation to magnetic and auroral disturbances. *Australasian Antarctic Expedition 1911–1914, Scientific Reports*, series B, vol II, part IV, pp. 445–534. Sydney.

Wright, C. S. & Priestley, R. E. (1922). *Glaciology*. *British* (Terra Nova) *Antarctic Expedition 1910–13*. Harrison & Sons, London.

Wright, I. P. & Grady, M. M. (1989). Meteoritics: Antarctic differences of opinion. *Nature*, **341**, 691.

Wüst, G. (1928). The origin of the deep waters of the Atlantic. Translated from 'Der Ursprung der Atlantischen Tiefenwasser' in *Sonderabdruck aus dem Jubiläums-Sonderband 1928 der Zeitschrift der Gesellschaft für Erdkunde zu Berlin*, 1928, pp. 506–34. In *Oceanography: Concepts and History*, ed. M. B. Deacon, pp. 65–88. Dowden, Hutchinson & Ross, Stroudsberg, Pennsylvania, 1978.

Wynn-Williams, D. D. (1985). The Signy Island terrestrial reference sites: XVI. Peat O_2-uptake in a moss turf relative to edaphic and microbial factors. *British Antarctic Survey Bulletin* no. 68, 47–59.

Yarrington, M. R. & Wynn-Williams, D. D. (1985). Methanogenesis and the anaerobic micro-biology of a wet moss community at Signy Island. In *Antarctic Nutrient Cycles and Food Webs*, ed. W. R. Siegfried, P. R. Condy & R. M. Laws, pp. 229–33. Springer-Verlag, Berlin.

Yoshida, M., Ando, H., Omoto, K., Naruse, R. & Ageta, Y. (1971). Discovery of meteorites near Yamato Mountains, East Antarctica. *Antarctic Record*, **39**, 62–5.

Young, A. (1821). Notice of the voyage of Edward Barnsfield [sic], master of His Majesty's Ship *Andromache*, to New South Shetland. *Edinburgh Philosophical Journal*, **4**, 345–8.

Young, E. C. (1970). The techniques of a skua-penguin study. In *Antarctic Ecology*, vol. 1, ed. M. W. Holdgate, pp. 568–84. Academic Press, London.

Young, W. (1980). On the debunking of Captain Scott – a critique against myths, errors, and distortions. *Encounter*, **54**, 8–19.

Zain-Azraai (1987). Antarctica: the claims of 'expertise' versus 'interest'. In *The Antarctic Treaty Regime: Law, Environment and Resources*, ed. G. D. Triggs, pp. 211–17. Cambridge University Press, Cambridge.

Zillman, J. W. (1972). Solar radiation and sea-air interaction south of Australia. In *Antarctic Oceanology. II. The Australian–New Zealand Sector*, ed. D. E. Hayes, pp. 11–40, *Antarctic Research Series* no. 19. American Geophysical Union, Washington, D.C.

Ziman, J. M. (1983). The Bernal Lecture, 1983. The collectivization of science. *Proceedings of the Royal Society of London* B, **219**, 1–19.

Zimmer, C. (1913). Untersuchungen über den inneren Bau von *Euphausia superba* Dana. *Zoologica, Stuttgart*, **26** (67), 65–128.

Zoller, W. H., Gladney, E. S. & Duce, R. A. (1974). Atmospheric concentrations and sources of trace metals at the South Pole. *Science*, **183**, 198–200.

Index

Printed in the United States
By Bookmasters